Springer Proceedings in Energy

The series Springer Proceedings in Energy covers a broad range of multidisciplinary subjects in those research fields closely related to present and future forms of energy as a resource for human societies. Typically based on material presented at conferences, workshops and similar scientific meetings, volumes published in this series will constitute comprehensive state-of-the-art references on energy-related science and technology studies. The subjects of these conferences will fall typically within these broad categories:

- Energy Efficiency
- Fossil Fuels
- Nuclear Energy
- Policy, Economics, Management & Transport
- Renewable and Green Energy
- Systems, Storage and Harvesting
- Materials for Energy

eBook Volumes in the Springer Proceedings in Energy will be available online in the world's most extensive eBook collection, as part of the Springer Energy eBook Collection. Please send your proposals/inquiry to Dr. Loyola DSilva, Senior Publishing Editor, Springer (loyola.dsilva@springer.com).

More information about this series at http://www.springer.com/series/13370

Ion Visa · Anca Duta
Editors

Solar Energy Conversion in Communities

Proceedings of the Conference for Sustainable Energy (CSE) 2020

Springer

Editors
Ion Visa
R&D Centre Renewable Energy
Systems and Recycling
Transilvania University of Brașov
Brașov, Romania

Anca Duta
R&D Centre Renewable Energy
Systems and Recycling
Transilvania University of Brașov
Brașov, Romania

ISSN 2352-2534 ISSN 2352-2542 (electronic)
Springer Proceedings in Energy
ISBN 978-3-030-55756-0 ISBN 978-3-030-55757-7 (eBook)
https://doi.org/10.1007/978-3-030-55757-7

This Springer imprint is published by the registered company Springer Nature Switzerland AG
The registered company address is: Gewerbestrasse 11, 6330 Cham, Switzerland

Organization

Scientific Committee CSE 2020

Viorel Badescu, Politehnica University of Bucharest and Romanian Academy
Mugur Balan, Technical University of Cluj-Napoca, Romania
Viorel Bostan, Technical University of Moldova, Chisinau, Republic Moldova
Octavian Buiu, National Institute for R&D in Microtechnologies, Bucharest, Romania
Bogdan Burduhos, Transilvania University of Brasov, Romania
Marcian Carstea, Anglia Ruskin University, Cambridge, UK
Ana Maria Dabija, Ion Mincu University of Architecture and Urbanism, Bucharest, Romania
Ioan Doroftei, Technical University Gheorghe Asachi, Iasi, Romania
Luis Duarte, Laboratório Nacional de Energia e Geologia, LNEG, Lisboa, Portugal
Neven Duic, University of Zagreb, Croatia
Valeriu Dulgheru, Technical University of Moldova, Chisinau, Republic Moldova
Anca Duta, Transilvania University of Brasov, Romania
Laurentiu Fara, Politehnica University of Bucharest and The Renewable Energy Commission of the Romanian Academy
Istvan Farkas, Szent Istvan University and Hungarian ISES President
Maria Gavrilescu, Technical University Gheorghe Asachi, Iasi, Romania
Brigitte Hasewend, European Sustainable Energy Innovation Alliance, Brussels, Belgium
Codruta Jaliu, Transilvania University of Brasov, Romania
Valeriu Jinescu, Politehnica University of Bucharest, Romania
Soteris Kalogirou, Cyprus University of Technology, Cyprus
Sixto Malato, Plataforma Solar de Almería, Spain
Nikolay Mihailov, Angel Kanchev University of Ruse, Bulgaria
Macedon Moldovan, Transilvania University of Brasov, Romania
Mircea Neagoe, Transilvania University of Brasov, Romania
Nicolae Olariu, Valahia University of Targoviste and SUN-E Association, Romania

Preface

Formulated in its contemporary use almost 50 years ago (1972), the sustainable development concept continuously evolved and is now well considered as the path towards the future. Going beyond the concept phase, various actions are considered for implementing this concept, going beyond the demonstration phase up to a broad, global scale.

One of the key elements supporting the sustainable development is the energy; the energy production, distribution and consumption are the main issues to be addressed and to be solved as our current economic development relies on energy that is (still) produced using mainly fossil fuels.

As it is well known, burning fossil fuels represent the main source of greenhouse gases (GHGs) that are directly related to the global heating. About 44% of the GHGs are the result of using coal as an energy source in our world; thus, the 7th UN Sustainable Development Goal to "ensure access to affordable, reliable, sustainable and modern energy for all" is well justified.

Through the Paris Agreement (2015), the stakeholders focused on the emissions' reduction, by using renewable energy sources, increasing the energy efficiency and electrifying sectors such as the buildings or the transportation sectors.

This represents a major change (as only 18% of the global energy consumption was in the form of electricity in 2016) that asks for conjugated efforts in implementing coherent actions involving the energy producers, distributors and users. However, significant results are possible to be obtained while these parts are considered as (sub)parts of legal entities that give coherence to all their efforts towards well-defined scopes.

This is why the renewable energy communities are considered important by the recast Renewable Energy Directive (RED-II) that aims at European level to stimulate the formation of *renewable energy communities* in all EU member states where the "citizens take ownership of the energy transition".

Solar energy, as a widely available energy resource, represents a significant part of the future energy scenario but, due to the differences between the solar radiation (and energy) variability and the pattern of the current energy demand, in 2018 solar energy accounted less than 2% of the total electricity production. However, it is expected that by 2050 the use of solar energy will raise to 16% and future decentralized solar energy conversion systems for electricity production systems are expected to feed communities worldwide.

The European Green Deal Investment Plan was presented by The EU Commission on January 2020. This represents a strategic response considered by EU for the development through the next decade aiming at supporting public and private investments required by the transition to a climate-neutral, green, competitive and inclusive economy. Thus, the conceptual phase is considered to be ended as it is also the demonstrator phase. The Green Deal asks for specific projects focusing on developing a green economy to be financed as "no one is left behind"; thus, a just transition is targeted. The aim is to have EU as a global leader.

The specific targets considered are related to: increasing the EU's climate ambition for 2030 and 2050; supplying clean, affordable and secure energy; mobilizing industry for a clean and circular economy; building and renovating in an energy and resource efficient way; a zero-pollution ambition for a toxic-free environment; preserving and restoring ecosystems and biodiversity; from "farm to fork": a fair, healthy, environmentally friendly food system; and accelerating the shift to sustainable and smart mobility.

Obviously, these targets require new knowledge and solutions for specific issues that are and will be further identified; therefore, "mobilizing research and fostering innovation" represents an important goal of the EU Green Deal.

These important trends in the near-future energy scenario are well mirrored in the 2020 edition of the Conference for Sustainable Energy (CSE) focusing on the main topic of *Solar Energy Conversion in Communities*.

The CSE takes place once every three years, starting with 2005, organized by the R&D Centre Renewable Energy Systems and Recycling (RESREC). Each edition was dedicated to a core subject, identified according to the sustainable energy pre-requisites: sustainable energy (2005), solar energy conversion (2008), sustainable communities (2011), nearly zero energy buildings (2014) and nearly zero energy communities (2017).

The 6th CSE edition runs in the R&D Institute of the Transilvania University of Brasov, designed as a sustainable community heavily relying on solar energy conversion systems (photovoltaic systems and solar-thermal systems). The institute was developed during 2009–2013; thus, we can now discuss the efficiency and feasibility of the installed solar energy conversion systems based on a real case study.

It is important to notice that the last two proceedings volumes (2014 and 2017) were published by Springer and reached over 40000 downloads for the 2017 and over 58000 downloads for the 2014 volume, thus proving their impact on the R&D community. The high level of the presentations is also preserved in this volume that contains the results of the latest research done on solar energy conversion in communities.

Brașov, Romania
October 2020

Ion Visa
Anca Duta

Contents

Part I Solar Energy Conversion Systems in the Built Environment

1 **Challenges in Implementing Solar Energy Conversion Systems in the Built Environment** 3
Ion Visa and Anca Duta

Part II Solar–Thermal Energy in Communities

2 **Simple Modeling of the Solar Seasonal Thermal Storage Behavior** 21
Paul Daniel Hiris, Florin Bode, Octavian Gabriel Pop, and Mugur Ciprian Balan

3 **Solar Solutions for Housing** 35
Dragoş Ştefan Gherghescu

4 **Outdoor Performance of Triangle Solar Thermal Collectors for Facades Integration** 51
Macedon Moldovan, Ion Visa, and Anca Duta

5 **Alternative Energy Sources Onboard Warships. Case Study: The Use of Solar-Thermal Systems on a Military Support Vessel** 65
Alexandru Cotorcea, Ion Visa, and Andrei Pocora

Part III Solar Electricity in Communities

6 **Agri.q: A Sustainable Rover for Precision Agriculture** 81
Giuseppe Quaglia, Carmen Visconte, Luca Carbonari, Andrea Botta, and Paride Cavallone

7 **A Comparative Analysis of the Solar Energy Receiving Share Using Four Tracking System Types at Mid-Latitude Regions** 93
Mircea Neagoe, Bogdan Gabriel Burduhos, Fazel Mohammadi, and Nadia Ramona Cretescu

8 **Reliability Assessment of PV Modules** 111
Divine Atsu, Istvan Seres, and Istvan Farkas

9 **Comparative Analysis of Software Accuracy in Photovoltaic Energy Estimation for a Temperate Mountain Climate** 125
Bogdan Gabriel Burduhos, Ion Visa, Mircea Neagoe, Mirjana Devetakovic, and Nadia Ramona Cretescu

10 **Hybrid RES Mobile Innovative System Optimized for DC-DC Applications** 141
Bogdan-Alexandru Onose, Ion Murgescu, and Ştefan-Adrian Sontea

11 **Emerging Issues and Challenges with the Integration of Solar Power Plants into Power Systems** 157
Fazel Mohammadi and Mircea Neagoe

Part IV Sustainable Energy Mixes in Communities

12 **Decarbonisation of Heating—Towards 2050** 177
Nevem Duić and Goran Krajačić

13 **Target for 100% Renewable Energy Systems Use in Cyprus for Electricity Production** 179
Rafaela A. Agathokleous and Soteris A. Kalogirou

14 **Vertical Axis Wind Turbines: The Behavior of Lift and Drag Airfoils** 195
Bostan Viorel, Bostan Ion, Rabei Ivan, Dulgheru Valeriu, and Ciuperca Rodion

15 **Vertical Axis Wind Turbines. Optimal Positioning of the Blades Defined by Asymmetrical Airfoils** 207
Bostan Viorel, Bostan Ion, Rabei Ivan, Gutu Marin, and Dulgheru Valeriu

16 **Hybrid Solar—Biomass System Design for Communities with Collective Houses** 215
Adrian Ilie and Ion Visa

17 **Analysis of the Water—Power Nexus of the Balkan Peninsula Power System** 235
Goran Stunjek, Antun Pfeifer, Goran Krajačić, and Neven Duić

18 **An Algorithm for the Design of a Stand-Alone Hybrid System** 259
Ciobanu Daniela, Săulescu Radu, and Jaliu Codruţa

19 **Renewable Energy Production in Decommissioned Power Plant Sites for Sustainable Cities—The Case Study of Romania** 267
Adrian Ciocănea, Mircea Scripcariu, Ştefăniţă Pluteanu, and Bogdan Tofan

20 Sustainable Biorefinery and Bioenergy Models Based on Seaweeds, as a Potential Feedstock for the Black Sea Coastal Area 283
Sandor Bartha, Florbela Carvalheiro, Patricia Moniz, Luis C. Duarte, and Santino Di Berardino

Part V New Trends in the Use of Solar Energy in Communities

21 Complex Electro-Intra-Conversions Within Rotaxanes' Molecular Machines 301
Mirela I. Iorga, Nicoleta A. Dudaş, and Mihai V. Putz

22 Quantum Metrological Matrices for Sustainable Graphentronics 315
Doru L. Buzatu, Paula Ianasi, and Mihai V. Putz

23 Influence of the Cd^{2+} Ions on the Crystal Violet Dye Adsorption from Aqueous Solutions by Activated Charcoal Obtained from Pine Cones 327
Nicoleta Popa, Maria Visa, and Silvia Patachia

24 TiO_2—Graphene Oxide VIS-Active Composites for Advanced Photocatalytic Wastewater Treatment 359
Anca Duta, Ioana Tismanar, Alexandru Cosmin Obreja, and Octavian Buiu

25 Challenges in the Up-Scaled Deposition of Self-Cleaning Thin Films on PV Modules 371
Maria Covei, Anca Duta, and Ion Visa

26 Hydrophilicity Variation of TiO_2—Graphene Oxide Composite Thin Films for Photocatalytic Applications 387
Ioana Tismanar, Alexandru Cosmin Obreja, Octavian Buiu, and Anca Duta

27 Correlation Between the Growth Process and Film Properties of RF-PECVD Grown Nanocrystalline Graphite/Graphene 399
Octavian-Gabriel Simionescu, Elena Anghel, Oana Tutunaru, Cristina Pachiu, Raluca Gavrilă, Andrei Avram, Octavian Buiu, and Gheorghe Dinescu

28 Multi-Temperature Model for Ultrafast Laser Experiments on Single Layered Graphene 411
Mihai Oane, Bogdan Alexandru Sava, Lucica Boroica, Rareş Medianu, Ileana Cristina Vasiliu, Mihail Elisa, Ana Violeta Filip, and Catalin Marius Dinca

29 Graphene Oxide—Glass Nanocomposite Obtained by Sintering and Melting 423
Bogdan Alexandru Sava, Lucica Boroica, Ileana Cristina Vasiliu, Mihail Elisa, Alexandra Trefilov, Ana Violeta Filip, Andreea Matei, Catalin Marius Dinca, Oana Gherasim (Fufa), and Mihai Eftimie

30 Thin Films Based on ZnO-Graphene Oxide Heterostructures for Self-Cleaning Applications 435
Dana Perniu, Cristina Bogatu, Silvioara Gheorghita, Maria Covei, and Anca Duta

Part VI Sustainable Communities

31 Renewable Energy Systems: Current Status and Prospects 451
Soteris A. Kalogirou

32 How the European Green Deal Promotes Sustainable Energy Research and Innovation 455
Brigitte Hasewend and Tijana Jokic

33 Environmental Impact of a Residential House Using Cardboard Waste as Construction Material 457
Teofil Câmpean, Nicolae Peiu, Dan-Alexandru Gavrilescu, and Maria Gavrilescu

34 National Mechatronic Platform—The Foundation of a Smart and Sustainable Society Development 471
Vistrian Mătieș, Călin Rusu, and Vlad Tămaș

35 Rethinking Architectural Spaces for Solar Energy Better Use 487
Anda-Ioana Sfintes and Radu Sfintes

36 From Theory to Practice and All the Way Back. Designing with The Sun 501
Ana-Maria Dabija

37 A Framework for Implementing Sustainable Design Strategies in Romanian School Buildings Projects 513
Oana Mihăescu

Correction to: Emerging Issues and Challenges with the Integration of Solar Power Plants into Power Systems C1
Fazel Mohammadi and Mircea Neagoe

Part I
Solar Energy Conversion Systems in the Built Environment

Chapter 1
Challenges in Implementing Solar Energy Conversion Systems in the Built Environment

Ion Visa and Anca Duta

Abstract Implementing the solar energy conversion systems has currently overpass the statement level. For a broad scale implementation, the pre-requisites has to be carefully formulated and fulfilled, according to the implementation site. The built environment represents a particularly challenging option, considering that besides functionality (and efficiency) the solar energy conversion systems have to fulfill various other conditions related to architectural acceptance and are facing, during operation, particular restrictions as e.g. shadowing. These aspects are discussed for the mostly used solar energy conversion systems implemented in the built environment (photovoltaic and solar-thermal systems) and emergent solutions are presented and analyzed.

Keywords Solar energy conversion systems · Building integrated photovoltaics · Solar-thermal systems in the built environment

1.1 Introduction

The Earth is considered to be formed about 4.53 billion years ago, [1]. However, life as we know it now-a-days occurred much later and evolved through centuries from human muscle power to horse, further on from horse power to railway while the past century witnessed the evolution from coal fired railway to the airplane and electric engines, [2].

Thus, the first industrial revolution (1750–1840) was driven by coal and supported the iron, textile, cement, chemicals, etc. industries along with a significant increase in the world population. The second industrial revolution (1840–1914) was launched by the invention of electricity and the discovery of oil that supported the development of rubber, glass and, further on, of various machine tools and the construction of

I. Visa (✉) · A. Duta
Transilvania University of Brasov, Brasov, Romania
e-mail: visaion@unitbv.ro

A. Duta
e-mail: a.duta@unitbv.ro

I. Visa and A. Duta (eds.), *Solar Energy Conversion in Communities*,
Springer Proceedings in Energy, https://doi.org/10.1007/978-3-030-55757-7_1

thermal power plants and of hydropower stations (as the Niagara Falls, completed in 1896), [2].

The broad use of computers and ICTs became well represented after 1970 being known as the third industrial revolution and we are currently living the grand energy transition from conventional fuels to renewable energy sources. This last energy transition is mainly the result of the use of fossil fuels that have accelerated the industrial development but has also increased the pollution, responsible for the climate change.

The industrial revolution allowed the development of health supporting processes and materials and this was mirrored in a continuous increase in the global population number. As the results in Fig. 1.1 show, a significant increase in the world's population was observed in the past 70 years along with a restructuring of this population in regions with factories. This led to an increase of the share of people living in urban areas, from 30% (in 1951) to 56% (in 2019), according to the data delivered by the United Nations, Department of Economic and Social Affairs, [3].

By 2100 the world population is projected to reach 10.9 billions with a much lower annual growth rate (0.1%) as compared with the current rate. However, by the end of our century, the population increase is projected to nearly stop as result of falling in the values of the global fertility rates from 2.5 (today) to 1.9 births per woman, according to the United Nations report, [4]. The urban population is expected to nearly double by 2030, increasing from 2.84 billions in 2000 to 4.9 billions by then [5].

The extended industrial development asked for larger amounts of energy while the extended world population asks for larger amounts of energy for direct use. As the data in Fig. 1.2 show, the primary world energy consumption steadily increased, even after the need for a wise (and reduced) energy use was officially recognized.

Currently, about 80% of the world's energy is obtained using fossil fuels, mainly oil, coal and gas, [7]. Natural gas generates more energy for the same amount of emissions as coal, thus the increase in the use of gas contributes in decreasing the carbon intensity and emissions in the last decade, in China, [8].

The share of the energy consumption used in buildings ranges between 20 and 40% in developed countries and is estimated to increase by 34% in the next 20 years, according to the review presented in [9].

The fossil fuels use is also linked with the increased amounts of greenhouse gases (GHG, mainly CO_2) emissions that did not exceed a concentration of 300 ppm in the atmosphere up to the industrial revolution but exceed now 400 ppm as result of fossil fuels burning, [6] Thus, to cover the energy demand and have a liveable Earth, in an affordable way, a set of measures have to be considered to mitigate the energy demand (in any process, particularly in buildings) and to produce this energy using renewable energy sources that are not emitting GHG during their energy production processes. In terms of GHG, during operation the renewable energy sources can be divided into: zero (wind, solar and hydro), low (geothermal) and neutral (biomass) resources, [8].

Moreover, *the energy returned to energy invested* (EROI) represents an indicator of the viability of the energy business, [2], thus the energy consumed for obtaining

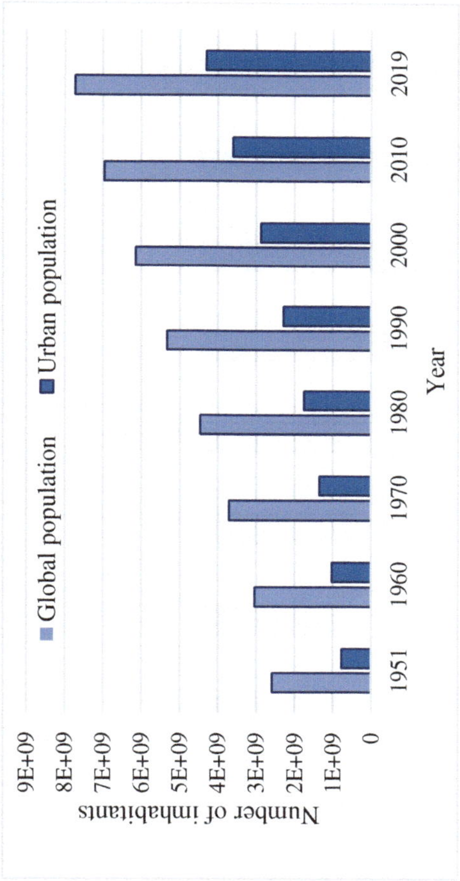

Fig. 1.1 Evolution of the global world population and of the urban population, [3]

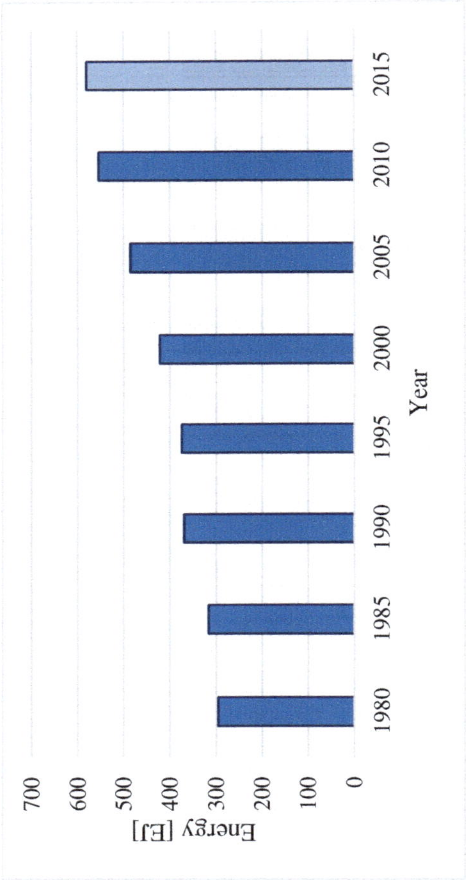

Fig. 1.2 Primary world energy consumption, [6]

the installations/systems using renewable energy sources have to be well watched to reach sustainability.

The state and the governmental policies are supporting, in different degrees, the development of energy-efficient buildings (and products). Starting with 2010, Sustainable Energy Action Plans were developed by EU countries aiming at CO_2 emissions mitigation and at decreasing the buildings energy demand, [10]. This is why relevant documents are currently regulating the development of the built environment as the Energy Performance of Buildings Directive (EPBD) in EU, the ASHRAE standard in US and the Design Standard for Energy Efficiency of Public Buildings in China, as synthesized in the review presented in [11]. In China, in 23 provinces, under the guidance of local and national authorities, a mandatory policy for implementing solar water heating systems in buildings with less than 12 floors, was launched since 2007 and resulted, after 10 years in reaching a share of 24% of solar water heating in the urban domestic area, [12]. Following the legal frame, the nearly Zero Energy Building (nZEB) concept was extended towards the nearly zero-energy district (nZED), [13], or nearly zero-energy community (nZEC), [14]. Further on, EU is exploring routes targeting a zero emissions economy by 2050 looking at the role played by the buildings and related industries, focusing on GHG emissions, resources efficiency, water use, health and comfort, resilience and adaptation to climate change, [15].

1.2 Decreasing the building's Energy Demand

The energy consumed in buildings in developed countries has surpassed the energy consumed in transportation and industrial sectors, [16, 17]. The energy is consumed for heating and/or cooling, for lighting and for powering other home appliances. The energy demand of a building has to consider the building's performance strongly depending on the form, compactness, orientation and thermo-physical properties of the building's envelope and shading. Besides these, the urban context where the building is implemented affects the building's performance e.g. by shadowing and reflection of the solar radiation from the surrounding buildings. Further on, for multi-stored buildings, the energy demand can vary about 10% according to the level where the apartment is.

The heat control entering/exiting the buildings represents a serious issue considering the heat gain during the warm season and the heat loss during cold(er) seasons. Thermal insulations represents the obvious solution to mitigate the effects caused by this issue. For the opaque part of the building envelope (walls, floor, roof) the usual insulation technologies are based on slabs or blocking elements (made of mineral wool, cellular rubber, saw dust, etc.) or on blanket insulation (cotton, wood fibers, etc.), insulating boards, etc. Insulations that are more efficient are currently under investigation as the thermally switchable insulation that can reversibly improve or decrease the heat transfer, [18]. However, the largest heat loss or gain is observed through glazing as outlined in [17] and in [19] for China where 40–50% of the winter heat losses are through the glazing.

The current trend is to develop solar facades that are satisfying the common requirements for facades and the possibility to reflect, absorb and reuse the solar energy. This can be done by implementing special glazing in these solar facades. According to the review presented in [20], the chromic devices can be electrochromic, gasochromic, photochromic, or thermochromic with the last two responding directly to light or temperature changes. The mostly often cited thermochromic materials are metal-oxide based, among which VO_2 is the most promising alternative where switching occurs between a low temperature conducting phase (low IR absorption thus large IR throughput) and a high(er) temperature, metal-like phase, with high IR reflectance, [21].

It has to be noticed that the buildings in the urban area are required to face additional issues as the urban heat island effect. This effect is the consequence of urban density that supports reaching temperatures with several degrees higher than the non-urban surroundings. Therefore, the use of thermochromic glazing (using VO_2/TiO_2 coating to prevent VO_2 oxidation, [17]) for buildings in the urban area represents a possible solution. To mitigate the heat island effect various solutions were proposed as the *cool pavements* that represent a regular asphalt pavement cooled down by a cooling liquid that flows through pipes inserted inside the pavement. This system resembles a solar-thermal collector but care should be taken as this type of system cannot face high traffic or heavy-vehicles loading. Obviously, the cooling liquid (e.g. anti-freeze) can be further used in various applications in the built environment, [22].

Besides these alternatives, the heat loss (thus the energy consumed for heating) can be mitigated by replacing the double-plane glazing with a triple- or quadruple-plane glazing that can lead to up to 67% energy savings, [17, 23].

These concepts were implemented when developing the R&D Institute of the Transilvania University of Brasov, as a sustainable built environment dedicated to the research community in the university. The construction has 12 laboratory buildings that are interconnected by two spine areas and a central atrium, Fig. 1.3.

The buildings are designed according to the Low Energy Buildings (LEB) status and the entire Institute functions as a Nearly Zero Energy Community, considering the usual energy consumers (not the high energy consuming research equipment). The thermal energy demand estimated for this community was of 57.8 kWh/(m^2 year) for heating, 5.75 kWh/(m^2 year) for cooling and 3.14 kWh/(m^2 year) for domestic hot water, [14] while the electricity consumption for lighting and other appliances (except the laboratory equipment) was 12.3 kWh/(m^2 year). This gives an overall energy demand of 79 kWh/(m^2 year) for the Institute and qualifies the construction in the Nearly Zero Energy Building standard.

The average electrical energy mix, for lighting, common appliances and for powering the thermal systems will eventually consist of photovoltaic systems and small wind turbines and will use the grid as backup source. The PV systems can be mounted on the top terrace of each building (117,535 kWh/year) and nearby the

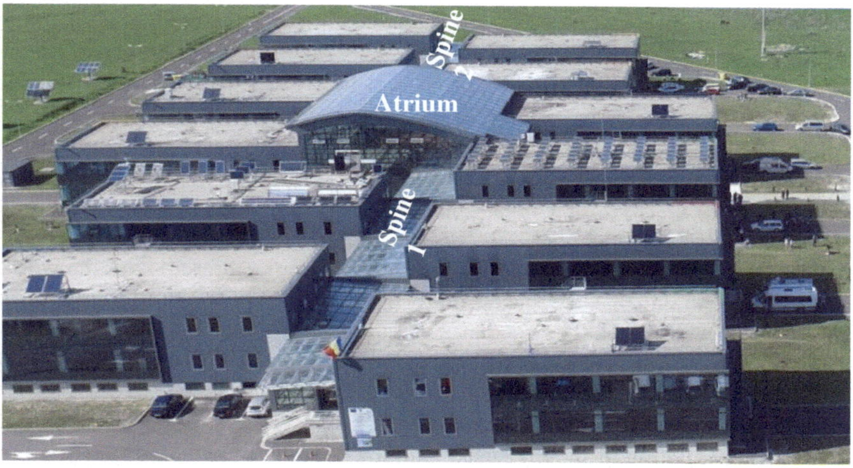

Fig. 1.3 The R&D Institute of the Transilvania University of Brasov, Romania

buildings as a 1 MWp tracked PV park, while the small wind turbines can be installed on the rooftops (34,672 kWh/year). Currently the components of the energy mixes are installed and monitored on two laboratory buildings, Fig. 1.3

1.3 Photovoltaic Systems in the Built Environment

Currently, solar energy is considered as a viable energy source (in the built environment) when using conversion systems to produce electricity (photovoltaic systems) and thermal energy (solar-thermal systems).

It is currently reported that the PV market approaches 134 billion USD in 2020, [2]. For the use at buildings level, a preliminary calculation shows that if a normal house consumes on average 4–6 kWh/day in an implementation location with an average solar insolation of 3.5–7 kWh/m^2/day, about 4–5 m^2 of photovoltaic modules are required. Obviously, this estimation can be more reality-related when specifying the climatic profile of the implementation location and the type of PV modules to be implemented. According to the review presented in [24], there are currently about 24 solar cells technologies and among these, about nine apply to transparent solar cells.

The solar to power energy conversion efficiency depends on the PV material and on the PV module construction (as e.g. opaque or semi-transparent). The highest conversion efficiency reported by the National Renewable Energy Laboratory in U.S. (NREL) corresponds to the four-junctions, concentrator solar cell (47.1%), followed by the four-junctions cell without concentrator (39.2%), single junction GaAs cell

(35.5%), two-junction cell (32.9%), single crystal silicon solar cell (26.7%), multi-crystalline silicon solar cells (21.2%), CIGS (23.4%), CdTe (22.1%), amorphous silicon (14%), [24]. These results correspond to PVs under research. The best commercial PV modules in 2019 were the mono-crystalline silicon modules with an expected lifetime for the materials of 25 years and a conversion efficiency of 22.8%, manufactured by SunPower, followed by the modules manufactured by LG and the REC Group (21.7%), China Sunergy (21.17%), Solaria (20.5%), Panasonic (20.3%), etc. [25].

In 2007 China (Taiwan) became the largest PV manufacturer all over the world, exceeding Europe, Japan, USA and currently producing about two thirds of the world total PVs production, [2]. The second-generation solar cells of CdTe (16.5%), or CIGS cells (18.4%), poly-crystalline and amorphous silicon solar cells are covering about 20% of the solar cell market. Other solar cell types, as the dye-sensitized solar cells (DSSC) are at laboratory testing stage.

The implementation of the PV systems in the built environment is supported by the regional strategies to develop (nearly) Zero Energy Buildings. These systems are investigated as Building Attached PVs (BAPV) or as Building Integrated PVs (BIPV) depending on their effect on the building structure: while BAPV are added to the building and have no direct effect on the structure's function, the BIPV can be integrated in the building envelope as in the roof or in the façade. The largely reported BIPV systems have as active elements Si-PV modules, [26].

Most of the BIPV modules have a lower efficiency comparing with the modules in-field implemented, mainly because of the temperature effect. As well known, the PVs register a drop in the efficiency when the temperature increases. The crystalline silicon solar cells have a temperature coefficient, T_c, of −0.5%/°C that means that a 10 °C increase in temperature can lead to a 5% power loss. If the power output is measured in standard testing conditions (STC), at 25 °C, any temperature on the module higher than this standard temperature will lead to a power loss; e.g. for a 250 W PV module that has 65 °C on the surface, the solar module power loss will be of 50 W, thus the maximum output that can be expected is of 200 W. Therefore, self-ventilation or self-heat dissipation mechanisms are proposed in the design stage supported by natural ventilation and buoyant force, [20]. Further on, mounting rows of PV modules (e.g. on the roof) can also consider to support the air-flow as a natural ventilation option.

The amount of collected solar energy can be increased by tracking. Tracking can be easily applied in the built environment to PV rows (single axis tracking), to platforms implemented on the flat roofs or in the close vicinity of the building (double axis tracking). Well-designed tracking algorithms may increase the amount of collected solar energy by about 30% [27, 28]; if this is not followed by heating the modules, the power output will be increased with a similar percentage. In-field studies, supported by an infrastructure consisting of tracked platform on-ground installed, near the buildings, and of tracked PV rows installed on the flat roof terrace of a building are developed in the R&D center Renewable Energy Systems and Recycling (RESREC), in the Transilvania University of Brasov, Romania, as presented in Fig. 1.4.

(a) (b)

Fig. 1.4 PV systems installed in the RESREC center: **a** single axis tracked PV rows and **b** dual axis PV tracked platforms installed on ground

Moreover, the amount of solar radiation that can reach the PV material can be increased by depositing a self-cleaning layer on the glazed surface of the PV module. This self-cleaning layer can be either super-hydrophobic or super-hydrophilic depending on the expected mechanism for cleaning. These are currently emergent technologies that are under industrial implementation, [29].

Besides their performance, the BIPV has to meet additional requirements for building integration. These requirements are related to aesthetic benefits and lack of shadowing in the implementation locations that has to be considered at the implementation moment and further on.

Research and in-filed studies outlined that for BIPV an average efficiency and performance degradation of 1% can be observed for each year of operation in low insolation areas, [30].

A research option intensively investigated for integrating PVs in buildings is related to the transparent or semi-transparent modules. These represent a real challenge considering that a PV cell has to absorb photons with energies in the VIS (and UV) part of the solar spectrum thus can hardly be considered transparent. The (semi)transparent solar cell have to absorb radiation in the NIR and in the UV spectral range and are used for integration in the glazing part of a building, suitably oriented. These cells are expected to reach transparency up to 80% and efficiencies high enough to justify their implementation.

A synthesis of the (semi) transparent solar cells currently investigated is presented in Table 1.1. For these cells, the best transparencies (Vis transmittance) are below 80% even for the laboratory best products, while the efficiencies are up to 8%, Table 1.1. The cost of these modules depends on the deposition technology and on the components, mainly the fillers in the structure.

The broad scale implementation of PVs for developing a sustainable built environment will eventually consider the entire building's energy consumption *vs.* the energy demand for manufacturing the PV system. The Energy Return on Energy Invested, EROI (or ERoEI) represents thus an indicator on how sustainable a PV system is. Obviously, this indicator depends on the type of PV system (type of materials used for manufacturing the PV modules) and on the insolation in the implementation location.

A study on the EROI of PVs implemented in low insolation locations (Germany or in Switzerland) is presented in [29]. Based on accurate and realistic calculations in the selected locations it was outlined that the PV implementation is far from the EROI value of 5 required for sustainability. Actually, the EROI value estimated for a building integrated PV system was less than 1 (0.82), thus the PVs broad implementation represents a clean way to produce electrical energy but is not a fully sustainable one. Moreover, the cost associated to these systems are currently subsidized but are high; the solar electricity price is over 10 cents/kWh that is double considering as reference the cost of 5 cents/kWh corresponding to solar-thermal power plants, [2]. Most of these solar-thermal power plants are located in USA and in Spain, [2].

Table 1.1 (Semi)transparent PV modules for building integration

Type of PV module	Efficiency (%)	Transmittance (%)	Film thickness	References
Thin film photovoltaics	7.8–8.4	20–23	6–12 μm	[30, 31]
Near infrared transparent solar cells	1.3–0.1%(using IR radiation)	Over 65		[31, 32]
VIS-transparent polymer solar cells	4	66		[31, 33]
	7.56	25		[31, 34]
	5.6	30		[31, 35]
Transparent luminescent solar concentrator	0.4	86		[31]
Semi-transparent Perovskite solar cell	6.4	29	40 nm	[31, 36]
	7.3	22	40 nm	[31, 37]
	12.7	77		[31]
DSSC obtained using the electrophoretic deposition	7.1	55	5 μm	[31]
DSSC deposited via dip coating	8.22	70		[31, 37]
Quantum dot solar cells	3.88	22		[31]
	5.4	24.1		

Further research will therefore be directed to increase the photovoltaic conversion efficiency and decrease the cost of the materials involved in the system (PV modules, inverter, etc.).

1.4 Solar-Thermal Systems in the Built Environment

The use of solar energy to produce heat is well known since 1896 when Clarence Kemp proposed the first system that used solar energy to produce hot water. During the first half of the twentieth century, in California and Florida (U.S.) solar powered hot water systems were installed in almost 30% of the households, [28]. The key component in a solar-thermal system is the solar energy collector that can be mainly developed as flat plate collector or as evacuated tube collector.

When installing solar-thermal systems in the temperate or cold(er) climates, one major issue that has to be faced is represented by the lack of full compliance between the periods when the solar energy is available and the periods when heat is needed. This is why, solar-thermal energy systems are mainly used to cover the hot water demand and less the space heating energy demand in the built environment. For space heating, usually the major RES component is a heat pump and solar-thermal heating systems are recommended for pre-heating or as supplementary heating source, as proved by the Solar Decathlon buildings, [38].

Additionally, the heat storage solutions are highly investigated as these solutions have to be implemented in buildings without asking for excessive storage space. Heat storage is proposed to be done using phase change materials (PCM) that can store a large amount of heat as latent heat, as e.g. microencapsulated paraffin (60%) within an ethylene based copolymer (40%), [20]. PCMs have large energy densities and may have different melting temperatures, [38].

To compensate the seasonal fluctuations in the amount of solar energy, thermochemical materials (TCM) represent a possible alternative, investigated in the review presented in [39]. Considering the boundary conditions, related to the heat storage capacity, energy density of the material, the hydration and the de-hydration temperatures, safety, chemical stability, corrosive action it was found that the material that mostly satisfies the pre-requisites is K_2CO_3 but all these pre-requisites were not fully satisfied by any of the over 500 investigated reactions and materials.

Another heat storage option is represented by the Trombe walls that represent a thermal mass installed on the Southern building envelope that regulates the heat flow into the building. In the case study discussed in [40] the system contains a 40 cm concrete structure with plaster on the inner side and a single glazed window in an aluminum frame on the outer/exterior side.

Besides their functionality, the solar-thermal systems installed in the built environment have to comply with additional requirements as aesthetics and safety. Considering these, flat plate solar-thermal collectors were developed by Visa and the group he coordinates with various shapes (trapeze, triangle), sizes and colors, [41, 42] for implementation on the building facades. These PV modules grouped in variously formed arrays can well cover the opaque part of the optimally oriented facades, as presented in Fig. 1.5.

The size of these type of modules can be redesigned according to the particular needs of the façade and various colors can be employed for the absorber plates allowing a diversity that might be required for the urban landscape [43, 44]. This topic was the core subject of a COST project, Action TU-1205 BIST that developed a network active in modelling and performance analysis of building integrated solar-thermal systems, [45].

1.5 Conclusions

A detailed overview of the solar energy conversion systems implemented in the built environment was recently developed, [43], and outlines that the use of solar energy in developing sustainable cities has to be carefully considered and implemented in a coherent way based on efficient affordable solutions. A better collaboration frame between architects, constructors and renewable energy systems designers has to be implemented from the very beginning when developing new dwellings. Urban retrofit represents another viable option for extending the solar share in the built environment

Fig. 1.5 Outdoor testing rig for the solar-thermal collectors in the R&D Institute of the Transilvania University of Brasov

by implementing and operating systems that well consider functionality and the specific features of the implementation location at the installment moment up to the end of the lifetime projected for these systems.

Sustainable (urban) communities represents the path to be followed starting with the near future; it requires the support of the inhabitants (users), of the stakeholders and it is advisable to have representatives of all the interested parties in a Community Company for Sustainability, working for the welfare of the community, based on well-informed acceptance and support, [46–48].

References

1. N. Kitadai, S. Maruyama, Origins of building blocks of life: a review. Geosci. Front. **9**, 1117–1153 (2018)
2. N. Khan, E. Kalair, N. Abas, A.R. Kalair, A. Kala, Energy transition from molecules to atoms and photons. Eng. Sci. Technol. Int. J. **22**, 185–214 (2019)
3. United Nations—Department of Economic and Social Affairs—Population Division, World Population Prospects (2019)
4. A. Cilluffo, N.G. Ruiz, World's population is projected to nearly stop growing by the end of the century, FactTank News in the Numbers, 19 June 2019. https://www.pewresearch.org/fact-tank/2019/06/17/worlds-population-is-projected-to-nearly-stop-growing-by-the-end-of-the-century/

5. R.J. Smithers, K.J. Doick, A. Burton, R. Sibille, D. Steinbach, R. Harris, L. Groves, M. Blicharska, Comparing the relative abilities of tree species to cool the urban environment. Urban Ecosyst. **21**, 851–862 (2018)
6. H. Ritchie, M. Roser, Energy. Published online at OurWorldInData.org (2020). Retrieved from: https://ourworldindata.org/energy
7. P. Lotfabadi, Solar considerations in high-rise buildings. Energy Build. **89**, 183–195 (2015)
8. W. Strielkowski, E. Volkova, L. Pushkareva, D. Streimikiene, Innovative policies for energy efficiency and the use of renewables in households. Energies **12**, 1392 (2019)
9. Z.B. Liu, L. Zhang, G.C. Gong, H.X. Li, G.F. Tang, Review of solar thermoelectric cooling technologies for use in zero energy buildings. Energy Build. **102**, 207–216 (2015)
10. E. Saretta, P. Caputo, F. Fronitini, A review study about energy renovation of building facades with BIPV in urban environment. Sustain. Cities Soc. **44**, 343–355 (2019)
11. D.H.W. Li, L. Yang, J.C. Lam, Zero energy buildings and sustainable development implications—a review. Energy **54**, 1–10 (2013)
12. J. Huang, J. Fan, S. Furbo, L. Li, Solar water heating systems applied to high-rise buildings—lessons from experiences in China. Energies **12**, 3078 (2019)
13. A.R. Amaral, E. Rodrigues, A.R. Gaspar, A. Gomes, Review on performance aspects of nearly zero-energy districts. Sustain. Cities Soc. **43**, 406–420 (2018)
14. I. Visa, A. Duta, M. Moldovan, Nearly zero energy community—a transition concept towards sustainability. J. Eng. Sci. Innov. **2**(3), 69–79 (2017)
15. E. Commission, *LEVEL(S) Taking Action on the Total Impact on the Construction Sector* (Publications Office of the European Union, Luxembourg, 2019)
16. L. Pérez-Lombard, J. Ortiz, C. Pout, A review on buildings energy consumption information. Energy Build. **40**, 394–398 (2008). https://doi.org/10.1016/j.enbuild.2007.03.007
17. S.D. Rezaei, S. Shannigrahi, S. Ramakrishna, A review of conventional, advanced, and smart glazing technologies and materials for improving indoor environment. Solar Energy Mater. Solar Cells **159**, 26–51 (2017)
18. H. Cui, M. Overend, A review of heat transfer characteristics of switchable insulation technologies for thermally adaptive building envelopes. Energy Build. **199**, 427–444 (2019)
19. S.Z. Wang, M. Guan, Windows—the key of building energy efficiency. Constr. Glas. Glas. Ind. **4**, 3–8 (2005)
20. C.M. Lai, S. Hokoi, Solar façades: a review. Build. Environ. **91**, 152–165 (2015)
21. C.G. Granqvist, G.A. Nik, Thermochromic oxide-based thin films and nanoparticle composites for energy-efficient glazings. Buildings **7**, 3 (2017)
22. Y. Qin, A review on the development of cool pavements to mitigate urban heat island effect. Renew. Sustain. Energy Rev. **52**, 445–459 (2015)
23. M. Arıcı, H. Karabay, M. Kan, Flow and heat transfer in double, triple and quadruple pane windows. Energy Build. **86**, 394–402 (2015)
24. A.A.F. Hussain, W.Z.W. Hasan, S. Shafie, M.N. Hamidon, A review of transparent solar photovoltaic technologies. Renew. Sustain. Energy Rev. **94**, 779–791 (2018)
25. NREL, Best Research-cell Efficiencies, (2019). https://www.nrel.gov/pv/assets/pdfs/best-research-cell-efficiencies.20200311.pdf
26. EnergySage, What are the best solar panels on the market? (2020). https://news.energysage.com/best-solar-panels-complete-ranking/
27. E. Biyik, M. Araz, A. Hepbasli, M. Shahrestani, R. Yao, L. Shao, E. Essah, A.C. Oliveira, T. del Caño, E. Rico, J.L. Lechón, L. Andrade, A. Mendes, Y.B. Atlı, A key review of building integrated photovoltaic (BIPV) systems. Eng. Sci. Technol. Int. J. **20**, 833–858 (2017)
28. I. Visa, Renewable energy systems case study: solar energy conversion systems, in *The 11 IFToMM Symposium*, vol. 18 (Springer Publishing House, 2014), pp. 31–49
29. M. Neagoe, I. Visa, B.G. Burduhos, Increasing the tracking efficiency of photovoltaic systems, sustainable energy in the built environment—steps towards nZEB, CSE 2014 (Springer Science and Bussiness Media, 2014), pp. 443–461
30. M. Covei, A. Duta, I. Visa, Self-cleaning metal oxide based thin films for photovoltaic modules, in *ASTR Conference Proceedings* (2019)

31. I. Visa, A. Duta, M. Moldovan, Nearly Zero Energy Community—a transition concept towards sustainability. J. Eng. Sci. Innov. **2**(3), 69–79 (2017)
32. S. Hore, C. Vetter, R. Kern, H. Smit, A. Hinsch, Influence of scattering layers on efficiency of dye-sensitized solar cells. Sol. Energy Mater. Sol. Cells **90**, 1176–1188 (2006). https://doi.org/10.1016/j.solmat.2005.07.002
33. F. Ferroni, R.J. Hopkirk, Energy return on energy invested (ERoEI) for photovoltaic solar systems in regions of moderate insolation. Energy Policy **94**, 336–344 (2016)
34. R.R. Lunt, V. Bulovic, Transparent near-infrared organic photovoltaic solar cells for window and energy-scavenging applications. Appl. Phys. Lett 98 (2011). https://dx.doi.org/10.1063/1.3567516
35. C. Chen, L. Dou, R. Zhu, C. Chung, T. Song, Y.B. Zheng, et al., Cells produced by solution processing. [120712084458005]. ACS Nano (2012). https://dx.doi.org/10.1021/nn3029327
36. K.-S. Chen, J.-F. Salinas, H.-L. Yip, L. Huo, J. Hou, A.K.-Y. Jen, Semi-transparent polymer solar cells with 6% PCE, 25% average visible transmittance and a color rendering index close to 100 for power generating window applications. Energy Environ. Sci. **5**, 9551 (2012)
37. R. Betancur, P. Romero-Gomez, A. Martinez-Otero, X. Elias, M. Maymó, J. Martorell, Transparent polymer solar cells employing a layered light-trapping architecture. Nat. Photon. **7**, 995–1000 (2013). https://dx.doi.org/10.1038/nphoton.2013.276
38. A. Bahramian, D. Vashaee, In-situ fabricated transparent conducting nanofibershape polyaniline/coral-like TiO2 thin film: application in bifacial dye-sensitized solar cells. Sol. Energy Mater. Sol. Cells **143**, 284–295 (2015). https://dx.doi.org/10.1016/j.solmat.2015.07.011
39. The Renewable Energy Hub, History behind solar thermal and solar water heating technology (2017). https://www.renewableenergyhub.us/solar-thermal-information/the-history-of-solar-thermal-technology.html
40. Z. Ma, H. Ren, W. Lin, A review of heating, ventilation and air conditioning technologies and innovations used in solar-powered net zero energy solar decathlon houses. J. Clean. Prod. **240**, 118–158 (2019)
41. P.A.J. Donkers, L.C. Sogutoglu, H.P. Huinink, H.R. Fisher, O.C.G. Adan, A review of salt hydrates for seasonal heat storage in domestic applications. Appl. Energy **199**, 45–68 (2017)
42. C. Lamnatou, D. Chemisana, R. Mateus, M.G. Almeida, S.M. Silva, Review and perspectives on life cycle analysis of solar technologies with emphasis on building-integrated solar thermal systems. Renew. Energy **75**, 833–846 (2015)
43. I. Visa, A. Duta, Innovative solutions for solar thermal systems implemented in buildings. Energy Procedia **85**, 594–602 (2016)
44. I. Visa, A. Duta, M. Comsit, M.D. Moldovan, D. Ciobanu, R. Saulescu, B.G. Burduhos, Design and experimental optimization of a novel flat plate solar thermal collector with trapezoidal shape for facades integration. Appl. Therm. Eng. **90**, 432–443 (2015)
45. I. Visa, M. Moldovan, M. Comsit, M. Neagoe, A. Duta, Facades integrated solar-thermal collectors—challenges and solutions. Energy Procedia **112**, 176–185 (2017)
46. D. Lamnatou, I. Farkas, Z. Aggelos, A. Buonomano, A. Palombo, D. Chemisana, *Building Integrated Solar Thermal Systems—Design and Application Handbook*
47. I. Visa, A. Duta, M. Moldovan, B. Burduhos, M. Neagoe, *Solar Energy Conversion Systems in the Built Environment* (Springer, Green Energy and Technology, 2020)
48. I. Visa, A. Duta, Community company for sustainability—an innovative tool for implementing renewables, in *The Xth edition of the ASTR Conference* (2015), pp. 56–61

Part II
Solar–Thermal Energy in Communities

Chapter 2
Simple Modeling of the Solar Seasonal Thermal Storage Behavior

Paul Daniel Hiris, Florin Bode, Octavian Gabriel Pop, and Mugur Ciprian Balan

Abstract The study presents a preliminary numerical investigation on the thermal behavior of a ground buried cylindrical solar thermal energy storage tank, operating in a solar district heating network. The proposed analytical model is based on energy balance and on the finite difference method. Due to the symmetry of the system, the model was considered one dimensional. A thermal solar field and an auxiliary gas boiler were considered as the thermal energy sources of the investigated system. The system serves a residential area situated in Zaragoza, Spain, with 1000 dwelling considered as heat consumer. The hourly variation of the operating conditions with different solar radiation and different ambient temperature were taken into account. The mathematical model was validated by comparison with results available in the literature and was further used in a case study, to prove the capability of this simple mathematical model to correctly describe the thermal behavior of a real solar district heating system with seasonal storage and to correctly adjust the storage tank volume with the solar field size and with the heating load. The solar fraction of the investigated system was determined in the range of (55.66–64.5%) depending on the size of the storage tank.

Keywords Seasonal thermal storage · District heating · Solar energy · Mathematical model

2.1 Introduction

In the context in which solar energy is accepted to be the most promising renewable alternative to the fossil fuels [1], its use in district heating systems appears logical and the first large-scale solar district heating system (SDHS) was inaugurated in Sweden in 1978 [2, 3]. Despite the advantage of providing energy without environmental pollution [4], the solar radiation also presents the disadvantage of important diurnal and seasonal variations [5]. Mainly because there is a difference between the district

P. D. Hiris · F. Bode · O. G. Pop · M. C. Balan (✉)
Technical University of Cluj-Napoca, Cluj-Napoca, Romania
e-mail: mugur.balan@termo.utcluj.ro

I. Visa and A. Duta (eds.), *Solar Energy Conversion in Communities*,
Springer Proceedings in Energy, https://doi.org/10.1007/978-3-030-55757-7_2

heat demand and the availability of solar radiation, seasonal thermal storage (STS) is required in all SDHS [3–5]. In such systems, solar energy is usual coupled with other heat sources [6, 7].

SDHS with STS are usually designed for solar fractions of approximate 50% and usually supply more than 100 flats [8]. The storage period can vary from a few hours for diurnal storage cycles, to many months for seasonal cycles [9].

Planning, design and realization of long-term STS is challenging because of the following issues: very long storage period, huge quantity of thermal storage materials, large thermal losses, important temperature variations at charging and discharging, etc. Over time, four different types of long term STS were considered: water tank, pits, boreholes, and aquifers [1, 10].

Due to its high specific heat, water is the most common storage medium in STS. Most STS use water for thermal storage and are located either inside or outside the buildings or underground. The sizes of the tanks used vary from a few hundred liters to a few thousand cubic meters [11].

A large variety of materials like steel, concrete and fiberglass, are used to make STS. The STS are frequently insulated with a thick layer of different materials due to the large surface area resulted from a seasonal storage [1, 11]. STS can operate with different degrees of stratification: highly stratified, moderately stratified and fully mixed or un-stratified [12].

Between the STS sizing methods, those based on climatic data and heat demand data are preferred because of low required computing efforts. Even the simple computing methods allow frequently the correct design of the main components and correct estimation of the STS behavior during long periods of time [13–15].

Numerical simulations performed to investigate different aspects of water storage systems, such as geometry, location of the diffuser, temperature variation, etc., are presented in [16]. Simplified dynamic models of different complexity are: zero-dimensional, quasi-one-dimensional and one-dimensional. Comparison with the experimental data presented in [17] showed that even the one-dimensional model is capable of predicting the performance, the thermal stratification and any other relevant aspects of the storage behavior with sufficient accuracy.

The goal of the study is to present and validate the accuracy of a simple analytical model of the thermal behavior of STS in a SDHS. The model is based on energy balance and on the finite difference method. After validation, the model was used in a case study of a system located in Zaragoza, Spain. The influence of the STS volume on the temperature variation and on the solar fraction was investigated, under the typical meteorological year (TMY) climatic data of the considered location.

2.2 Materials and Methods

2.2.1 Configuration, Location and Energy Components of the SDHS

The investigated SDHS is the same as the one considered in [13]. The scheme of the SDHS is presented in Fig. 2.1.

The SDHS is considered located in Zaragoza, Spain at the coordinates 41.652 °N and 0.881 °E, at the altitude of 218 m. The residential area (1) is considered of 1000 dwellings in multifamily buildings, each of 100 m^2, located in Zaragoza (Spain) as in [13, 18] and is characterized by the required heat demand (Q_d [kWh]) that can be calculated as function of the temperature difference between the inside temperature (t_i) considered constant at 21 °C like in [13] and the exterior temperature (t_e [°C]). In this study only the heating of the residential area is considered. The heating load required for preparing the domestic hot water (DHW) was not considered, because the first goal of the study is to validate the proposed mathematical model by comparing

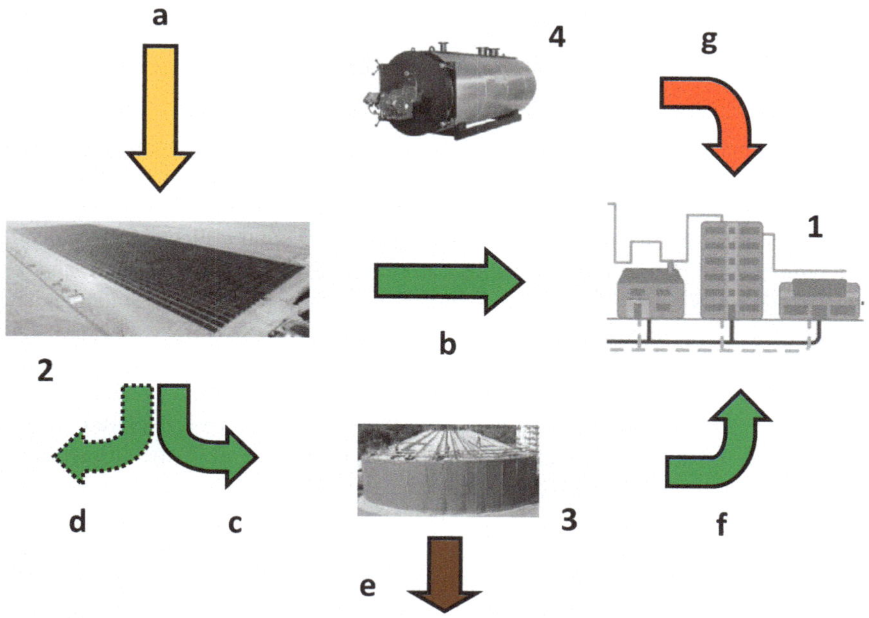

Fig. 2.1 Principle scheme of the solar district heating system: 1—residential area; 2—solar field; 3—seasonal storage tank; 4—gas boiler; a—solar energy (Q_r); b—solar heat for heating (Q_{sh}); c—STS charging solar storage heat (Q_{st}); d—excess solar heat (Q_{sx}); e—heat losses (Q_l); f—STS discharging solar storage heat for (Q_{ss}); g—gas heat (Q_g)

the results with the ones of the considered reference case presented in [13], where DHW heat load was not investigated.

The heat demand can be calculated as:

$$Q_d = k \cdot S \cdot (t_i - t_e) \cdot \Delta\tau \tag{2.1}$$

where $k \cdot S = 0.12$ MW/K was considered the thermal characteristic of the heated buildings (compatible with [13]) and $\Delta\tau = 1$ h is the time step considered in the calculations. If $t_i < t_e$ or if $t_e \geq 12$ °C $=> Q_d = 0$.

The solar energy (Q_r [kWh]) represents the global solar energy on the horizontal plane, available at the considered location and represents the local potential of the solar radiation. The solar field (2) provides the solar heat (Q_s [kWh]) that during the operation of the SDHS is distributed directly to the consumers (Q_{sh} [kWh]) or is accumulated into the cylindrical STS (3) (Q_{st} [kWh]).

The STS is buried underground because it requires a high mounting area and the landscape in the vicinity of residential areas is usually expensive. In fact, only one of the first 11 SDHS systems built in Europe contains an over ground STS [2].

The solar heat provided by the solar field (Q_s [kWh]) depends on the thermal efficiency of the solar collectors (η_{th} [-]) that can be calculated using:

$$\eta_{th} = \eta_0 - k_1 \frac{\Delta t}{I_{gt}} - k_2 \frac{\Delta t^2}{I_{gt}} \tag{2.2}$$

where η_0 [-] is the optical efficiency of the collectors, k_1 [W/m^{2}K] and k_2 [W/m^{2}K^2] are correction factors taking into account thermal losses and I_{gt} [W/m^2] is the global solar radiation on the tilted plane of the collectors, considered of 30°.

In this study, Arcon-Sunmark HT-SolarBoost 35/10 solar collectors are considered with the following characteristics: $\eta_0 = 0.838$; $k_1 = 2.46$ W/m^{2}K and $k_2 = 0.0197$ W/m^{2}K^2, determined by Solartechnic Prüfung Forschung independent laboratory. The aperture area of this collector type is of 12,555 m^2.

The global solar radiation on the tilted plane of the collectors was calculated using a validated algorithm previously used in [19–21]. The total number of collectors was considered 230, in order to obtain almost the same collectors area considered in [13].

In the case of no heat demand or of very low heat demand, if the storage tank is already at its upper temperature limit, the exceeding solar heat (Q_{sx}) must be dissipated.

The heat demand of the residential area (Q_d [kWh]), can be covered either directly by the solar field (Q_{sh} [kWh]) or by the heat provided by the solar storage tank (Q_{ss} [kWh]) or by the gas boiler (Q_g [kWh]) or by any combination of the three heat components.

The situations that can be reached during the operation of the SDHS as a function of the tank temperature (t_{tank} [°C]) relative to the lower temperature limit ($t_{lim,min}$) and upper temperature limit ($t_{lim,max}$), but also to the relative values of the heat demand

Table 2.1 Situations that can be reached during the operation of the SDHS

$t_{tank} < t_{lim,min}$		$t_{lim,min} \leq t_{tank} \geq t_{lim,max}$		$t_{tank} \geq t_{lim,max}$	
A	B	C	D	E	F
$Q_d \geq Q_s$	$Q_d < Q_s$	$Q_d \geq Q_s$	$Q_d < Q_s$	$Q_d \geq Q_s$	$Q_d < Q_s$
$Q_{sh} = Q_s$	$Q_{sh} = Q_d$	$Q_{sh} = Q_s$	$Q_{sh} = Q_d$	$Q_{sh} = Q_s$	$Q_{sh} = Q_d$
$Q_{st} = 0$	$Q_{st} = Q_s - Q_d$	$Q_{st} = 0$	$Q_{st} = Q_s - Q_d$	$Q_{st} = 0$	$Q_{st} = 0$
$Q_{sx} = 0$	$Q_{sx} = 0$	$Q_{sx} = 0$	$Q_{sx} = 0$	$Q_{sx} = 0$	$Q_{sx} = Q_s - Q_d$
$Q_{ss} = 0$	$Q_{ss} = 0$	$Q_{ss} = Q_d - Q_{sh}$	$Q_{ss} = 0$	$Q_{ss} = Q_d - Q_{sh}$	$Q_{ss} = 0$
$Q_g = Q_d - Q_{sh}$	$Q_g = 0$	$Q_g = 0$	$Q_g = 0$	$Q_g = 0$	$Q_g = 0$

Table 2.2 Description of the operating cases

Heat source	A	B	C	D	E	F
Solar	Only heating	Also storage	Only heating	Also storage	Only heating	Exceeding heat
Tank	Stay cold	Self-heating	Provides heating	Self-heating	Provides heating	Stay hot
Gas	Gas heating	No gas heating	No gas heating	No gas heating	No gas heating	No gas heating

(Q_d [kWh]) and of the solar heat provided by the solar field (Q_s [kWh]) are presented in Table 2.1. In this case study $t_{lim,min} = 45$ °C and $t_{lim,max} = 90°$.

The description of the operating cases is presented in Table 2.2.

2.2.2 *The Mathematical Model of the STS Thermal Behavior*

The proposed mathematical model of the STS thermal behavior is one dimensional and is based on the finite difference method. The transition from the real model of the storage tank filled with water to the numerical model is suggested in Fig. 2.2.

The real STS (Fig. 2.2a) was divided into n layers, or volumes, thus the discretization scheme is obtained (Fig. 2.2b). Each of these volumes is considered fully-mixed, meaning that the water temperature is uniform. A similar hypothesis is considered in [22]. Due to the assumed uniformity of the water temperature, the heat transfer between the elementary volumes can be considered one-dimensional and the whole mass of the water in each volume can be considered concentrated in equivalent nodes (Fig. 2.2c).

Such simple models are usually used for fully mixed STS and only the average water temperature is expected to be accurately predicted [13], even if the water temperature is computed in all the nodes.

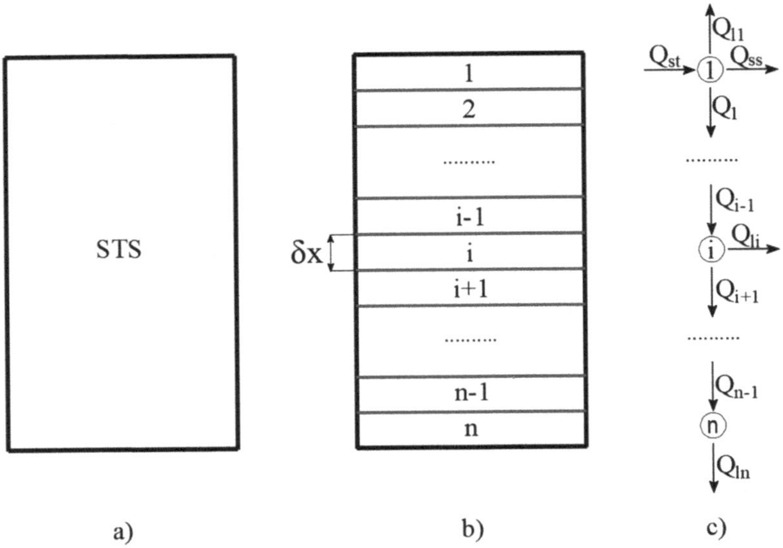

Fig. 2.2 Transition from the real model to the mathematical model, **a** real STS; **b** discretized STS; **c** numerical model of the STS

There are three types of nodes in the considered model.

The upper node (1) represents the energy interface with the environment: in this case the solar field and the heated residential area. The energy balance for this node and for each considered time step (of 1 h) can be calculated as:

$$Q_{a1} = Q_{st} - Q_{ss} - Q_{l1} - Q_1 \tag{2.3}$$

where:

Q_{a1} [kWh] represents the heat accumulated in the node and can be calculated as:

$$Q_{a1} = m_i \cdot c \cdot \Delta t_1 \cdot \Delta\tau \tag{2.4}$$

where: m_i [kg] is the mass of water in each elementary volume, c = 4.18 kJ/kgK is the specific heat capacity of the water and Δt_1 [°C] is the water temperature variation in the upper node (1);

Q_{st} [kWh] represents the share of the heat provided by the solar field that is stored into the STS and can be computed as presented in Table 2.1;

Q_{ss} [kWh] is the heat discharged from the STS into the heating system of the residential area and can be computed as presented in Table 2.1;

Q_{l1} [kWh] is the partial heat loss into the environment (ground), corresponding to this node and can be calculated as:

$$Q_{l1} = k_l \cdot S_1 \cdot (t_1 - t_g) \cdot \Delta\tau \tag{2.5}$$

where k_l [kW/m^{2}K] is the global heat transfer coefficient between the water and the ground, S_l [m^2] is the heat transfer surface, corresponding to the upper elementary volume, t_l [°C] is the water temperature in this node at the previous time step and t_g [°C] is the ground temperature at the previous time step;

Q_l [kWh] is the heat transferred through conduction to the adjacent node and can be calculated as:

$$Q_1 = \frac{\lambda}{\delta x} \cdot S_t \cdot (t_1 - t_2) \cdot \Delta\tau \quad (2.6)$$

where $\lambda = 0.6$ W/mK is the thermal conductivity of the water, δx [m] is the width of each elementary volume, S_t [m^2] is the area of the tank cross section, t_l [°C] is the water temperature in this node at the previous time step and t_2 [°C] is the water temperature in the adjacent node (2) at the previous time step.

The intermediate nodes (i) are the nodes between 2 and n − 1, for which the thermal energy balance for each considered time step can be calculated as:

$$Q_{ai} = Q_{i-1} - Q_{i+1} - Q_{li} \quad (2.7)$$

where Q_{i-1} [kWh] and Q_{i+1}[kWh] are the heat transferred through conduction to the adjacent nodes and can be calculated with Eq. (2.6), by adapting the nodes temperatures, and Q_{li} [kWh] is the partial heat loss to the environment (ground), corresponding to this node and can be calculated with Eq. (2.5) by adapting the heat transfer surface and the node temperature.

The lower node (n) is the node corresponding to the lower elementary volume, for which the thermal energy balance for each considered time step can be calculated as:

$$Q_{an} = Q_{n-1} - Q_{ln} \quad (2.8)$$

where Q_{n-1}[kWh] is the heat transferred through conduction from the adjacent node (n − 1) and can be calculated with Eq. (2.6), by adapting the nodes temperatures and Q_{ln} [kWh] is the partial heat loss to the environment (ground), corresponding to this node and can be calculated with Eq. (2.5) by adapting the heat transfer surface and the node temperature.

Mathematical models based on similar principles were also investigated [23, 24].

2.3 Results and Discussions

2.3.1 Validation

The validation of the presented mathematical model of the STS thermal behavior was realized by comparing the obtained results with similar data available in the

Table 2.3 Input data for the Lunde model

	Jan.	Feb.	Mar.	Apr.	May	Jun.	Jul.	Aug.	Sep.	Oct.	Nov.	Dec.
Q_d	1309	865	632	366	80	0	0	0	0	142	807	1287
Q_s	162	210	279	296	349	330	341	294	186	128	69	99
Q_g	1147	655	353	70	0	0	0	0	0	0	0	606

literature. The Lunde model [25] with its results presented in [13] was considered as reference case, where all the available results are monthly.

The input data corresponding to the Lunde method are presented in Table 2.3.

The comparative values of the monthly average water tank temperature (t_{avg} [°C]) and monthly heat losses (Q_l [MWh]) are presented in Figs. 2.3 and 2.4, respectively.

The monthly accumulated heat (Q_a [kWh]) was also calculated and compared with the corresponding values presented in [13].

The relative errors are situated for the water temperature between (0.0–6.43) %, for the heat losses between (0.22–6.07) % and for the accumulated heat between (0.3–6.06%). Similar results were obtained by comparing the determined values with the presented mathematical model with the results obtained by the methods BKS and GLS, which are also available in [13]. Taking into account the small differences, the presented model can be considered validated.

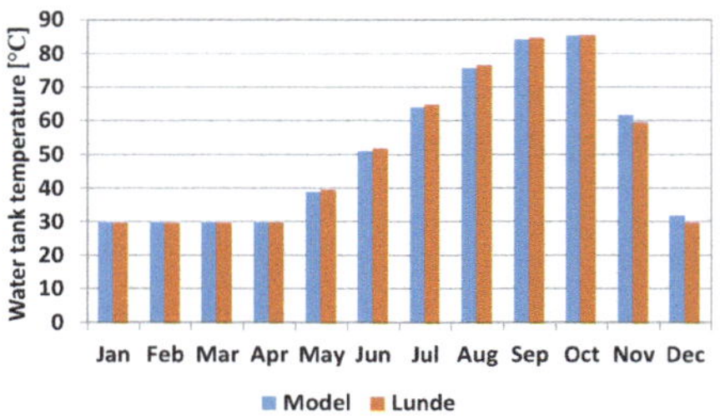

Fig. 2.3 Comparison for water temperature

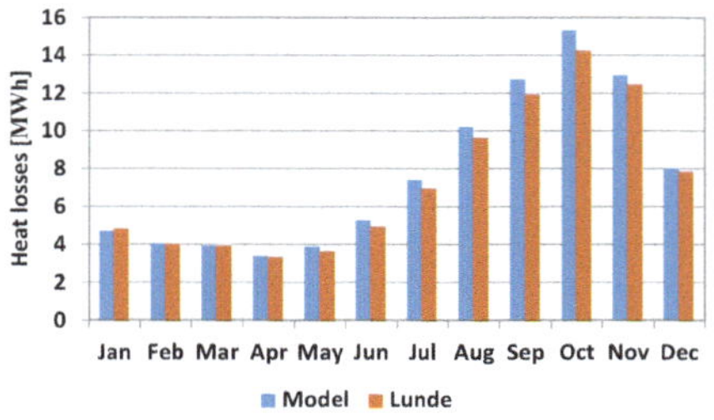

Fig. 2.4 Comparison for heat losses

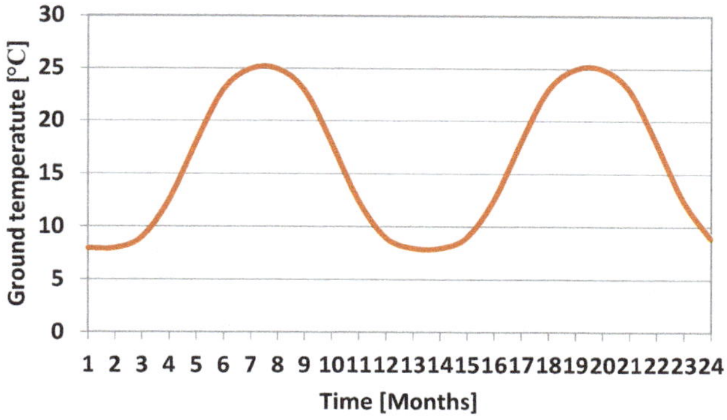

Fig. 2.5 The ground temperature variation

2.3.2 *Case Study*

The proposed model was also used in a case study, in order to evaluate its capability to estimate the thermal behavior of the STS and of the entire SDHS.

The same SDHS was considered, and the simulation was realized based on the hourly variation of the solar radiation and the ambient temperature, available from TMY for Zaragoza.

The considered ground temperature variation for a period of two years is presented in Fig. 2.5.

The yearly average ground temperature is of 15.9 °C and it is in agreement with the available values for different locations in Europe [26].

The upper limit of the average water temperature ($t_{lim,max}$) was set at 90 °C like in the reference case and the lower limit of the water temperature ($t_{lim,min}$) was set at 45 °C, which is different from the value of 30 °C considered in the reference case. The value of 45 °C is considered as the minimum supply temperature for the SDHS and corresponds to a low temperature heating system inside the buildings, considered crucial for the overall efficiency of the SDHS systems [2].

The initial temperature of the STS, at the beginning of the year, corresponding to the middle of the heating season, was considered 55 °C.

The influence of the STS water volume was investigated, considering the following values: 0 m^3 (no STS); 23,000 m^3 (near 22,829 m^3 in the reference case where 50% of the solar fraction was targeted); 30,000 and 36,000 m^3.

The solar fraction of the SDHS (SF [%]) represents the share of solar heat introduced in the heating system of the considered residential area, directly from the solar field or from the STS and is calculated as:

$$SF = \frac{Q_{sh} + Q_{ss}}{Q_d} \cdot 100 \quad (2.9)$$

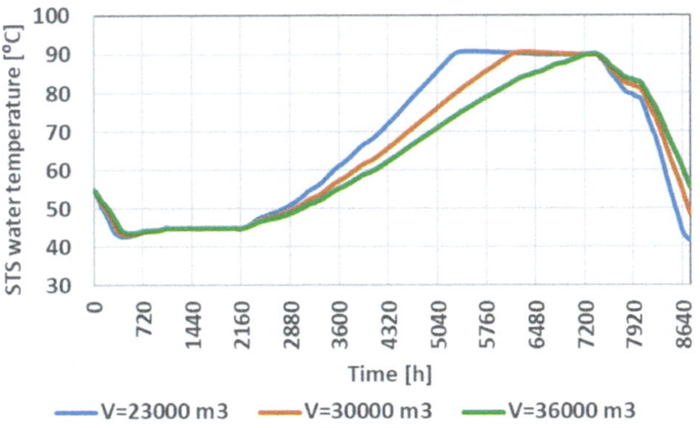

Fig. 2.6 The influence of the STS volume on the average water temperature

The following values of the SF were determined for the considered values of the STS volume: 4.68%; 55.66%; 62.93% and 64.50%, respectively.

The influence of the STS volume on the average water temperature inside the STS, is presented in Fig. 2.6.

It can be observed that the average water temperature varies between the upper limit and the lower limit. Values below the lower limit occur only in a single period when the charging of the STS is not possible because the heat demand is higher than the heat provided by the solar field and in this period heat losses still exist.

The comparative variation curves of the heat components of the SDHS energy balance (Q_d; Q_{sh}; Q_{st}; Q_{ss}; Q_{sx}; Q_g) are presented in Figs. 2.7, 2.8, 2.9, 2.10, 2.11 and 2.12, respectively.

The heat demand (Q_d) is important during late autumn, winter and early spring, but it is null during summer, when solar radiation is high. The share from solar heat that can be directly used for heating (Q_{sh}) is low because solar radiation is low when heat demand is high.

Solar heat available for storage (Q_{st}) is higher in periods when heat demand is low, but when the heat demand increases, the solar heat from the storage (Q_{ss}) becomes available.

Fig. 2.7 The variation of the heating demand (Q_d)

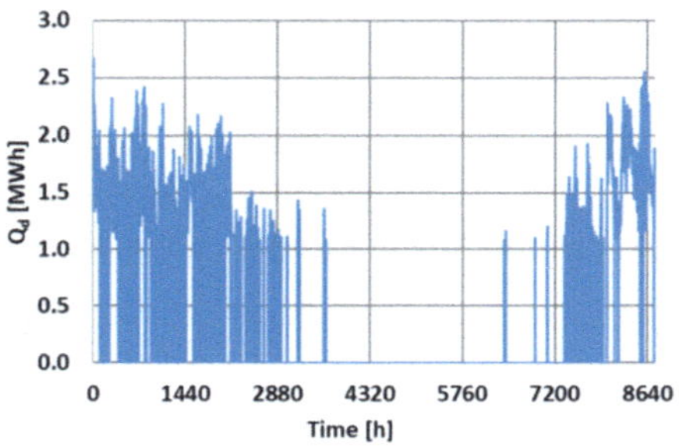

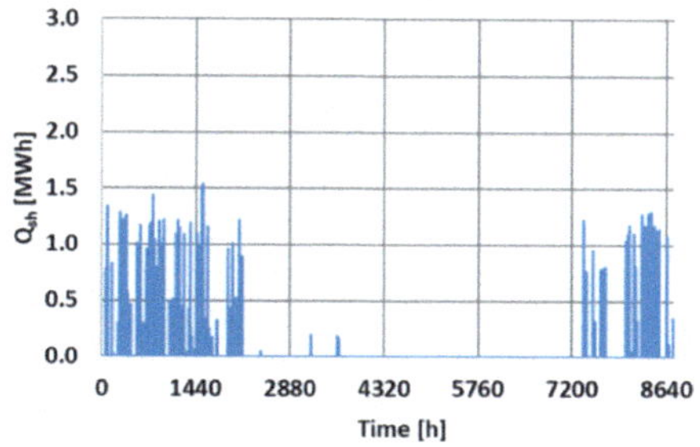

Fig. 2.8 The variation of the solar energy for heating (Q_{sh})

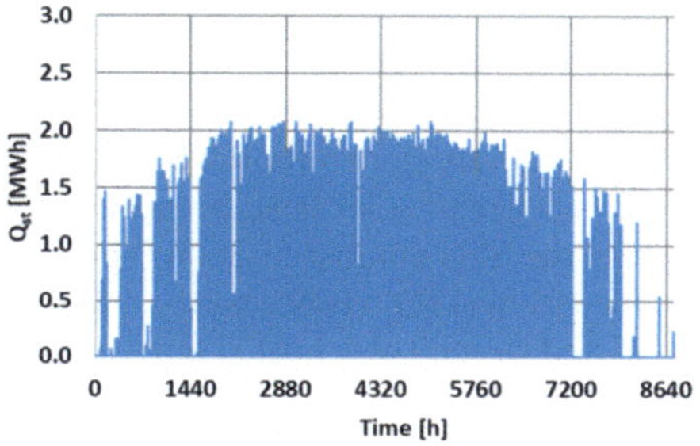

Fig. 2.9 The variation of the solar energy for storage (Q_{st})

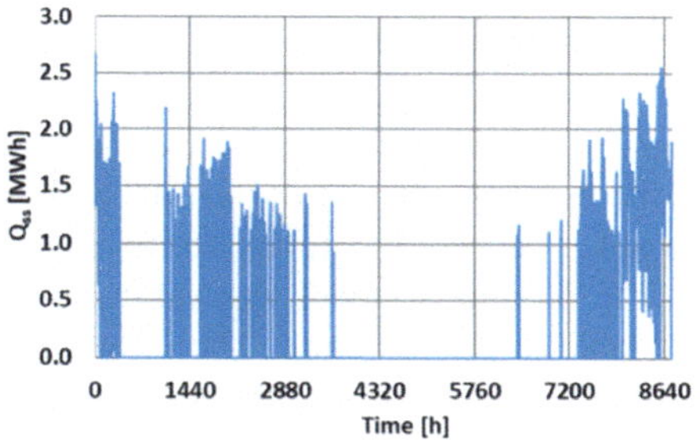

Fig. 2.10 The variation of the heat from STS (Qss)

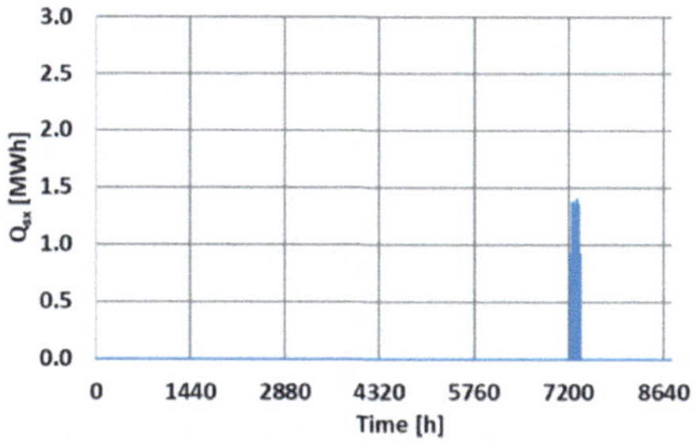

Fig. 2.11 The variation of the excess solar energy (Q_{sx})

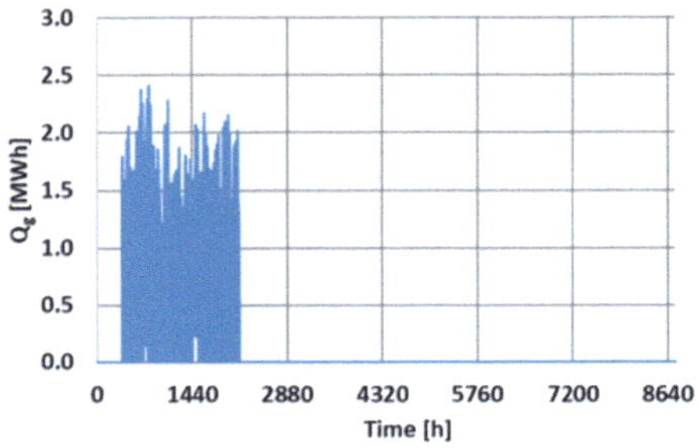

Fig. 2.12 The variation of the gas heat (Q_g)

There is only a single short period with excess solar heat that must be rejected. For lower volumes of the STS this component of the energy balance is much higher.

All charts are drawn for the STS volume of 36,000 m^3. This volume corresponds to the best use of solar heat, because the excess solar heat (Q_{sx}) is lower. In fact $Q_{sx} = 0$ for the STS volume of 36,500 m^3.

The gas heat (Q_g) is necessary only in the last period of the heating season, when direct and stored solar heat is not available.

2.4 Conclusions

The study presents a simple one dimensional mathematical model for the simulation of the thermal behavior of the STS and even of the SDHS. Such simple models are capable to correctly estimate the average water temperature inside the storage tank.

The presented model was validated by comparing the results with similar data available in the literature. The presented validation includes the comparative data with the Lunde method, and the maximum relative differences are of 6.43% for the water tank temperature, of 6.07% for the heat losses and of 6.06% for the accumulated heat. These low differences obtained for a very long simulation period of one year, confirm the correct calculation capability of the proposed method. Moreover, similar comparative results were also obtained for other two different calculation methods.

The mathematical model was also used to investigate the thermal behavior of the SDHS that includes a STS. It was proved that the model can be used to adjust the size of the STS in order to obtain the best use of the available solar radiation. If the size of the solar field is also considered variable, the model can be used to adjust the relative sizes of the two major components of the SDHS system: solar field and STS.

Acknowledgements The research activities were partially supported by the project POCU/380/6/13/123927—ANTREDOC, "Entrepreneurial competencies and excellence research in doctoral and postdoctoral study programs", project co-funded from the European Social Fund through the Human Capital Operational Program 2014–2020.

References

1. A. Dahash, F. Ochs, M.B. Janetti, W. Streicher, Advances in seasonal thermal energy storage for solar district heating applications: a critical review on large-scale hot-water tank and pit thermal energy storage systems. Appl. Energy **239**, 296–315 (2019)
2. International Energy Agency (IEA), Solar heating and cooling program. Task VII. Central solar heating plants with seasonal storage—status report (Stockholm, Sweden, 1990)
3. Z. Tian, S. Zhang, J. Deng, J. Fan, J. Huang, W. Kong, B. Perers, S. Furbo, Large-scale solar district heating plants in Danish smart thermal grid: developments and recent trends. Energy Convers. Manage. **189**, 67–80 (2019)
4. S.A. Kalogirou, *Solar Energy Engineering. Processes and Systems*, 2nd edn (Elsevier, Amsterdam, The Netherlands, 2014)
5. D.Y. Goswami, *Principles of Solar Engineering*, 3rd edn. (Taylor & Francis, USA, 2015)
6. H. Lund, S. Werner, R. Wiltshire, S. Svendsen, J.E. Thorsen, F. Hvelplund, B.V. Mathiesen, 4th generation district heating (4GDH): integrating smart thermal grids into future sustainable energy systems. Energy **68**, 1–11 (2014)
7. H. Lund, P.A. Østergaard, M. Chang, S. Werner, S. Svendsen, P. Sorknæs, J.E. Thorsen, F. Hvelplund, B.O.G. Mortensen, B.V. Mathiesen, C. Bojesen, N. Duic, X. Zhang, B. Möller, The status of 4th generation district heating: research and results. Energy **164**, 147–159 (2018)
8. D. Bauer, R. Marx, J. Nußbicker-Lux, F. Ochs, W. Heidemann, H. Müller-Steinhagen, German central solar heating plants with seasonal heat storage. Sol. Energy **84**, 612–623 (2010)
9. I. Dincer, M.A. Rosen, *Thermal Energy Storage: Systems and Applications*, 2nd edn. (Wiley Ltd., United Kingdom, 2011)
10. E. Guelpa, V. Verda, Thermal energy storage in district heating and cooling systems: a review. Appl. Energy **252**, 113474 (2019)
11. O. Ercan Ataer, Storage of thermal energy, in energy storage systems, in *Encyclopedia of Life Support Systems (EOLSS)*, Developed under the auspices of the UNESCO (Eolss Publishers, Oxford, UK, 2006)
12. M.Y. Haller, C.A. Cruickshank, W. Streicher, S.J. Harrison, E. Andersen, S. Furbo, Methods to determine stratification efficiency of thermal energy storage processes—review and theoretical comparison. Solar Energy **83**, 1847–1860 (2009)
13. M. Guadalfajara, M.A. Lozano, L.M. Serra, Comparison of simple methods for the design of central solar heating plants with seasonal storage. Energy Procedia **48**, 1110–1117 (2014)
14. M. Guadalfajara, M.A. Lozano, L.M. Serra, A simple method to calculate central solar heating plants with seasonal storage. Energy Procedia **48**, 1096–1109 (2014)
15. M. Guadalfajara, M.A. Lozano, L.M. Serra, Simple calculation tool for central solar heating plants with seasonal storage. Sol. Energy **120**, 72–86 (2015)
16. L. Kocijel, V. Mrzljak, V. Glažar, Numerical analysis of geometrical and process parameters influence on temperature stratification in a large volumetric heat storage tank. Energy **194**, 116878 (2020)
17. J. Raccanello, S. Rech, A. Lazzaretto, Simplified dynamic modeling of single-tank thermal energy storage systems. Energy **182**, 1154–1172 (2019)
18. M. Guadalfajara, M.A. Lozano, L.M. Serra, Evaluation of the potential of large solar heating plants in Spain. Energy Procedia **30**, 839–848 (2012)
19. M.C. Balan, L. Jäntschi, S.D. Bolboacă, M. Damian, Thermal solar collectors behaviour in Romania. Polish J. Environ. Stud. **19**(1), 231–241 (2010)
20. P.V. Unguresan, R.A. Porumb, D. Petreus, A.G. Pocola, O.G. Pop, M.C. Balan, Orientation of façades for active solar energy applications in different climatic conditions. J. Energy Eng. **143**(04017059), 1–11 (2017)
21. A.C. Abrudan, O.G. Pop, A. Serban, M.C. Balan, New perspective on performances and limits of solar fresh air cooling in different climatic conditions. Energies **12**, 2113 (2019)
22. J.E. Braun, S.A. Klein, J.W. Mitchell, Seasonal storage of energy in solar heating. Sol. Energy **26**, 403–411 (1981)

23. S. Scolan, S. Serra, S. Sochard, P. Delmas, J.-M. Reneaume, Dynamic optimization of the operation of a solar thermal plant. Sol. Energy **198**, 643–657 (2020)
24. K.M. Powell, T.F. Edgar, An adaptive-grid model for dynamic simulation of thermocline thermal energy storage systems. Energy Convers. Manage. **76**, 865–873 (2013)
25. P. Lunde, Prediction of the performance of solar heating systems utilizing annual storage. Sol. Energy **22**, 69–75 (1979)
26. K. Ochsner, *Geothermal Heat Pumps. A Guide for Planning and Installing*, 1st edn. (Earthscan, UK and USA, 2008)

Chapter 3
Solar Solutions for Housing

Dragoş Ştefan Gherghescu

Abstract Since the first types of housing appeared in history, they were built in such matter that benefit from the Sun. Further in history, in Ancient Greece, the right to Sun was specified in the Civil Law, which gave each house the right to have direct sunlight. Today, in the context of environment protection for lowering the emissions of CO_2, new ways are being searched for consuming less energy by living in a building and for producing renewable energy. We find the Sun once again as an infinite resource. By benefiting of its light, during the day we can obtain electricity and heat that we can use in our homes. The manufacturing industry for the active solutions offers a large variety of products that can be applied on building. The technological support being assured, the architects have the final task of finding new ways of integrating the solutions in the architecture of the building. Due its limitations of material and colours, the active solutions require special architectural approaches for obtaining a final coherent image of the building where the products blend in with the construction materials. While the active solutions are integrated in the building's façade, the passive solutions have to be integrated in the volume of the building in a more unitary way. This paper analyses the newest active and passive solar systems and modern architectural solutions of integrating them in the building's architecture.

Keywords Architectural integration · Passive solution · Active solutions · Solar architecture · Solar solutions

3.1 Introduction

The increasing level of the pollution in the last two centuries have visible effects even today, by increasing of the annual mean temperature which leads to the ice melting. In the last year the permafrost thawing [1] represented a trigger to the all sectors on the subject of changing the way of consuming and produce energy and in the way of protecting the natural environment.

D. Ş. Gherghescu (✉)
University of Architecture and Urbanism "Ion Mincu", 010014 Bucharest, Romania
e-mail: dragos.gherghescu@uauim.ro

I. Visa and A. Duta (eds.), *Solar Energy Conversion in Communities*,
Springer Proceedings in Energy, https://doi.org/10.1007/978-3-030-55757-7_3

The building landscape represents the biggest footprint of the human civilization on Earth, thus the way the building looks and the way it is preserved has a high impact on the human lives. In the same rhythm as the technology evolved, the demand for energy to fuel the factories and the logistics increased. The construction sector is a high consumer of energy and a waste generator; only itself produces one third of the world's waste.

In the last decades many specialists from all the research fields came up with great solutions for reducing the energy consumption, by creating alternative materials for many large-scale fabrication products and solutions for harvesting and producing alternative energy. In architecture, various styles emerged, that integrate the nature as an energy resource. Since the Ancient period the wind and the Sun were used for the comfort of the interior space only by modeling the building with no active systems whatsoever. Many of the passive methods used before in the Ancient History are adopted today along with biophilic, sustainable, solar and bioclimatic architecture styles.

The solar energy represents one of the nature most important infinite resource that the man can use for ensuring the comfort of the interior space. The house has, not only the largest presence on the building environment but represents one of the oldest preoccupations of the man. It always had a special attention and dedication from the architect. For a house to be well integrated to its natural environment, it should have well designed both the active and the passive solutions in respect with its climate, orientation and natural surroundings. The solar power is a natural resource from which a house can benefit both in passive and active way. Comparing with the wind power or the water power, the Sun is present all around the globe during the whole year, the only factors that affect the power of the solar radiation being the clouds and the changes of the Earth's axis inclination.

3.1.1 Sun a Provider, More Than a Star

For the man the Sun represented an important resource since the beginning of agriculture, and thus, the beginning of the first types of houses. Because agriculture required prolonged work of the soil, the need of a property and a permanent shelter appeared. With the sedentariness came the need of a more durable home, well adapted to its environment. From this point forward, the man started to look upwards, differently to the Sun, as a dependable resource for the crops. In time, after the people started to spread around the world, new territories were discovered and, with it, different types of climate.

Even if the Sun is viewed in different ways, many religions see it as the center of all thing being represented as the symbol of a Divinity, the science sees it as the center of our solar system. As for the day to day living point of view, the people see it in a unitary way, representing the beautiful light on the sky that warms us in the cooler days, the light that brings joy and wellbeing when appearing from the clouds, the symbol of the day and the nourishment of the nature. It is the presence on the sky that every man rises his head to, with joy and hope, even in the cloudiest days.

The sunlight, besides assuring the physical and psychological condition, both for the nature and the man, also represents an infinite resource of energy that can be collected and used as renewable clean energy, instead of the fossil fuels. Even if the core of the Sun would stop emitting energy, its successive layers would continue to radiate energy into the space for millions of years. The amount of Sun's energy that reaches the top of the Earth's atmosphere in one hour is equivalent with the amount of energy demand worldwide during a whole year. Once the sunlight entered into the atmosphere, the following phenomenon happens: scattering and absorption by the air particles, reflection by the clouds and conversion to low temperature heat by the soil, oceans and air [2].

A beam of solar light is in the same time both a wave and a particle. The speed of light is the physical limit for a particle to reach such speed, beyond it any particle would dematerialize in a wave. This paradox of the light makes possible for the solar beam radiation to produce in the same time heat and electricity. By harvesting the solar light in a collector, it can be transformed in hot air or hot water and in electricity. In Ancient History, besides warming the interior space by the direct solar gain, the Sunlight was also collected and stored in building elements with high thermal inertia like stone floors or stone walls.

3.2 Sun Power in Architecture

The house being a construction that the man built for living purposes for both day and night, it must be equipped to offer the comfort that the man needs, including a series of natural factors. If possible, a great living space would be a replica to the natural image with a controlled environment of the temperature, humidity, light and smell. No matter of the architectural styles adopted for a house, either modern or traditional, with artificial or natural materials used as finishing, the need of the occupant would at least still be reduced to a natural factor, the presence of the Sun in the interior space. By definition, in the Romanian legislation [3, 4] a space for living must have a minimum of natural sunlight in the winter period.

The presence of the Sun is not only a human necessity, but it is also a construction requirement, because a healthy construction is a sunny and well ventilated one. There are multiple factors that can influence the solar gain: (1) the different climates on the planet, (2) different possible geographical positions of the future site and (3) the Sun and Earth movement. After taking into consideration all the three factors mentioned above, it can be proper designed a solar control which represents an important part in the set of the passive architectural methods applied on the volume of the house. Having a good solar control, the house will receive sunlight when is cool outside and will be shaded when it is very hot, to maintain a constant warm interior temperature.

The passive methods have their limitations in the temperate or cold climates, from which point forward the active solutions can substitute. A house can be designed to protect itself from the Sun or to naturally ventilate and cool the interior space, but when it comes to heating the interior temperature to +21 °C despite the outside below freezing temperature, certain active systems powered by electricity must function. Also, heating the domestic house water (DHW) is a high energy consuming process, even if it is powered by electricity produced by the solar radiation. So, additional systems to heat the DHW are needed to be installed.

Each solution has its advantages and its limitations, therefore it is necessary to realize a thorough analysis of the site and to determine all the climate and geographical conditions in order to apply the best methods that fit to the location for the future building and so, to design a proper architectural integration of the solar solutions to the volume of the house.

3.2.1 Solar Passive Solutions

Solar passive solutions are methods that can be applied to the architectural volume of the building with the purpose of obtaining the thermal comfort of the interior space using no energy or no mechanical parts beside the natural power following the principles of thermodynamics.

The basic passive approach of the sun power is by having control on three factors: when, how and where. *When* during year, *how* much and *where* should the sunlight enter the house.

When. Due to the planet movement around its own axis, Earth has the day-night sequence. In this cycle, the Sun has a continuous path on the sky, it starts from the East, from the horizon, uprising to a maximum point in the South direction. That specific point is associated with a certain unique day, which repeats exactly each year. Each day has the maximum point at a different position than its previous or the following day. After that point the Sun goes downwards disappearing on the horizon, on the West.

On its path, the solar radiation has different intensity, different color and most important different angle with the Earth's surface.

The movement around the Sun in an elliptical path that takes one year, during which the inclination of the Earth's axis is changing, producing variations in temperatures on the surface of the planet, therefor the existence of the seasons. The summer and the winter are caused by the effect of the beam radiation travel through space. As further is a surface from the Sun, as less solar power it gets. Due to the inclination of the Earth's axis, when a hemisphere is closer to the Sun, the other is further. That means that when the north hemisphere it is closer to the Sun, it has the hot season, while the south hemisphere is further from it, having a cold season.

The changes of the sun path on the sky during a day and during a year requires a special design of a house. Starting from the outside to the inside of the building, orientation is the first major step in the designing process. Usually the rooms preferred to be oriented to the south are the living rooms related to the day-time activities, to be able to benefit as much as possible of the sunlight.

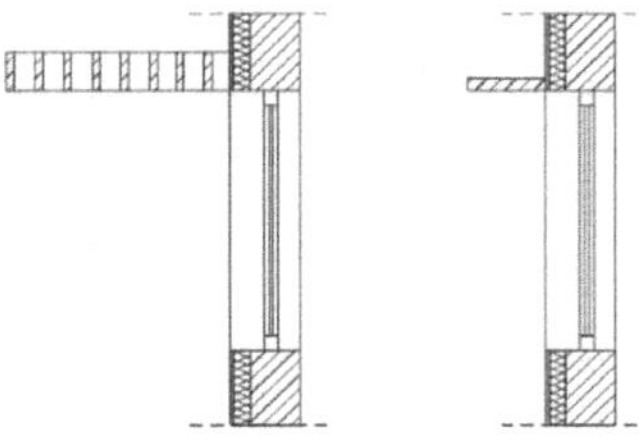

Fig. 3.1 Shading devices: horizontal fins above the window—vertical section through a window. The concept can be applied the same for the vertical fins by looking at the drawing as a horizontal section

How to limit the amount of light that enters the house is by having a solar control. By using elements that block the sunlight at certain moments of the day, the shading of the interior space is assured. The main attention in the designing process is focused on the architectural integration of the shading elements with the façade or even with the whole volume.

Where is best to allow the sunlight inside depends much on the project theme determined by the client needs. But it can be generally accepted that the common living spaces need to receive as much as possible natural light, direct or indirect, the amount of each type is determined by the when factor.

Solar control—Applied shading devices. The applied shading devices represent a solution for equipping the new or the old buildings with minimum intervention on the existing volume. The solution aims to create the solar control by protecting the glazing surfaces with system applied directly on the façade or as structure in front of it, serving also to other purposes (Fig. 3.1).

The basic concept of the sun shading has a fin to block the sun rays from the desired direction. From the East and from the West, the solar beams are at a low angle which require vertical fins to block them, while solar beams from the South direction require horizontal fins. The difference from a simple sunshade to a shading device is given by the design of the fins. The design of the length, the number and the distance between the fins offer the actual solar control.

Starting from the geographical position of the existing or the future house position, the Sun path diagram can be calculated, revealing the position of the Sun on the sky at any hour of the day, during any day of the year. Using this data and the client needs, it can be designed a personalized shading system to allow the sunlight to enter the house only at a certain time, different by each season. The purpose is to block the solar radiation to warm the interior air during the hot summer days and to allow it to enter during the winter days to use the solar energy to warm the air (Fig. 3.2).

Solar control—Integrated solutions. The integrated solutions of the solar control represent an important part of the passive and bioclimatic architectural styles. It is an

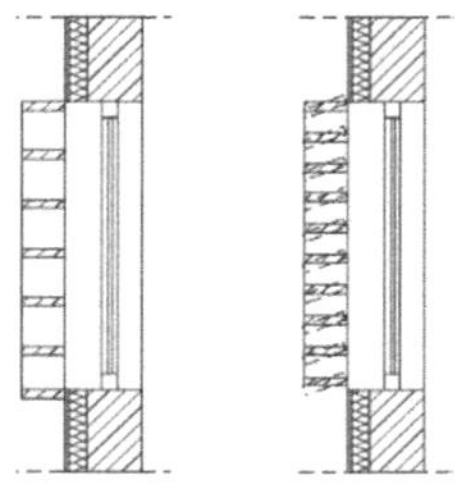

Fig. 3.2 Shading devices: horizontal fins in front of the window—vertical section through a window. The concept can be applied the same for the vertical fins by looking at the drawing as a horizontal section

approach that express more sincere expression of the house architecture. All its parts are integrated in its volume and function, as a whole. Using the building elements as a shading device, the need of maintenance is minimized. By using this method, it is obtained a great wide visibility through the glazing, with no obstacle on the exterior. The only limitation of the integrated solution compared with the applied shading devices is the lack of mobility. Once designed to have a fixed solar control, it can't be changed to receive or block the sun rays on other periods of time (Fig. 3.3).

The design of the building elements, to be a part of the solar control, focus on tilted or horizontal elements in front of the glazing. Same data is collected as in the previous solution, by using the Sun path diagram. If the fins system, by placing the elements at different distances and with different tilting, are able to choose which range of sun paths to be blocked, the passive methods of shading does not allow to select multiple ranges. For example, if is planned to have solar gain from 10 to 11 a.m. and from 4 p.m. to 6 p.m. in July by using fins system it can be done. Using passive method of solar control, it can be designed to block all the beams with angles bigger than 50° and allow all below 50°. Most of the solutions work with the balcony or top floor slab. The most workable element is the roof, whose large variety of structure types allows to be used as a continuous shade element or as a form that replicate the geometry of parallel elements, same as the horizontal fins.

Shading the house from excessive radiation might be enough for the interior comfort, but for the comfort of the outside yard building elements might be over expensive and might affect the natural landscape no matter the size of the yard. If the climate and the soil allow, the use of vegetation can have a substantial effect on the outside comfort. Each climate influence differently and requires a specific type of plants according to the seasons and soil composition.

Solar control—Vegetation. For the temperate-continental climate the vegetation can be used both for the cold and hot climate, due to the change of wind direction during the seasons. On the South direction, deciduous trees are fit to be planted due to the high density of crown leaves and due to the fact, that during the winter, after the leaves are falling, the bare branches allow the sun rays to pass. The distance of the

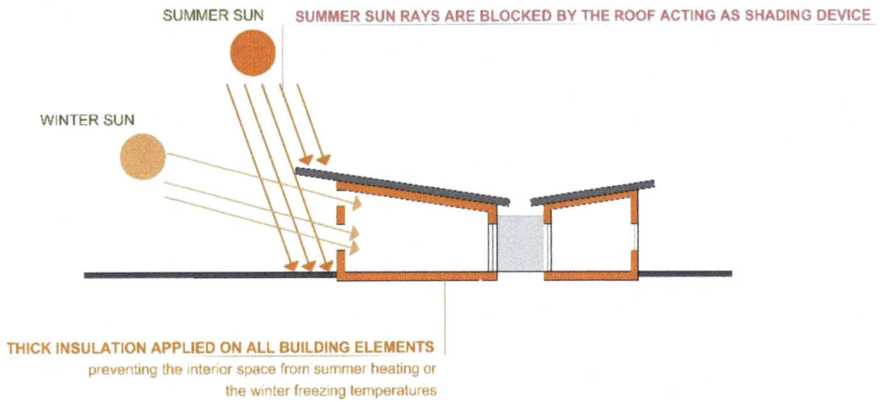

Fig. 3.3 Solar gain scheme

trees from the house must be determined by their future height for allowing to pass, if the case, the desired sun rays. On the North direction of the yard, evergreen trees are best suited having green leaves through the whole year offering great protection against the winds (Fig. 3.4).

Solar gain—Direct and Indirect gain. The amount of solar radiation that a house receives and uses for improving the comfort of the interior space, is called solar gain. By having the passive methods applied to the house, the building volume can gain solar radiation directly through windows raising the temperature or indirectly, the heat being transferred by another building element.

Offering great visibility and creating sunny rooms, the direct solar gain method is associated with shading devices. A direct exposure of the interior to the sunlight without a solar control affects both the temperature ambiance and the living comfort. Due to the direct sunlight many objects inside the house can be affected by long solar exposure and a glazing UV treatment might be required. The effect produced by the prolonged exposure to the sunlight of the plastic, fabrics, photographs and other objects are decolorization or even cracking. So, a special approach for the interior design is required in order to have a long time well behavior of the finishing and decorations.

For a full benefit of the solar power gained during the day, high thermal inertia materials are used for storing the heat. Thermal inertia is a method often adopted by regions from the hot climates where there are big day-night differences and from the continental climate during the winter days. The areas where the solar radiation can be stored are next to the glazing. The floor slab or finishing next to the windows can be best designed for this purpose. Their quality stands in absorption the radiation during the day when the interior air heating is assured by the direct solar gain, and giving it back the heat stored during the night when the Sun is absent, through the conductive process.

There are situations when, due to the urban environment, high reflectiveness or privacy issues might occur and makes it impossible to open the house to the outside for gaining solar radiation. The Trombe Wall is a concept developed to gain and store solar heat while protecting the interior space. By building a buffer room between the living room and the exterior, with glazing on the façade on one side and a thick wall on another side, it acts like a solar collector gaining the radiation inside the buffer room and storing heat. Being very efficient in the sunny days during the winter, its efficiency it's dropping during the cloudy days. During the summer the glazing in front of it needs to be shaded for avoiding overheating the interior, [5].

This concept uses the solar radiation indirectly through a solar collector that can be from a variety of materials with high thermal inertia like masonry, concrete, stone or even a water tank. A dark color is used for the collector surface for enhancing the radiation absorption. The energy stored it is transferred to the interior either conductive by the wall directly, or by convection using openings for the air movement. For improving the air exchange, a ventilated system can be installed and it increases the overall efficiency by 10%.

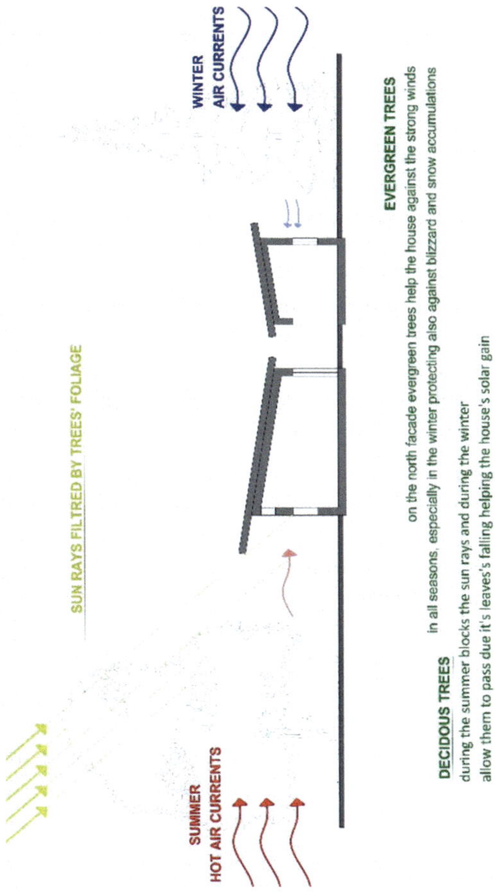

Fig. 3.4 Use of vegetation as a passive method

3.2.2 Solar Active Systems

The solar active systems represent a set of devices which collect the solar radiation and transform it in a usable form of electricity or heat. The active systems are composed of collectors, distribution parts, and in some cases, storage units. The limitations of the active systems are due to the need of direct sunlight; thus, it requires a certain orientation and inclination of the solar collectors for them to be as much as possible efficient. The different inclination of the beam radiations from Winter to Summer periods, require either movable panels, which imply very high costs, or either a specific tilting that, during a whole year, assure the highest mean efficiency. Compared with the passive solutions, solar active systems require a special design, additional installing and maintenance which results in higher costs.

At this moment there are available two concepts of transforming the solar energy, in electricity and in thermal energy. Each collector is mounted outside in the same way on the building. Even though the technical requirement is the same, to have full access to the sunlight, there are two different types of equipment available to be installed on a building. The first type is the oldest on the market, called Building Applied solar panels, they are mounted directly on the building, or even on the ground, stand-alone best suited for the existing buildings. Building Integrated solar panels represent the second type of solar panels, which are designed to be perfectly integrated in the building volume as finishing, lighter, bendable, being able to offer great possibilities in order to achieve a unique architectural design (Fig. 3.5).

Building Applied Systems. The first models of solar panels whose concept and design are still used at this moment represents stand-alone solar panels that are separate from the building materials with the possibility of mounting anywhere on it due to the necessity of its own structural frame. It will always exist a separation of the panel from the building finishing.

Building applied Photovoltaics (BAPV). The photovoltaic cells (PV) produce direct current from sunlight, which after passing through an inverter is converting it in alternative current, and, if needed stored in batteries for night time loads. A photovoltaic or solar cell represents the basic unit which produces electricity through silicon-based semiconductors. A single solar cell represents a module that is multiplied on the glass support to form a panel which is mounted on a metal structure on the ground or on the building, directly on the roof (Fig. 3.6).

According to a study on multiple PV manufacturers developed by the author, each panel produces in an hour at 100% efficiency between 250 Wh (the polycrystalline cells), and up to 450 Wh (the monocrystalline cells). These are mean values that depends on the manufacturer and on the number of the modules on a panel, which are at this moment available, for the residential use, with 60 or 72 modules, the difference of the 12 modules are arranged on the longitudinal axis of the panel, the width being the same, respectively 99 cm. The difference of the two types of the PV panels is small regarding a residential system. The monocrystalline solar cell requires more energy to be produced but it has a slightly higher efficiency. In the end, the amount of energy required to manufacture the monocrystalline solar cell, is

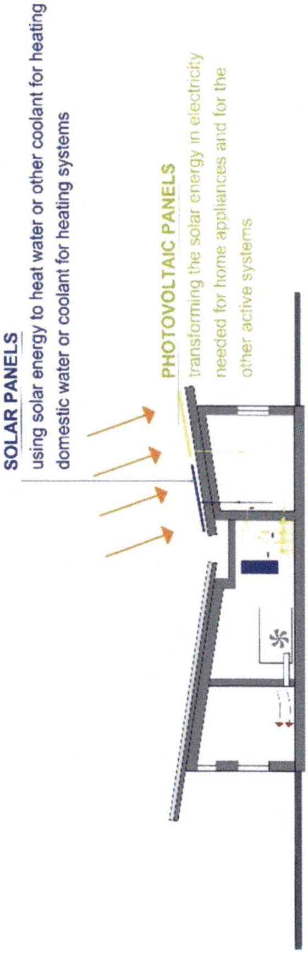

Fig. 3.5 Active methods scheme

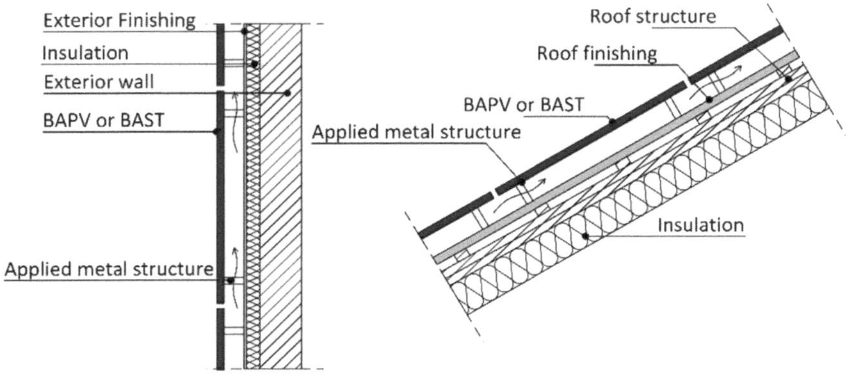

Fig. 3.6 Vertical section through a wall and roof. Relation between building element and applied solar equipment

equal with the amount of energy they produce in their operational life, which makes the polycrystalline solar cells to be better in a long-term approach.

The number of the panels on a building is determined by calculating a few scenarios of the loads a house can have. By adding the consumption of the equipment's that must run in the same time for the comfort of the house, a mean load measured in kWh is obtained which will represent the power of the PV's. Overloading power can be supplied either by a stack of batteries, in the case of an off-grid solution, or by the grid. The automatization of the solution is made by the inverter which can function in off-grid mode, on-grid mode or hybrid mode, the last solution would offer power from the PV, the batteries and the grid in the same time.

Roof mounting PV panels represent a practical approach due to the large applicability on all the houses, old or new, with inclined roof or with a terrace, south oriented or not. By using a metal frame, at all costs, neglecting the image of the building, it can be assured the orientation and the inclination needed for the best efficiency. Besides the orientation factor, the surrounding of the site can affect the efficiency of the PV's or even the possibility of installing them in the first place. The shading that may come from a mountain, or a tall building, will limit the possibility of receiving the solar beams only for a few hours, or not even at all. By using the sun path diagram to study the shadows casted by the surroundings can anticipate the efficiency of the future PV panels.

Building applied Solar Thermal panels (BAST). The solar thermal panels function by very old technique with the most basic concept: heating a liquid under the sunlight. By putting a large tank of water under the Sun it will get warm. The heating process is related to the amount of sun power that can heat up a liquid during the limited time of a day (Fig. 3.7).

For eliminating the issue regarding the time, it takes to a water tank to be heated up, small pipes have been developed that run water or a mix of glycol–water which collects the solar radiation through a dark colored surface. The purpose of the small

Fig. 3.7 Photo of a BAST on the roof of a house marking the lack of compatibility between the equipment and the building due to the wrong solar orientation of the house

pipes is that it requires less energy, and so less time for the liquid to achieve the desired high temperature.

The solar thermal panels collect the energy and transfer it to a liquid that circulates through pipes, well insulated to maintain the high temperature. The high temperature achieved fast in the small amount of liquid is then transferred to a heat exchanger to store the thermal energy, [2]. The energy stored can be used in multiple ways by adding specialized equipment for each purpose: DHW, heating by air ventilation or if installed in larger quantity for conductive heating by radiating panels.

Two types of solar thermal panels exist: pressurized or unpressurized. The main difference is that the pressurized panel have, as the name suggests, liquid under pressure in the piping, which allows it to produce heat even during the winter which increases its efficiency. Another advantage of the pressurized panel is that the heat exchanger can be placed inside a utility room, and so the panels design will close to the BAPV, not having a big cylinder above it.

BAST are widely spread at this moment, being a solution well implemented on the market in the last 20 years, especially in the hot climates, Greece for example has one of the most developed solar thermal panel market since 1970's, by 2002 had installed more than 25% of the systems in the EU-15 countries, [6].

Building Integrated solar panels. The BIPV's offer new ways of equipping a building with solar systems on the vertical surfaces of the buildings. If the BAPV's are limited to the horizontal surfaces or to the opaque areas on a façade, the new technology makes possible for the architects to design the façade and integrate the solar panels without compromising the desired image of the building. Building integrated solar panels represent a different solution for a building, instead of applying them on the roof, new materials can be placed instead of the finishing materials of the façade, roof and glazing. The integrated solar panels become one with the building offering the possibility to interplay them during the design process, in accordance with the architecture style (Fig. 3.8).

The integration in the façade extends the way of designing building, offering the possibility to install in each building solar equipment enough to produce at least the

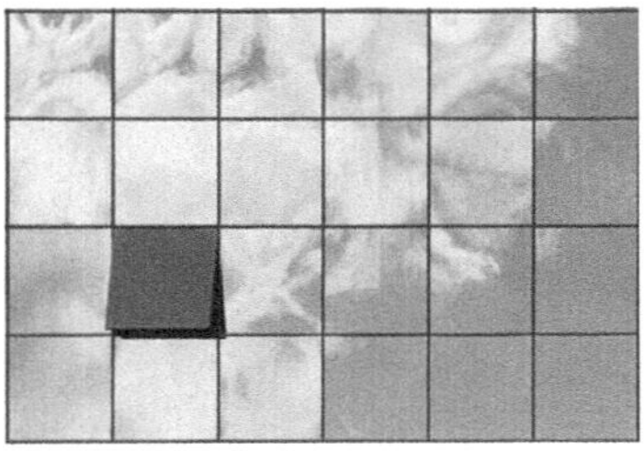

Fig. 3.8 Photo of a Curtain façade highlighting the large possibilities of integrating the solar panels in the façade architecture (free images online library www.pexels.com)

amount of energy needed by it, and even maybe enough to contribute to the grid and to be able to have self-sufficient cities.

The importance of the building integrated solar panels in housing. For a building, the surface available for installing the solar equipment decreases with its volume. A house need of electricity in Europe goes from 1639 kWh/dw, where "dw" stands for dwelling, in Romania up to 17,666 kWh/dw in Norway in the year 2017, [7], the big difference being given by the fact that countries like France and Norway use electricity as source for heating, where Romania uses natural gas. Considering that the mean house area in Europe is 91 m^2 of all new and old houses, [8] and given the need of electricity, it can be determined that the surface of a house roof is enough for mounting PV, even for Norway, where it's needed approximately 48 kWh/dw for each day. If it would be to design solar thermal panels, the amount of electricity needed would obviously decrease. So, the need of use the BIPV or BIST for housing is different from the situation of the urban tall building. The integrated solar panels represent a tool very useful for the architect to achieve the perfect architectural image of the house, exceeding the technical need.

Building integrated Photovoltaics (BIPV). The new technology of producing photovoltaic cells makes the PV to be lighter and plastic allowing it to be used as finishing on the roof or on the façade. New possibilities appear also due to the use of colors, in the façade design the game of colour improve the integration of the PV for a better architectural result (Fig. 3.9).

Building integrated Solar Thermal panels (BIST). The integration of the solar thermal panels has been experienced before the BIPV. By placing the tubes on the façade, with proper insulation on the building side it acts like a BIST (Fig. 3.10).

The modern approaches involve placing the dark tubes as shading devices, or as decorative panels on the façade. Flat dark collectors are an option for smother finishing of a building. Many possibilities exist for the BIST, mainly because the architecture of the façade already includes solutions that makes it seamlessly to integrate the panels in its metal rail system.

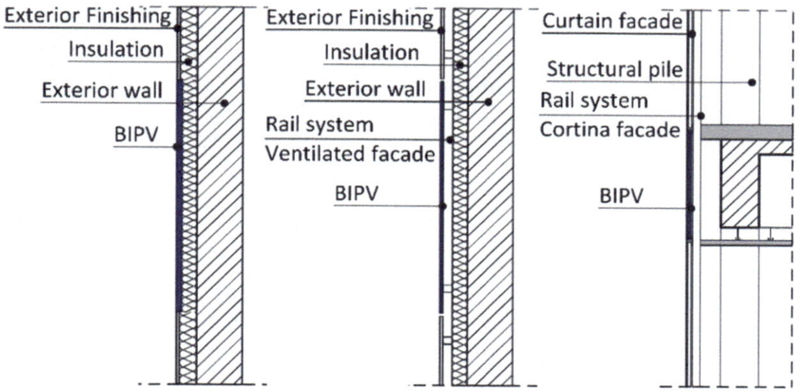

Fig. 3.9 Vertical section through exterior wall presenting various situations of how BIPV can be integrated in a building façade. From the left: 1. integrated in façade plaster like finishing, 2. installed in a ventilated façade, 3. installed in a Curtain façade

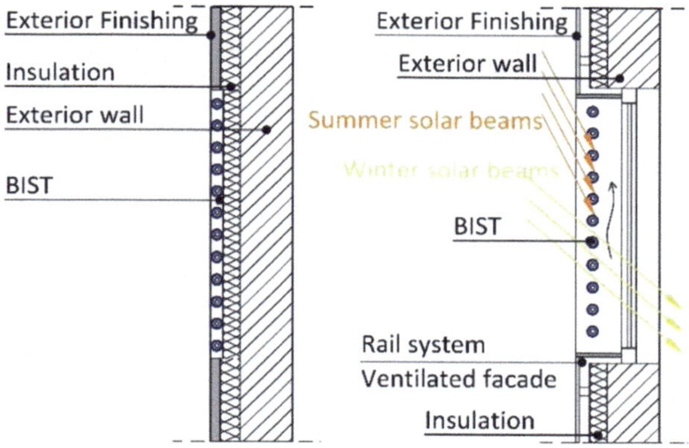

Fig. 3.10 Vertical section through exterior wall presenting various situations of how BIST can be integrated in a building façade. From the left: 1. integrated in façade, 2. installed in a ventilated façade either as a module or as shading device in front of a window or a balcony

3.3 Conclusions

The house represents the smallest individual unit of a building type the man uses for its day to day needs. Fulfilling one of the basic needs of the man, the housing, receives and deserves all the attention and the dedication of the future inhabitant, the client, and of the architect. Being spread all around the world, according to the high number of populations, the housing sector represents an important factor in the high energy consumption that happens nowadays. If any other natural resource on the planet depends on the actions of the man, the Sun is the only independent and infinite resource available. It is necessary to preserve the natural environment in order to achieve a sustainable future. By starting to use clean solar energy and to transform the houses into devices that are able to use natural sources in order to obtain interior comfort it is one important step. No matter where the energy comes from, the responsible consumption of it has an equal part.

References

1. E. Struzik, *How Thawing Permafrost Is Beginning to Transform the Arctic* (2020). https://e360.yale.edu/features/how-melting-permafrost-is-beginning-to-transform-the-arctic. Last accessed 2020/04/15
2. P.J. Axaopoulos (ed.), *Solar Thermal Conversion* (Active Solar Systems, Symmetria, 2011)
3. Romanian Health Minister Order no. 119 from 4th February 2014
4. Romanian Normative regarding the safety of the residential buildings (revision NP 016-96) Indicative NP 057-02
5. P. Lapithis, *Bioclimatic Architecture and Cyprus*. Pantheon Cultural Association (2018)
6. A. Argiriou, S. Mirasgedis, The solar thermal market in Greece—Review and perspectives. Renew. Sustain. Energy Rev. **7**, 397–418 (2003). https://doi.org/10.1016/S1364-0321(03)00064-9
7. Electricity consumption per dwelling. https://www.odyssee-mure.eu/. Last accessed 2020/04/22
8. International house sizes. https://demographia.com/db-intlhouse.htm. Last accessed 2020/04/05

Chapter 4
Outdoor Performance of Triangle Solar Thermal Collectors for Facades Integration

Macedon Moldovan, Ion Visa, and Anca Duta

Abstract The European Commission presented the European Green Deal in December 2019, aiming at making Europe the first climate-neutral continent by 2050. An Investment Plan was released in January 2020, mobilizing at least €1 trillion of sustainable investments over the next decade, to reach the ambitious climate and energy targets. The need for clean, secure and affordable energy supply is recognized as one of the main policy areas. Following this international context, one of the key research objectives in the Renewable Energy Systems and Recycling R&D Centre of the *Transilvania* University of Brasov focuses on the increase of the solar energy share in meeting the thermal energy demand of the buildings. Thus, novel solar thermal collectors were developed, with nonrectangular shapes (trapeze and triangle) and non-traditional colours (green, red, orange), to increase the architectural acceptance of the facade integrated solar thermal collectors. After indoor testing, four triangle solar thermal collectors have been installed on the southern facade of an outdoor testing rig for a long-term testing procedure. The collectors are parallelly interconnected and equipped with temperature sensors at each inlet and outlet and with digital flow meters that allow evaluating the thermal power output of each collector. The conversion efficiencies for each solar thermal collector are discussed in this paper for four representative clear sky days during the monitoring period. These efficiencies are compared with the values previously obtained on the indoor testing rig, outlining the behaviour of the novel solar thermal collectors in the outdoor working conditions.

Keywords Triangle solar thermal collector · Coloured absorber plate · Efficiency

M. Moldovan (✉) · I. Visa · A. Duta
Transilvania University of Brasov, Brasov, Romania
e-mail: macedon.moldovan@unitbv.ro

I. Visa
e-mail: visaion@unitbv.ro

A. Duta
e-mail: a.duta@unitbv.ro

I. Visa and A. Duta (eds.), *Solar Energy Conversion in Communities*,
Springer Proceedings in Energy, https://doi.org/10.1007/978-3-030-55757-7_4

4.1 Introduction

The European Commission goal of making Europe the first climate-neutral continent by 2050 was stated in the European Green Deal released in December 2019 [1]. Ambitious climate and energy targets were established for the next decade backed up by an Investment Plan released in January 2020, mobilizing at least €1 trillion for sustainable investments [2]. The need for clean, secure and affordable energy supply is recognized as one of the main policy areas, in line with the United Nations 2030 Agenda [3] and the Paris Agreement [4].

In this context, one of the research objectives of the Renewable Energy Systems and Recycling R&D Centre (RESREC) in the *Transilvania* University of Brasov focuses on the increase of the renewable energy share in meeting the energy demand of the buildings [5], aiming to comply with the Nearly Zero Energy Building standard, as defined in the European Directive 2010/31 [6].

Solar energy is a promising alternative to fossil fuels, with an expected increase of the implemented solar thermal collectors at global level from 686 million m^2 reported for 2018 to 2000 million m^2 by 2030 and 5800 million m^2 by 2050 [7]. To reach these targets, the number of the building integrated solar thermal systems [8] must be significantly increased and thus, large available surfaces will be required. Moreover, the architectural acceptance of the solar thermal collectors is an issue, especially for those installed on the visible surfaces of the buildings, such as the building facades [9, 10] that require specific policies [11] and planning processes [12]. The solar facades are particularly targeted due to theirs better exposure to the sun during the cold season allowing thus to be used for space heating [13].

Considering the high share (71%) of the flat plate solar thermal collectors installed in Europe [14], a novel concept of this type of collectors with nonrectangular shapes (trapeze and triangle) and non-traditional colors (green, red, orange) was developed in RESREC to increase the architectural acceptance of the façade integrated solar thermal collectors [15]. Firstly, novel trapeze solar thermal collectors with small aperture surface (0.63 m^2) were designed and developed at demonstrator level [16] and the optimized prototype was further implemented on an outdoor facade to assess its performances in real implementation conditions [17]. To further increase the coverage degree of the available surfaces on the building facades, novel triangle solar thermal collectors with smaller surface (0.067 m^2) were developed [18]. After indoor testing on a vertical testing rig [19], four triangle solar thermal collectors were installed on the southern facade of an outdoor testing rig for a long-term testing procedure. The collectors are parallelly interconnected and equipped with temperature sensors and digital flowmeters that allow evaluating the thermal power output for each collector. The conversion efficiencies for each solar thermal collector are presented in this paper for four representative clear sky days during the monitoring period. These efficiencies are compared with the values previously recorded on an indoor testing rig, outlining the behaviour of the novel solar collectors in the outdoor working conditions.

4.2 Methodology

Due to the seasonal variability of the solar irradiance in the implementation location (Brasov, Romania, 45.67 °N, 25.55 °E), the output of the novel triangle solar thermal collectors was evaluated in the sunniest days during the monitoring period (August 2018–July 2019), close to the four representative days: spring and fall equinoxes, and summer and winter solstices. The sunniest day in a given period of time is considered the day with the highest amount of the daily available global solar energy that is calculated using:

$$E_G = \frac{1}{n}\sum_{i=t_{sr}}^{t_{ss}} G \tag{4.1}$$

where: t_{sr} and t_{ss} are the sunrise and sunset time of the day, G is the available global solar irradiance at the time i and n is the number of the measurements during one hour.

The available global solar irradiance is calculated using:

$$G = B + D \tag{4.2}$$

where B and D are the available direct and diffuse solar irradiance.

The available direct solar irradiance was measured by the pyrheliometer installed on a Solys2 sun tracker (Kipp and Zonen) while the available diffuse solar irradiance was calculated using:

$$D = D_h \frac{(1 + \sin\alpha)}{2} \tag{4.3}$$

where: D_h is the available diffuse solar irradiance measured in the horizontal plane by the shaded pyranometer installed on the Solys2 sun tracker and α is the solar altitude angle calculated with well-known equations as in [20].

The performance of the novel solar thermal collector is evaluated through its thermal efficiency calculated using:

$$\eta = \frac{p_t}{G_n} 100 \tag{4.4}$$

where: p_t is the specific thermal power output of the solar thermal collector and G_n is the global solar irradiance received in the plane of the solar thermal collector.

The specific thermal power output of the solar thermal collector is calculated based on the measured mass flow rate of the fluid through the solar thermal collector ($\dot{m}$), on the specific heat at constant pressure of the antifreeze thermal fluid (c), on the outlet and inlet temperatures (t_o and t_i) and on the absorber surface (S_a) using:

$$p_t = \frac{\dot{m}c(t_o - t_i)}{S_a} \tag{4.5}$$

The global solar irradiance received in the plane of the solar thermal collector is calculated using:

$$G_n = B_n + D_n \tag{4.6}$$

where B_n and D_n are the direct and diffuse components of the solar irradiance received by the solar thermal collector.

The direct solar irradiance received in the plane of the solar thermal collector is calculated based on the measured direct solar irradiance (B) and on the incidence angle (ν) using:

$$B_n = B\cos\nu \tag{4.7}$$

The incidence angle, defined as the angle between the sun ray and the normal to the solar thermal collector [21], is calculated using:

$$\nu = \cos^{-1}(\cos\alpha\cos\alpha_{\mathrm{n}}\cos(\psi - \psi_{\mathrm{n}}) + \sin\alpha\sin\alpha_{\mathrm{n}}) \tag{4.8}$$

where: α and ψ are the altitude and the azimuth angles that define the solar ray direction calculated with well-known equations as in [21] while α_{n} and ψ_{n} are the altitude and the azimuth angles of the solar thermal collector. In the particular case of the solar thermal collectors vertically installed on the southern façade of the outdoor testing rig ($\alpha_n = 0°$ and $\psi_n = 0°$), the Eqs. (4.8) and (4.7) become:

$$\nu = \cos^{-1}(\cos\alpha\cos\psi) \tag{4.9}$$

$$B_n = B\cos\alpha\cos\psi \tag{4.10}$$

The diffuse solar irradiance received in the plane of the solar thermal collector is calculated based on the measured horizontal diffuse solar irradiance (D_h) and on the altitude angle of the solar thermal collector (α_{n}) using:

$$D_n = D_h\frac{(1 + \sin\alpha_n)}{2} \tag{4.11}$$

In the particular case of the solar thermal collectors vertically installed on the southern façade of the outdoor testing rig ($\alpha_n = 0°$), the Eq. (4.11) becomes:

$$D_n = \frac{D_h}{2} \tag{4.12}$$

The daily received global solar energy is calculated using:

$$E_{G_n} = \frac{1}{n} \sum_{i=t_{sr}}^{t_{ss}} G_n \tag{4.13}$$

4.3 Experimental Setup

The novel triangle solar thermal collectors developed in the RESREC Centre in the *Transilvania* University of Brasov [18] were installed on the southern façade of the outdoor testing rig starting with August 2018. The outdoor testing rig is installed in the R&D Institute of the *Transilvania* University of Brasov (temperate continental climate) and it is equipped with commercial flat plate (1) and evacuated tubes (2) solar thermal collectors used as references during the simultaneous tests of the newly developed solar thermal collectors of trapeze (3) and triangle (4) type (Fig. 4.1).

The hydraulic scheme of the solar thermal system including four triangle solar thermal collectors is presented in Fig. 4.2. The four triangle solar thermal collectors (1) are parallelly interconnected to a storage tank (2) with a capacity of 500 L. A mix of water and propylene glycol is circulated through the solar circuit by the pumping group (3) driven by the controller (4). The volume variation of this antifreeze fluid is compensated by an expansion vessel (5). An electric heater (6) is used to increase the temperature in the storage tank when a higher inlet temperature is required for testing of the solar thermal collectors. The temperature is measured using Pt1000 sensors (T_1–T_{12}), with an accuracy of ±0.1 °C, installed at each inlet and outlet of the solar thermal collectors as well as in four positions in the storage tank. The mass flow rate through each solar thermal collector is measured using IFM-SM4000 flow meters (Q_1–Q_4) with an accuracy of ±0.5%. The measured parameters are stored, each minute, in a data logger.

After implementing the novel triangle solar thermal collectors on the southern façade of the outdoor testing rig, the mass flow rate was adjusted to 0.001 kg/s through each collector, according to the standard for solar thermal collector testing [22]. During the testing period, there was no consumption of hot water and the pump was continuously running the antifreeze solution through the solar thermal system.

The meteorological parameters were monitored using a DeltaT weather station (Fig. 4.3a) and a Solys2 sun tracker (Fig. 4.3b) installed on the rooftop of a nearby building in the R&D Institute of the *Transilvania* University of Brasov. The DeltaT weather station is equipped with a RHT2 sensor to measure the outdoor air temperature and relative humidity, with a ±0.1 °C temperature precision and an accuracy of 2% for the relative humidity; the station also consist of an AN4 anemometer to measure the wind speed with an accuracy of ±0.5 m/s, a WD4 wind vane to measure the wind direction with an accuracy of ±2% and a RG2 rain gauge to measure the precipitation with an accuracy of ±2%. The Solys2 sun tracker is equipped with a CHP1 pyrheliometer to measure the direct solar irradiance (B) with an accuracy of 1%, a CMP22 pyranometer to measure the horizontal global solar irradiance (G_{h})

Fig. 4.1 Triangle solar thermal collectors installed on the southern façade of the outdoor testing rig

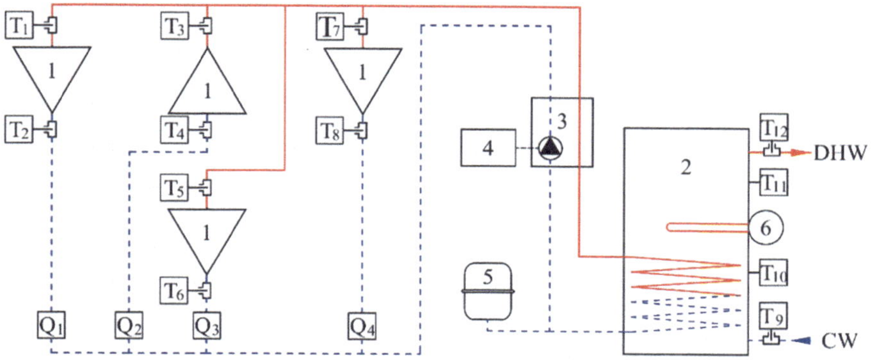

Fig. 4.2 Hydraulic scheme of the solar thermal system with four triangle solar thermal collectors

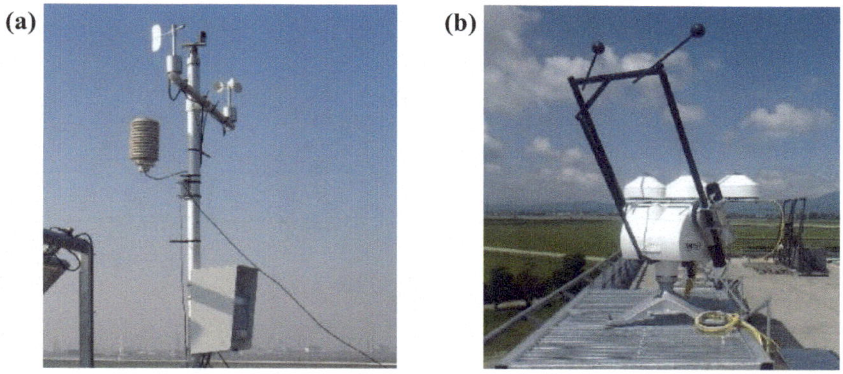

Fig. 4.3 Monitoring system of the meteorological parameters: **a** Delta T weather station. **b** Solys 2 sun tracker

with an accuracy of 0.5% and a shadowed CMP22 pyranometer to measure the horizontal diffuse solar irradiance (D_h) with an accuracy of 0.5%.

4.4 Results and Discussions

The results obtained by processing the parameters recorded during the one year monitoring period are outlined. The sunniest days close to the equinoxes and solstices were selected and the solar irradiance measured during these days is comparatively presented in Fig. 4.4.

Based on the measured values of the solar irradiance, the direct (B_n), diffuse (D_n) and global irradiance (G_n) received by the vertically installed, south oriented solar

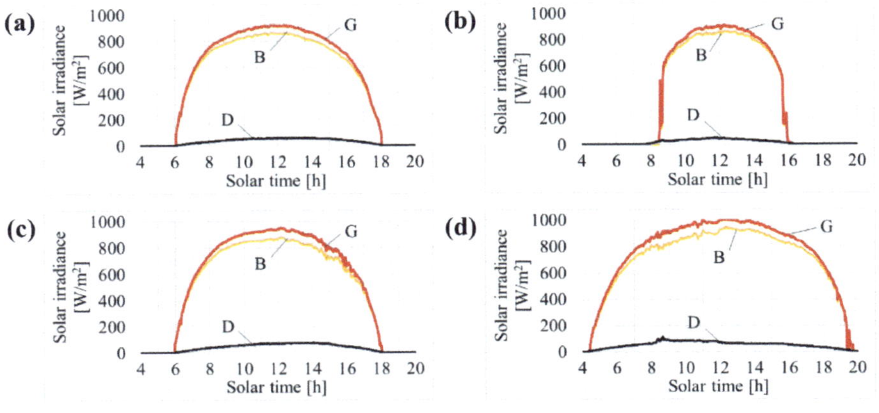

Fig. 4.4 The available global (G), direct (B) and diffuse (D) solar irradiance measured during the selected days: **a** 20.09.2018, **b** 05.01.2019, **c** 24.03.2019 and **d** 26.06.2019

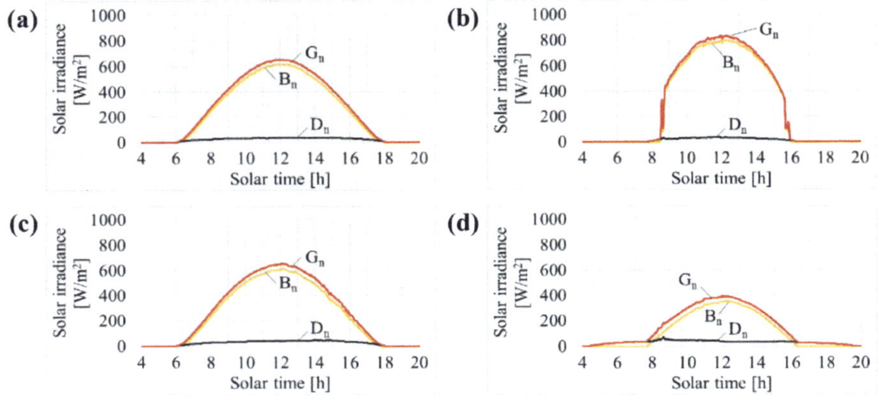

Fig. 4.5 The received global (G_n), direct (B_n) and diffuse (D_n) solar irradiance calculated for the selected days: **a** 20.09.2018, **b** 05.01.2019, **c** 24.03.2019 and **d** 26.06.2019

thermal collectors are calculated using Eqs. (4.10), (4.12) and (4.6), respectively. The results are comparatively presented in Fig. 4.5.

Similar profiles can be observed in Figs. 4.4 and 4.5 for the days close to the equinoxes for both the available and the received solar irradiance. The maximum solar irradiance reaches 1013 W/m^2 only at summer day noon, while for the other representative days it reached only about 900 W/m^2, even in the winter day. Due to the vertically installed, south oriented solar thermal collectors, the received solar irradiance is significantly reduced (392 W/m^2) during the summer day (Fig. 4.5d) when the maximum available solar irradiance is the highest while for the winter day the received solar irradiance reaches a maximum of 836 W/m^2 (Fig. 4.5b) representing

Table 4.1 The daily available and received global solar energy in the selected days

Day	Daily available global solar energy (kWh/m^2)	Daily received global solar energy (kWh/m^2)
20.09.2018	8.77	4.61
05.01.2019	5.78	4.86
24.03.2019	8.88	4.55
26.06.2019	11.98	2.41

91.7% of the available solar irradiance (912 W/m^2). These results show that the solar irradiance of 1000 W/m^2, required in the standard for solar thermal collector testing [22], is never reached on the southern façade in the implementation location.

The values of the daily available and received global solar energy are included in Table 4.1 for the selected days. When comparing the values of the daily available global solar energy, a large variation can be observed, ranging from 5.78 to 11.98 kWh/m^2, mainly due to the length of the daytime. For the daily received global solar energy, the maximum amount resulted in the winter day, closely followed by the fall and the spring days, indicating an almost constant distribution of the received solar energy during the cold season; this makes the solar facades good candidates for both space heating and domestic hot water preparation. The minimum received solar energy (almost half comparing to that in the winter day) resulted in summer. This represents an advantage in terms of protection to overheating that may appear in summer due to the reduced thermal load (only for hot water preparation).

Further on, the specific thermal power output of each collector was calculated for the selected days based on the measured mass flow rates and inlet/outlet temperatures. Similar results were observed for the two orange absorbers therefore, the results will be presented only for one of them. The specific thermal power output (p_t) is plotted in Fig. 4.6 along with the calculated global solar irradiance received in the vertical plane (G_n). Once again, similar profiles result for the days close to the equinoxes (Fig. 4.6a, c). The highest specific thermal output power is obtained during the winter day (Fig. 4.6b), despite the lowest outdoor air temperatures, compensated by the higher received solar irradiance. For the summer day (Fig. 4.6d), because of the high solar altitude angles the incidence angle is the worst and so is the received solar irradiance generating the lowest specific thermal output.

The efficiencies of the solar thermal collectors are comparatively plotted in Fig. 4.7 for the time interval close to noon of each selected day. Almost similar variations result, except the winter when the lower outdoor temperature slightly decrease the efficiency.

As expected, the colour of the absorber plate has a significant influence on the conversion efficiency, with the highest values (close to 50%) for the black plate and the lowest values (close to 30%) for the orange absorber plate. Despite these differences, it can be also observed that the sensitivity to the outdoor conditions is similar for all the colours, as the results in Fig. 4.7b show for the winter day.

Finally, the outdoor efficiencies variation with the reduced temperature difference is plotted (using dots) in Fig. 4.8 and compared with the efficiencies of the same

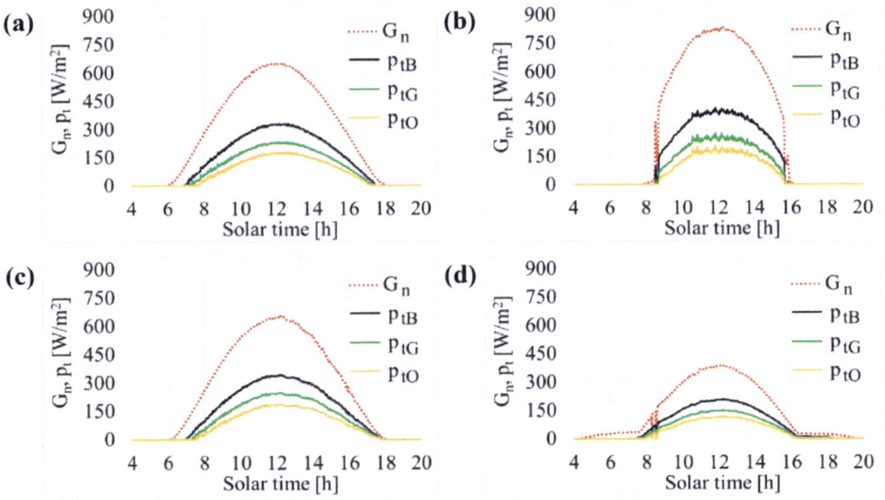

Fig. 4.6 The received global solar irradiance (G_n) and the specific thermal output power of the black (B) green (G) and orange (O) solar thermal collectors in the selected days: **a** 20.09.2018, **b** 05.01.2019, **c** 24.03.2019 and **d** 26.06.2019

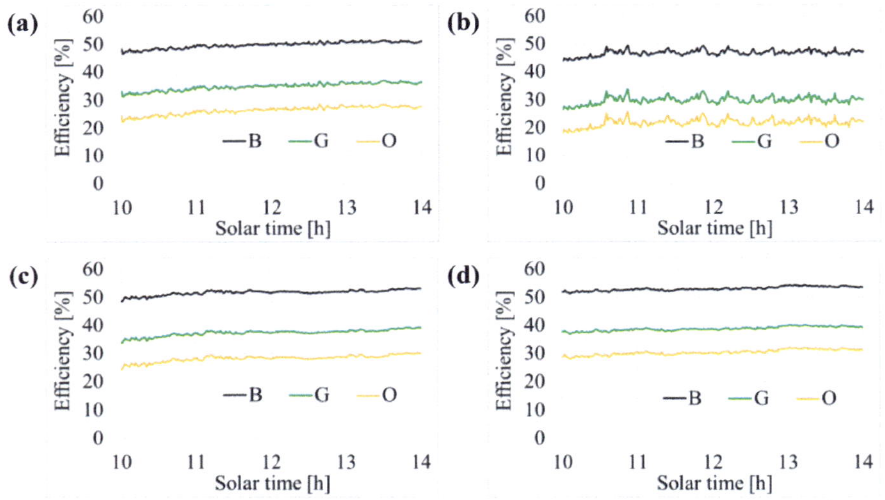

Fig. 4.7 The efficiency of the black (B) green (G) and orange (O) solar thermal collectors in the central part of the selected days: **a** 20.09.2018, **b** 05.01.2019, **c** 24.03.2019 and **d** 26.06.2019

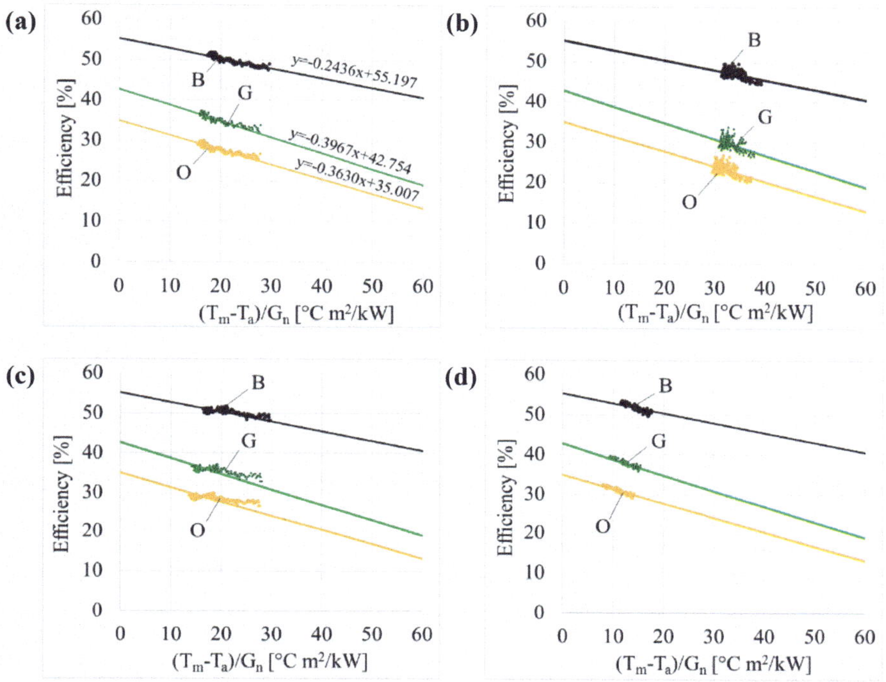

Fig. 4.8 Experimental efficiency obtained on the indoor (lines) and outdoor (dots) testing rigs for the black (B) green (G) and orange (O) solar thermal collectors in the selected days: **a** 20.09.2018, **b** 05.01.2019, **c** 24.03.2019 and **d** 26.06.2019

solar thermal collectors previously recorded on an indoor testing rig [18] plotted using lines in Fig. 4.8. Similar values of the root mean squared deviation (RMSD) ranging between 0.45 and 1.03% result for the spring and fall days (Fig. 4.8a, c). The influence of the low outdoor temperatures during the winter day can be observed in Fig. 4.8b with RMSD ranging between 0.83 and 1.20%. The best fit is observed during the summer day (Fig. 4.8d) when the RMSD value ranges between 0.30 and 0.37%. However, RMSD values lower than 1.2%, as observed during the testing period, can be considered mirroring a good agreement between the outdoor and the indoor experimentally assessed efficiencies.

4.5 Conclusions

The thermal efficiency is assessed in this paper for three solar thermal collectors having triangle shape, small surface (0.067 m^2) and differently colored absorber plates (black, green and orange), installed on the southern façade of an outdoor testing rig during a monitoring period of one year (August 2018–July 2019).

The results obtained for four sunny days, close to the equinoxes and solstices in the monitoring period, are comparatively presented and discussed.

The vertical implementation of the solar thermal collectors on the southern façade has as main advantages the increase of the received solar energy during winter sunny days (that is beneficial for the space heating systems) and the decrease during summer, protecting the solar thermal system against overheating. For the location of the outdoor testing rig (Brasov, mountain area with temperate continental climate) the solar energy received on the southern façade during the winter sunny day is two times higher than during the summer sunny day, opposite to the available solar energy that in the winter day is half of the summer day.

Accordingly, the specific thermal power output of the solar thermal collectors resulted the highest during the winter day and the smallest in the summer day.

The thermal efficiencies obtained for the time interval close to noon during the four selected days (10:00–14:00) are in good agreement with the results previously obtained on an indoor testing rig, showing a root mean squared deviation lower than 1.2%, with the highest values during the winter day (between 0.83 and 1.20%) and with the best fit during the summer day (between 0.30 and 0.37%).

Future research will focus on the influence of the mass flow rate on the thermal efficiency of the novel solar thermal collectors and on the influence of the serial interconnections among them.

Acknowledgements We hereby acknowledge that the infrastructure required for this work was supported by a grant of the Romanian National Authority for Scientific Research and Innovation CNCS/CCCDI-UEFISCDI, project no. PN-III-P-2.2.1-PED-2016-0338, within PNCDI III.

References

1. European Commission, The European Green Deal, Communication from the Commission to the European Parliament, the European Council, the Council, the European Economic and Social Committee and the Committee of the Regions, COM 640, Brussels, 2019.
2. European Commission, Sustainable Europe Investment Plan, European Green Deal Investment Plan, Communication from the Commission to the European Parliament, the European Council, the Council, the European Economic and Social Committee and the Committee of the Regions, COM 21, Brussels, 2020
3. United Nations, *Transforming Our World: the 2030 Agenda for Sustainable Development*, A/RES/70/1, United Nations General Assembly, 2015
4. United Nations Framework Convention on Climate Change, *Paris Agreement*, United Nations Treaty Collection, 2016
5. I. Visa, M. Moldovan, M. Comsit, A. Duta, Improving the renewable energy mix in a building toward the nearly zero energy status. Energy Build. **68**, 72–78 (2014)
6. European Union, Directive 2010/31/EU on the energy performance of buildings, *Official Journal of the European Union*, 18.6.2010 L153/13–35, 2010
7. International Renewable Energy Agency, *Global Renewables Outlook: Energy Transformation 2050* (International Renewable Energy Agency, Abu Dhabi, 2020)
8. S.A. Kalogirou, Building integration of solar renewable energy systems towards zero or nearly zero energy buildings. Int. J. Low Carbon Technol. **10**(4), 379–385 (2015)

9. R. O'Hegarty, O. Kinnane, S.J. McCormack, Review and analysis of solar thermal facades. Sol. Energy **135**, 408–422 (2016)
10. A. Buonomano, C. Forzano, S.A. Kalogirou, A. Palombo, Building-facade integrated solar thermal collectors: energy-economic performance and indoor comfort simulation model of a water based prototype for heating, cooling, and DHW production. Renew. Energy **137**, 20–36 (2018)
11. M.C. Munari Probst, C. Roecker, Criteria and policies to master the visual impact of solar systems in urban environments: The LESO-QSV method. Sol. Energy **184**, 672–687 (2019)
12. G. Lobaccaro, S. Croce, C. Lindkvist, M.C. Munari Probst, A. Scognamiglio, J. Dahlberg, M. Lundgren, M. Wall, A cross-country perspective on solar energy in urban planning: lessons learned from international case studies. Renew. Sustain. Energy Rev. **108**, 209–237 (2019)
13. M. Dehghan, C. Pfeiffer, Modelling and control of collecting solar energy for heating houses in Norway. J. Sustain. Dev. Energy, Water Environ. Syst. **5**(3), 359–376 (2017)
14. W. Weiss, M. Spork-Dur, Solar heat worldwide. Global market development and trends in 2018. Detailed market figures 2017, in *International Energy Agency Solar heating and Cooling Programme* (2019)
15. I. Visa, M. Comsit, A. Duta, M. Neagoe, M. Moldovan, B. Burduhos, D. Perniu, A. Enesca, L. Isac, M. Cosnita, I. Totu, I. Savvides, C. Vassiliades, Novel solar-thermal collectors/array with increased architectural acceptance for building integration, in *Building Integration of Solar Thermal Systems Design and Applications Handbook*, pp. 373–391 (2017)
16. I. Visa, A. Duta, M. Comsit, R. Saulescu, M. Moldovan, D. Ciobanu, B.G. Burduhos, Design and experimental optimisation of a novel flat plate solar thermal collector with trapezoidal shape for facades integration. Appl. Therm. Eng. **90**, 432–443 (2015)
17. I. Visa, A. Duta, M. Moldovan, Outdoor performance of a trapeze solar thermal collector for facades integration. Renew. Energy **137**, 37–44 (2019)
18. I. Visa, M. Moldovan, A. Duta, Novel triangle flat plate solar thermal collector for facades integration. Renew. Energy **143**, 252–262 (2019)
19. M. Moldovan, I. Visa, Development of an indoor testing rig for façade integrated solar thermal collectors, in Book Series: *E3S Web of Conferences*, vol. 85, pp. 04005 (2019)
20. J. Duffie, W. Beckmann, *Solar Engineering of Thermal Processes*, 4th edn. (Wiley Interscience, New York, 2013)
21. I. Visa, A. Duta, M. Moldovan, B. Burduhos, M. Neagoe, *Solar Energy Conversion Systems in the Built Environment* (Green Energy and Technology, Springer Nature Switzerland AG, 2020)
22. International Organization for Standardization, *Solar energy—Solar Thermal Collectors—Test Methods, ISO 9806:2017* (2017)

Chapter 5
Alternative Energy Sources Onboard Warships. Case Study: The Use of Solar-Thermal Systems on a Military Support Vessel

Alexandru Cotorcea, Ion Visa, and Andrei Pocora

Abstract In an oil market with prices per barrel less than half than few years ago, the use of classic fuels remains the economically viable and affordable solution. This aspect has a negative impact on the design, development and implementation of solutions based on the use of alternative energy sources in all sectors of activity. The present paper reviews the efforts made and the results obtained in order to provide energy to warships from sources other than the classical ones. Implementation of alternative sources of energy on warships is even more difficult, because on such a vessel there is generally no free space needed to place the systems. In this context, an analysis of the installation of solar thermal systems on military support vessels, which are not directly involved in combat and which are built like a merchant vessel, is presented. Also, three solar thermal systems are proposed for the preparation of DHW onboard vessels.

Keywords Military support vessels · Solar thermal system · Naval boiler

5.1 Introduction

Shipping is an energy-efficient mean of transportation compared to air and road transport, due to large carrying capacity of vessels and low fuel consumption per ton transported [1]. However, in this industry as well there are measures needed in order to reduce the fuel consumption from ships and, thus, reduce the carbon footprint of the marine industry [2]. International Maritime Organization (IMO) naval machineries efficiency requirements for existing ships include a Ship Energy Efficiency Management Plan and International Energy Efficiency Certificate and are enforce starting with January 1, 2013 [3]. Also, IMO prepared an initial strategy

A. Cotorcea (✉) · A. Pocora
Mircea Cel Batran Naval Academy of Constanta, Constanta, Romania
e-mail: alexandru.cotorcea@anmb.ro

I. Visa
Transilvania University of Brasov, Brasov, Romania
e-mail: visaion@unitbv.ro

I. Visa and A. Duta (eds.), *Solar Energy Conversion in Communities*,
Springer Proceedings in Energy, https://doi.org/10.1007/978-3-030-55757-7_5

to reduce greenhouse gases (GHGs) emissions from worldwide fleet. The strategy includes a 50% reduction in GHGs by 2050, compared to 2008 levels, and complete elimination of GHG's as soon as possible [2]. Among the methods of reducing dependence of fossil fuels IMO has identified use of alternative fuels, wind power and solar energy [4]. Issuing energy efficiency policies and technological leadership in the shipping industry leads to the idea that the environmental impact of energy use (renewable energy) has and will improve even more [5].

The use of alternative sources of energy onboard vessels is intended to minimize the consumption of diesel oil or heavy fuel oil. Recent studies, based on experimental data, consider hybrid energy systems for naval application [6–8]. As example, according to an analysis of the use of a solar energy hybrid power system, the fuel consumption is reduced by 4.02% and carbon dioxide emissions by 8.55% a year [6]. Also, the concern about impact of emissions on neighboring communities [9] during anchoring in port, conduct to regulations for the use of shore-energy, known as cold-ironing and onboard clean technologies [7, 8].

Regarding the warships, the regulations issued by IMO do not fully apply, due to the relative reduced impact on the environment (the traffic of military vessels is less intensive than that of commercial ones). Thus, opportunities for energy efficiency and renewable energy as a means of cutting costs, emissions and risks associated with reliance on traditional energy sources have been identified in this area of activity [10]. Consequently, USA Defense Department (DoD) allocate an important budget for increasing energy efficiency of military facilities, use of renewable energy and reduction of petroleum products for non-tactical vehicles. Also, the program Great Green Fleet, with standards and solutions in the field of alternative-energy, is intended to increase fuel efficiency and save millions of dollars [11].

In this context, the utility and opportunity of installing solar-thermal conversion systems onboard a military support vessel are further analyzed. The amount of thermal energy onboard any ship is high [12] and is used to heat the main engine, the fuel from different tanks, the main engine lubricating oil before entering the separator and producing the domestic hot water (DHW) needed by the crew.

In order to identify the benefits of using solar thermal collectors onboard the ship, the following steps were completed:

- description of the vessel, its main characteristics and identification of the areas available for the location of the collectors;
- choosing and evaluating from the point of view of the energy gain of a relevant trip, taking place in the two hemispheres and under extreme astronomical conditions;
- identification of the way to use the thermal energy obtained through conversion.

Also, three systems, of which two hybrids (used to produce thermal energy in combination with the ship's boiler), are proposed for implementation onboard the ship.

5.2 Determination of the Amount of Energy Produced by Solar-Thermal Conversion

As a military logistic support vessel, the Albatros vessel was chosen from the endowment of the Romanian Naval Forces. The vessel is intended for the transport of general goods, bulk goods (excluding ores), timber under decks or on decks and containers. It is useful for any type of navigation area with a range of 8000 nautical miles (Nm), which can be increased to 12,000 Nm by using mixed fuel tanks.

The vessel has a 2-stroke reversible diesel engine, engine running on heavy fuel oil while on open sea and diesel fuel oil during entry or leaving ports maneuvers, as well as in high-risk areas. The main engine is of type SULZER 5RD68 with an effective power $P_e = 6100\ HP = 4550$ kW and a specific fuel consumption $c_e = 0.165\left[\frac{\text{kg}}{\text{kWh}}\right]$. The fuel used has a lower calorific value $Q_i = 42,700\left[\frac{\text{kJ}}{\text{kg}}\right]$. Table 5.1 shows the main technical characteristics and capabilities for the Albatros vessel.

In order to obtain the thermal energy, the Albatros vessel is equipped with a naval boiler CAVNO1000 type. This type of boiler is intended to produce saturated steam at a maximum pressure of 7 bar and works with heavy fuel oil or diesel oil and has a nominal steam flow rate of 1000 kg/h. The Albatros can take 604 tons of heavy marine fuel, and according to the legislation in force 10% of this quantity must be permanently in conditions of immediate use. The vessel consumes, at an economic speed of 7 knots, a quantity of fuel of 500 kg/h [13].

The possibilities of placing renewable energy systems (photovoltaic modules, solar thermal collectors etc.) on a vessel are usually limited, because the design of the vessels is realized with the purpose of minimizing unnecessary spaces. In the present analysis for the military logistic support vessel, the following areas were considered in the initial analysis as being available for the location of solar thermal

Table 5.1 Main technical characteristics and capabilities of the Albatros vessel

Maximum length	130.86 m	Oil (Castrol CDX 30)	40 t
Width	17.70 m	Drinking water	55 t
Construction height	10.20 m	Warehouse cover surface 1	10.2 m^2
Draft	8.10 m	Warehouse cover surface 2	136 m^2
Displacement	12,350 t	Warehouse cover surface 3	136 m^2
Deadweight	8750 t	Warehouse cover surface 4	136 m^2
Heavy fuel	604 t	Full speed	15 knots
Diesel fuel	265 t	Standard/maximum crew	20/44

collectors: storage lid with an area of 136 m^2, the compass deck with a surface of 105.24 m^2 and the aft deck with an area of 96 m^2.

The analysis of the areas initially identified for the installation of solar-thermal collectors led to the following conclusions:

- the installation of the collectors on the covers of the warehouses, although they have the largest surface, can affect the manipulation of the goods and can damage the collectors during the activities; also, the positioning of the collectors in these areas will make impossible the transport of goods with arrangement on the covers of the warehouses;
- the arrangement of collectors in the stern area may affect the crew's ability to act during mooring or towing maneuvers; also, there may be shading of this area at certain times (when the Sun is positioned towards the bow of the vessel), thus resulting in a lower efficiency of the collectors;
- the compass deck has the advantage of being the area best exposed to the Sun (being the highest point of the vessel); therefore, communication and positioning equipment are installed in this area, which in this context is a disadvantage; however, this disadvantage can be eliminated by adopting modern, powerful and small equipment in a retrofitting and modernization process of the vessel.

From the aspects identified in the analysis of the areas for the installation of solar thermal collectors it can be concluded that the compass deck is the most suitable and, consequently, this area will be considered further. The aim is to use at maximum the available space and the way of disposing of the collectors (horizontal or inclined). Considering the area previously identified for the installation of the collectors, in Fig. 5.1 there are presented the main dimensions and distances between the solar-thermal collectors that can be installed on the compass deck.

As it can be seen in Fig. 5.1 on the surface of the compass deck, 23 solar thermal collectors can be installed, which represents from the point of view of the absorber surface a total of 59.11 m^2.

In order to evaluate the optimal option (in which the thermal energy obtained by the system has the highest value), three scenarios were considered, namely:

- Scenario 1: the collectors are horizontally arranged;

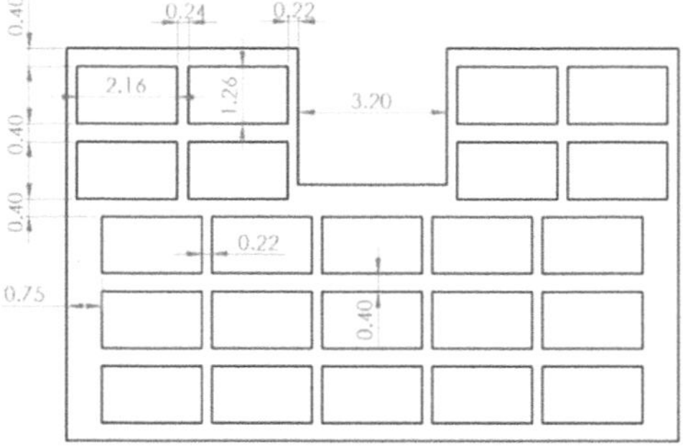

Fig. 5.1 The main dimensions and distances between the solar-thermal collectors on the compass deck

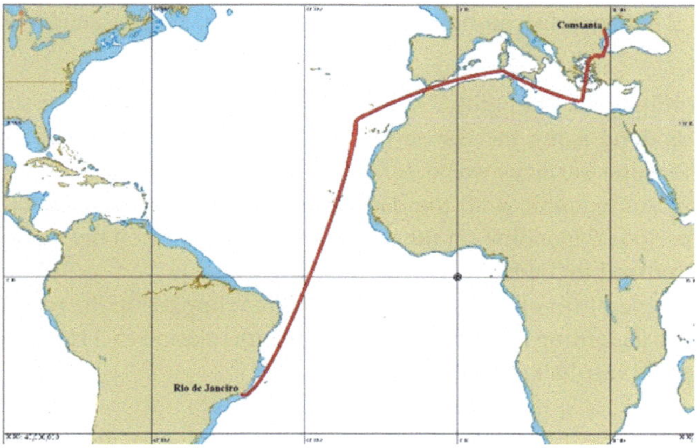

Fig. 5.2 Voyage sketch on the Constanta—Rio de Janeiro route (overview)

- Scenario 2: the collectors are tilted with the optimum angle depending on the season and latitude and permanently oriented towards the South/North;
- Scenario 3: the collectors are tilted at the optimum angle and permanently follow the Sun (daily movement in steps).

All three scenarios consider the use of 23 solar thermal collectors, the main purpose being for the evaluation of energy according to the absorber surface, without consideration of the effect of self-shadowing.

For the carried-out study, a long route was chosen, covering the two hemispheres, the route Constanta (Romania)—Rio de Janeiro (Brazil), whose general map is shown in Fig. 5.2. The mapping was carried out in Transas Navi-Sailor ECDIS program, software used on most vessels, with the support of specialized personnel from the Ship-handling Simulator within the Naval Academy of Constanta, Romania.

The route has a length of 6482 nautical miles and the duration of the trip is 926 h; on this route the latitude varies between 44.08 °N and 22.51 °S. The speed of the vessel was considered constant at 7 knots, this being an economic speed that respects the current trends in the field.

The simulation of the voyage, from the point of view of the deployment period, was performed by considering two relevant options, corresponding to totally opposite astronomical phenomena:

- Option 1 envisaged that during the summer solstice in the northern hemisphere (respectively the winter solstice in the southern hemisphere), June 21, the vessel would be in the middle of the voyage. Thus, the voyage takes place between June 2 (day of the year n = 153) at 08.00 and July 10 (n = 191) at 22.00;
- Option 2 was chosen so that during the winter solstice in the northern hemisphere (respectively the summer solstice in the southern hemisphere), December 21 (n = 355), the vessel would be as in the previous case at mid-voyage. Thus, the voyage

takes place between December 2 (n = 336) at 08.00 h and January 9 (n = 9) at 22.00 h.

The determination of the global solar radiation in the plane of the solar-thermal collectors was done using the Bason model [14, 15]. The turbidity (Linke) factor specific to the water surfaces was determined considering latitude and season, cold or warm [16]. An example with the daily values of the turbidity factor during the two voyage periods depending on the average latitude at which the vessel is located daily are centralized in Table 5.2.

Optimum angle of tilt of the collectors, with great impact on the received energy, was determined according to latitude and season with relation (5.1) [16]. An example with the obtained results is presented in Table 5.3.

Table 5.2 Linke factor values by average daily latitude and day of the year

Average daily latitude (°)	Option 1		Option 2	
	Day of the year	Linke factor	Day of the year	Linke factor
43.05	153	3.25	336	2.33
35.75	158	3.52	341	2.60
37.06	163	3.48	346	2.55
31.63	168	3.64	351	2.76
23.80	172	3.80	355	3.06
13.86	176	3.85	359	3.41
0.45	181	3.68	364	3.74
−12.79	186	3.31	4	3.83
−23.03	191	2.97	9	3.80

Table 5.3 The optimum tilt angle according to the average daily latitude and the day of the year

Average daily latitude (°)	Option 1		Option 2	
	Day of the year	Optimum tilt angle (°)	Day of the year	Optimum tilt angle (°)
43.05	153	8	336	70
35.75	158	2	341	63
37.06	163	3	346	64
31.63	168	−2	351	59
23.80	172	−9	355	52
13.86	176	−17	359	43
0.45	181	−29	364	31
−12.79	186	−40	4	16
−23.03	191	−49	9	7

Table 5.4 Main technical specifications of the TitanPower-ALDH 29 collector

Absorber surface	2.57 m^2	Collector efficiency	82.0%
Total area	2.72 m^2	Maximum working pressure	11 bar
Length	2160 mm	Absorption factor	95%
Width	1260 mm	Emission factor	5%
Depth	105 mm	Stagnation temperature	200 °C
Mass	55.2 kg	Price per piece	599 $

Source www.siliconsolar.com

$$\chi_{opt} = 0,919 \cdot \varphi \pm 17,448 \tag{5.1}$$

In relation (5.1), the "+" sign will be adopted between October and March, and the sign "−" between April and September.

The negative values of the optimum tilt angle in Table 5.3 indicate the north orientation of the solar collectors (in the southern hemisphere).

For the conversion of solar energy into thermal energy a flat-plate collector of the type TitanPower-ALDH29 was considered, whose technical specifications are given in Table 5.4.

Knowing the efficiency of the collectors based on the characteristics emitted by the manufacturer (82%) and considering the heat losses from the installation (approximately 40%), the efficiency of the whole solar-thermal system will be ≈50%.

Further, the three scenarios of arrangement of solar-thermal collectors are compared in terms of thermal energy produced during the two voyages. Following the analysis, the optimal arrangement of the solar thermal collectors will be determined onboard the vessel.

The determination of the thermal energy produced by the solar-thermal system during the voyages on the Constanta—Rio de Janeiro route was made considering position of the vessel from hour to hour, hourly estimated global solar radiation and an efficiency of the solar-thermal collectors of 82% and an overall efficiency of the system of $\eta = 50\%$.

The thermal energy resulting from the conversion of solar energy (E_{STC}) was determined based on the relationship:

$$E_{STC} = E \cdot A_{abs} \cdot \eta \tag{5.2}$$

where E is solar energy in Wh/m^2 and A_{abs} is the absorber area in m^2.

The amount of thermal energy produced in the three scenarios is presented in Table 5.5. As can be seen the maximum energy corresponds to scenario 3 (tilted with optimum angle and diurnal tracking in steps), and the minimum one corresponds to scenario 1 (collectors arranged horizontally). Also, Table 5.5 shows the average

Table 5.5 The thermal energy produced during the voyages in the three scenarios

Period of voyage	The thermal energy produced during voyage (kWh)		
	Scenario 1	Scenario 2	Scenario 3
2 June–10 July	8709.02	9274.98	11,179.31
2 December–9 January	5967.06	8200.81	9845.88
Average energy	7338.04	8737.90	10,512.60

Table 5.6 Percentage differences between scenario 1 and scenarios 2 and 3

	2 June–10 July		2 December–9 January	
	Thermal energy (kWh)	Difference from scenario 1 (%)	Thermal energy (kWh)	Difference from scenario 1 (%)
Scenario 1	8709.02	–	5967.06	–
Scenario 2	9274.98	+6.5	8200.81	+37.4
Scenario 3	11,179.31	+28.4	9845.88	+65.0

values over the two periods. This average, given that the chosen periods are astronomically extreme, can be considered as a reference value regardless of the period of voyage of a Constanta—Rio de Janeiro route.

Another relevant aspect is the percentage differences that appear between the most unfavorable case (horizontal arrangement of collectors—scenario 1) and cases of tilted collectors (scenarios 2 and 3), Table 5.6.

From the analysis of the results presented in Tables 5.5 and 5.6 the following can be stated:

- the arrangement with an tilt angle $\chi = 0$ (horizontal) of the solar-thermal collectors (scenario 1) represents the least favorable option, regardless of the period analyzed; the differences are significant especially during the winter period when the contribution of the solar radiation is lower, being between 37.4% compared to scenario 2 and 65.0% compared to scenario 3;
- scenario 2, which implies the daily tilt of the collectors with the optimum angle depending on the latitude and their orientation towards the south/north, becomes efficient compared to scenario 1, especially in winter, in summer the difference is 6.5%;
- scenario 3, which considers the permanent track of the Sun, as well as the tilt of the collectors with the optimal angle depending on latitude, represents the most advantageous option regardless of the season; thus, at an annual level, there is a difference of 65.0% compared to scenario 1 and 20.0% compared to scenario 2.

5.3 Methods of Using the Energy Generated by the Solar-Thermal System

The use of solar-thermal collectors leads to the reduction of the fuel consumption used to produce thermal energy onboard the vessels and implicitly to the reduction of the greenhouse gases resulting from the burning of this quantity of fuel. In this case, the fuel used onboard the vessel (and generally onboard the commercial vessels) is heavy fuel oil with a lower calorific value $Q_i = 42{,}700$ (kJ/kg). The quantity of fuel m saved by using the thermal energy delivered by the solar-thermal system considered, considering that 1 kWh = 3600 kJ, is given by:

$$m = \frac{E_{STC} \cdot 3600}{Q_i} [\text{kg}] \tag{5.3}$$

where: E_{STC} represents the energy produced by the solar thermal system in kWh, and Q_i the calorific power of the fuel, in kJ/kg.

Knowing the thermal energy obtained in the most favorable case (scenario 3 in summer time) as being 11,179.31 kWh and using the above relationship, will result in a fuel saving during the voyage of 942.52 kg, which leads to an average of 24.8 kg/day. Compared to the total fuel consumption during the journey which is 463 tons (taking into account an hourly consumption of 500 kg and a duration of 926 h), the amount saved represents 0.20%, but as will be shown further, the thermal energy obtained by the solar-thermal conversion covers 100% of the energy needed to prepare DHW, and when stationed at anchorage it substantially reduces pollution.

The thermal energy obtained onboard the vessel from the considered solar-thermal system is intended to be used for the purpose of providing the DHW needed by the crew. The DHW is used in the installations serving the sanitary system, as well as in the installations that ensure the functioning of the kitchens. Also, the use of the solar-thermal system during the stationing of the vessel in the port or vicinity of the port for the purpose of obtaining domestic hot water, can lead to the following benefits:

- reduces the burden of facilities commonly used for hot water preparation, installations that use fossil fuels for operation; thus, the installations will function less (for the realization of other functionalities), which leads to fuel savings and to increase the life of the subcomponents of the installation;
- by reducing fuel consumption, greenhouse gas emissions will also be reduced, with direct favorable impact on the communities near the ports.

To identify this potential for use it is necessary to know the necessary DHW of the crew and the average daily amount of energy delivered by the solar thermal system (to determine the amount of water that can be heated with this energy). The number of sailors in the crew according to the main characteristics of the Albatros can be between 20 (standard crew) and 44 (maximum crew). The daily requirement of DHW for each crew member is 50–100 L, depending on the activities carried out. This results in a daily requirement of 1000–2000 L for the standard crew and 2200–4400 L for the maximum crew. The average daily amount of energy delivered by the solar-thermal system is 276.6 kWh, considering a total of 10,512.6 kWh during the 38 days of the trip. The daily amount of DHW that can be obtained with this energy is determined according to the relation:

$$E_{avgdaily} = m \cdot c \cdot \Delta t \rightarrow m = \frac{E_{avgdaily}}{c \cdot \Delta t} = \frac{276,6}{0.001161 \cdot 40} = 5956.1[\mathrm{kg}] = 5956.1[\mathrm{l}] \quad (5.4)$$

where: $c = 4,18\left[\frac{\mathrm{kJ}}{\mathrm{kg \cdot K}}\right] = 0,001161\left[\frac{\mathrm{kWh}}{\mathrm{kg \cdot K}}\right]$ is the specific heat of the water, $\Delta t =$ 40 °C is the temperature difference between the DHW temperature required for consumption (50 °C) and the water temperature of the storage tanks (10 °C), and $E_{avgdaily}$ is the average daily energy produced by the solar thermal system.

The obtained results make it possible to state the following observations:

- the average thermal energy during the trip obtained from the considered solar-thermal system satisfies the need for DHW preparation for the whole crew, exceeding even the maximum required regardless of the number of sailors in the crew;
- the surplus of DHW produced on a clear day (if the obtained thermal energy is higher than the average thermal energy considered in the present study) can be used on cloudy days; also, this surplus can be transferred to naval boilers, thus reducing their load by delivering water at 50 °C instead of 10 °C as usual, which also leads to a reduction in the amount of fuel used by boilers;
- the surplus of thermal energy obtained during sunny days can also be used for periodically raising of the DHW temperature above 60 °C, in order to prevent the development of bacteria on distribution facilities and storage tanks [17, 18];
- in case of consecutive cloudy days, when the DHW requirement cannot be provided by the solar thermal system, the water heating can be done in the classical way through the boiler, thus creating a combined DHW production system.

The utility of solar-thermal systems onboard vessels should also be looked at from the point of view of the environmental impact. As previously shown, solar energy conversion systems offer clean energy-saving solutions for periods when vessels are in coastal areas. In addition, these solutions are more likely to be implemented on cruise vessels or pleasure boats where, together with immediate utility (DHW production), another important application, that of drinking water production, through desalination process, can be considered. For this type of application, the implementation of solar-thermal collector systems must be considered from the design stage of the vessel, with surfaces dedicated to their positioning.

5.4 Design of Solar-Thermal Systems for the Military Support Vessel

Based on the results obtained previously regarding the amount of DHW that can be prepared by using the solar energy, further there are proposed three types of systems that can be adapted onboard the vessel (Figs. 5.3, 5.4 and 5.5). Within the proposed systems, the conversion of solar energy into thermal energy, the transfer of energy

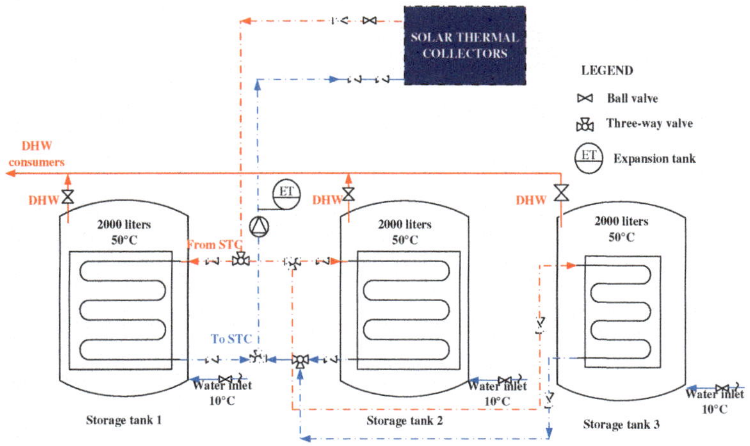

Fig. 5.3 Solar-thermal system for DHW preparation

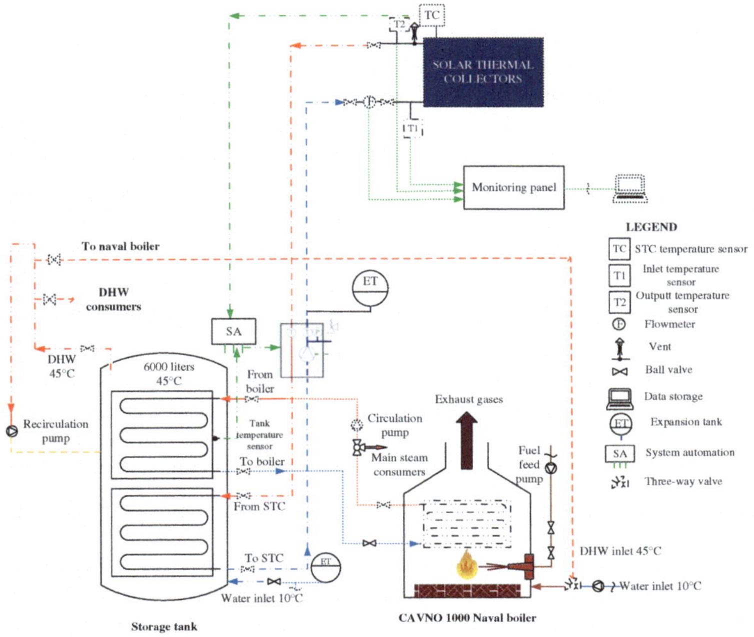

Fig. 5.4 Combined solar—boiler system for DHW preparation

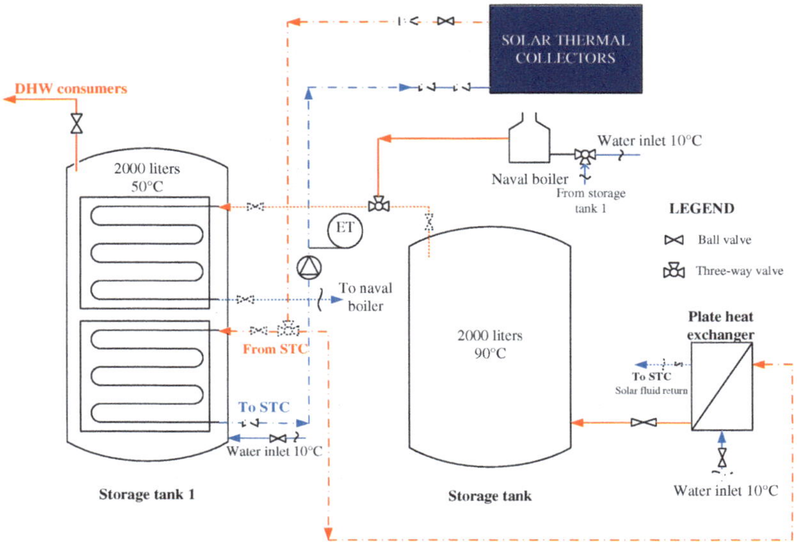

Fig. 5.5 Combined solar-boiler DHW preparation system with heat exchanger

necessary for the preparation of DHW, as well as its storage and/or distribution to consumers, is carried out.

The system shown in Fig. 5.3 is a classic solar-thermal system composed of collectors, storage tanks and distribution system. The storage tanks have been sized in such a way as to allow the preparation and storage of the quantity of DHW previously determined based on the average daily thermal energy delivered by the collector system. Three 2000 L storage tanks were included, the design being made so that one of them could be insulated in the event of a malfunction or a low supply of thermal energy.

The systems in Figs. 5.4 and 5.5 are designed in such a way as to ensure a close connection between the classic system of hot water production onboard (boiler) and the solar thermal system. Thus, in the first proposed combined system (Fig. 5.4), a storage tank of 6000 L is designed, a tank that takes the thermal energy from the collectors but can receive energy when needed from the boiler. Additionally, to avoid overheating the collectors if the hot water in the storage tank is not used by the consumers, it can become an auxiliary source of water for the boiler. The main advantage of this use is that in the boiler the water is already preheated, and the amount of fuel needed to obtain steam will be lower.

The system in Fig. 5.5 is designed with two 2000 L storage tanks and a plate heat exchanger. The role of the storage tank that can store hot water at 90 °C is to reduce the space required onboard for the installation of the system (having a smaller capacity) compared to the system proposed in Fig. 5.3 and to allow storage and use at need of hot water with a high temperature (thus limiting the losses that can occur during the days with an increased radiation intake).

The systems proposed in this case can be adopted onboard the logistic support vessel but can also be viable solutions for any vessel. Two of the basic principles to consider when choosing the right system are the dependence on the available space (in the case of vessels already in service) and the additional weight that appears onboard, weight that is intended to be as lower as possible, as well as the quantity of cargo that can be transported and consequently the revenue from it.

5.5 Conclusions

An analysis of a logistical support military vessel from the point of view of the functionalities of the thermal energy and of the implementation of the solar thermal systems onboard vessel was carried out. The analysis took into account a voyage on the Constanta—Rio de Janeiro route (38 days, 12,000 km), made during different periods of the year and astronomically extreme in order to determine the thermal energy delivered by a solar-thermal system. Further, there were identified methods of using the energy obtained by solar-thermal conversion, as well as the quantity of fuel saved. Results showed that a maximum of 11,179.31 kWh is produced on the chosen voyage, using tracking systems during summer time, and the reduction in terms of fuel and emissions were 886 kg (which represent 0.2% of total amount of fuel), respectively 2760 kg of CO_2. Also, three types of solar-thermal systems that can be adapted on vessels for the direct delivery of DHW and/or for the storage of thermal energy and its subsequent use have been proposed.

References

1. E.A. Bouman, E. Lindstad, A.I. Rialland, A.H. Strømman, State-of-the-art technologies, measures, and potential for reducing GHG emissions from shipping—a review. Transp. Res. Part D: Transp. Environ. **52**(Part A), 408–421 (2017)
2. International Maritime Organization, *MEPC.304 (72)—Initial IMO Strategy on Reduction of GHG Emissions from ships*. London, England (2018)
3. E. Eastlack, et al.,Zero emission super-Yacht, in *2019 Fourteenth International Conference on Ecological Vehicles and Renewable Energies (EVER)*, Monte-Carlo, Monaco, pp. 1–8 (2019)
4. P. Nuttall, A. Newell, B. Prasad, J. Veitayaki, E. Holland, A review of sustainable sea-transport for Oceania: providing context for renewable energy shipping for the Pacific. Marine Policy **43**, 283–287 (2014)
5. V. Koilo, Sustainability issues in maritime transport and main challenges of the shipping industry. Environ. Econ. **10**(1), 48–65 (2019)
6. Y.P. Yuan, J.X. Wang, X.P. Yan, Q. Li, T. Long, A design and experimental investigation of a large-scale solar energy/diesel generator powered hybrid ship. Energy **165**(Part: A), 965–978. https://doi.org/10.1016/j.energy. 2018.09.085, Published: DEC 15 2018
7. T. Ruoli, W. Zhou, L. Xin, Optimal operation of photovoltaic/battery/diesel/cold-ironing hybrid energy system for maritime application. Energy **162**(1), 697–714 (2018)
8. T. Ruoli, L. Xin, L. Jingang, A novel optimal energy-management strategy for a maritime hybrid energy system based on large-scale global optimization. Appl. Energy **228**(15), 254–264 (2018)

9. J. Corbett, J. Winebrake, E. Greean, P. Kasibhatla, V. Eyring, A. Lauer, Mortality from ship emissions: a global assessment. Environ. Sci. Technol. **41**, 8512–8518 (2007)
10. H.P. Cullom, A.D.M. Rear, Navy task force energy, in *Climate & Energy Symposium*, chap. IV., Energy Imperatives, Part II (2010)
11. A.M. Chambers, S.A. Yetiv, The great green fleet—The U.S. Navy and fossil-fuel alternatives. Naval War Coll. Rev. **64**(3) (2011)
12. M. Manzan, A. Pezzi, E. Zandegiacomo de Zorzi, A. Freni, A. Frazzica, B.M. Vaglieco, Z. Lucio, D. Claudio, Potential of thermal storage for hot potable water distribution in cruise ships. Energy Procedia **148**, 1105–1112 (2018)
13. A. Cotorcea, I. Vișa, Study of adaptability of solar thermal systems on merchant vessels, in *Springer Proceedings in Energy, Proceedings of the Conference for Sustainable Energy (CSE)*, pp. 221–228 (2014). ISBN 978-3-319-09706-0
14. F. Bason, Diffuse solar irradiance and atmospheric turbidity, in *Proceedings EuroSun Conference*, Freiburg, Germania (2004)
15. F. Bason, Solar irradiance measurements from the Danish Galathea 3 expedition, in *ISES Solar World Conference*, Beijing (2007)
16. A. Cotorcea, Adaptation and implementation of solar energy conversion systems into thermal energy for merchant vessels, Ph.D. thesis, Brasov (2017). (in Romanian language)
17. MGN 397 (M+F), Guidelines for the Provision of Food and Fresh Water on Merchant Ships and Fishing Vessels
18. World Health Organization, International Health Regulations—Guide to Ship Sanitation, 3rd edn. (2007)

Part III
Solar Electricity in Communities

Chapter 6
Agri.q: A Sustainable Rover for Precision Agriculture

Giuseppe Quaglia, Carmen Visconte, Luca Carbonari, Andrea Botta, and Paride Cavallone

Abstract In this paper, an innovative mobile and sustainable robot for precision agriculture, named "Agri.q", is presented. Characterized by a peculiar mechanical architecture and provided with specific sensors and tools, the Agri.q is able to operate in unstructured agricultural environments in order to fulfill several tasks as mapping, monitoring, and manipulating or collecting small soil and leaf samples. In addition, the rover is equipped with a top platform covered with solar panels, whose orientation can be exploited both to maximize the sunrays collection during the auto-charging phase and to permit a drone landing over a horizontal surface, regardless of the ground inclination. A particular attention to energy consumptions and sustainability has driven the mechanical design of the Agri.q powertrain: the weight reduction results into a limited number of small size locomotion motors, enhancing the importance of the harvested solar energy on the energy balance of the whole system. In this paper, all these characteristics are described and analyzed in detail. Moreover, some preliminary tests aimed at evaluating the energetic behaviour of the rover under different working and weather conditions are presented.

Keywords Mobile robot · Agri.q · Service robotics · Precision agriculture · PV solar panels

G. Quaglia (✉) · C. Visconte · L. Carbonari · A. Botta · P. Cavallone
Politecnico di Torino, 10129 Torino, Italy
e-mail: giuseppe.quaglia@polito.it

C. Visconte
e-mail: carmen.visconte@polito.it

L. Carbonari
e-mail: luca.carbonari@polito.it

A. Botta
e-mail: andrea.botta@polito.it

P. Cavallone
e-mail: paride.cavallone@polito.it

I. Visa and A. Duta (eds.), *Solar Energy Conversion in Communities*,
Springer Proceedings in Energy, https://doi.org/10.1007/978-3-030-55757-7_6

6.1 Introduction

Precision agriculture aims at improving the sustainability of agricultural production by taking advantage of specialized technological equipment. As a side effect, also profitability and efficiency turn conveniently enhanced [1]. In a few words, the whole precision agriculture paradigm can be summarized in applying 'the right treatment in the right place at the right time' [2]. Many relevant studies demonstrated in the recent past the beneficial effect of such paradigm on environment and economics [3, 4]. The interest demonstrated by private producers and governments is then well understandable. In this technological transition, robotics and automation perform a leading role, which can profit of the advancements of decades of research. Those technologies will produce a significant thrust in the modernization of the agricultural production. Many examples of re-adaptation of agricultural machines are available in the literature: Wang et al. [5], and later Zaman et al. [6], focused on the use of advanced navigation systems; in [7] a cooperative system of aerial and ground vehicles is used for monitoring tasks; Khaliq et al. [8] approached the problem under the point of view of trajectory planning using multispectral imagery. Aside such interesting advancements in the field of control, few examples are available of machines designed to fulfill some specific requirements. Among others, [9] and [10] present two innovative ground robotic units for automated harvesting, while in [11] an aerial robot for mango harvesting is proposed. As a matter of fact, one of the most important features that an autonomous robot should have is to work without human intervention as long as possible. Then, the use of solar panels can allow the robot to have a high exploration capability in open spaces. Many studies focused on motion planning and control to reduce the energy consumption [12, 13], better than increasing the stored energy. At present, two main strategies are in use for battery recharging, based on board solar panels [14] and off board stations [15].

In this scenario, the researchers of the Politecnico di Torino developed a novel wheeled UGV named Agri.q (shown in Fig. 6.1) tailored on precision agriculture tasks and specifically designed for monitoring and sampling of crops and soil [16–18], especially for vineyard cultures. The rover is a small size (approximately 1 m $\times$ 2 m $\times$ 0.7 m) electric vehicle designed to operate in unstructured environments, to move through the rows of grapes and to cooperate with aerial drones. Moreover, it is equipped with a robotic arm to accomplish sampling tasks. The Agri.q mobile rover has been equipped with an orientable platform covered with solar panels, with the aim of enhancing the battery runtime and therefore the duration of the missions executable by the rover.

6.2 Prototype Description

The actuators of the modular locomotion systems have been carefully chosen considering, as the most burdening condition, a slope of about 15° to be overcome with only

Fig. 6.1 Prototype of the Agri.q mobile robot

two front traction motors (namely the two front driving units) at a maximum speed of about 5 km/h. To achieve such goal, the transmission of each unit is composed by an electric brushless DC motor, connected to a planetary gearbox. A further speed reduction is then provided by a chain system between the gearmotor and a pair of wheel (front left or front right). Some details on the adopted mechanical components are provided in Table 6.1. Each gearmotor unit is also provided with an electric parking brake.

In addition, the rover is equipped with two rear locomotion units that are used only when the slope exceeds 15% or when the adherence condition requires to pass from the 2 to the 4 driving units architecture; in this second case, the power transmission is distributed to the eight wheels in contact with the soil (Fig. 6.2).

The powertrain layout was completed by a battery able to feed the adopted locomotion systems, and by a solar panel whose dimension was reasonably chosen to ensure the recharge of the battery during the idle time between missions. Details about the rechargeable battery are provided in Table 6.2.

The solar panel was chosen to meet the needs deriving from the mobile application.

In particular, two flexible light-weighted panels were installed, able to guarantee an adequate recharging capability without affecting the robot dynamics with an excessive weight. Some technical details are shown in Table 6.3 and Fig. 6.3. The data refers to a perpendicular solar irradiation of 1000 W/m^2 at the working temperature of 25°.

Table 6.1 Front and rear driving modules functional parameters

		Front driving units	Rear driving units
Motor nominal power	$P_{M,Nom}$	120 W	120 W
Motor nominal torque	$C_{M,Nom}$	0.38 Nm	0.38 Nm
Motor nominal speed	$\omega_{M,Nom}$	3000 rpm	3000 rpm
Motor nominal current	$I_{M,Nom}$	10 A	10 A
Motor maximum current	$I_{M,\text{Max}}$	20 A	20 A
Motor torque constant	K_I	0.055 Nm/A	0.055 Nm/A
Gearbox reduction	τ_1	1 : 15.88	1 : 28.93
Chain system reduction	τ_2	1:3	1:1
Total reduction	τ	1:47.64	1 : 28.93
Wheels radius	r_w	0.200 m	0.200 m

Fig. 6.2 Usage of the orientable upper panel during solar recharge (**a**) and for safe drone landing (**b**)

Table 6.2 Battery technical specifications

Battery voltage	29.7 V	Maximum current	128 A
Battery capacity	56 Ah	Total stored energy	4.84 MJ

Table 6.3 Solar panel technical specifications

Peak power	154 W	Maximum tension	18.2 V
Dimensions	1046 mm × 996 mm	Maximum current	8.5 A
Mass	2.40 kg	Working temperature	−45/ + 85 °C

6.3 Photovoltaic Potentiality of the Agri.q

As well known, solar tracking strategies can substantially increase the efficiency of the energy harvesting capability of a solar panel (up to 40%). In general, the idea is to optimize the direct absorption of the solar radiation by reorienting the panels so that the sunrays turn to be perpendicular to them [19].

In order to get an effective solar tracking, the rover can take advantage of two expressly conceived degrees of freedom able to orientate the upper platform around two independent axes. In particular, the pitch and the roll angles of the solar panel can be modified (as shown in Fig. 6.4a), so that the reference frame $\xi - \psi - \zeta$ solid with the panels assume a non-identical rotation matrix with respect to the rover reference frame $x - y - z$.

As shown in [16–18], the Agri.q upper plane pitch mechanism can provide a rotation of $\gamma = -5°/+40°$. The plane can also roll of $\alpha_R = -20°/+20°$. Such degrees of freedom can orientate the solar panels so that the perpendicular unit vector ζ is:

$$\zeta = R_y(\gamma) R_x(\alpha_R) \begin{bmatrix} 0 & 0 & 1 \end{bmatrix}^T \tag{1}$$

where $R_y(\gamma)$ and $R_x(\alpha_R)$ represent two rotation matrices respectively around the y and the x axes of the reference frame $x - y - z$. These ranges of rotation allow the rover to point the solar panels towards a relevant section of the sky (red spherical cap in Fig. 6.4b). The minimum pointing angle $\varphi_{\min}$ that the platform can perform is $\varphi_{\min} = 46.04°$.

Moreover, thanks to its maneuverability in the plane $x - y$, the rover can assume whatever orientation around the axis z, so increasing the number of possible poses of the platform. Figure 6.3b shows through the gray spherical cap the reachable orientations of the solar panel.

6.4 Consumptions Estimation

The average power consumption can be estimated monitoring the mechanical power absorbed by the front left (W_{FL}) and right (W_{FR}) traction motors in different conditions. In the following, four tests are analysed (see Fig. 6.5): Test 1 to 3 refer to straight trajectories on non-sloped surfaces at different velocity levels (details are shown in Table 6.4), while Test 4 refers to a constant curvature trajectory. Aside the

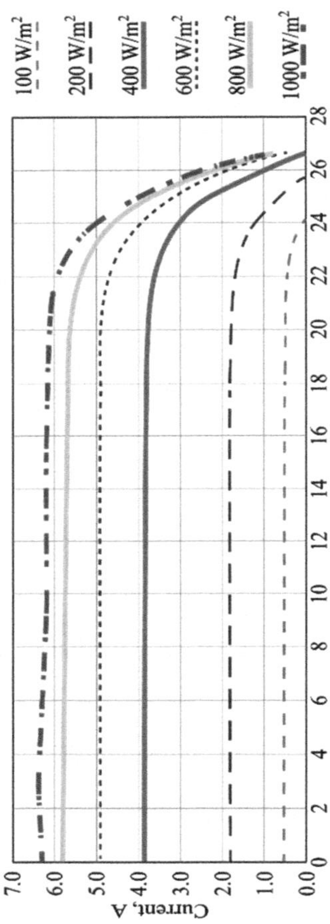

Fig. 6.3 Current/voltage characteristic of the photovoltaic panels at different levels of solar incident power

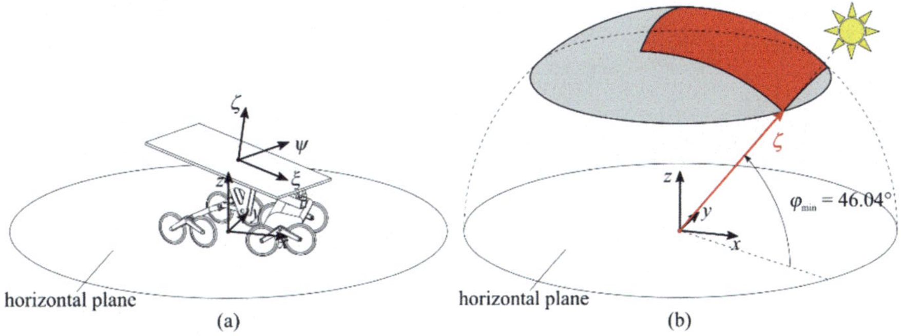

Fig. 6.4 **a** Reorientation of the Agri.q upper platform and **b** minimum inclination of the panels perpendicular with respect to the horizontal plane

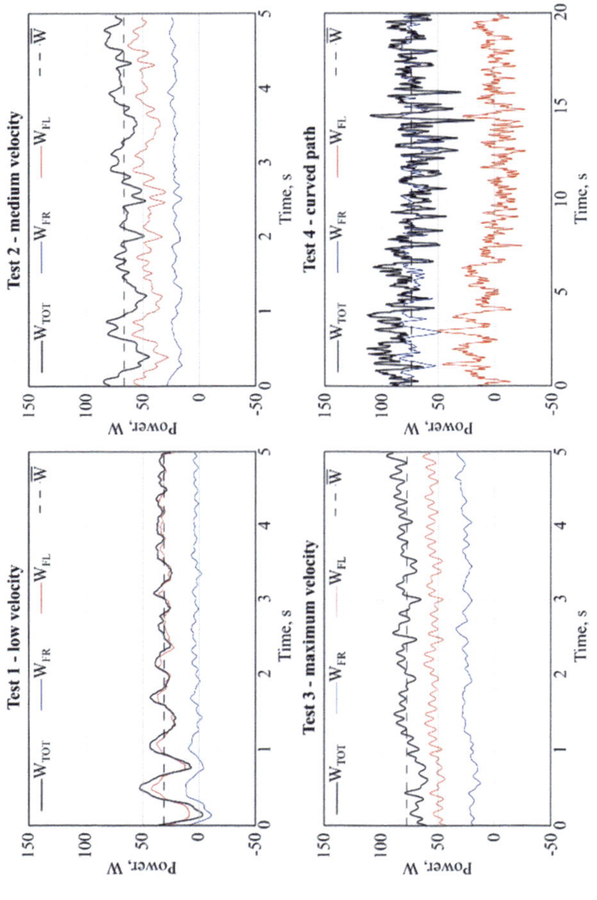

Fig. 6.5 Mechanical power measured in different use conditions

Table 6.4 Velocity, torque and power measurement for tests of Fig. 6.5

	Test 1	Test 2	Test 3	Test 4
Velocity	2.15 km/h	3.2 km/h	4.7 km/h	2.00 km/h
Mean motors torque	0.91 Nm	0.93 Nm	0.94 Nm	0.25 Nm
Mean motors power	30.94 W	66.28 W	78.19 W	74.14 W

mechanical power of the left and right driving units, also the global power absorption W_{TOT} and the mean value $\overline{W}$ are shown. A combination of such experimental tests can be used to estimate the average energy consumption of the rover during an actual monitoring mission. By simply mediating the registered powers, a mean value of 62.4 W is obtained. Such a value should be properly increased to take into consideration slopes, different possible levels of terrain roughness, and even the case in which all the four driving units (both front and rear) are needed. Moreover, the re-orientation actuators and the electronic hardware are expected to increase the power absorption too. As also evicted in the following, an average consumption of about 90 W is an acceptable approximation.

As a further consumption estimation proof, some numerical data obtained during a two hours mission of the rover are reported. The tested duty cycle included almost all the working conditions reasonably foreseeable during a standard monitoring mission: paths on non-sloped uneven terrains, curves, slopes (of about 10° approached both with only the front locomotion units activated and with all the four units active), reorientations of the solar panels, etc. This wide use scenario resulted in the energy consumption reported in Table 6.5. A total consumption of 179.5 Wh, corresponding to 0.65 MJ, was recorded. Taken into account the tested mission duration, it corresponds to a power absorption of approximately 89.75 W; therefore, the 13.35% of the whole battery energy was used. A similar result can also be inferred from data on the used battery capacity: 6.54 Ah on a total capacity of 56 Ah, corresponding to the 11.71%. Even in terms of peak current, the powertrain turns to be well sized on the rover needs: the maximum value, recorded along a slope of 10° approached at the maximum rover velocity with all the four driving units active, is 24.19 A, which is well lower than the maximum battery current (128 A).

Table 6.5 Energy and electric parameters recorded during a two hours mission

Time	Voltage	Actual current	Actual power	Used charge	Used energy
0 h 00′	28.95 V	0 A	0 W	0 Ah	0 Wh
0 h 30′	28.50 V	0.97 A	27.9 W	3.86 Ah	106.8 Wh
1 h 00′	28.36 V	3.15 A	89.3 W	5.33 Ah	145.3 Wh
1 h 30′	28.47 V	0.94 A	26.7 W	5.44 Ah	148.8 Wh
2 h 00′	28.34 V	0.86 A	24.3 W	6.56 Ah	179.5 Wh
Peak current: 24.19 A		Peak power: 674.1 W		Average tension: 27.76 V	

6.5 Concluding Remarks

The exploitation of robotics technologies for precision agriculture applications is increasingly catching the attention of both research and industrial community, for it representing a smart and sustainable way to enhance profitability and productivity. Moreover the sustainable design approach adopted in this research represents a method to address the technological evolution to the achievement of some Sustainable Development Goals, fixed by United Nations in 2015 [20]. The most relevant SDGs, in this case are 7, affordable and clean energy, 13, Climate action, 15, life on land. The researchers of the Politecnico di Torino developed a novel wheeled UGV, named Agri.q, designed for unmanned monitoring and sampling tasks of crops and soil. This paper focuses on the sustainability features which characterize the robot, and in particular on its capability to exploit solar renewable energy. The installed set of photovoltaic panels, in fact, makes available an effective source of energy that increases the autonomy of Agri.q during operating missions. The panels can recharge the rover battery in the idle time between the missions, avoiding it to come back to base for such a reason. The tests showed the behavior of the robot in the most common use conditions, such as straight and curved paths, and allowed verifying the effectiveness of the chosen powertrain. Aside that, some details are also provided about the energy consumption during a reasonable mission-like usage.

Acknowledgements Authors' acknowledgement goes to the PIC4SeR (PoliTO Interdepartmental Centre for Service Robotics) which gave support and assistance to this research.

References

1. A.M. Adrian, S.H. Norwood, P.L. Mask, Producers' perceptions and attitudes toward precision agriculture technologies. Comput. Electron. Agric. **48**(3), 256–271 (2005)
2. R. Gebbers, V. Adamchuk, Precision agriculture and food security. Science **327**(5967), 828–831 (2010)
3. J. Lowenberg-DeBoer, I.Y. Huang, V. Grigoriadis, S. Blackmore, Economics of robots and automation in field crop production. Precis. Agric. 1–22 (2019)
4. A.P. Barnes, I. Soto, V. Eory, B. Beck, A. Balafoutis, B. Sánchez, M. Gómez-Barbero, Exploring the adoption of precision agricultural technologies: A cross regional study of EU farmers. Land Policy **80**, 163–174 (2019)
5. Y. Wang, Y. Lan, Y. Zheng, K. Lee, S. Cui, J.-A. Lian, A UGV-based laser scanner system for measuring tree geometric characteristics, in *5th International Symposium on Photoelectronic Detection and Imaging*, (Beijing, China, 2013).
6. S. Zaman, L. Comba, A. Biglia, D.R. Aimonino, P. Barge, P. Gay, Cost-effective visual odometry system for vehicle motion control in agricultural environments. Comput. Electron. Agric. **162**, 82–94 (2019)
7. P. Tokekar, J.V. Hook, D. Mulla, V. Isler, Sensor planning for a symbiotic UAV and UGV system for precision agriculture. IEEE Trans. Rob. **32**(6), 1498–1511 (2016)
8. A. Khaliq, L. Comba, A. Biglia, D.R. Aimonino, M. Chiaberge, P. Gay, Comparison of satellite and UAV-based multispectral imagery for vineyard variability assessment. Remote Sens.**11**(4) (2019)

9. A.P. Chatzimichali, I.P. Georgilas, V.D. Tourassis, Design of an advanced prototype robot for white asparagus harvesting, in *International Conference on Advanced Intelligent Mechatronics* (2009), pp. 887–892
10. A.A. Aljanobi, S.A. Al-Hamed, S.A. Al-Sushaibani, A setup of mobile robotic unit for fruit harvesting, in *19th International Workshop on Robotics in Alpe-Adria-Danube Region RAAD* (Budapest, Hungary, 2010)
11. S. Konam, Agricultural aid for mango cutting (AAM), in *International Conference on Advances in Computing, Communications and Informatics (ICACCI)* (2014), pp. 1520–1524.
12. A. Barili, M. Ceresa, C. Parisi, Energy-saving motion control for an autonomous mobile robot. Proc. IEEE Int. Symp. Ind. Electron. **2**, 674–676 (1995)
13. Z. Sun, J Reif, On energy-minimizing paths on terrains for a mobile robot. In: *Proceedings of the 2003 IEEE International Conference on Robotics & Automation* (2003), pp. 3782–3788
14. T. Kaur, D. Kumar, Wireless multifunctional robot for military applications, In *2nd International Conference on Recent Advances in Engineering and Computational Sciences* (2016)
15. G.B. Parker, R.S. Zbeda, Learning navigation for recharging a self-sufficient colony robot, In *Proceedings of the 2007 IEEE International Conference on Systems* (2007)
16. G. Quaglia, C. Visconte, L.S. Scimmi, M. Melchiorre, P. Cavallone, S. Pastorelli, Design of the positioning mechanism of an unmanned ground vehicle for precision agriculture, in, *IFToMM World Congress on Mechanism and Machine Science* (2019), pp. 3531–3540
17. G. Quaglia, C. Visconte, L.S. Scimmi, M. Melchiorre, P. Cavallone, S. Pastorelli, Robot arm and control architecture integration on a UGV for precision agriculture, in *IFToMM World Congress on Mechanism and Machine Science* (2019), pp. 2339–2348.
18. G. Quaglia, P. Cavallone, C. Visconte, Agri_q: Agriculture UGV for monitoring and drone landing, In *IFToMM Symposium on Mechanism Design for Robotics* (2018), pp. 413–423
19. G. Quaglia, S.L. Maurino, Solar. q_1: A new solar-tracking mechanism based on four-bar linkages. *in Proceedings of the Institution of Mechanical Engineers* (2017), pp. 2855–2867
20. https://sustainabledevelopment.un.org/

Chapter 7
A Comparative Analysis of the Solar Energy Receiving Share Using Four Tracking System Types at Mid-Latitude Regions

Mircea Neagoe, Bogdan Gabriel Burduhos, Fazel Mohammadi, and Nadia Ramona Cretescu

Abstract This paper presents a comparative study of the share of direct solar radiation received by flat solar converters (photovoltaic modules, solar thermal collectors) using four tracking systems of equatorial, azimuth, pseudo-equatorial, and pseudo-azimuth type. The study is developed for the Brașov, Romania location (45.65° N, 25.60° N), but its results can be extended for most mid-latitude regions. The effect of dual-axis vs. single-axis tracking is studied in the first part, highlighting high shares of received solar energy using single-axis diurnal tracking with an optimum fixed tilt angle. The four tracking systems with unequal steps tracking algorithms are compared with a fixed tilted system as reference. The paper outlines the ability of single-axis tracking in receiving high shares of direct solar radiation. This methodology allows the designers to rank the four tracking system types, for mid-latitudes, according to the criteria of solar energy receiving share and tracking angular strokes.

Keywords Solar energy · Direct solar radiation receiving share · Solar tracking systems

M. Neagoe · B. G. Burduhos · N. R. Cretescu (✉)
Renewable Energy Systems and Recycling Research Center, Transilvania University of Brașov, Brașov, Romania
e-mail: ncretescu@unitbv.ro

M. Neagoe
e-mail: mneagoe@unitbv.ro

B. G. Burduhos
e-mail: Bogdan.burduhos@unitbv.ro

F. Mohammadi
Electrical and Computer Engineering (ECE) Department, University of Windsor, Windsor, ON N9B 1K3, Canada
e-mail: fazel@uwindsor.ca

I. Visa and A. Duta (eds.), *Solar Energy Conversion in Communities*,
Springer Proceedings in Energy, https://doi.org/10.1007/978-3-030-55757-7_7

7.1 Introduction

The main role of a solar tracking system is to increase the share of the solar energy received by the conversion devices, i.e. Photovoltaic (PV) modules or Flat Plate Solar Thermal Collectors (FPSTC), thus increasing their output.

A high-precision PV tracking system can support an annual increase in the electrical energy production up to 45% compared to a fix-tilted PV array, depending on the latitude and weather profile of the implementation location, the tracking system complexity (single-axis or dual-axis), and type of the tracking algorithm. This significant increase represents an important advantage in the case of solar energy conversion systems integrated into the built environment, considering the limited space available for implementation. However, this advantage also has several drawbacks compared to the fixed tilted PV systems [1, 2]. Fixed solar systems, tilted at an annual optimal angular value, can receive a maximum of 70–75% of the available solar energy during clear sky periods, therefore their output is at least 25% lower than the possible one. The yearly output gain when using solar trackers can vary from 20% in unsteady weather regions up to 40–45% in sunnier areas, compared to fixed, optimally tilted PV modules. The daily gain may range from almost zero (during the cloudiest days) to nearly 100% (during the clear sky days) [3]. Thus, solar tracking can be a feasible solution to significantly increase the output of solar energy conversion devices [4–6].

Tracked solar energy conversion systems have significant advantages compared to the stationary systems as they produce more energy for the same installed capacity or require a smaller number of PV modules/FPSTCs to obtain the same output and there is a current mature market that offers reliable and affordable tracking systems. However, the balance between these advantages and drawbacks of the higher initial investment and operational costs represents a critical point in the design process of a solar energy conversion system, highly dependent on the climatic profile of the implementation location, the available space for implementation, and the solar tracker type.

Starting from these considerations, this paper proposes a study on the capabilities of various tracking systems to increase the share of the direct solar radiation received at mid-latitude locations. Therefore, the methodology proposed for computing the the share of the received solar energy is presented in the first part, focusing on modeling the solar angles (Sect. 7.2), describing the four tracking system main types (Sect. 7.3), modeling the direct solar radiation (Sect. 7.4) along with the solar tracking algorithms and simulation scenarios (Sect. 7.5). The scenarios are simulated for the Brașov location, Romania, and numerical results are discussed in Sect. 7.6, along with the final conclusions formulated in Sect. 7.7.

7.2 Solar Angles

The solar radiation has two main components: the *direct radiation* that has the maximum irradiance value on the sunray's direction, and the *diffuse radiation*, with its maximum irradiance measured on the horizontal plane of the observer. The solar radiation profile with the preponderant direct component is mostly suitable for tracking applications and consequently, the design of solar tracking systems starts with modeling the angles of the sunray vector.

Considering an implementation location Q (Fig. 7.1) defined by the latitude angle φ, the solar angles associated to the sunray vector can be defined both in the global coordinate system of the Earth ($OX_0Y_0Z_0$), denoted also as the equatorial coordinate system (further symbolized by E_q), and in the observer coordinate system ($QX_qY_qZ_q$). The point O is the Earth's center, the Z_0 axis is the South (S)—North (N) rotation axis of the Earth, the X_0 axis is perpendicular to the local meridian plane and oriented toward the local East (E) direction, and the Y_0 axis is given by the intersection of the equatorial plane with the local meridian plane. The axis Z_q corresponds to the local vertical (zenith) axis, while Y_q to the N direction and X_q to the E direction.

The solar angles defined in the equatorial coordinate system (E_q) defined in Fig. 7.1a are:

- The *hour angle* (ω_s) is the angle between the sunray projection in the equatorial plane and the ($-Y_0$) axis and changes by 15°/hour. Thus, it can be calculated considering:

$$\omega_s = 15^\circ(12 - t_s) \tag{7.1}$$

where t_s is the solar time expressed in hours.

- The *declination angle* (δ_s) is the angle between the sunray and the equatorial plane, ranging between -23.45° and $+23.45^\circ$ during the year as the Earth axis is tilted at $23^\circ 26' 21''$ from the normal vector of its orbit plane. Its variation during 24 h is about 0.25° and hence, it is considered constant over one day in practical applications.

$$\delta_s = \sin^{-1}\left(\sin 23.45^\circ \sin\left(\frac{360^\circ(N - 80)}{365}\right)\right) \tag{7.2}$$

where N denotes the day number of the year.

The Sun crosses the local meridian at solar noon ($t_s = 12$ and $\omega_s = 0^\circ$). The hour angle is positive toward East (E) and negative towards West (W). The sunray

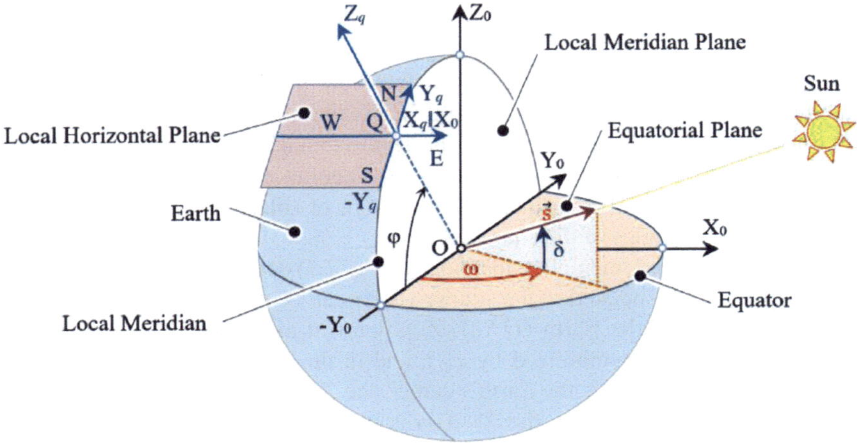

(a)

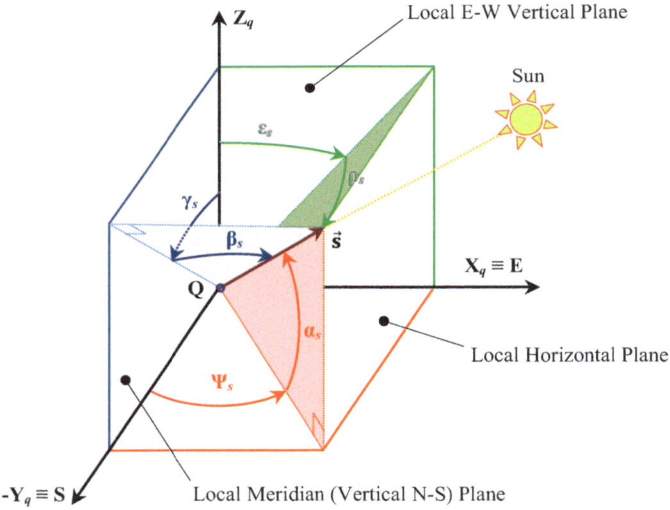

(b)

Fig. 7.1 The sunray angles: **a** equatorial solar angle pair (ω_s, δ_s) defined in the global coordinate system $OX_0Y_0Z_0$, **b** azimuth solar angle pair (Ψ_s, α_s), pseudo-azimuth solar angle pair (ε_s, ρ_s) and pseudo-equatorial solar angle pair (β_s, γ_s) defined in the local coordinate system $QX_qY_qZ_q$

vector can be projected on the three orthogonal planes of the local coordinate system $QX_qY_qZ_q$ (Fig. 7.1b) and the following solar angle pairs result [7].

- (Ψ_s, α_s) *Azimuth* (Az) solar angles, where QX_qY_q is the projection plane (the local horizontal plane).
- (ε_s, ρ_s) *Pseudo-Azimuth* (PAz) solar angles, where QX_qZ_q is the projection plane (the vertical East–West plane).
- (β_s, γ_s) *Pseudo-Equatorial* (PEq) solar angles, where QY_qZ_q is the projection plane (the vertical North–South plane).

The solar angles Ψ_s, ε_s, and β_s are *diurnal angles* and describe the relative motion of the Sun from the East (sunrise) toward the West (sunset). The elevation of the Sun on the sky is modeled by the solar angles α_s, ρ_s, and γ_s. In the particular case of the azimuth system, the solar angle Ψ_s is named *azimuth* and α_s*–altitude*. The daylight corresponds to $\alpha_s > 0$. Thus, the sunrise time t_{sr} and at sunset time t_{ss} $\alpha_s = 0$.

The diurnal angles Ψ_s of the Az system and ε_s of the PAz system are measured relative to a fixed axis (i.e. $-Y_0$ and Z_0, respectively), while the solar angle β_s of the PEq system has a time variable axis used as a reference. Generally, the stroke of the diurnal angles is higher than the stroke of the elevation angles and the diurnal motion has a bigger impact on the direct solar radiation collected by a tracked PV or FPSTC surface. The Sun path on the sky changes every day its parameters (sunrise and sunset positions on the horizontal plane, the height at noon). Thus, solar angles strokes are continuously modified during the year. The longest Sun trajectory during daylight occurs at the summer solstice.

$(N = 172)$. As a result, the summer solstice is typically the key day for evaluating the maximum strokes of the solar angles.

In the assumption of parallel sunray falling on the Earth, the expressions of the solar angle pairs of Az, PAz, and PEq types can be established in relation with the Eq solar angle pair, as follows [7]:

$$\begin{cases} \alpha_s = \sin^{-1}(\cos\omega_s \cos\delta_s \cos\varphi + \sin\delta_s \sin\varphi) \\ \Psi_s = \operatorname{sgn}(\omega_s)\cos^{-1}\left(\frac{\sin\alpha_s \sin\varphi - \sin\delta_s}{\cos\alpha_s \cos\varphi}\right) \end{cases} \tag{7.3}$$

$$\begin{cases} \rho_s = \sin^{-1}(\cos\omega_s \cos\delta_s \sin\varphi - \sin\delta_s \cos\varphi) \\ \varepsilon_s = \operatorname{sgn}(\omega_s)\cos^{-1}\left(\frac{\cos\omega_s \cos\delta_s \cos\varphi - \sin\delta_s \sin\varphi}{\cos\rho_s}\right) \end{cases} \tag{7.4}$$

$$\begin{cases} \gamma_s = \tan^{-1}\left(\frac{\cos\omega_s \cos\delta_s \sin\varphi - \sin\delta_s \cos\varphi}{\cos\omega_s \cos\delta_s \cos\varphi + \sin\delta_s \sin\varphi}\right) \\ \beta_s = \operatorname{sgn}(\omega_s)\cos^{-1}\left(\frac{\cos\omega_s \cos\delta_s \cos\varphi + \sin\delta_s \sin\varphi}{\cos\gamma_s}\right) \end{cases} \tag{7.5}$$

The incidence angle ν of the sunray on a plane surface represents the angle between the unit vector $\vec{s}$ of the Earth-Sun direction and the normal unit vector $\vec{n}$ of the surface, Fig. 7.2.

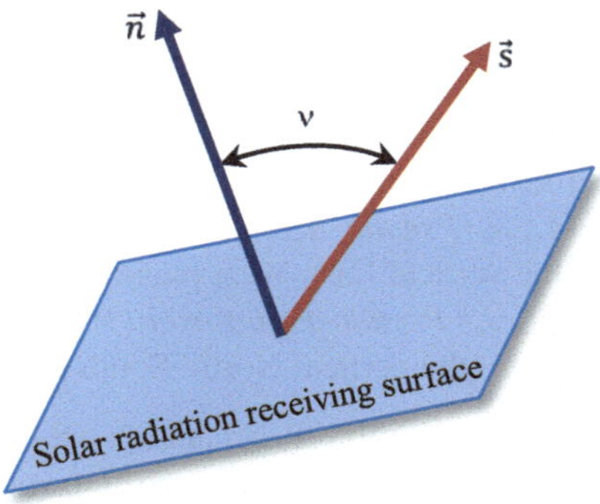

Fig. 7.2 The incidence angle of the sunray on a receiving surface

The angle ν is important in evaluating the direct solar radiation received by the PV modules or FPSTC and can be calculated based on the angle pairs of the two-unit vectors ($\vec{s}$ and $\vec{n}$), expressed in the same coordinate system.

$$\begin{aligned}\cos\nu = \vec{s}\cdot\vec{n} &= \cos\delta_s\cos\delta_n\cos(\omega_s-\omega_n)\\ &+\sin\delta_s\sin\delta_n = \cos\alpha_s\cos\alpha_n\cos(\Psi_s-\Psi_n)\\ &+\sin\alpha_s\sin\alpha_n = \cos\rho_s\cos\rho_n\cos(\varepsilon_s-\varepsilon_n)\\ &+\sin\rho_s\sin\rho_n = \cos\beta_s\cos\beta_n\cos(\gamma_s-\gamma_n)\\ &+\sin\beta_s\sin\beta_n\end{aligned} \tag{7.6}$$

7.3 Solar Tracking Systems

Considering the four solar angle pairs so far described, four types of dual-axis solar tracking systems can be defined (Fig. 7.3):

- *Equatorial* (or *polar*) type, Fig. 7.3a, with the diurnal axis parallel to the Earth polar axis [8–10].
- *Azimuth* (or *azimuth-altitude*) type, Fig. 7.3b, where the diurnal axis is parallel to the zenith (local vertical) axis [9–13].
- *Pseudo-Azimuth* type, Fig. 7.3c, with a fixed horizontal rotation axis parallel to the North–South direction [7, 14, 15].
- *Pseudo-Equatorial* type, Fig. 7.3d, with a fixed horizontal rotation axis parallel to the East–West direction [7, 14, 15].

A dual-axis tracking system contains two revolute joints with perpendicular axes. The first (primary) axis is fixed to the ground, while the other (secondary) is perpendicular to the first axis. The primary axis is used for the daytime (diurnal) motion, and the secondary axis for the elevation motion, except the PEq system where the role of the two motion axes is reversed.

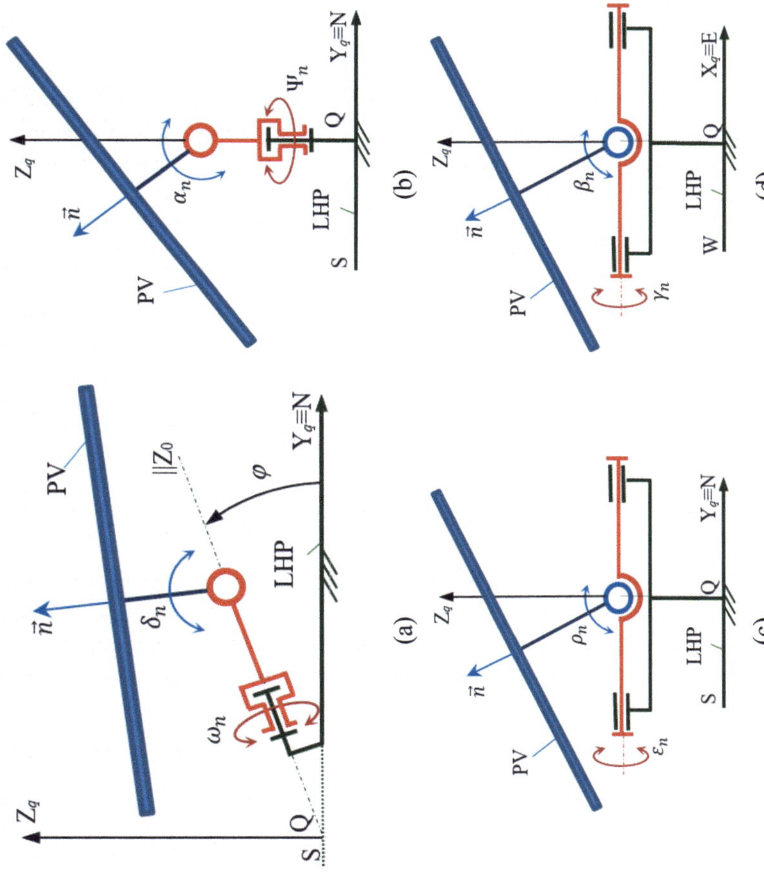

Fig. 7.3 Dual-axis tracking systems of: **a** Equatorial type, **b** Azimuth type, **c** Pseudo-Azimuth type, and **d** Pseudo-Equatorial type (LHP—Local Horizontal Plane)

The single-axis tracking systems result from dual-axis systems typically by removing the elevation axis or, by exception, the diurnal axis where the tilt angle has to be periodically adjusted. The maximum values of the solar angles and, particularly, of the diurnal angles (i.e. the angular strokes) are important input data in the design process of optimal tracking systems suitable for a given location.

7.4 Received Direct Solar Radiation

Under clear sky conditions, neglecting the diffuse component of the global solar radiation can be an acceptable hypothesis when designing solar tracking systems. Thus, the solar energy received depends both on the PV or FPSTC surface position relative to the sunrays and on the available direct solar radiation, variable in time according to the location on the Earth and the local weather conditions.

The irradiance of the received direct solar radiation (B_n) on a PV or FPSTC surface with the normal unit vector $\vec{n}$ (Fig. 7.2) can be obtained using the Lambert law [16]:

$$B_n = B_s \cos \nu \tag{7.7}$$

where the incidence angle ν is set according to Eq. (7.6) and B_s is the available direct solar radiation measured on a plane perpendicular to the sunray at the ground level under clear sky assumption, expressed as follows [16]:

$$B_s = I_0\left[1 + 0.0334 \cos\left(\frac{\pi}{180}(0.9856N - 2.72)\right)\right] e^{-\frac{T_R}{0.9+9.4 \sin \alpha_s}} \tag{7.8}$$

where $I_0 = 1367\ \mathrm{W/m^2}$ is the solar constant, N is the day number in the year and T_R is the turbidity (Linke) factor.

The share of received direct solar radiation is expressed as an instantaneous value by the ratio between the irradiances of the received and of the available direct solar radiation, as follows:

$$\eta = \frac{B_n}{B_s} = \cos \nu \tag{7.9}$$

or as an average value on a time interval T (e.g. a day, a month, a season, a year), as follows:

$$\eta_T = \frac{E(B_n)}{E(B_s)} = \frac{\int_{t_s \in T} B_n dt_s}{\int_{t_s \in T} B_s dt_s} \tag{7.10}$$

where $E(B_n)$ and $E(B_s)$ represent the the energy of received and available direct solar radiation during the time interval T.

7.5 Solar Tracking Algorithms and Scenarios

The *solar tracking algorithms* are aiming at optimizing (typically at increasing) the solar energy collected by the conversion devices. The tracking performance depends on the solar energy converter type (concentrating or non-concentrating the solar radiation) and on their configuration (platforms, strings), along with the specific meteorological and shading conditions of the implementation location [7, 17–19]. Since continuous tracking is not practically feasible, stepwise tracking represents a feasible and affordable option.

For a given application, the steps of the tracking algorithm are designed to control the tracking mechanism motions, that cannot be continuous for technical and economic reasons [7]. Consequently, stepwise tracking algorithms are commonly used in practice, i.e. a sequence of commands (t_s, u) for each tracking axis, whereby the motion from the current angular position to the new position u is triggered at the time t_s.

Various solar tracking algorithms can be implemented considering constant or variable time steps, bigger or smaller number of steps, the size of the angular strokes, and the duration of a tracking season, i.e. the continuous period of time when a tracking algorithm remains unchanged, e.g. one month [14]. The share of the received solar energy increases as the number of tracking seasons is higher, the number of steps in a day is higher and steps with variable optimized durations are used while the angular stokes of the tracking system are closer to the strokes of the solar angular parameters. The tracking algorithms can be designed aiming at increasing the receiving share of different solar radiation components:

- The irradiance of the direct solar radiation B_n, by following as closely as possible the Sun's trajectory in the sky. Since the diffuse component is neglected, these algorithms are recommended for sunny areas with large numbers of clear sky days during the year.
- The direct or diffuse components of the solar radiation at a certain time, under variable environmental conditions [20–22].
- The global solar radiation, as proposed in [14, 23, 24].

The stepwise solar tracking algorithms can be divided into the categories [24]:

- *Predefined algorithms* that calculate and follow the solar angles, aiming at maximizing the irradiance of the received direct or global solar radiation.
- *Adaptive algorithms*, aiming at identifying the brightest point on the sky by a 3-facet PV sensor and a logic controller [25]. These are recommended in sunny regions, for accurate tracking using appropriate movements, each several minutes [26].

A comparative analysis proposed in the paper considers the city of Brasov, Romania (45.65° N, 25.5° E) as implementation location, the four tracking system types (Eq, Az, PAz, PEq) and the following scenarios of predefined tracking algorithms, in the clear sky assumption, aiming at increasing the share of the received direct solar radiation during the year.

- Scenario 1: Fix-tilted system at the annual optimal tilt angle, used as a reference for the tracking scenarios.
- Scenario 2: Single-axis elevation tracked system with fixed diurnal angle set at 0° and optimal elevation angle adjusted monthly (manually or automatically), aiming at investigating the impact of the tilt angle on collecting the solar energy during various seasons by using less complex mechanical systems.
- Scenario 3: Single-axis diurnal tracked system considering a stepwise diurnal tracking algorithm with a small number of optimized unequal steps (2, 4, and 6 steps) and annual optimally fixed tilted angle.
- Scenario 4: Dual-axis tracked system using stepwise tracking algorithms with a small number of optimized unequal steps (2, 4, and 6 steps) for both diurnal and elevation motions.

The tracking procedure can be optimized by identifying the optimal values (t_s, u) of each step, aiming at maximizing the received solar energy [7]. A four-steps tracking algorithm, optimized for the Az system in mid-June, is depicted in Fig. 7.4 outlining an anti-symmetrical design in relation to the solar noon and four motions (2 steps in the morning and 2 steps in the afternoon) between sunrise to sunset.

7.6 Numerical Results and Discussions

The numerical simulations aim at identifying the impact of various solar tracking systems and algorithms on the share of the received direct solar radiation compared with the fix-tilted case. The methodology is applied to a mid-latitude location (Brașov, Romania), in a mountain temperate climate on the northern hemisphere, by neglecting the diffuse radiation (i.e. in the clear sky hypothesis). According to the four simulation scenarios, the numerical results of the receiving share of the direct solar radiation are obtained as monthly and yearly average values by using Eqs. (7.1)–(7.10), where the altitude angle is φ is equal to 45.65° and the values of the turbidity factor are presented in Table 7.1.

The four scenarios are simulated in a temperate climate for three representative months: December (Fig. 7.5a), March (Fig. 7.5b), and June (Fig. 7.5c), when the 4-steps tracking algorithms are only involved in Scenarios 3 and 4. The following remarks can be formulated.

- A South-facing fixed surface, tilted at an annual 40° optimal angle will receive between 59.5% (in June) and 75.3% (in December) of the available direct solar radiation; the yearly average receiving share is of approximatively 69%. Despite its reduced energy performance, this solution is commonly preferred by designers

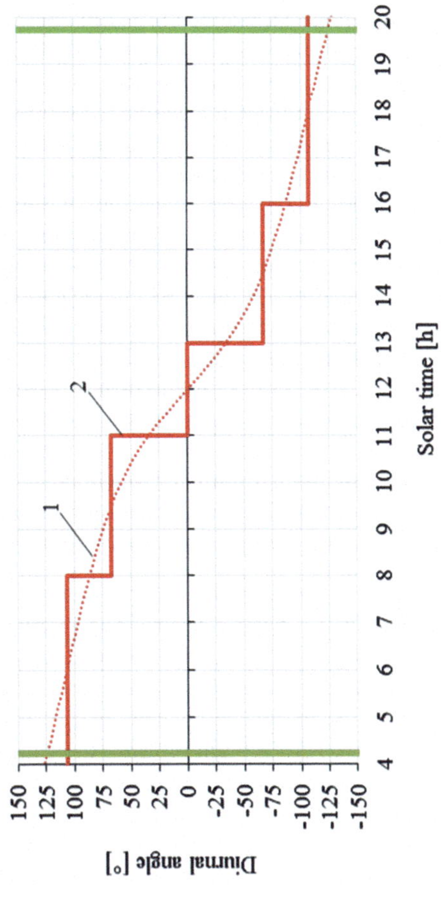

Fig. 7.4 Optimized 4-steps tracking algorithm for an azimuth type single-axis tracking system in mid-June, in the Brașov location: the variation of the solar (1) and tracking (2) diurnal angles

Table 7.1 Monthly values of the turbidity (Linke) factor TR for Braşov, Romania [15

Jan	Feb	Mar	Apr	May	Jun	Jul	Aug	Sep	Oct	Nov	Dec
2.8	2.9	3	3	3	3.2	3.2	3.2	3	2.9	2.8	2.8

due to its simplicity and because it requires lower costs for the mechanical structure and for maintenance.

- By monthly adjustment at an optimal tilt angle, a single-axis elevation tracking system (Scenario 2) allows receiving an average monthly energy amount of the direct solar radiation ranging between 67.7% (in June) and 89% (in December). The system receives about 74% of the yearly available radiation, with only 5% more than the fixed-tilted system. Therefore, the increased complexity and the cost of the system by introducing one mobile axis to adjust the tilt angle periodically (e.g. monthly) is an important factor for designers.
- A more efficient strategy is the single-axis diurnal tracking with a smaller number of unequal steps and fixed tilt angle at the yearly optimal value, as considered in Scenario 3. The 4-steps optimized tracking algorithm brings a significant monthly average share of over 84%, depending on the tracking system type. The yearly value of the solar radiation receiving share ranges between 91.8% (PAz system) and 94.5% (Az system) as presented in Table 7.2. The Eq and PEq systems also allow high receiving shares. Thus, the single-axis diurnal tracking allows an increase with 20% and 25% of the receiving share compared to Scenario 2 and Scenario 1.
- Moreover, the dual-axis tracking with optimized unequal 4-steps algorithms implemented on both motion axes supports the collection of approximately 98% of the available direct solar radiation, irrespective of the month, or system type.

The results of Scenarios 2 and 4 are independent of the tracking system type and they are depicted over one year in Fig. 7.6, using Scenario 1 as a reference. The higher values of the receiving share are observed during the winter season and a decreased performance results during the summer months due to the higher strokes of the solar angles. Compared to the reference (Scenario 1), the solar energy collected by using dual-axis tracking (Scenario 4) is over 50% higher during the summer season and significant (~30%) during the transient and winter seasons.

The number of unequal steps, optimally set, has a little relevance during the cold season, but has a significant impact in the warm season (Figs. 7.7 and 7.8).

The 4-steps tracking algorithm is the main candidate for implementation in a single-axis tracking system, Fig. 7.7, as it allows an increase by ~2% of the yearly receiving share compared to the 2-steps algorithm, and ~0.6% difference from the 6-steps algorithm. Similar conclusions can be drawn from Fig. 7.8 for a dual-axis tracking system, the 4-steps tracking algorithm can bring over 2% than the 2-steps algorithm. The 6-steps algorithm proved only a slight increase (less than 0.7%) in the share of the collected solar energy compared to the 4-steps case.

A global overview of the receiving share using the four tracking system types, analyzed over one year, is developed in Fig. 7.9 for Scenario 3.

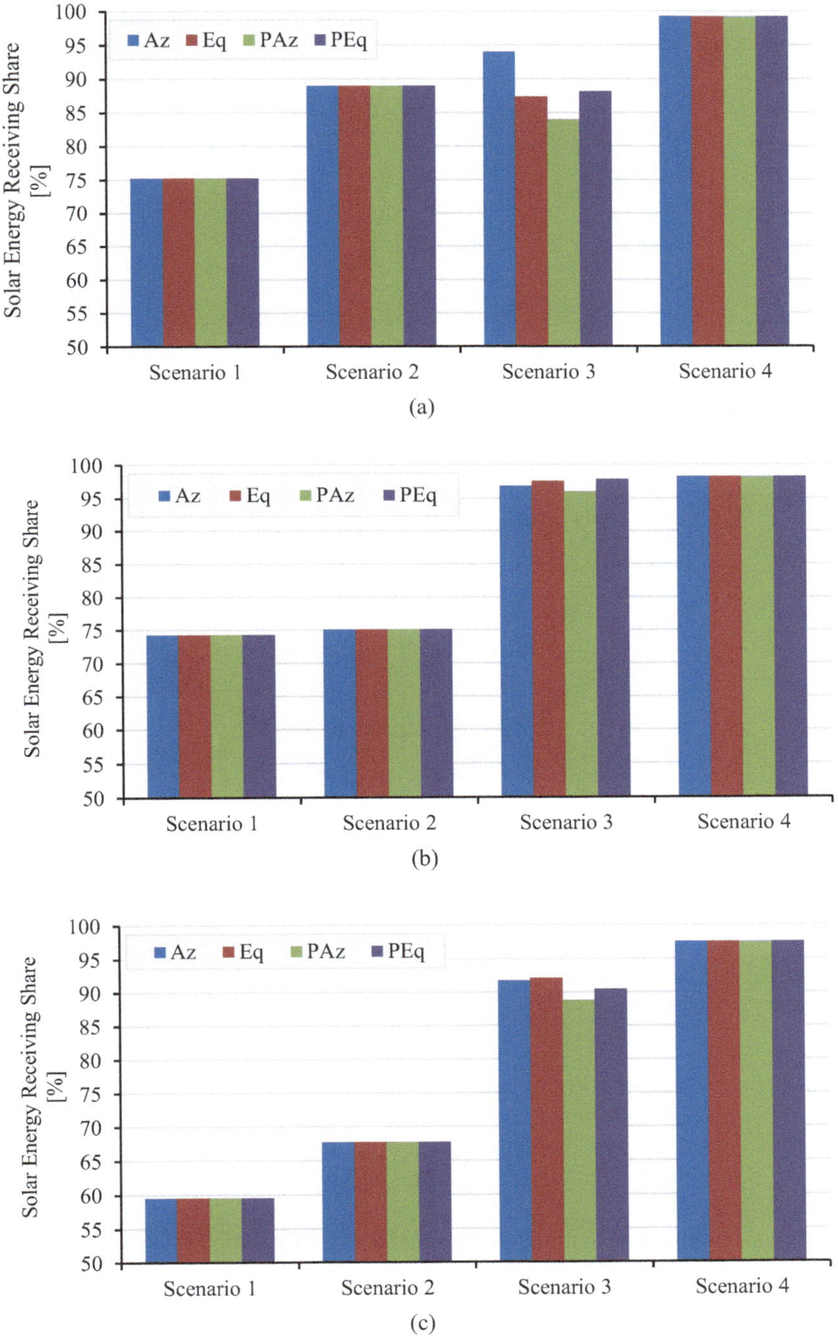

Fig. 7.5 Monthly average receiving share of direct solar radiation in Brașov using various scenarios in **a** December, **b** March, **c** June

Table 7.2 Yearly average values of the share of the received direct solar radiation (η_y) and the maximum angular strokes (Δ_{max}) for Brasov, Romania

Scenario No	Tracking system type	$\eta_y[\%]$	Tracking axis type	$\Delta_{max}[°]$
Scenario 1	–	68.9	–	–
Scenario 2	All	74.0	Elevation	61.8
Scenario 3 (4-steps single-axis)	Az	94.5	Diurnal	214.0
	Eq	94.2	Diurnal	180.0
	PAz	91.8	Diurnal	168.0
	PEq	94.1	Diurnal	154.0
Scenario 4 (4-steps dual-axis)	Az	98.1	Diurnal/Elevation	214.0/62.3
	Eq	98.1	Diurnal/Elevation	180.0/46.9
	PAz	98.1	Diurnal/Elevation	168.0/85.2
	PEq	98.1	Diurnal/Elevation	154.0/121.7

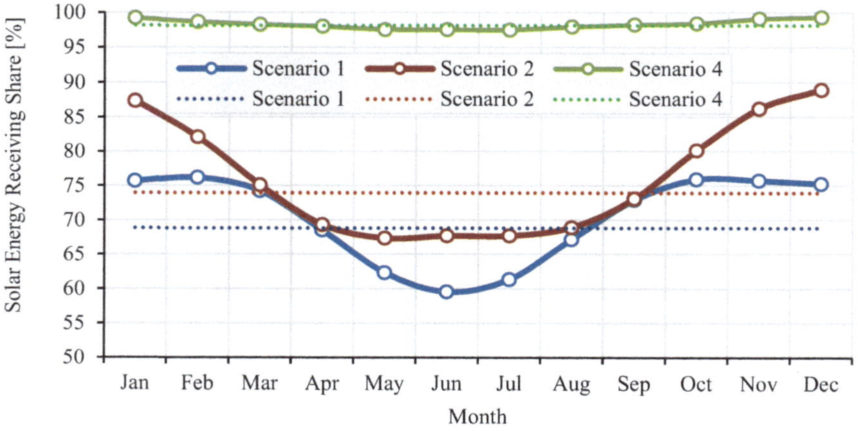

Fig. 7.6 Monthly (continuous line) and yearly (dotted line) average receiving share of the direct solar radiation in Brașov for Scenarios 1, 2, and 4 (in the 4-steps algorithm)

The results depicted in Fig. 7.9 show the higher performance when using the Az, Eq, and PEq tracking systems, and slightly lower performance for the PAz system. Therefore, considering the implementation difficulties raised by the Eq system at mid- and high-latitudes, the Az and PEq tracking systems can be recommended for mid-latitude regions.

Another important design parameter in selecting a tracking system type is represented by the angular tracking strokes required by the implemented stepwise tracking algorithms. According to the data in Table 7.2, the lowest diurnal angular stroke is required by the PEq system (154°), while the Az system needs the largest diurnal stroke (214°). It is worth to mention that the elevation stroke required by the Eq

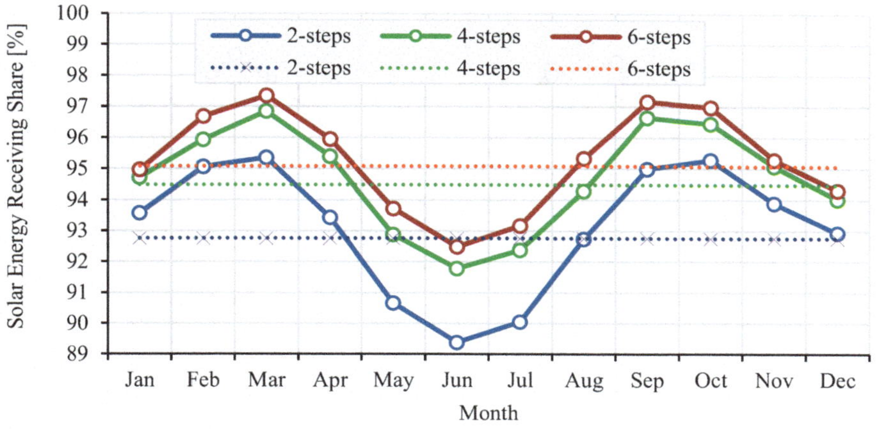

Fig. 7.7 Monthly (continuous line) and yearly (dotted line) average receiving share of the direct solar radiation in Brașov for Scenario 3 using 2-steps, 4-steps and 6-steps algorithms for single-axis Az tracking

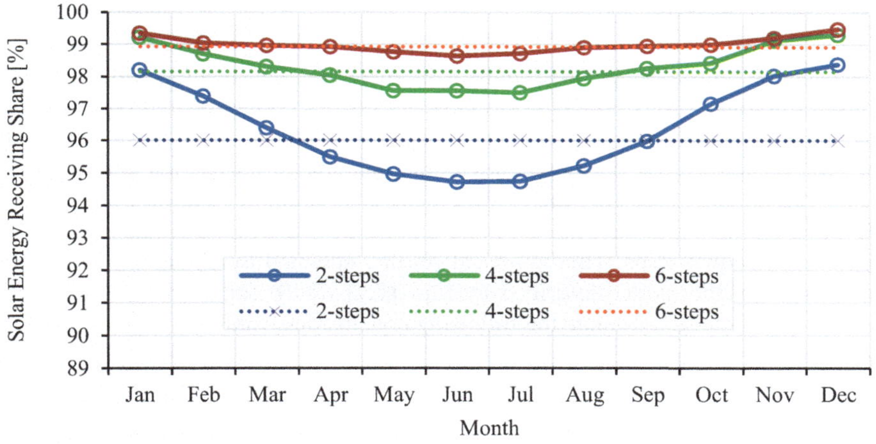

Fig. 7.8 Monthly (continuous line) and yearly (dotted line) average receiving share of the direct solar radiation in Brașov in case of Scenario 4 (dual-axis tracking) using 2-steps, 4-steps and 6-steps algorithms for both diurnal and elevation axes of the Az tracking system

system is almost double than that of the Az system, but in the range accepted for simple tracking mechanism of triangle type [15]. As a result, single- or dual-axis PEq tracking systems with 4-steps tracking algorithms, with optimized unequal steps, can be considered as an optimal solution for the Brasov implementation location, and generally, for mid-latitude regions.

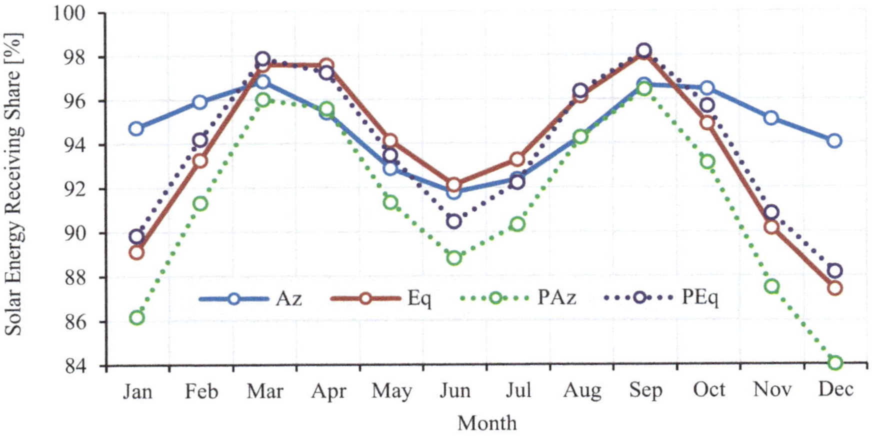

Fig. 7.9 Monthly average receiving share of direct solar radiation in Brașov in Scenario 3 using a 4-steps algorithm for Az, Eq, PAz, and PEq single-axis tracking systems

7.7 Conclusions

The paper presents a comparative study of the share of direct solar radiation received considering four types of tracking systems and various scenarios used typically for tracking non-concentrating solar converters versus fixed South-facing system tilted at the yearly optimal angle. Two strategies for the solar tracking scenarios were applied, which are monthly adjusting the elevation angles at optimal values for single-axis elevation trackers and stepwise tracking algorithms based on a small number of optimized unequal steps. The numerical results, obtained by simulating over one year based on the model described by Eqs. (7.1)–(7.10) for the Brasov location (Romania), allow to formulate the following conclusions, valid to a large extent for mid-latitude regions:

- The monthly adjustment of the tilt angle allows increasing the share of the received solar radiation by approximately 5% compared to the fixed system ($\eta_y \approx 69\%$), while this requires a more complex and expensive single-axis mechanical structure.
- A significant increase of the receiving share (up to 95%) can be obtained using single-axis tracking systems of diurnal type and optimized stepwise tracking algorithms with lower diurnal tracking strokes.
- Dual-axis tracking systems lead to smaller energy gains compared to the single-axis diurnal tracking (~3% more); therefore, they are less recommended for non-concentrating solar applications due to the significant increase of the system complexity and cost.
- Compared with the classical stepwise tracking algorithms using equal steps, the tracking algorithms with a smaller number of optimized steps bring the advantages

of shorter controller programs, smaller tracking angular strokes, and a higher share of the received solar radiation. The choice of the 4-steps tracking algorithms seems to be a good compromise between simplicity and efficiency.

- The single-axis diurnal tracking systems with optimized unequal steps algorithms reaches different receiving shares and requires different tracking angular strokes. Considering these evaluation criteria, the PEq system type is the best recommended for mid-latitude locations, followed by the Az system type.

Solar tracking supports the output increase of the solar energy converters by increasing the share of the collected solar energy. The feasibility of a tracked solar application is analyzed considering certain technical and economic criteria, such as the available space for project implementation, the energy demand, the overall lifetime cost of the conversion system. The decreasing cost of the solar energy conversion devices and tracking systems, along with the efficiency increase of the modern solar energy technologies, indicates that the solar tracking systems can be used as an affordable option for applications with limited available implementation area, as the built environment, or under the requirement of high energy shares produced using renewables.

References

1. E. Paris, K. Beisner, *Fixed-Tilt Versus Tracker: Why a One-Size-Fits-All Approach Can Limit Solar Production. SPW* (2018)
2. Greentech Media Site, *Solar Balance-of-System: To Track or Not to Track*, Part I, https://www.greentechmedia.com/articles/read/solar-balance-of-system-to-track-or-not-to-track-part-I, Last Accessed 05 Apr 2020.
3. Tracstar site, *Small Power Systems Has Built Trackers for Over 25 Years*, https://www.helmholz.us/smallpowersystems, Last Accessed 05 Apr 2020
4. S. Abdallah, The effect of using sun tracking systems on the voltage-current characteristics and power generation of flat plate photovoltaics. Energy Convers. Manage. **45**(11–12), 1671–1679 (2004)
5. M.D. Moldovan, I. Visa, B.G. Burduhos, Energetic autonomy for a solar house. Environ. Eng. Manag. J. **10**(9), 1283–1290 (2011)
6. C. Sungur, Sun-tracking system with PLC control for photo-voltaic panels. Int. J. Green Energy **4**(6), 635–643 (2007)
7. I. Visa, A. Duta, M. Moldovan, B. Burduhos, M. Neagoe, *Solar Energy Conversion Systems in the Built Environment*, 1st edn (Springer International Publishing, Berlin, 2020)
8. P.A. Davies, Sun-tracking mechanism using equatorial and ecliptic axes. Sol. Energy **50**(6), 487–489 (1993)
9. C.Y. Lee, P.C. Chou, C.M. Chiang, C.F. Lin, Sun tracking systems: a review. Sensors **9**(5), 3875–3890 (2009)
10. W. Nsengiyumva, S.G. Chen, L. Hu, X. Chen, Recent advancements and challenges in solar tracking systems (STS): A review. Renew. Sustain. Energy Rev. **81**, 250–279 (2018)
11. M. Koussa, M. Haddadi, D. Saheb, A. Malek, S. Hadji, Sun tracking mechanism effects on flat plate photovoltaic system performances for different step time and main parameters affecting the obtained gains: case of North Africa and mediterranean site. Energy Procedia **18**, 817–838 (2012)

12. L. Miloudi, D. Acheli, A. Chaib, Solar tracking with photovoltaic panel. Energy Procedia **42**, 103–112 (2013)
13. I. Visa, Mechanisms in building integrated renewable energy systems: case study-solar energy conversion systems, in *The 11th IFToMM International Symposium on Science of Mechanisms and Machines* 18, (2014), pp. 31–49
14. M. Neagoe, I. Visa, B. Burduhos, Increasing the tracking efficiency of photovoltaic systems. Sustain. Energy Buil. Environ. Steps Towards nZEB, 443–460 (2014).
15. I. Visa, C. Jaliu, A. Duta, M. Neagoe, M. Comsit, M. Moldovan, D. Ciobanu, B. Burduhos, R. Saulescu, *The Role of Mechanisms in Sustainable Energy Systems*. Transilvania University of Brasov Publishing House (2015)
16. M. Meliss, *Regenerative Energiequellen: Praktikum*, 1st edn. (Springer-Verlag, Berlin Heidelberg, 1997)
17. N.H. Helwa, A.B.G. Bahgat, A.M.R. El Shafee, E.T. El Shenawy, Maximum collectable solar energy by different solar tracking systems. Energy Sources **22**(1), 23–34 (2000)
18. F. Mohammadi, G.A. Nazri, M. Saif, A bidirectional power charging control strategy for plug-in hybrid electric vehicles. Sustainability **11**(16), 4317 (2019)
19. C. Mehdipour, F. Mohammadi, Design and analysis of a stand-alone photovoltaic system for footbridge lighting. J. Solar Energy Res. **4**, 85–91 (2019)
20. S. Armstrong, W.G. Hurley, A new methodology to optimise solar energy extraction under cloudy conditions. Renew. Energy **35**(4), 780–787 (2010)
21. N.A. Kelly, T.L. Gibson, Increasing the solar photovoltaic energy capture on sunny and cloudy days. Sol. Energy **85**(1), 111–125 (2011)
22. G.Y. Liu, Y. Ren, S.K. Nguang, Simulated photovoltaic array systems under a changing environment for temporal performance. Int. J. Green Energy **9**(7), 673–684 (2012)
23. B.G. Burduhos, I. Visa, M. Neagoe, M. Badea, Modeling and optimization of the global solar irradiance collecting efficiency. Int. J. Green Energy **12**(7), 743–755 (2015)
24. B.G. Burduhos, M. Neagoe, A. Duta, Adaptive stepwise orientation algorithm for non-concentrated dual-axis solar tracking systems, in *Proceedings of the 14*th *World Congress in Mechanism and Machine Science* (Taipei, Taiwan, 2015)
25. A. Toloei, M. Zahednamazi, R. Ghasemi, F. Mohammadi, A comparative analysis of star identification algorithms. Astrophys. Space Sci. **365**(4), 63 (2020)
26. Deger, MLD-Sensor Site: https://www.degerenergie.de/en/mld-sensor.html, Last Accessed 05 Mar 2020

Chapter 8
Reliability Assessment of PV Modules

Divine Atsu, Istvan Seres, and Istvan Farkas

Abstract Degradation and reliability evaluation of solar photovoltaic (PV) modules is very critical to enhance the durability and lifespan of PV systems in operation. This study assesses the reliability and rate of degradation of a twelve-year-old ground-mount monocrystalline silicon (c-Si) solar PV modules exposed to the tropical climate of the sub-Saharan. The methods used for the assessment are visual inspection, current–voltage (I–V) curve characterization and thermal imaging. The process of partial shading was applied to determine the state of the bypass diodes. Results revealed a decrease in short circuit current (I_{sc}) ranging from 7% to 16.4% whiles the reduction in open-circuit voltage (V_{oc}) was between 11.4 and 17.1%. The decline in Fill Factor (FF) and maximum power were respectively between 11.3% to 24.2% and 34.5% to 41.4% with 3.19% per year decline in power output. It was found that V_{oc} had the most significant impact on the decrease in the performance of the modules.

Keywords Degradation rate · Reliability · Solar photovoltaic

8.1 Introduction

Solar PV market has been growing very fast over the last decade, recording an unprecedented yearly increase of 107.3 GW in 2018, which is about 34.3% over the preceding year's addition. The total global capacity of solar PV as of 2018 stands at 509.3 GW [1]. The growth of solar PV in the share of the global power market is as a result of several factors such as the favourable international and regional policies for

D. Atsu (✉)
Mechanical Engineering Doctoral School, Szent István University, Páter K. utca 1., Gödöllő H-2100, Hungary
e-mail: atsud22@yahoo.com

I. Seres · I. Farkas (✉)
Department of Physics and Process Control, Szent István University, Páter K. utca 1., Gödöllő H-2103, Hungary
e-mail: farkas.istvan@gek.szie.hu

I. Visa and A. Duta (eds.), *Solar Energy Conversion in Communities*,
Springer Proceedings in Energy, https://doi.org/10.1007/978-3-030-55757-7_8

their dissemination, reduction in the cost of solar modules, its environmental friendliness and the growing awareness of the negative impact of conventional sources of energy. However, the increasing growth in PV is not without the associated reliability and efficiency challenges encountered in operation, which are site-specific.

Warranty provided by manufacturers is to offer the assurance of reliability for the period. The number of years given as warranty by most manufacturers for PV modules is 25 years of 80% peak power output. These warranties provide for 90–95% of power output in the first ten years and then 80–87% peak power output from the tenth year until the 25 years [2]. Manufacturers give two types of warranties for PV modules. Performance warranty which guarantees the performance output and equipment warranty for non-failure over the given years [3, 4].

The lifespan and the reliability of solar modules in outdoor operation depend on the degradation and failure modes which are controlled by several factors such as, defects incurred during the production process, transportation, installation and the issues that arise from climate-specific conditions such as solar irradiation and ambient temperature, humidity, wind, water ingress and ultraviolet (UV) intensity [5].

To guarantee the performance of PV modules over their lifetime, they are taken through rigorous quality qualification test procedures as per the IEC standards. These tests are intended to reveal any design and manufacturing flaws which may instigate the early failure of PV modules than expected. The accelerated ageing test has also been included in the test requirements, which has led to the improvement in technology and reliability in the last decade. These tests, as mentioned above, however, are incapable of revealing all degradation modes and the reliability issues that may exist whiles in outdoor operation with varying climatic influences. Testing for the degradation whiles in operation is a purposive approach to ascertain the problems that show forth whiles in real operating conditions [6].

The main mechanisms that lead to efficiency and performance loss in solar PV cells are mainly due to corrosion, light-induced degradation, decrease in stability of contacts and cracked cells, cell delamination, moisture ingress, and the discoloration of the encapsulant material [5]. At the module level, the loss in power is found to be as a result of failures from individual cells and compounded. It is also, from breakage of cover glass, delamination, failure of the busbar, front surface discoloration, diode failure, hotspots, loss in interlayer adhesion, and broken interconnects [7]. A study conducted by NREL [8] presents module degradation rates to reach as high as 4% per year while the average degradation rate is estimated to be an 0.8% reduction in power output per year. Data available for the studies on PV module reliability in the sub-Saharan tropical climates, which has been proven to have the worst impact on PV modules is rather insignificant [9]. This study investigates the reliability and degradation of performance output of a twelve (12) year ground mount monocrystalline solar PV modules exposed to the tropical climate of the sub-Saharan of Ghana.

8.2 Materials and Methods

The location of the study is Koforidua in the Eastern Region of Ghana. According to the Köppen-Geiger climate classification, the climate of Koforidua is classified as 'Aw' which is savanna with dry winter [10]. The average temperature of the site is 25.9 °C and experiences about 1407 mm of precipitation annually, as shown in Fig. 8.1. The average maximum temperature of the site is 30.58 °C, while the average minimum temperature is 24 °C (Climate-data, 2019). The annual average horizontal irradiation for the site is 1733.75 $kWhm^{-2}$ d with an average value of 4.75 $kWhm^{-2}$ d. The solar radiation map of Ghana is presented in Fig. 8.2. The monthly global horizontal radiation figures for the location of the study range from 4.19–5.37 $kWhm^{-2}$ d. Minimum and maximum wind speed are 2.40 and 4.19 ms^{-1}, respectively. The average monthly humidity ranges from 76.5 to 88.3%. Koforidua experiences two major seasons, The dry season and the wet season. The dry season usually starts from December to March while the wet season starts from April to November.

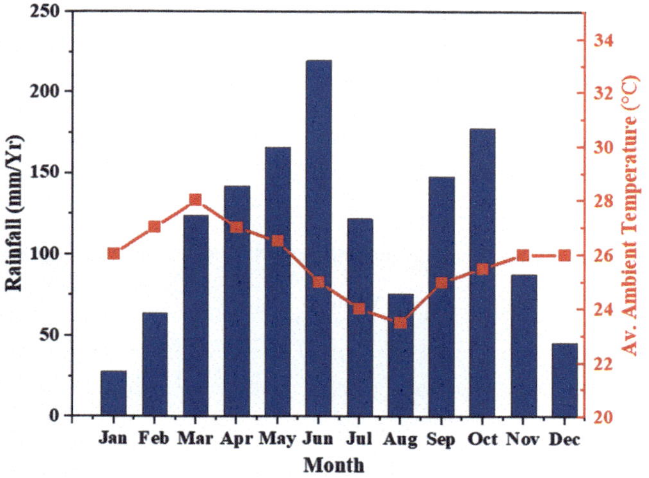

Fig. 8.1 Rainfall pattern and temperatures [11]

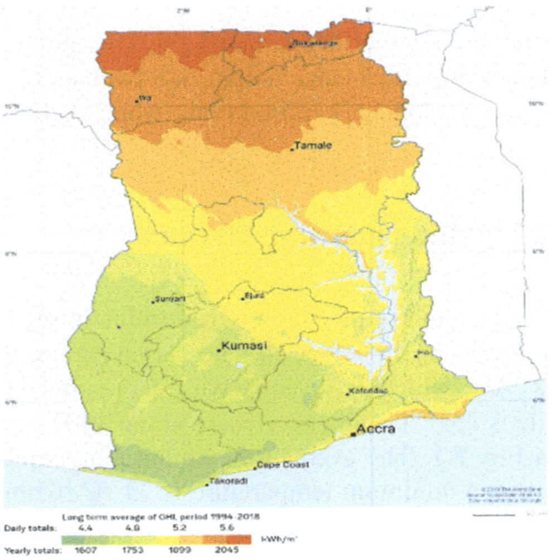

Fig. 8.2 Global Horizontal Radiation (GHI) map of Ghana [12]

8.2.1 System Description

The system for the study is an off-grid PV system located at the forecourt of the demonstration workshop (Energy House) of the Renewable Energy Department of the Koforidua Technical University in Ghana. It is installed for demonstration purposes to aid students' practical training. It also serves as a backup for the power requirements of the Energy House. It is situated on the 6.062545 N, −0.266001 W at an elevation of 173 m above sea level. The ground-mounted system was installed in 2007 with the modules fixed on a structure made of aluminium angle bars and bolted onto metal bars, as shown in Fig. 8.3.

The array consists of four monocrystalline modules with specifications outlined in Table 8.1. The modules are CE Europe, TÜV and ESTI certified. The warranty

Fig. 8.3 General view of the installation

Table 8.1 Specification of the PV module at STC given by the manufacturer

Parameters	Value
Maximum power (P_{max})	50 Wp
Short circuit current (I_{sc})	3.16 A
Open circuit voltage (V_{oc})	21.6 V
Voltage at Pmax (V_{mp})	17.60 V
Current at Pmax (I_{mp})	2.9 A
NOCT	43 ± 2°
Power tolerance	±10%

given by the manufacturer is 25 years for 80% power output.

The EKO MP–170 I–V curve plotter was used to obtain the module parameters under the experimental conditions. It has an accuracy of (±1% for 0.1–10 A) for current (I_{dc}), and (±1% for 10 –1000 V) for Voltage (V_{dc}). Solar radiation was measured using the Kimo solarimeter LSL 200 (resolution 1 Wm^{-2}, accuracy *5%)* at the plane of the modules. The Voltcraft infrared thermometer *(IR260—8S)* with a measuring range of −30 to 260 °C(±2 °C), resolution and emissivity of 0.1 °C and 0.95 respectively were employed in the measurement of the temperature of the modules. Quantitative analysis of the captured infrared (IR) images was performed with the aid of the Report Generator Lite software. IR images were obtained using the (NEC Avio H2640) camera having specifications as follows: temperature range, −40 to 500 °C, Spectral range, 8–13 μm, Emissivity, 0.1–1.00 with an accuracy of ±2% *or* ±2 °C.

Measurements were done on a bright sunny day to avoid significant variations in curated data which consequently will affect the translated results at low levels of solar radiations [13]. The modules were disconnected from each other and tested individually. The inbuilt logger stored measured data in the I–V curve plotter. The characteristics of the modules were obtained simultaneously with the acquisition of data on irradiation and module temperatures.

8.2.2 Translation of Experimental Data to STC Values

A uniform translation procedure developed by the Joint Research Centre (JRC) of the European Commission in Ispra is adopted. The method is convenient to apply and could be used even when specific module parameters under a particular temperature and irradiation are not known. Input parameters can be deduced from experimental measurements. Application of this procedure has a translation accuracy of 4% achieved for one array measurement [14]. The default values applied are dimensionless temperature coefficient of I_{sc} (α) = 0.0045, dimensionless temperature coeffcient of V_{oc} (b) = 0.06 and Dimensionless irradiance correction factor (a) = 0.06. These default values are valid for crystalline silicon modules.

Equations 8.1 to 8.4 provides a less complex, practicable, convenient and accurate procedure of translation or correction of the parameters of a PV module in operation without having to determine any constant at predetermined ambient and irradiation conditions which are not achievable with modules in operation [15].

$$I_{sc,2} = I_{sc,1}[1 + \alpha(T_2 - T_1)]\frac{G_2}{G_1} \tag{8.1}$$

$$V_{oc,2} = V_{oc,1}\left[1 + aln\frac{G_2}{G_1} + b(T_2 - T_1)\right] \tag{8.2}$$

$$I_2 = I_1\left(\frac{I_{sc,2}}{I_{sc,1}}\right) \tag{8.3}$$

$$V_2 = V_1 + \left(V_{oc,2} - V_{oc,1}\right) + R_s(I_1 - I_2) \tag{8.4}$$

where $I_1(A)$ and $V_1(V)$ are the experimentally acquired current and voltage, respectively. I_2 and V_2 are the pair of the corrected module current and voltage characteristics. G_1 and G_2(W/m^2) are the experimental and the reference irradiance values, respectively. $I_{sc,1}$ (A) and $V_{oc,1}$ (V) are the measured short circuit current, and open-circuit voltage while $I_{sc,2}$ and $V_{oc,2}$ are the corrected short circuit current and open-circuit voltage. α is the dimensionless temperature coefficient of short circuit current, b is the dimensionless temperature coefficient of open-circuit voltage, a is the dimensionless correction factor for irradiance, with a default value of *0.06*. R_s is the series resistance [15]. The default values are valid for crystalline silicon.

8.2.3 Module Degradation Factor (MDF)

MDF is an estimation of the total degradation, which indicates the decline in Isc as a result of an increase in Rs. Since module degradation is directly related to Isc, the MDF is defined as the level of degradation in a module and is obtained by the relation presented by Eq. 8.5 [16].

$$\%MDF = \left(1 - \frac{I_{sc(degraded\,module)}}{I_{sc(ideal\,module)}}\right) \times 100 \tag{8.5}$$

8.3 Results and Discussions

In this section, the results of the various observations and measurements conducted on the modules are presented, discussed and compared with results in other climatic conditions.

8.3.1 *Visual Inspection*

The fastest but most effective and less complicated method for finding defects and failures in solar modules is by visual inspection [17]. According to [13], the appropriate period for undertaking visual inspection is under the illumination of more than 1000 lx during the visual inspection and apply to defects detectable with the unaided eye. The visual inspection of the PV modules and the entire measurements were conducted at irradiation levels close to 1000 W/m^2 to satisfy the above condition. Table 8.2 presents the results of the visual inspection.

The IEC 61215 standard of testing for degradation requires only total irradiation of 15 kWh/m^2 of UV energy. With this requirement, modules pass the test. However, under real operation conditions, some modules have the potential of 5% or more power degradation and browning of encapsulant material just under UV irradiation of 150 kWh/m^2 or more [17]. The annual average horizontal radiation for the site is 1733.75 kWh/m^2. Thus, there is a high possibility of browning of the studies modules. Table 8.2 shows the level of browning or discoloration of EVA, covering about 70% of each cell area. EVA browning, which is a commonly reported defect is directly proportional to the Ultraviolet (UV) light, leading to transmittance loss hence drop in performance output [17, 18]. Another defect observed during the visual inspection was the browning of the interconnect metallization. However, it was not as prevalent as the EVA browning.

8.3.2 *Bypass Diode Test*

Partial shading method was employed to assess the state of the bypass diodes. Modules are partially shaded one at a time whiles tracing the I–V and P–V curves. In cases where the bypass diodes are malfunctioning, there is a decrease in the current flow in the shaded string of cells when they are not protected. However, in the presence of healthy diodes, the decline is transformed to inflection points. It is observed from the output that all modules deviate from the standard I–V curves with a drastic decrease in current to the extent of being negligible, as shown in Fig. 8.4 which presents the output for module 4. The I–V curves and P–V curves under shading conditions for all four modules are similar to Fig. 8.4 in magnitude and shape. The voltage, however, remains unaffected. This indicates an open circuit performance for

Table 8.2 Results of visual inspection

PV module component	Observation/remark	Image
Front of the PV module	Shows no sign of delamination, or browning. Glass still feels smooth. Except for accumulated dirt in the lower end and the edges	
PV cells	All cells uncracked and not broken. Each cell shows some level of discoloration of encapsulant (>70%). EVA discoloration was observed in all the modules	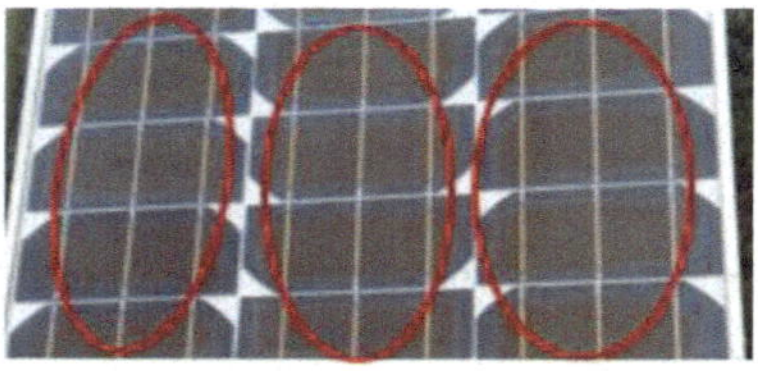
Cell metallization	Shows no sign of burns or having been oxidized except for browning or colouration of cell interconnects ribbons	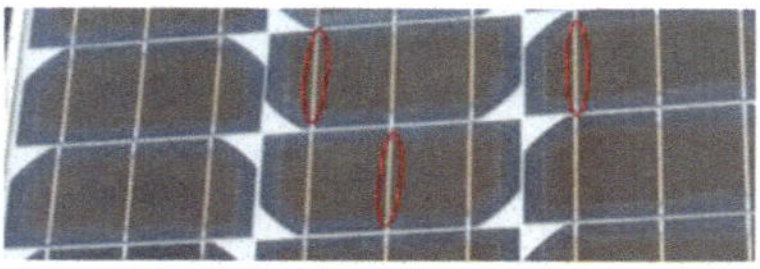
Module frame	Well intact with no scratches or broken parts and not askew. Some discoloration as a result of the accumulation of dirt and water mostly at the bottom half of the module	
Juntion box	Intact and all well closed but corrosion observed at portions when opened. This was seen in the junction boxes for all four modules at similar spots. This may be due to moisture ingress. It can also be as a result of the material used which may have a high affinity for water vapour	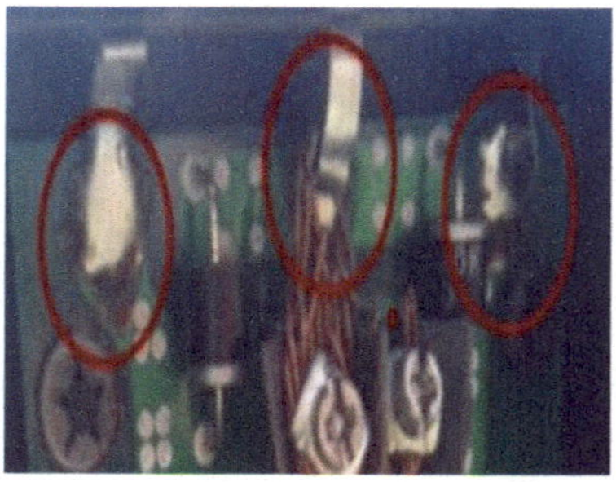

(continued)

Table 8.2 (continued)

PV module component	Observation/remark	Image
	Evidence of bad wiring was observed. (Module wire touching conducting components within the box). This may lead to internal arcing in the box. This was seen for only module 4. This could be a result of students' activities on the system	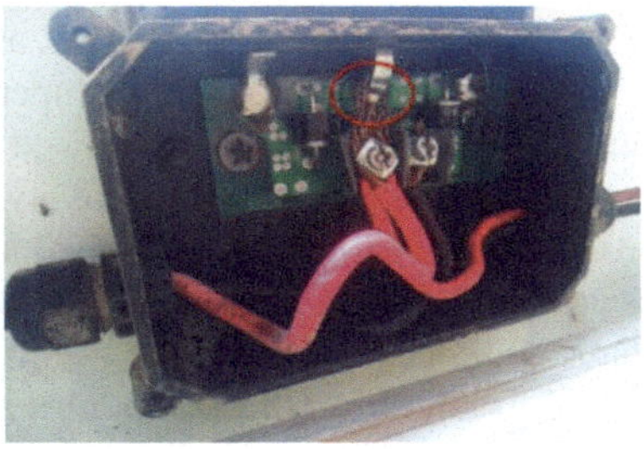

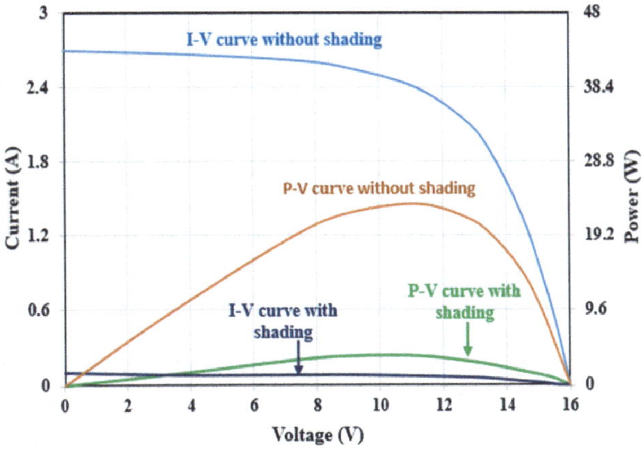

Fig. 8.4 Measured characteristic I–V, P–V curves (unshaded and partially shaded) of module 4

the modules; a state in which an insignificant amount of current or no current flows through the circuit. This condition may be the result of defective bypass diodes or soldering disconnection between the bypass diode and the metal contact inside the junction box [14].

8.3.3 Temperature Measurement

Thermography was utilized to determine the operating temperatures of the modules for the period of the study. The report lite software was used for the quantitative analysis of the captured IR images. The expanse of the modules was captured during the process. Results show that module 3 had the highest temperature of 59.5 °C whiles the lowest temperature of 45.9 °C was measured for module 4 during the period. The highest mean temperature of 56.3 °C was obtained by module 1, with module 4 recording the least mean temperature of 52.4 °C. Modules 2 and 3 recorded average temperatures of 55.3 °C and 55.03 °C respectively. Module 4 had the most significant temperature dissimilarity of 10.7 °C, as well as the lowest mean temperature. Module 3, 2 and 1 respectively recorded temperature dissimilarity of 10.0, 6.5 and 4.2 °C. The standard deviation calculated showed values of 1.5905, 2.8557, 3.8676, and 4.7545 for modules 1, 2, 3 and 4, respectively. The average temperatures were 56.28, 55.25, 55.03 and 52.38 for modules 1, 2, 3 and 4 respectively. The temperature variations observed for the four modules are as a result of the different absorption rates of solar radiation caused by the non-uniform discoloration of EVA in each module. The uneven discoloration of the modules also causes dissimilar electrical defects which lead to mismatch losses and consequently causing the modules to generate varying temperatures on their surfaces.

8.3.4 Degradation Rates

Correction of module parameters to the corresponding values at standard test conditions (STC) was executed by adopting the JRC approach. Results show that module 2 recorded the highest Isc (2.94 A) and the closest to the nameplate value. The lowest Isc (2.6 A) was observed for module 1. The average Isc for the array was 2.78 A, whiles the mean decline in Isc was determined as 11.7%. Module 1 had the highest reduction of 16.4% in Isc and the lowest drop of 7% was observed for module 2, as illustrated in Fig. 8.5. The annual average decline in Isc is 0.98%.

The average V_{oc} determined for the array is 18.4 V which is 14.8% lower than the nameplate value of 21.6 V, as shown in Fig. 8.6. The highest V_{oc} of 19.4 V was recorded by module 2, and the least V_{oc} of 17.91 V determined for module 3. The most significant decline of 17.1% in V_{oc} was observed for module 3. The average decrease in V_{oc} was 14.8% which presents an annual decline in V_{oc} of 1.23%. The average fill factor determined for the array was 60.05. The highest drop in fill factor of

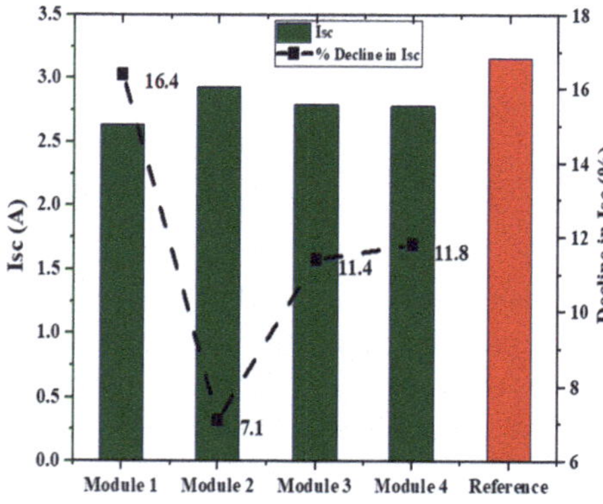

Fig. 8.5 Translated short circuit current

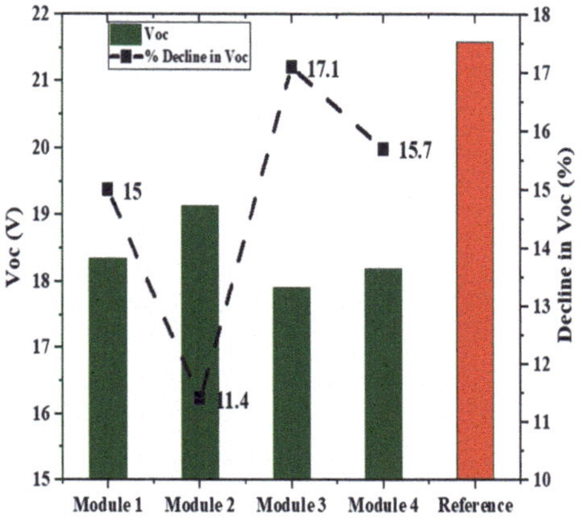

Fig. 8.6 The Translated open-circuit voltage

24.2% was determined for module 3, while the least decline of 11.3% was recorded by module 1, as illustrated by Fig. 8.7. The average decline in FF was 18.0%. This shows an annual average decline in FF to be 1.5%.

Module 2 recorded the highest Vmp of 13.76 V whiles the lowest Vmp of 12.4 V was observed for module 3 and the average for the array was found to be 13.08 V. The average decline in Vmp was found to be 25.7% with an annual decline being 1.09 V as shown in Fig. 8.8. The decrease in *Imp* was relatively lower than is recorded by Vmp, which was as a result of the significant reductions in the observed V_{OC}. The highest observed Imp was 2.40 A for module 4 and the least Imp of 2.30 A recorded for module 1. The mean Imp for the studied array was found to be 2.36 A, as shown in Fig. 8.9. The average decline in Imp was 18.6%, with an annual decline of 1.55%.

The average power output observed for the modules was 30.86 W. The highest and lowest power of 32.72 and 29.3 Wp were determined for modules 2 and 3, respectively. The significant decline in power output of 41.4% was recorded by module 3. The lowest drop was found to be 34.6% for module 2, as depicted in

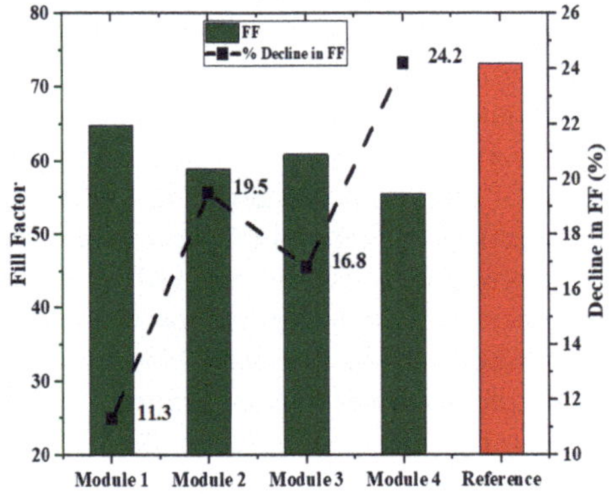

Fig. 8.7 Fill factor of modules at STC

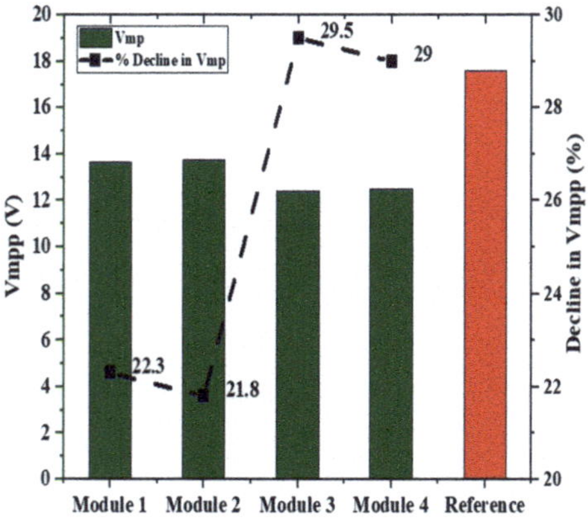

Fig. 8.8 Translated maximum power point voltage

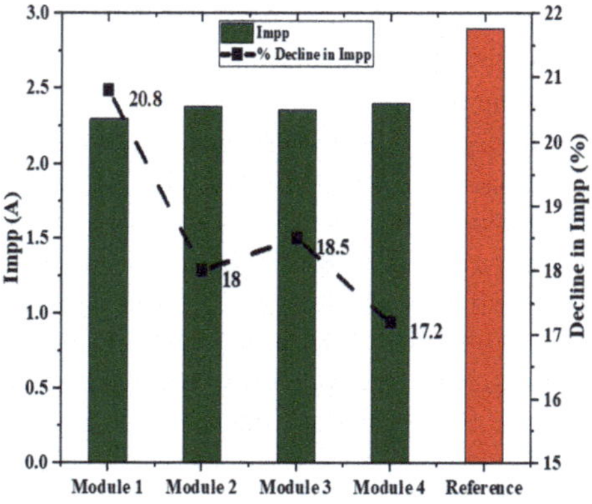

Fig. 8.9 Translated maximum power point current

Fig. 8.10. The average decline in the nominal power for the array was 3% which indicates an average annual decline of 3.19% for the 12 years in operation. Similar studies conducted by Carr et al. in Australia reveal a decline of 2%/year in nominal power of c-Si modules after 16 months in operation [19]. Also, a 2%/year decline in power was observed for a-Si modules in the tropical climate of Singapore [20]. The most significant contributing factors are found to be the V_{oc} and the FF, considering their average declines. According to [21], the order of impact of failure modes is the potential induced degradation, failure of bypass diodes and discoloration of EVA, the encapsulant material of the module. The reason given for this is that there are not appropriate tests yet approved for these failure mechanisms. It is therefore not unexpected the high degradation levels of the modules which demonstrated acute EVA discoloration and bypass diode failures. As it is corroborated by [22], the degree of discoloration of EVA determines the change in the transmittance of the light reaching the solar cells. Thus, the decrease in the power generated [22]. The estimated

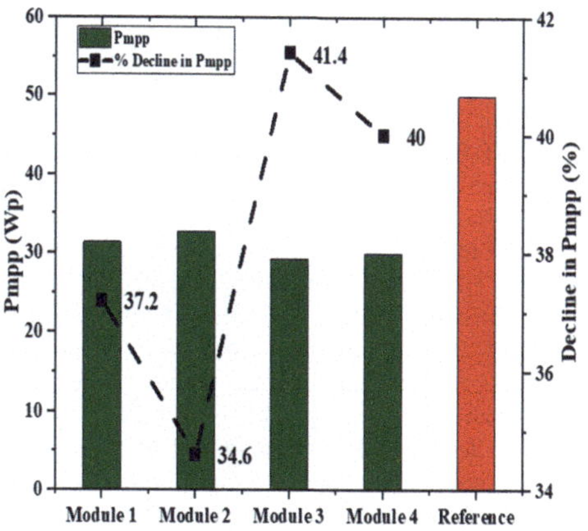

Fig. 8.10 Translated nominal power

MDF of the modules is 16.46%, 6.96%, 11.39% and 11.71% for modules 1, 2, 3 and 4 respectively.

8.4 Conclusion

This work presented the failures and degradation levels in a 12-year ground mount solar PV array in the tropical climate of the sub-Saharan. The power decline observed ranges from 34.6 to 41.4% with an average decrease in power output per year being 3.19%. Compared with the manufacturer's values of nominal power 50 Wp, Isc 3.16 A, V_{OC} 21.6 V, the performance decline of the measured parameters for the 12 years of outdoor exposure are Isc, 7.1–16.4%, V_{OC}, 11.4–17.1%, FF, 11.3–24.2%. All studied modules had malfunctioning bypass diodes. The modules showed high levels of EVA discoloration, covering about 70% of the cell area. The thermal evaluation of the modules revealed temperature dissimilarities ranging from 4.2 to 10.7 °C. This work provides further insights into the varying impacts of specific climates on the long-term performance of solar PV modules and the sources of degradation of PV modules in outdoor operation in the tropical environment.

Acknowledgements This work was supported by the Stipendium Hungaricum Programme and the Mechanical Engineering Doctoral School, Szent István University, Gödöllő, Hungary.

References

1. https://www.fi-powerweb.com/Renewable-Energy.html. Last Accessed 19 Feb 2020
2. J. Solórzano, M.A. Egido, Automatic fault diagnosis in PV systems with distributed MPPT. Energy Convers. Manage. **76**, 925–934 (2013)

3. M. Vázquez, I. Rey-Stolle, Photovoltaic module reliability model based on field degradation studies. Prog. Photovoltaics Res. Appl. **16**, 419–433 (2008). https://doi.org/10.1002/pip.825
4. Energy Sage Inc., *What to Know About a Solar Panel Warranty* (Energysage Inc., Boston, MA, USA, 2015)
5. M.M. Rahman, M. Hasanuzzaman, N.A. Rahim, Effects of various parameters on PV module power and efficiency. Energy Convers. Manage. **103**, 348–358 (2015)
6. V. Sharma, S.S. Chandel, Performance and degradation analysis for long term reliability of solar photovoltaic systems: a review. Renew. Sustain. Energy Rev. **27**, 753–767 (2013)
7. C.E. Chamberlin, M.A. Rocheleau, M.W. Marshall, P.A. Lehman, Comparison of PV module performance before and after 11 and 20 years of field exposure, in *Proceedings of the 37th IEEE Photovoltaic Specialists Conference* (2011).
8. D.C. Jordan, S.R. Kurtz, Photovoltaic degradation rates-an analytical review. Progress Photovoltaic Res Appl **21**, 12–29 (2013)
9. D.C. Jordan, T.J. Silverman, J.H. Wohlgemuth, S.R. Kurtz, K.T. VanSant, Photovoltaic failure and degradation modes. Prog. Photovoltaics Res. Appl. **25**, 318–326 (2017)
10. M.C. Peel, B.L. Finlayson, T.A. Mcmahon, Updated world map of the Köppen-Geiger climate classification. Hydrol. Earth Syst. Sci. Discuss. Eur. Geosci. Union **11**, 1633–1644 (2007)
11. Weather of Koforidua, https://en.climate-data.org/africa/ghana/eastern-region/koforidua-52344/. Last Accessed 20 Oct 2019
12. GHI of Ghana, https://solargis.com/maps-and-gis-data/download/ghana. Last Accessed 31 Mar 2020
13. S.S. Priya, O.S. Sastry, B. Bora, A. Kumar, Comparison of curve correction procedure of current and voltage as per IEC 60891 for thin film technology, in *2015 IEEE 42nd Photovoltaic Specialist Conference PVSC 2015*, (2015). https://doi.org/10.1109/PVSC.2015.7355843.
14. IEA-PVPS, *Review of Failures of Photovoltaic Modules* (OECD/IEA, Paris, France, 2014)
15. D. Quansah, M. Adaramola, G. Takyi, I. Edwin, Reliability and degradation of solar PV modules—Case study of 19-year-old polycrystalline modules in Ghana. Technologies **5**(2), 22 (2017)
16. A. Dhoke, R. Sharma, T.K. Saha, PV module degradation analysis and impact on settings of overcurrent protection devices. Sol. Energy **160**, 360–367 (2018)
17. M.C.C. de. Oliveira, A.S.A.C. Diniz, M.M. Viana, V.D.F.C. Lins, The causes and effects of degradation of encapsulant ethylene vinyl acetate copolymer (EVA) in crystalline silicon photovoltaic modules: A review. Renew. Sustain. Energy Rev. **81**, 2299–2317 (2017)
18. A. Ndiaye, C.M. Kébé, A. Charki, P.A. Ndiaye, V. Sambou, A. Kobi, Degradation evaluation of crystalline-silicon photovoltaic modules after a few operation years in a tropical environment. Sol. Energy **103**, 70–77 (2014)
19. A.J. Carr, T.L. Pryor, A comparison of the performance of different PV module types in temperate climates. Sol. Energy **76**, 285–294 (2004)
20. J.Y. Ye, T. Reindl, A.G. Aberle, T.M. Walsh, Performance degradation of various PV module technologies in tropical Singapore. IEEE J. Photovoltaics **4**(5), 1288–1294 (2014)
21. M. Köntges et al., *Assessment of Photovoltaic Failures in the field* (IEA PVPS Task 13, Subtask 3, 2017)
22. M.A. Munoz, M.C. Alonso-García, N. Vela, F. Chenlo, Early degradation of silicon PV modules and guaranty conditions. Sol. Energy **85**(9), 2264–2274 (2011)

Chapter 9
Comparative Analysis of Software Accuracy in Photovoltaic Energy Estimation for a Temperate Mountain Climate

Bogdan Gabriel Burduhos, Ion Visa, Mircea Neagoe, Mirjana Devetakovic, and Nadia Ramona Cretescu

Abstract The paper analyses four of the mostly used photovoltaic (PV) simulation software tools (PVsyst, PV*SOL online, PV-Watts and PVGIS) from the produced electrical energy point of view. Based on the monthly and yearly estimated performances that are compared with the 6-years-long experimental values recorded from a PV system, recommendations are formulated on the accuracy and setup flexibility of each tool in order to replicate the infield PV application. The experimental data are recorded using a fixed tilted 2 kWp PV platform installed in Brasov, Romania (a mountain temperate climate location), containing two types of PV modules (m-Si and p-Si). The obtained results are useful for the prefeasibility studies of PV systems installed in similar conditions.

Keywords PV system · Electrical energy production · Simulation software · Experimental result · Temperate mountain climate · Validation

9.1 Introduction

Renewable energy systems, able to convert local energy resources (solar, wind, hydro, etc.) represent an important component of the energy mix in sustainable communities

---

B. G. Burduhos (✉) · I. Visa · M. Neagoe · N. R. Cretescu
Transilvania University of Brasov, Eroilor Blvd. 29, 500036 Brasov, Romania
e-mail: bogdan.burduhos@unitbv.ro

I. Visa
e-mail: visaion@unitbv.ro

M. Neagoe
e-mail: mneagoe@unitbv.ro

N. R. Cretescu
e-mail: ncretescu@unitbv.ro

M. Devetakovic
Faculty of Architecture, University of Belgrade, Bulevar kralja Aleksandra 73/2, 11000 Belgrade, Republic of Serbia
e-mail: mirjana.devetakovic@gmail.com

I. Visa and A. Duta (eds.), *Solar Energy Conversion in Communities*,
Springer Proceedings in Energy, https://doi.org/10.1007/978-3-030-55757-7_9

that will also diminish the conflict between energy-rich and -poor regions. Especially the solar energy converters (photovoltaic and solar-thermal systems) will play a key role, as solar radiation is the only resource available anywhere on Earth [1].

The accurate estimation of the solar and photovoltaic energy potential [2] is an important issue in selecting the optimal energy solutions for the specific conditions of a region. Various tools are available for this purpose, out of which the following are mainly used: PVsyst, PVGIS, PVWatts, PVSol, SolarGIS, Homer, SISIFO, etc.[3]. The use of such tools can generate recommendations on the installation of large-scale PV plants in different countries [4, 5].

The different estimation accuracy of the simulation tools requires for each type of implementation location/climate a comparative analysis between the available tools (e.g. Ujjain, India [6]), or between estimated and experimental results (e.g. Ramagundam, India [7]; Khatkar-Kalan, India [8]; Valencia, Spain [9]; Port Elizabeth, South Africa [10]; Brasov, Romania [11]).

Since most literature studies compare the software estimated with the infield results mainly for locations from the tropical and subtropical climates, this paper focuses on the accuracy of the simulation tools for mono- and poly-crystalline PV systems installed in the European temperate climate.

One offline software (PVsyst) and three online web-based simulation tools (PVWatts, PVGIS and PV*SOL online) are used; their photovoltaic estimations are compared with the experimental measurements recorded from a 2 kWp fixed tilted PV platform with both mono- and poly-crystalline PV modules, installed at the outskirts of Brasov, Romania, located in the temperate climate.

The paper has two parts; the first one describes the details of the experimental PV platform and the configuration possibilities of the four simulation tools. The second part is dedicated to the comparison between measured and estimated data and allows different recommendations formulated in the conclusions of the paper, for similar PV systems installed in the temperate climate conditions.

9.2 Experimental Setup

The PV system analysed in this paper is a 2 kWp PV platform, part of a photovoltaic infrastructure with 5 trackable platforms and a total nominal power of 10 kWp. The infrastructure was installed in 2011 near the buildings of the newly established R&D Institute of the Transilvania University of Brasov, Romania (Fig. 9.1). It was installed both for covering a part of the electrical energy demand of the institute with renewable energy but also for research purposes. In this study, the PV platform is tilted at 25° from the horizontal plane and oriented towards South.

The 2 kWp PV platform (Fig. 9.2a) has eight crystalline Silicon PV modules (250 Wp). Four modules are of m-Si type (model Heliene HEE215M) while the other four are of p-Si type (model LDK_DS_250P-20). For compatibility reasons, in order not to limit each other's power production, the two PV module models have very similar electrical parameters. Table 9.1 describes the main electrical parameters

Fig. 9.1 Grid-connected 2 kWp photovoltaic system installed in the Renewable Energy Systems and Recycling R&D Centre, near the Research Institute buildings of the *Transilvania* University of Brasov, Romania

Table 9.1 Electrical parameters under STC conditions of the two PV modules installed

Parameter module type	P_{mp} (Wp)	V_{mp} (V)	I_{mp} (A)	V_{OC} (V)	I_{SC} (A)
HEE215M	250	30.3	8.22	37.5	8.72
LDK_DS_250P-20	250	30.3	8.17	37.7	8.69

under *Standard Testing Conditions* (STC), where P_{mp} is the nominal power, V_{mp} is the voltage at maximum power point, I_{mp} is the current at maximum power point, V_{OC} is the open-circuit voltage and I_{SC} is the short-circuit current.

In 2011 the 8 PV modules were connected in a series string to a SMA Sunnyboy SB 1700 inverter that injected the produced electrical energy to the local grid. This solution allowed the monitoring of the overall platform performance, but was not able to provide information about each module in the system. However, the study required also to evaluate the performance of each individual PV module on the platform therefore, in May 2014, the system was upgraded with Solaredge OPI300-LV power optimizers (DC/DC converters). Each PV module was connected to a power optimizer, while all optimizers were serially connected to a compatible inverter (Solaredge SE 2200) as indicated in Fig. 9.3.

Besides the possibility to extract the maximum available power from each module, without any interferences from the other modules in the series, this solution also allows individual monitoring of the PV modules on a dedicated web interface, https://monitoring.solaredge.com (Fig. 9.2b).

The upgraded PV platform was fully operational and commissioned at the beginning of June 2014. In this paper the monthly energy production data since June 2014 until March 2020, extracted from the web interface for each PV module, are used.

The DC to AC size ratio of the SE 2200 inverter is close to 0.91 indicating that power cut-offs can hardly occur even during intervals with better conditions than STC.

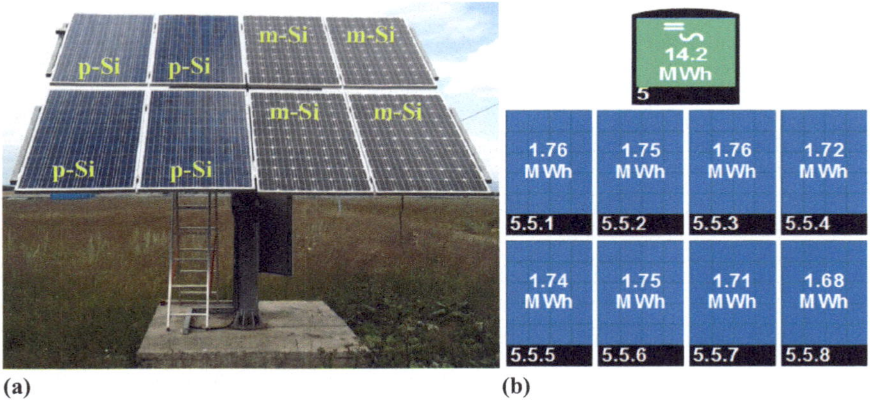

Fig. 9.2 Grid-connected 2 kWp photovoltaic platform with p-Si (LDK_DS_250P-20) and m-Si (Heliene HEE215M) photovoltaic modules: **a** infield installation; **b** Solaredge online monitoring platform

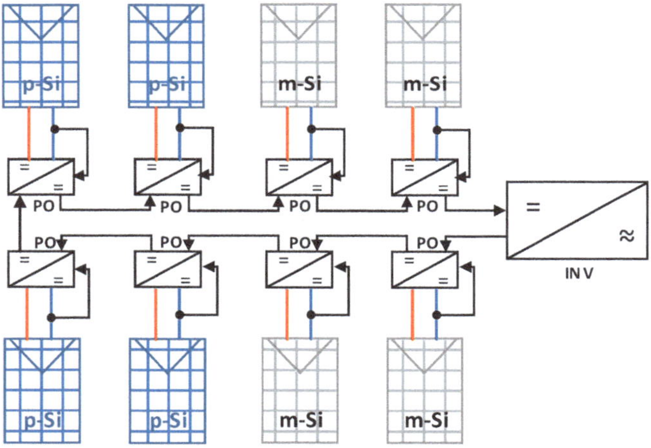

Fig. 9.3 Electrical connection and operation model of the Solaredge solution based on power optimizers (PO) [12]

9.3 Simulation Setup

Four common tools for simulating photovoltaic systems (PVsyst, PVGIS, PVWatts and PV*SOL online) were used to estimate the electrical energy produced by a PV system similar to the one above described and installed near the buildings of the R&D Institute of the Transilvania University of Brasov. Further the four simulation tools are briefly described.

PVsyst is a specialized offline software, available both as full and 1-month trial version. It is widely used for sizing, simulation and prefeasibility studies of PV systems and considered a reference tool in the PV industry.

The software was firstly developed by the Energy Group at the University of Geneva, Switzerland during the 1990s and updated regularly until the current version no. 6.86, available since November 1st, 2019 (https://www.pvsyst.com).

As input data for simulations, PVsyst uses the hourly values of the available solar radiation and several other meteorological parameters from the analysed location. These parameters can be obtained directly using the software from the Swiss company called Meteotest which develops the MeteoNorm software and from NASA (National Aeronautics and Space Administration), but they can also be imported to PVsyst if they are available from different other sources.

For a detailed replication of the PV system to be simulated, PVsyst offers a broad database with all types of components specific for photovoltaic systems (PV modules, inverters, charge regulators, rechargeable batteries, electrical loads, etc.), produced by different manufacturers. If a specific product is not available in the database, the user can define the product, add it to the database and use it further in the analysis.

The influence of shading on the PV modules can also be simulated in PVsyst, if the PV structure and the surrounding objects are 3D modelled in the dedicated interface within PVsyst (Fig. 9.4).

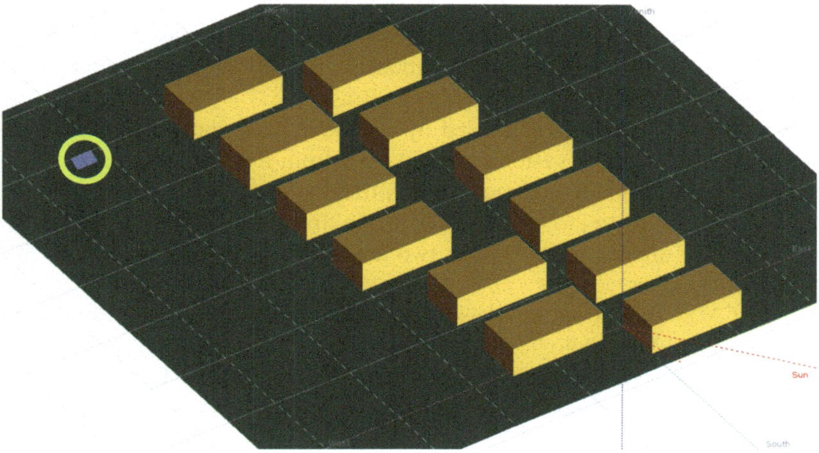

Fig. 9.4 3D perspective of the PV platform and the surrounding shading scene, implemented in PVsyst software

PV*SOL is the name of a simulation software bundle provided by the German company called Valentin. It can be found in three different formats, either as a paid full installable program (PV*SOL Premium and PV*SOL) that allows a detailed simulation of the PV system, or as a free online tool (https://pvsol-online.valentin-software.com) with only general proprieties that can be adjusted. The later was the format used in this paper to simulate the PV platform installed at the Transilvania University of Brasov.

This tool uses for worldwide locations input climate data as hourly values, originated mainly from Meteotest. Additionally climate data from the German Weather Service (DWD) are used for more precise analyses of systems located in Germany.

PVWatts is an online web application (https://pvwatts.nrel.gov) for estimating the electrical energy production of different photovoltaic system types using only few general parameters of the simulated PV system.

The input solar radiation used by this tool are hourly typical meteorological year (TMY) datasets from different sources:

- National Solar Radiation Database (NSRDB) developed by NREL for most locations on the American continent and Indian/Central Asian subcontinents;
- NREL International database, that contains, for locations outside the United States, data from the Solar and Wind Energy Resource Assessment Programme (SWERA), ASHRAE International Weather for Energy Calculations Version 1.1 (IWEC) and Canadian Weather for Energy Calculations (CWEC).

  PVGIS is also an online web-based simulation tool (https://re.jrc.ec.europa.eu/pvg_tools/en/tools.html) which provides free and open access to solar radiation resource and temperature, photovoltaic performance for different technologies/systems and full time series on a large geographical scale, on most of the Earth's surface.

  This tool has been developed since 2001, at the Joint Research Center of the European Commission (Ispra, Italy). The analysis in the current paper used PVGIS 5.1, last updated on December 17, 2019 in order to estimate, based on few general characteristics, the electrical energy produced by a similar PV system to the one installed at the Transilvania University of Brasov.

  The input solar radiation data used by PVGIS for simulating the photovoltaic performance of the PV systems are obtained from five different databases, but the number of available sources might differ for each location on the Earth's surface. According to the tool developers, the first three databases are calculated from satellite images while the last two are obtained by regional reanalysis for high-latitude areas:
- PVGIS-CMSAF, is a dataset developed by the European Organization for the Exploitation of Meteorological Satellites (EUMETSAT) and Satellite Application Facility on Climate Monitoring (CM SAF), which covers the European, African and parts of South-American continents. Since this operational dataset is not updated anymore in the PVGIS tool, its developers recommend using the PVGIS-SARAH database instead;

- PVGIS-SARAH is a dataset developed by the PVGIS team in collaboration with CM SAF, based on a new algorithm that also covers the Asian continent;
- PVGIS-NSRDB is a dataset provided by National Renewable Energy Laboratory (NREL) part of the National Solar Radiation Database, which covers only the American continents;
- PVGIS-ERA5 is a dataset developed through reanalysis by the European Centre for Medium-Range Weather Forecast (ECMWF), containing solar radiation data only for the European continent;
- PVGIS-COSMO dataset is also a reanalysis product taken from the Consortium for Small-scale Modelling Regional Reanalysis 6 (COSMO-REA6) database, covering also only the European continent.

Among the four simulation tools used in this paper, PVsyst allows the most detailed parameter adjustment of all PV components, electrical loads and nearby objects (see Table 9.2) so that the simulated PV system is very similar to the real implemented one.

In contrast to PVsyst the other three simulation tools only allow the adjustment of the main, general parameters of the implemented system, as described in Table 9.2.

Table 9.2 Parameters that can be adjusted in the four simulation tools

Parameter name	PVsyst	PV*SOL online	PVWatts	PVGIS
General parameters				
Location	x	x	x	x
PV technology	–	–	x	x
Model of PV module	x	x	–	–
Tilt/Inclination angle	x	x	x	x
Azimuth/Orientation angle	x	x	x	x
Installation/Mounting type	x	x	x	x
Overall system losses	–	–	x	x
Albedo	–	x	–	–
Shading	–	x	–	–
Soiling	–	x	–	–
Inverter efficiency/selection	x	x	x	–
DC to AC size ratio	x	x	x	–
Detailed parameters				
PV module editing	x	–	–	–
PV string configuration	x	–	–	–
Horizon editing	x	–	–	x
Detailed system losses	x	–	–	–
Inverter editing	x	–	–	–

All four simulation tools considered in the paper allowed the accurate selection of the PV platform location, installed near the R&D Institute of the Transilvania University in Brasov (45.6701°N, 25.5486°E).

The two Silicon based PV modules (mono- and polycrystalline silicon) installed on the platform were selected from the PVsyst database and all PV parameters were adjusted [13]. In PVWatts the simulation of both modules was done by selecting the same module type (standard, *premium* or thin-film), while in PVGIS only the PV technology could be selected (*crystalline silicon*, CIS and CdTe). This means that a single result could be obtained in PVWatts and PVGIS for both types of modules. In PV*SOL two module types (*Example mono* 200 W and *Example poly* 200 W) were selected and the simulation for each type was independently done.

The tilt/inclination and the azimuth/orientation angle were defined in all tools according to the ones of the real PV platform: 25° from the horizontal plane and facing the South direction.

The installation/mounting type was selected in all tools so that it was similar to the mounting of the experimental platform: in PVsyst the option *Free mounted modules with air circulation* was selected, in PVWatts the *Fixed (open rack)* option, in PVGIS the *Free-standing* option, while in PV*SOL the *Mounted—open space* option.

The losses of the PV platform were selected as follows in the three simulation tools: PVsyst allowed the adjustment of several losses sources (the module's *Aging* was considered to decrease the efficiency with 0.4%/year; the *Soiling* was only considered during December (15%) and January (20%) due to snow; the PV module *Efficiency loss* was considered −1.5% for p-Si modules and −1% for m-Si modules). In PVWatts and PVGIS the overall system losses were considered as predefined (14%). In PV*SOL the losses were adjusted using the *Albedo* (20%), the *Shadowing* (3%) and the *Soiling* (5%) parameters.

From the electrical configuration point of view, the real electrical connections of the module optimizers, inverter etc. used on the PV platform could be replicated in PVsyst. In PVWatts the *DC/AC size ratio* (2000 W/2200 W) and the *Standard inverter efficiency* (97.6%) were the only parameters which could be adjusted according to the experimental setup. In PVGIS and PV*SOL online, no details of the connection between PV module—optimizer—inverter could be detailed.

9.4 Results and Discussions

Further, the electrical energy produced in a temperate mountain area (RESREC R&D Centre, Brasov, Romania) by a photovoltaic platform is comparatively analysed, on a monthly and annual basis, with the simulation results from the four simulation tools above described (PVsyst, PVWatts, PVGIS and PV*SOL). The experimental results were recorded during June 2014–March 2020, from the 2 kWp platform with individually monitored m-Si and p-Si PV modules.

The validation of the experimental data was an important step in the analysis and showed that each four modules of the same type similarly performed throughout the analysed interval, with a maximum monthly relative deviation from the average value of 4.6% for m-Si and 10.35% for p-Si modules, as the monthly energy production of m-Si—Fig. 9.5a and p-Si modules—Fig. 9.5b show. Consequently, each set of four modules of the same type was considered as a separate 1 kWp unit whose values will be further analysed.

The monthly electrical energy produced by a PV unit of 1 kWp (Fig. 9.6a, b) indicates a high variation during all the analysed years, mainly during the winter and transient seasons. This high variation is rather normal in temperate mountain climates, as the one where the platform is installed, due to the high variability throughout the

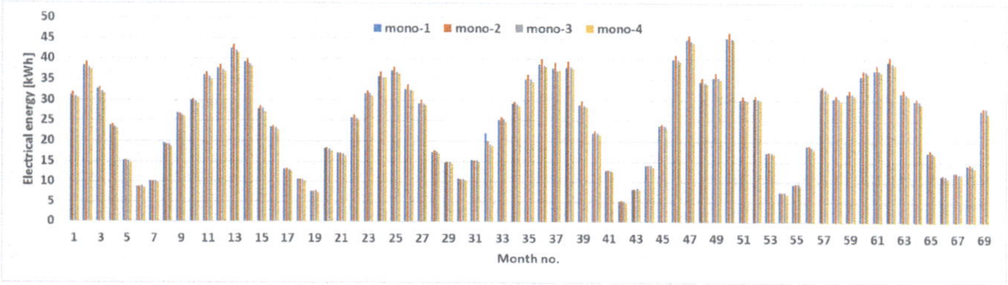

(a)

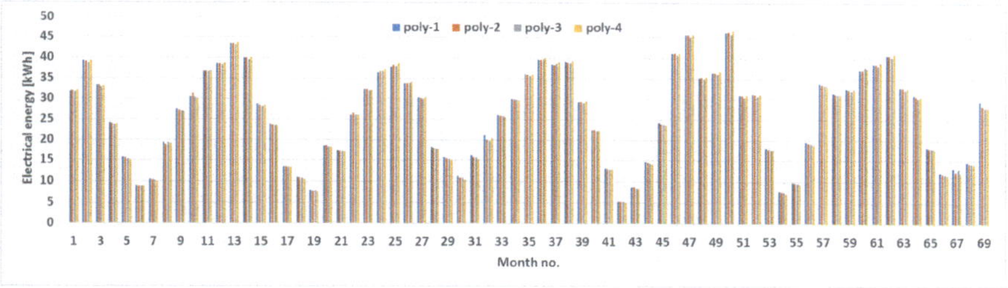

(b)

Fig. 9.5 Experimental monthly electrical energy produced during June 2014–March 2020 by the four PV modules of the: **a** 1 kWp m-Si and **b** 1 kWp p-Si photovoltaic units

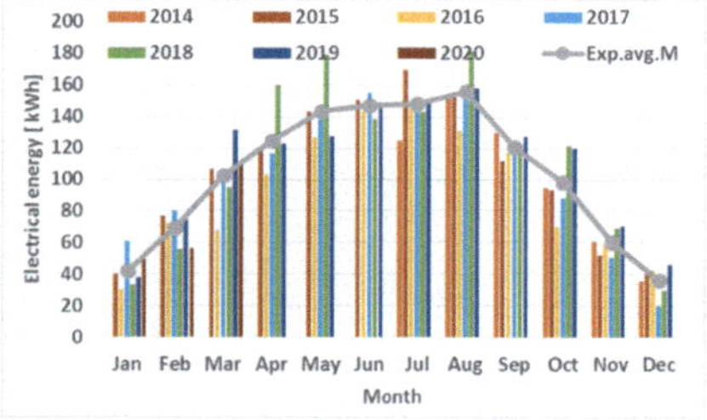

(a)

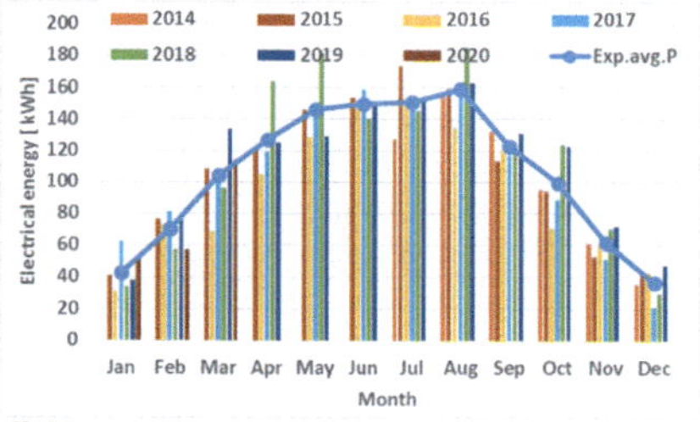

(b)

Fig. 9.6 Experimental monthly and average electrical energy produced during June 2014–March 2020, by: **a** a 1 kWp m-Si and **b** a 1 kWp p-Si photovoltaic unit

years of the meteorological parameters that influence the photovoltaic conversion (available solar radiation, air temperature and rainfall amount) [14].

The month with the highest relative deviation from the average monthly value was December 2017 (−42.92% for the m-Si unit and −43.06% for the p-Si unit), which can be attributed to the low solar irradiance conditions and heavy soiling of the PV modules due to snowfall.

The monthly electrical energy produced by a PV unit of 1 kWp (Fig. 9.6a, b) indicates a high variation during all the analysed years, mainly during the winter and transient seasons. This high variation is rather normal in temperate mountain climates, as the one where the platform is installed, due to the high variability throughout the years of the meteorological parameters that influence the photovoltaic conversion (available solar radiation, air temperature and rainfall amount) [14].

The month with the highest relative deviation from the average monthly value is December 2017 (−42.92% for the m-Si unit and −43.06% for the p-Si unit), which can be attributed to the low solar irradiance conditions and heavy soiling of the PV modules due to snowfall.

The two PV module types had a similar performance during the entire investigated period with the monthly average values presented in Fig. 9.7, p-Si performing slightly better than m-Si. The average yearly energy produced by the m-Si PV unit is 1,248.8 kWh and 1,272.8 kWh by the p-Si module indicating a better performance of the p-Si unit in the temperate mountain climate conditions (23.6 kWp increase).

Considering only the complete years, the highest relative deviation from the annual average value was observed during 2016 (−10.8%, for the m-Si unit and −10.44% for the p-Si unit), while the minimum relative deviation was during 2017 (−0.89%, for the m-Si unit and −1.14% for the p-Si unit).

The monthly average experimental values denoted with *Exp.avg.M* and *Exp.avg.P* in Figs. 9.6 and 9.7 will be further compared with the estimated values by the four simulation tools.

As already described, the PVsyst software is the most advanced tool among the four used in this paper. It allowed fine tuning of detailed parameters of the PV modules, optimizers, inverter etc. and consequently generated the most realistic simulation model. Several databases can be used in PVsyst for generating the weather

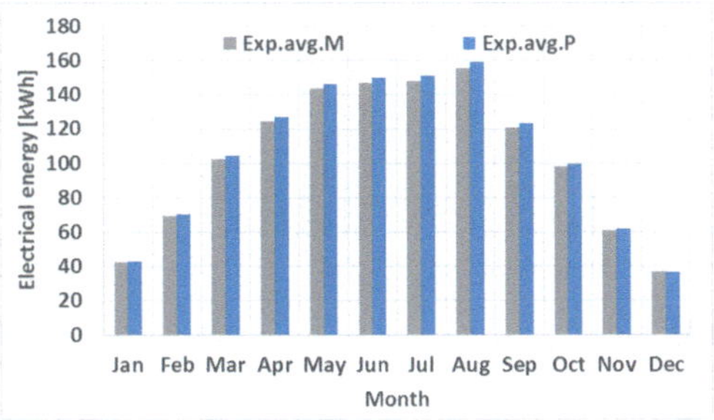

Fig. 9.7 Experimental electrical energy produced during an average month by the two 1 kWp photovoltaic units of m-Si and p-Si type

profile of a location, out of which the Meteonorm dataset showed the best correlation in the considered mountain climate location (Brasov, Romania) [11].

According to Table 9.3, the results indicate a very low annual relative deviation of the energy estimated compared to the infield energy produced (0.1% for the m-Si unit) and (−0.5% for the p-Si unit), but rather high monthly deviations during winter (December, January) and August. The highest deviation can be observed during January (15.9% for m-Si and 16.2% for p-Si). As indicated in Fig. 9.8a the increased performance of the p-Si unit compared to the m-Si unit can also be observed, in agreement with the experimental results.

Similar to PVsyst, PV*SOL allows the selection of the PV module type, but only few other general parameters of the PV system can be adjusted. As such, two simulations were generated with the results outlined in Fig. 9.10 and Table 9.3, that lead to low annual deviations of 0.6% (m-Si) and −2.5% (p-Si). Unfortunately the tool underestimates the energy production during the summer interval, when most of the solar energy is available (e.g. August: −23.9% for m-Si and −25.9% for p-Si) and greatly overestimates the production during the winter months (e.g. December: 83.1% for m-Si and 77.4% for p-Si). Moreover, according to Fig. 9.8b, this tool indicates a slightly better performance of the m-Si photovoltaic unit, contrary to the infield and the PVsyst results.

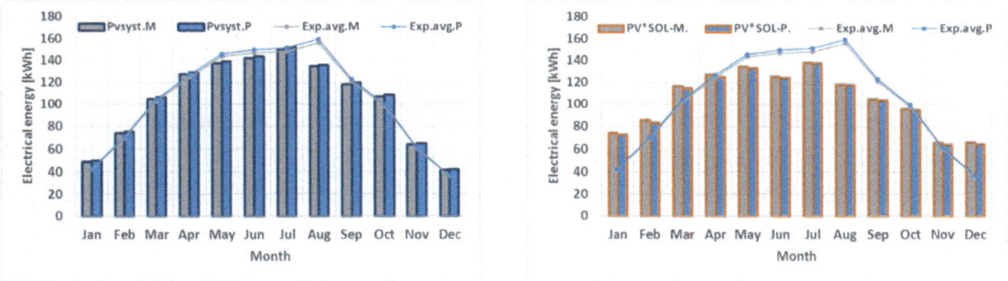

(a) (b)

Fig. 9.8 Monthly electrical energy produced by 1 kWp m-Si and p-Si photovoltaic units, estimated using: **a** the Meteonorm dataset in PVsyst software; **b** the PV*SOL online simulation tool

(a) (b)

Fig. 9.9 Monthly experimental and estimated electrical energy produced by 1 kWp m-Si and p-Si photovoltaic units, using online simulation tools: **a** PVWatts; **b** PVGIS

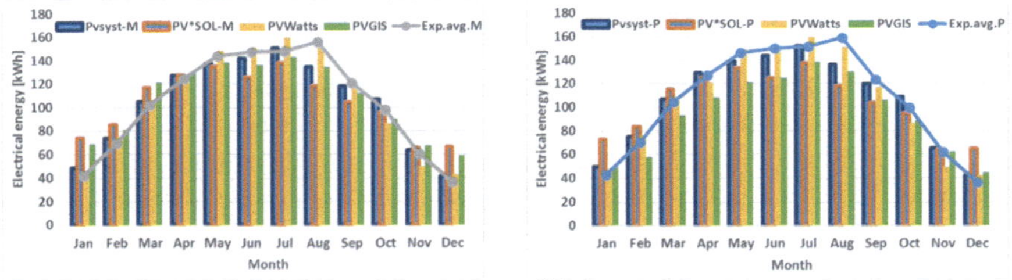

(a) **(b)**

Fig. 9.10 Monthly electrical energy produced versus estimated using PVsyst, PVWatts, PVGIS and PV*SOL online simulation software for: **a** the 1 kWp m-Si and **b** the 1 kWp p-Si photovoltaic unit

In contrast to the two tools analysed above, in PVWatts only the simulation of a generic PV system is allowed, with no possibility to define the exact PV module type or any other electric component.

As such, PVWatts generated a single set of simulation results (Fig. 9.9a) that was compared with the energy produced by the 1 kWp m-Si and p-Si experimental units. Despite this limitation, the results indicate a good match between real and simulated data (Table 9.3), with a yearly relative deviation of 0.7% (m-Si) and −1.2% (p-Si) respectively. The maximum monthly relative deviation observed is 18.3% during December for the m-Si unit, and 19.5% during November for p-Si modules.

The same limitations in simulating a PV system are valid for PVGIS (Fig. 9.9b), where also a single set of simulation results was obtained. According to the tool developers, the recommended dataset for the implementation location is PVGIS-SARAH, because PVGIS-CMSAF is not updated anymore, while PVGIS-ERA5 and PVGIS-COSMO are datasets intended for high latitude locations (over 65°).

The correlation in this case between the real and simulated results is not as accurate as in the previous case (Table 9.3), especially during the summer interval (May–August), when the electricity production was underestimated with about 15%. The yearly relative deviation is −9.9% for m-Si and −11.6% for p-Si modules. Also, the maximum monthly relative deviations are higher and were observed during December, with values of 24.3% (m-Si) and 23.2% (p-Si).

Table 9.3 Monthly and yearly relative deviations (%) of the estimated electrical energy compared to the infield experimental system

	Jan	Feb	Mar	Apr	May	Jun	Jul	Aug	Sep	Oct	Nov	Dec	Year
mono-crystalline Silicon													
PVsyst versus Exp	15.9	6.5	2.5	2.3	−4.4	−3.6	1.7	−13.6	−2.0	9.1	5.6	14.4	**0.1**
PV*SOL versus Exp	76.2	23.2	14.4	2.4	−6.0	−14.4	−6.5	−23.9	−13.0	−1.6	9.0	83.1	**0.6**
PVWatts versus Exp	8.0	9.0	4.0	−2.5	3.6	3.2	8.2	−3.0	−3.1	−11.6	−17.8	18.3	**0.7**
PVGIS versus Exp	16.0	−17.1	−8.8	−13.4	−15.6	−15.0	−6.8	−16.4	−12.4	−9.5	3.5	24.3	**−9.9**
poly-crystalline Silicon													
PVsyst versus Exp	16.2	6.4	2.0	1.5	−4.9	−4.3	0.6	−14.5	−2.9	8.7	5.0	15.6	**−0.5**
PV*SOL versus Exp	70.0	18.4	10.8	1.3	−8.7	−16.7	−8.8	−25.9	−15.6	−4.7	4.3	77.4	**−2.5**
PVWatts versus Exp	6.4	7.2	2.0	−4.5	1.8	1.3	5.9	−5.1	−5.0	−13.1	−19.5	17.2	**−1.2**
PVGIS versus Exp	14.2	−18.4	−10.6	−15.2	−17.0	−16.5	−8.8	−18.2	−14.2	−11.0	1.4	23.2	**−11.6**

9.5 Conclusions

This paper comparatively analyses, using four common photovoltaic simulation software tools, the electrical energy that can be produced with a fixed tilted 2 kWp photovoltaic system installed in a temperate mountain climate, containing both mono- and polycrystalline silicone PV modules. The simulated data were compared with the experimental results recorded in Brasov, Romania during 2014–2020 using a PV system installed in the RESREC R&D Centre of the Transilvania University of Brasov. The comparative data analysis allowed to formulate the following conclusions:

- The results obtained using PVsyst indicate the best correlation with the experimental data. This aspect together with the fact that the software allows the most detailed setup of all PV system parameters, according to the real PV application, makes it an accurate solution especially if real local meteorological data are available in the simulation.
- Despite the fact that PVWatts is mainly recommended for preliminary studies on the performance of photovoltaic systems allowing only a general setup, the simulation results indicate the best agreement with the experimental results, among all the investigated simulation tools. Given the simplicity of this tool and the low yearly and acceptable monthly relative deviations, PVWatts can be considered a good simulation tool for similar PV systems installed in mountain areas with a temperate climate.
- Since PV*SOL online gives good results at annual level it can represent a good solution for yearly analyses of PV systems, but the high deviations during summer and winter months make the free online version of PV*SOL an unreliable solution for monthly analyses.
- The current version of PVGIS is the least reliable tool for electrical energy estimations of similar PV systems due to its high yearly and monthly deviations from the experimental data.

Acknowledgements We hereby acknowledge the COST Action CA16235, PEARL PV—Performance and Reliability of Photovoltaic Systems: Evaluations of Large-Scale Monitoring Data.

References

1. I. Visa, A. Duta, M.D. Moldovan, B.G. Burduhos, M. Neagoe, *Solar Energy Conversion Systems in the Built Environment* (Springer, 2020)
2. B.G. Burduhos, M. Neagoe, Improved model for forecasting global solar irradiance during sunny and cloudy days, annals of West University of Timisoara—Physics. J. West Univ. Timisoara **60**(1), 14 (2018)
3. J. Hofierka, J. Kanuk, Assessment of photovoltaic potential in urban areas using open-source solar radiation tools. Renew. Energy **34**, 2206 (2009)

4. T. Pavlovic, D. Milosavljevic, I. Radonjic, L. Pantic, A. Radivojevic, M. Pavlovic, Possibility of electricity generation using PV solar plants in Serbia. Renew. Sustain. Energy Rev. **20**, 201 (2013)
5. D.Z. Djurdjevic, Perspectives and assessments of solar PV power engineering in the Republic of Serbia. Renew. Sustain. Energy Rev. **15**, 2431–2446 (2011)
6. C. Dondariya, D. Porwal, A. Awasthi, A.K. Shukla, K. Sudhakar, S.R., M.M., Bhimte, A., *Performance simulation of grid-connected rooftop solar PV system for small households: A case study of Ujjain, India, Energy Report,* **4**, 546, 2018.
7. B. Shiva Kumar, K. Sudhakar, Performance evaluation of 10 MW grid connected solar photovoltaic power plant in India. Energy Reports **1**, 184 (2015)
8. V. Sharma, S.S. Chandel, Performance analysis of a 190 kWp grid interactive solar photovoltaic power plant in India. Energy **55**, 476 (2013)
9. B. Quesada, C. Sanchez, J. Canada, R. Royo, J. Paya, Experimental results and simulation with TRNSYS of a 7.2 kWp grid-connected photovoltaic system. Appl Energy **88**, 1172 (2011)
10. D. Okello, E.E. van Dyk, F.J. Vorster, Analysis of measured and simulated performance data of a 3.2 kWp grid-connected PV system in port Elizabeth, South Africa. Energy Convers Manage **100**, 10 (2015)
11. B.G. Burduhos, I. Visa, M. Neagoe, N.R. Cretescu, Simulated versus Produced electrical energy of a 9.6 kWp PV system installed in a temperate mountain climate. J. Sci. Arts **1**(50) (2020)
12. Solaredge, *Technical Note—SolarEdge Fixed String Voltage, Concept of Operation—9/2012.*
13. D.T. Cotfas, A.M. Deaconu, P.A. Cotfas, Application of successive discretization algorithm for determining photovoltaic cells parameters. Energy Convers. Manage. **196**, 545 (2019)
14. B.G. Burduhos, I. Visa, A. Duta, M. Neagoe, Analysis of the conversion efficiency of 5 types of photovoltaic modules during high relative humidity time periods. J. Photovoltaics **8:6**, 1716 (2018)

Chapter 10
Hybrid RES Mobile Innovative System Optimized for DC-DC Applications

Bogdan-Alexandru Onose, Ion Murgescu, and Ştefan-Adrian Sontea

Abstract The research focus was to develop an off-grid hybrid RES mobile containerized system, combining the energy from the photovoltaic and micro-wind turbine source in order to directly providing the energy via DC-DC power supply. The applicability of the system was intended for irrigation and fertigation, in semi-dry and dry-sub-humid arid climates. The project tackled the most common technologies used in solar photovoltaic and small wind turbine applications combining them in an autonomous hybrid RES system that can be used in agriculture to power directly DC power supply (without the need of DC–AC conversion). During the development of the project an experimental model of the hybrid RES mobile system was deployed using a 2.16 kWp photovoltaic and 600 Wp wind turbine power source combined with a 5.3 kWh battery pack for extended use period. For the photovoltaic panels a mobile mechanical system of the photovoltaic source was developed and patented (no. A00730/12-11-2019) in order to increase the flexibility and maximize power production in real life conditions. Further testing in real file conditions is planned for a yearlong irrigation period in order to validate and further improve the energy management algorithm for the for irrigation and fertigation periods.

Keywords Off-grid photovoltaic · Hybrid RES system · Micro-wind turbine · Energy management · Irrigation and fertigation

10.1 Introduction

Mobile, photovoltaic and wind power generators have entered history in the last half of the last century. In fact, the first photovoltaic generator used for practical purposes was a mobile system. In 1958, Hoffman Electronics created a photovoltaic generator

B.-A. Onose (✉) · I. Murgescu · Ş.-A. Sontea
Department of Renewable Energy Sources and Energy Efficiency, National Institute for R&D in Electrical Engineering ICPE-CA, Bucharest, Romania
e-mail: bogdan.onose@icpe-ca.ro

I. Murgescu
e-mail: ion.murgescu@icpe-ca.ro

I. Visa and A. Duta (eds.), *Solar Energy Conversion in Communities*,
Springer Proceedings in Energy, https://doi.org/10.1007/978-3-030-55757-7_10

with 9% efficiency, with a power of 0.1 W and a 100 cm^2 panel, for Vanguard I, the first solar-powered satellite [1]. Although today's photovoltaic mobile generators are much more common than wind turbines, which is easy to understand given that the sun sends its energy all over the planet's surface, the wind generator has outperformed the photovoltaic one in time. The first wind turbine to generate electricity was a battery charger installed in July 1887 by Scottish scholar James Blyth to illuminate his vacation home in Marykirk, Scotland [2].

Mobile electric generators are usually stand-alone generators being used in places without access to the electricity grid. The explosive technological development of the photovoltaic industry and the exponential fall in prices for photovoltaic modules in the first two decades of the twenty-first century have naturally led to the development of a large number of mobile applications in all areas of human activity [3].

Solar energy water pumps, one of the applications that could not be missed by the photovoltaic industry, work with zero cost electricity generated by the photovoltaic panels, replacing the diesel generators. Solar pumps are useful where the electric grid is not available and alternative sources (wind power) cannot provide enough energy [4]. When they do, even if it is less predictable than photovoltaic, hybrid applications are developed, photovoltaic-wind, which can improve the energy supply continuity.

10.2 Concept Definition

The irrigation and fertigation application, which represents the energy consumer in this case, dictates the type and size of the mobile hybrid electric generator. In our case, for the supply of the water pump for irrigation and fertigation of agricultural crops, the solution was to use a small containerized system with a mobile mechanical structure that will provide optimal photovoltaic placement regardless of the terrain configuration.

All the above must be achieved with minimum costs and with a maximum of 2 operators. Thus being the case, the irrigation and fertigation system characteristics are:

- A plantation with an area of 500 m^2 as presented in Fig. 10.1;
- The water requirement for the crop, maximum 4 m^3/day and a maximum of 8 h/day of irrigation;
- The pump will be positioned at a maximum of 10 m from the irrigation plantation are system (dynamic level), with a maximum of 4 bar pressure in the irrigation pipes, but not more than 5 m above the water level;
- Water temperature is not relevant to this case;

Solar irrigation pumps are built to operate without batteries as long as there is sun. To make irrigation possible during the morning and evening, an additional energy storage will be used to capture the excess energy from the hybrid photovoltaic-wind system.

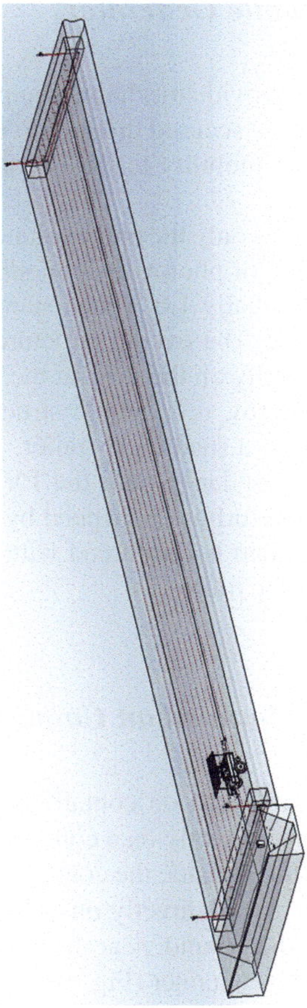

Fig. 10.1 The use case plantation area structure for irrigation. *Source* INMA Bucharest

10.3 Mechanical and Structural Analysis of Mobile Systems

In order to develop a new and innovative solution for the hybrid RES mobile system, a state-of-the-art analysis of the available systems on the market was made taking into account two important structural aspects: trailer or container mounted systems.

10.3.1 Trailer Mounted Mobile Generators

Trailer mounted mobile generators provide mechanical support for the photovoltaic modules and/or wind turbine, that are secured directly to the trailer structure. This approach provides a high degree of mobility to the overall system and can easily towed to the application site.

The most common option is to have only the photovoltaic system fixed on a trailer chassis (Fig. 10.2a) [5]. The number of photovoltaic modules may vary depending on the size of the application, but usually 1.6 × 1 m standard dimension are used with a weight of approx. 20 kg each. The overall structure is wind resistance with reinforced steel bars mounted vertically on the floor at the ends of the PV modules.

In the second case (Fig. 10.2b) [6], is a simple structure of a hybrid system (photovoltaic and wind) mounted on a road legal trailer. The foldable structure is made from aluminium profiles, in solidarity with the PV modules, similar to the structure of a solar tracker. The wind turbine is gripped by a folding pillar so that it does not exceed the transport limit and is reinforced with three steel pipe bars for anchorage that are also fastened to the trailer.

10.3.2 Mobile Generators Mounted on Containers

In the case of mobile generators mounted on containers, the mechanical support structure and equipment are mounted directly on a container. For transport ease, the structure is easily detachable and stored inside the container. The structure is built to be in solidarity with the container placed directly on the ground or on an arranged platform or is mounted directly on the ground, near the container, with the electrical equipment usually is fixed inside the container (Fig. 10.3) [7–9].

The above solution shows a variety of containerized systems available on the market, that can be either only photovoltaic systems, or hybrid photovoltaic-wind systems. To further improve the overall efficiency of this systems, energy storage solution was adopted, due also to the available space inside the container [10].

Container mobile structures can provide much higher powers, only depending on the size of the container in which they are transported. The systems can also be modularized with dedicated containers for electrical equipment and for mechanical

(a) (b)

Fig. 10.2 Trailer mount mobile generators **a** solar PV system and **b** hybrid solar PV and wind system. *Source* MobileSolar, GURU Mobil

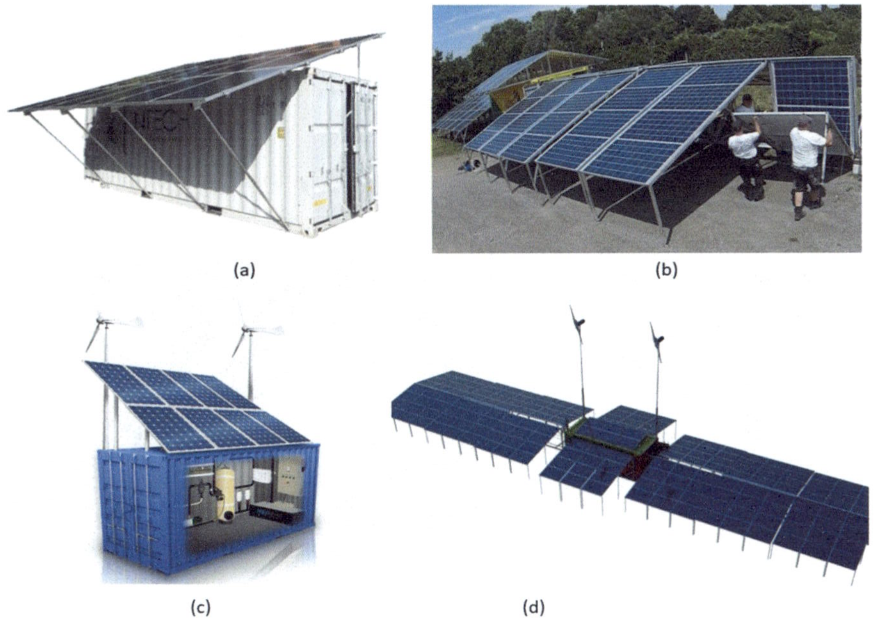

Fig. 10.3 Container or ground mount mobile generators **a** container fixed structure solar PV system, **b** ground foldable structure of solar PV system, **c** container fixed hybrid solar PV and wind system and **d** ground fixed hybrid solar PV and wind system. *Source* Power-MoveIT, Motive, EcosPowerCube

structures that can be installed on-site and in this case the power of this mobile generators becomes unlimited.

The disadvantage is that there is time and labor required for installation and uninstallation. Nevertheless, this drawback becomes more insignificant in relation to the length of the time that this mobile generator is used in a single location or application.

This solution represents the most common mobile solar photovoltaic and/or wind turbine solution because it can provide higher powers than the trailer and also offers greater safety conditions during transport and storage periods. The only down side is the reduced mobility compared with trailer solutions [11].

10.4 System Design and Component Sizing

10.4.1 DC–DC Water Pump

Using the concept definition data from Chap. 2 and the manufacturer simulation software we managed to identify the correct pump for this application. The solar surface water pumping systems resulted from the simulation has an ECDRIVE DC brushless motor, designed for solar, with over 90% efficiency, MPPT technology to maximize power use from PV modules, an inbuilt data logging and also a direct connect with AC connection options. A typical application using the solar surface water pumping system is outlined in Fig. 10.4 [12].

For optimum performance the manufacturer simulation software report [13] provided information about the pump technical data, with performance matching of the concept case with daily or monthly pump operation scenarios connected directly to the photovoltaic system.

When choosing the pump, it will be taken into account to ensure a greater need due to the increase of the irrigated surface in the future or the same performance for a greater depth in drought times. Also, in addition to the simulation, battery storage capacity will be added that will allow the pump to keep running even during the night or days without too much sun.

10.4.2 Hybrid Photovoltaic-Wind System with Energy Storage

During the development of the project an experimental model of the hybrid PV-wind DC pumping surface system was deployed using 8 × 270 Wp photovoltaic modules, a 600 Wp micro-wind turbine and a 5.3 kWh gel battery pack, to extend the system usage period during the period without sun (Fig. 10.5).

Fig. 10.4 Typical solar surface water pump (dc pump) system layout. *Source* Lorentz

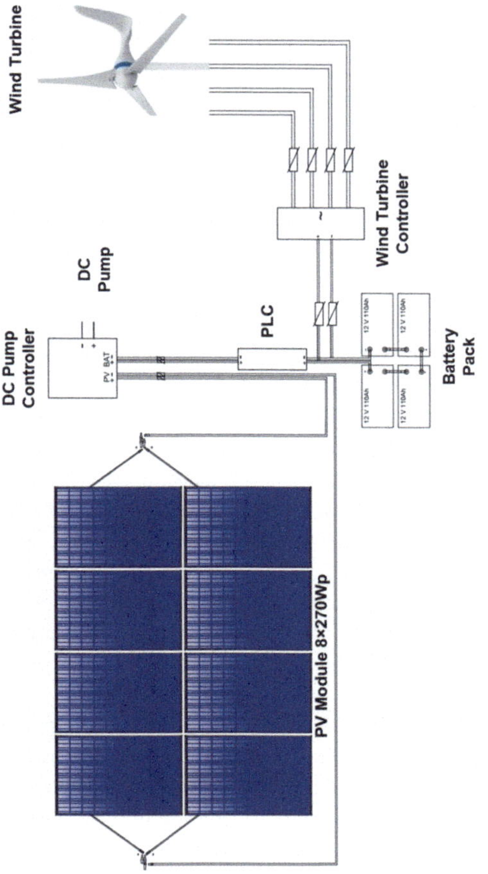

Fig. 10.5 Technical diagram of the hybrid PV-wind DC pumping system with battery pack

The photovoltaic modules will be grouped in 2 groups of 4 modules connected directly to the DC solar pump controller. The micro-wind turbine will be connected to the wind turbine charging controller and will provide the energy required to charge the 48 V battery pack. The DC solar pump controller is connected to the battery pack via a programmable source (PLC).

The DC pump controller is in contact communication with the solar surface water pump to determine the energy demand at any moment, thus making it the primary factor in the control logic of the hybrid PV-wind DC pumping system.

During the irrigation/fertigation hours, all energy from the PV system is supplied directly to the solar pump to cover the load. During periods with low irradiation levels, the energy demand will be covered from the wind turbine and/or the battery pack.

10.4.3 Mobile Container and Mechanical Photovoltaic System

The development of the new and innovative solution for the hybrid RES mobile system, was done with the purpose of solving the major disadvantages discovered in the already existing systems (detailed in Chap. 3).

For the structure of the photovoltaic panels a mobile mechanical system of the photovoltaic source was developed in order to increase the free movement flexibility in real life conditions. The designed folding support system is ideal for easy photovoltaic modules installation and positioning to maximize the energy captured from the sun, thus increasing the system energy production (Fig. 10.6 Diagram of the photovoltaic panels a mobile mechanical system structure (a) without PV modules and (b) with PV modules attached. The final result of the experimental model of the mobile container structure with hybrid PV-wind with photovoltaic modules mobile mechanical system structure and micro-wind turbine position is presented in Fig. 10.7).

The design of the mechanical photovoltaic support and folding system is very well adapted for a wide range of photovoltaic solar panels from the point of view of its size. The support and folding system for solar modules, from a constructive point of view, with a movable metallic structure that has multiple inclination angle adjustment for better capturing the solar rays. The azimuth adjustment is done using the wheels and position it on the ground based on the solar position. As a result of the development, a patent was registered to the State Office for Inventions and Trademarks for the mobile mechanical system of the photovoltaic source (no. A00730/12-11-2019).

The main system components are: the metal supports on which the solar panels are mounted, a fastening systems that allow the fastening and folding assembly of the system, suspension rods that allow adjusting the angle of exposure of the PV modules, clamping elements that ensure the fixing of the PV modules on the metallic structure, mobile wheels that allow this system free movement over the terrain, a hinge system

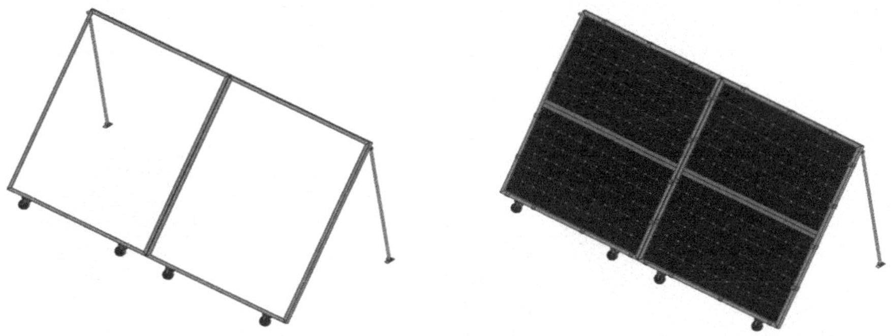

(a) (b)

Fig. 10.6 Diagram of the photovoltaic panels a mobile mechanical system structure **a** without PV modules and **b** with PV modules attached

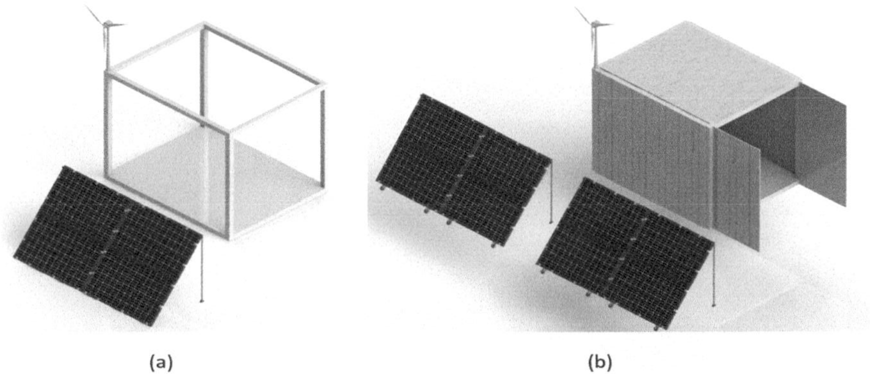

Fig. 10.7 Mobile container structure with hybrid PV-wind with photovoltaic modules mobile mechanical system structure and micro-wind turbine position

that ensure the folding of the 2 metal supports with solar panels, having as main advantage the diminution of the storage space and the ease of their transport.

The final result of the experimental model of the mobile container structure with hybrid PV-wind with photovoltaic modules mobile mechanical system structure and micro-wind turbine position is presented in Fig. 10.7.

The mobile container is custom made based on a predefined structure of 3 m × 2.4 m and a height of 2.54 m. In order to provide the necessary safety during transport and anti-vandal in the operation site, a metal structure was chosen, from special p zinc-plated profiles, the upper part (exterior) form double-zinc plated sheet, anti-condensation foil, wooden panels, mineral wool insulation and PVC board and the exterior walls made of cut sheet, fastened on rectangular support posts covered by undulated painted sheet.

10.5 Simulation and Results

Based on the simulations, the technical properties of the solar surface water pumping system required for the irrigation/fertigation were estimated. The period required for the irrigation/fertigation is between March–September, so this was the focus period of the simulation of the hybrid PV-wind energy source.

The simulation of the micro-wind turbine energy production was done based on monthly average data for the defined period of time, using the test site location real-life weather conditions.

The simulation of the photovoltaic system was done using PVSYST dedicated software in order to determine the performance of the system in real-life operation for the defined period of irrigation/fertigation (Fig. 10.8) [14].

The simulation results show that the off-grid hybrid RES mobile system has an overall energy availability from hybrid PV-wind of 2563.4 kWh/period and an user energy demand of 1605.0 kWh/period from the solar surface water pumping system used in the irrigation/fertigation period, as outlined in Table 10.1.

The results are based on an annual horizontal global irradiation of 1436 kWh/m^2 [15] for the selected location, an annual average wind speed at 10 m above ground of 3.42 m/s [16] and a defined irrigation/fertigation schedule of 8 h/day.

Thus, the results for used energy from the total hybrid system fraction is 1.60, with a photovoltaic solar fraction of 1.43. The defined algorithm for irrigation/fertigation uses the energy produced by the photovoltaic system is directly to cover the demand of the solar surface water pumping system and the micro-wind turbine energy to charge the battery pack. This control logic provides more flexibility to the hybrid PV-wind system in the order to cover the required demand by the irrigation/fertigation system, in periods without sun and offer autonomy to the system from 1 to 3 days.

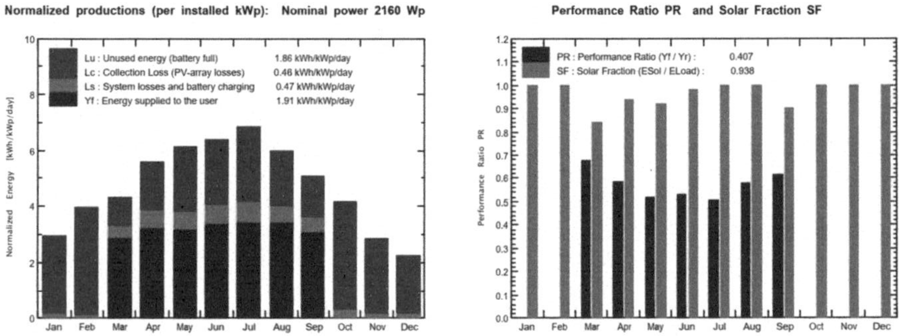

(a) (b)

Fig. 10.8 Energy production simulation of the off-grid photovoltaic system

Table 10.1 Energy produced by the hybrid PV-wind system during the irrigation/fertigation season

Month	PV system		Micro-Wind turbine		Total available energy	User energy demand	Total fraction
	Horizontal global irradiation	Available solar energy	Average wind speed	Available wind energy			
	[kwh/m^2]	[kWh]	[m/s]	[kWh]	[kWh]	[kWh]	[-]
March	111.3	262.2	3.7	40.7	302.9	232.5	1.30
April	151.0	320.1	3.7	40.5	360.6	225.0	1.60
May	184.4	354.0	3.3	36.6	390.6	232.5	1.68
June	191.3	352.3	3.3	36.3	388.6	225.0	1.73
July	208.0	387.5	3.2	35.4	422.9	232.5	1.82
August	172.5	339.9	3.2	34.6	374.5	232.5	1.61
September	128.6	285.3	3.5	38.0	323.3	225.0	1.44
TOTAL/period					**2563.4**	**1605.0**	**1.60**

10.6 Conclusions

The hybrid RES mobile innovative system optimized for DC-DC applications, resulted from the project has a mobile container custom structure 3 × 2.4 m, with a hybrid PV-wind system using 8 × 270 Wp photovoltaic modules, a 600 Wp micro-wind turbine and a 5.3 kWh gel battery pack. Also, a patent was registered to the State Office for Inventions and Trademarks for the mobile mechanical system of the photovoltaic source (no. A00730/12-11-2019).

For the irrigation/fertigation application a solar surface water pumping system is used, having a solar designed ECDRIVE DC brushless motors, with over 90% efficiency and MPPT technology to maximize power use from PV modules.

The simulation results show that the off-grid hybrid RES mobile system has an overall energy availability from hybrid PV-wind of 2563 kWh/period and a total hybrid system fraction of 1.60 and a photovoltaic solar fraction of 1.43, for a defined irrigation and fertigation schedule of 8 h/day.

Further testing in real life conditions at the Research and Development Facility for Plant Culture in Dăbuleni, Dolj county, Romania will be done for a yearlong irrigation period in order to validate and further improve the energy management algorithm for the irrigation and fertigation periods.

Acknowledgements The work was done under the *SmartIRRIG* project (no. 27PCCDI/2018), financed by the Romanian Ministry of Research and Innovation CCDI – UEFISCDI, under PNIII Program 1—Development of National Research and Development System, Subprogram 1.2—Institutions Performance Complex Projects in RDI consortia.

References

1. “Photovoltaic Dreaming 1875–1905: First Attempts at Commercializing PV". Retrieved 8 April 2017
2. Wikipedia, https://en.wikipedia.org/wiki/Timeline_of_solar_cells, last edited on 13 October 2019
3. Y. Hamakawa, Solar PV energy conversion and the 21st century's civilization. Sol. Energy Mater. Sol. Cells **74**(1–4), 13–23 (2002)
4. J. R. Fanchi, C. J. Fanchi, *Energy in the 21st Century* (4th Edition), World Scientific, November 2016
5. MobileSolar, https://www.mobilesolarpower.net/ (2019)
6. GURU Mobil, https://www.aselsan.com.tr/en/capabilities/energy-systems/hybrid-energy-systems/guru-mobil (2019)
7. Power-MoveIT, https://www.power-moveit.tech/mobile-solar-container/ (2019)
8. Motive, https://www.motive-offshore.com/renewable-hybrid-power-container (2019)
9. EcosPowerCube®, https://ecospheretech.com/powercube (2019)
10. Out-of-the-box, https://www.outofthebox.energy/ (2019)
11. A. Goetzberger, J. Luther, G. Willeke, Solar cells: past, present, future. Sol. Energy Mater. Sol. Cells **74**(1–4), 1–11 (2002)
12. Lorentz, https://www.lorentz.de/products-and-technology/surface-solar-pumps (2019)
13. Lorentz Compass, https://www.lorentz.com.au/lorentz-compass-pump-design-tool (2019)
14. PVSYST V6.82, https://www.pvsyst.com/
15. PVGIS, https://re.jrc.ec.europa.eu/pvg_tools/en/tools.html
16. NASA LaRC for the Forecast of World Energy Resources, https://power.larc.nasa.gov/data-access-viewer/ (2020)

Chapter 11
Emerging Issues and Challenges with the Integration of Solar Power Plants into Power Systems

Fazel Mohammadi and Mircea Neagoe

Abstract The characteristics of solar-generated electricity, including intermittency, uncertainty, and non-synchronous power generation, lead to some technical challenges to large-scale power grid integration. Each of those characteristics causes an economic challenge as well as reverse power flow, power quality issues, dynamic stability, and big data challenges. This paper aims to comprehensively investigate the existing challenges with the integration of high-penetration solar power plants, particularly Photovoltaic (PV) power plants, into power systems and corresponding solutions to improve the security, reliability, and resiliency of power systems.

Keywords Distributed Generations (DGs) · Photovoltaic (PV) power plants · Renewable Energy Sources (RESs) · Solar power plants

11.1 Introduction

The limited fossil fuel resources, global warming and environmental concerns, growth in the load demand, cyber-physical attacks, power shortage, and interconnection of new load types, such as Plug-in Hybrid Electric Vehicles (PHEVs), to power grids, have enforced the energy sector using Renewable Energy Sources (RESs) [1–6]. Conventional power systems are based on centralized and regulated power generation, transmission, and distribution systems and there is no flexibility for

The original version of this chapter was revised: In this version, URLs as CrossRef and Google Scholar of reference 1 is corrected and title of the paper, URLs as CrossRef and Google Scholar of reference 108 has been corrected. The correction to this chapter is available at https://doi.org/10.1007/978-3-030-55757-7_38

F. Mohammadi (✉)
Electrical and Computer Engineering (ECE) Department, University of Windsor, Windsor, ON N9B 1K3, Canada
e-mail: fazel@uwindsor.ca; fazel.mohammadi@ieee.org

M. Neagoe
Renewable Energy Systems and Recycling Research Center, Transilvania University of Brașov, Brașov, Romania
e-mail: mneagoe@unitbv.ro

I. Visa and A. Duta (eds.), *Solar Energy Conversion in Communities*,
Springer Proceedings in Energy, https://doi.org/10.1007/978-3-030-55757-7_11

Distributed Generations (DGs) [1, 2]. The interconnection of DGs to power systems requires control, communication, and computation systems to ensure efficient, stable, and reliable operation [7]. Solar power plants, particularly Photovoltaic (PV) power plants, are one of the fast-growing types of DGs being integrated into power systems in recent years. Solar power plants reduce operational costs to generate electricity and provide added value to customers and utilities. The share of solar power plants capacities is increasing by roughly 40% annually [8]. The most favorable characteristics of solar power plants are the availability of solar irradiation in most of the world sites and the fact that solar power plants can be installed in a variety of sizes from small-scale to very large-scale systems [9].

Most of the solar power plants are integrated with the low-voltage distribution grids. While the increase in the solar power plants penetration into power systems leads to many challenges, which all depend on the point of interconnection of the solar power plants to power systems and the state and performance of equipment that are already installed on power systems [10]. To integrate solar power plants into power systems, advanced inverters, anti-islanding capability, protection systems, forecasting technology, and smart metering and control systems are required [11].

This paper comprehensively reviews the challenges with the integration of solar power plants, specifically PV power plants, into power systems and explains some possible technical solutions to mitigate such challenges and improve the security, reliability, and resiliency of power systems.

The rest of the paper is organized as follows. Section 11.2 describes the existing challenges of solar power plants integration into power grids. Possible solutions for solar power plants integration into power grids are presented in Sect. 11.3. A summary of the existing challenges and possible solutions for solar power plants integration into power grids is given in Sect. 11.4. Finally, some brief conclusions are indicated in Sect. 11.5.

11.2 Existing Challenges of Solar Power Plants Integration into Power Grids

The integration of solar power plants into power systems requires to take the characteristics of solar-generated electricity and their corresponding challenges into account. Such characteristics along with their challenges are presented in this section.

11.2.1 Variability

The output power of conventional power plants, including some hydro and geothermal power plants, that run on fuel can increase and decrease on command. Therefore, they are dispatchable. However, DGs, such as solar power plants, generate electricity when the Sun is shining. Power systems operators do not control DGs; they accommodate them, which need some agility. From the grid operator point of view, the output power of DGs, particularly solar power plants, is functionally equivalent to a reduction in the load demand. This can cause fluctuations in the demand for dispatchable power [12–16]. Therefore, there should be rules and regulations that govern existing power systems infrastructure while the integration of DGs into power systems.

11.2.2 Uncertainty

The output power of solar power plants in day-ahead and/or day of forecast cannot be accurately predicted [17–19]. Hence, power systems operators should ensure having excess reserve running to meet the demand.

11.2.3 Non-Synchronous Power Generation

Conventional power generation units provide voltage support and frequency control to power grids. Solar power plants do not currently help to maintain grid frequency [20]. In the case of doing so, additional capital investment is required.

11.2.4 Location-Specificity and Low Capacity Factor

The solar irradiation is good and efficient in some places than others. Such places may not be close to power grids. Thus, power transmission infrastructure is required to transfer the power to where it is needed. The area occupied by solar power plants is directly related to the size of the plant, solar irradiance at specific locations, and the technology and efficiency of solar cells. Such issues adversely affect the agriculture industry and the environment. In addition, solar power plants operate when there is enough solar irradiation [21–24]. Hence, the average capacity factor, which is defined as generation relative to potential, for large-scale solar power plants is approximately 30%. Comparing to the average capacity factor of a typical nuclear power plant, which is ~93%, DGs are generating electricity at low capacity factors.

11.2.5 Hazardous Materials and Life-Cycle Emissions

Solar cells contain heavy metals, such as lead and cadmium, which may be hazardous when they are decommissioned. Solar cells are mainly made with thin-films cells containing harmful materials, such as indium, gallium, and arsenic. Considering the manufacturing process of silicon-based solar cells, the silicon dust is released that causes breathing problems when inhaled. In addition, for cleaning and purifying the semiconductor surface of solar cells, hazardous chemicals are involved. Moreover, the solar cells manufacturing process, their transportation, and installation and decommissioning cause aggregate life-cycle emission [25–27]. It can be concluded that the increase in solar power plants installation can increase the risk of exposure to hazardous material, as well as an increase in the life-cycle emission.

11.2.6 Power Grids Flexibility

Proper operation of power systems depends on continuously balancing power generation from different conventional and non-conventional sources with the load demand. In order to balance the electricity supply with the load demand, when solar power plants cannot operate, thermal and hydroelectric power plants provide operating reserves and supply the loads. Also, there is a need for storing energy during periods of low demand and injecting the stored energy when it is needed, as well as (1) sharing energy and capacity across particular regions, which needs both power transmission capacity and power market participants to trade electricity and (2) changing the load demand in response to power grids conditions [28–32]. Solar power plants can cause overgeneration conditions. This refers to a condition during which the aggregated supply of conventional power plants and solar power plants exceeds the demand. To avoid this, solar power plants generation should be curtailed by either reducing the output from the inverter or disconnecting the entire power plants from power systems. To do so, the physical control systems of the generation sources are required.

11.2.7 Capacity Value to Meet the Peak Demand

Solar power plants can provide capacity value by reducing the load demand that must be supplied by the conventional generation units during periods of high demand. In other words, capacity value shows how much additional load can be added with the addition of solar power plants [32–34]. The major issue with this is related to the reliability of solar power plants to continuously and adequately supply the load demand.

11.2.8 Big Data and Cybersecurity

The high penetration of DGs, such as solar power plants, causes an increase in the volume of data. The data includes power consumption pattern data, smart metering and control devices data, and operational data [35–40]. The data is used for real-time monitoring and control of the entire grid. To successfully monitor and control the system, there is a need for a highly efficient communication infrastructure. At high penetration of solar power plants, the massive data intrusion and cyber-attacks can affect the quality of service. However, the latency, adequate bandwidth, efficiency, and reliability of communication should be considered.

11.3 Possible Solutions for Solar Power Plants Integration into Power Grids

To tackle the challenges of solar power plants integration into power systems, several solutions are further proposed.

11.3.1 Minimum Power Injection Limit

The minimum power injection limit can be applied when the output power of the grid-connected solar inverters falls below a certain threshold value and accordingly, the relay should disconnect the grid-connected solar inverters from the power grid. In the case of an overload (>1.20 of the nominal load level), the relay should also disconnect the grid-connected solar inverters from the power grid [41, 42].

11.3.2 Reverse Power Flow Index

When the power flow from the power grid becomes zero or changes to the opposite direction, the relay should disconnect the grid-connected solar inverters from the power grid [43, 44]. To do so, the relay should continuously monitor the direction of power flow and send the trip signal to the corresponding breaker, when the reverse power flow is detected.

11.3.3 Loading Condition

The output power of the solar inverter should be dynamically controlled by the Diode Clamped Inverters (DCIs) and smart inverters and varies based on the loading condition. This helps the solar inverters to efficiently operate and mitigate the impact of reverse power flow. In addition, solar inverters can participate in the voltage regulation on the feeders [45–49].

11.3.4 Utilization of Energy Storage Systems

As mentioned in Sect. 11.2.1, variability is one of the biggest challenges to integrate the large-scale solar power plants into power grids. To mitigate the impacts of the variability of the output power of solar power plants, Energy Storage Systems (ESSs), such as battery banks and supercapacitors, can be utilized to smooth the output power and prevent the sudden power outage [50, 51].

11.3.5 Utilization of Solid-State Transformers

Solid-state transformers are capable of improving the power quality and protecting the system against faulty conditions [52, 53]. However, the efficiency of the solid-state transformers is lower than electromagnetic induction-based transformers and their protection scheme both at the low voltage and high voltage is complicated, with their help, large-scale solar power plants can be integrated into power grids while providing real-time control and monitoring of energy dispatch and improving the power quality.

11.3.6 Optimal Energy Dispatch

A practical solution to supply the load demand is to generate electricity using different types of DGs. However, it is difficult to achieve efficient energy dispatch. To overcome this issue, optimal energy dispatch algorithms and power electronics topologies should be investigated [54–59]. By implementing optimal energy dispatch algorithms with high penetration of DGs, particularly solar power plants, (1) balancing the load demand in an efficient way can be guaranteed, (2) the feeders power losses can be minimized, and (3) the stability of the power grid can be improved.

11.3.7 Utilization of Advanced Control and Distribution Management Systems

Using the advanced control and distribution management systems can optimize the overall performance of power systems while solar power plants are integrated and provide an automated outage restoration procedure. In addition, the advanced control and distribution management systems enable smooth integration of DGs, voltage regulation, and resiliency improvement [60–62]. However, it is difficult to achieve an optimal solution by combining the advanced control and distribution management systems and the existing power grids infrastructure.

11.3.8 Utilization of Advanced Communication Infrastructure and Intelligent Protection Systems

The integration of DGs into power grids impacts the reliability, security, and resiliency of power systems. Using the advanced communication infrastructure and intelligent protection systems can improve the reliability and resiliency of power systems. Dynamic monitoring and control of the grid parameters, such as voltage, current, frequency, through a fast and reliable communication infrastructure can cause the operation of the entire grid more efficient [63–74]. To deal with the dynamic nature of power systems with the integration of solar power plants, more intelligent protection systems with self-awareness, self-reconfiguration, and self-healing capabilities should be deployed.

11.3.9 Determining the Optimal Size and Allocation of Distributed Generators

Determining the optimal size and allocation of DGs, especially solar power plants, is a function of the feeder power losses, voltage profile, operating costs, line load ability, and the existence of other previous DGs installations. For solar power plants, the average solar irradiation for the candidate locations is important. Therefore, the immense benefits with high integration of solar power plants can be achieved, if the size and location of solar power plants, subject to the technical and non-technical constraints, are optimally determined [74–79].

11.3.10 Utilization of Hybrid Grid-Connected Distributed Generators

In order to improve the efficiency and reliability of power systems, the grid-connected DGs can be used in a hybrid form. This means different types of DGs with different power capacities can be combined with each other. In this case, by properly selecting the type and size of DGs, such as wind, ESSs, etc., and integrating them with solar power plants, and also scheduling of such power generation units, the overall performance of the grid improves [80–88].

11.3.11 Utilization of Flexible Conventional Power Generation Units

Power systems planners always consider more flexible conventional power generation units, such as natural gas and small-scale Combined Heat and Power (CHP) plants to deal with the variable nature of power generation by non-conventional generation units [89, 90]. It should be noted that the operating costs of conventional power plants can be smaller than fuel savings from the increased DGs penetration.

11.3.12 Demand Response Program and Demand-Side Management

It was before mentioned that power generation units must be capable of supplying the aggregated load demand and power losses associated with transferring power from the generation units to the load centers. By dynamically control the loads to match the power generation in real-time through demand response program and demand-side management, i.e., load shifting and load shedding/curtailment, the DGs can be more dispatch-able within certain limits [91–94]. This also causes overcoming the variable nature of power generation by DGs. For instance, if the majority of the customers in a certain area participate in demand response program and demand-side management, the generated power by the concentrated solar power plants that have molten-salt storage becomes available for 24 h of the day. In addition, demand response program and demand-side management lead to a reduction in power generation and operating costs to balance the load demand, reduction in power market price variations, improving the reliability of the grid, managing the grid congestion, and improving power systems security [93, 94].

11.3.13 Interconnected Power Transmission Grids

If different DGs, e.g., solar power plants and wind farms, are aggregated across a broader region, the generated power by such DGs becomes less variable. In this case, if the geographical area linked up by power transmission grids is big enough, this is highly possible that the Sun is shining and the wind is blowing somewhere within that specific area [95].

11.3.14 Cloud Computing and Artificial Intelligence

To deal with big data and cybersecurity challenges, the collected data from all nodes in power systems should be stored in high capacity and high-speed data storage systems. One solution is to use big data streaming frameworks, such as Storm, Spark, Flink, Kafka, and Samza suggested. The other solution is to use cloud computing, which enables achieving more level of flexibility and efficiency in data management [96–99]. Some cloud computing services are Software as a Service (SaaS), Infrastructure as a Service (IaaS), Platform as a Service (PaaS), Data as a Service (DaaS), Communication as a Service (CaaS), and Monitoring as a Service (MaaS). However, the major challenges in cloud computing are data security and latency. In order to improve the security of cloud computing services, scalability, and visibility, and reducing the bandwidth and latency, the post-cloud computing frameworks, such as fog computing, Mobile Edge Computing (MEC), and dew computing, are proposed [100–102]. It should be noted that the deployment of artificial intelligence along with cloud computing and post-cloud computing helps to efficient integration of solar power plants into power systems by analyzing the new and historical data and weather forecasting, and optimal controlling of the system, estimating the state of the system, and diagnosing faults in power systems [103–114].

11.4 Summary of the Existing Challenges and Possible Solutions for Solar Power Plants Integration into Power Grids

Table 11.1 gives a summary of the existing challenges of solar power plants integration into power systems and possible solutions that can address such challenges. In addition, suggested future solutions are mentioned.

According to Table 11.1, the integration of small-scale and large-scale solar power plants into power grids requires to develop more advanced control, protection and communication systems to improve the reliability, security, and resiliency of the power systems.

11.5 Conclusions

This paper presents different challenges with the integration of solar power plants into power systems and their consequences. In addition, considering the future of solar power plants integration, various existing and practical solutions are presented. As the existing solutions need to be further improved, some suggested future solutions are indicated in this paper. It should be noted that technical, operational, and environmental challenges with the integration of solar power plants into power grids

Table 11.1 Solar power plants penetration problems and suggested future solutions

Challenges	Existing solutions	Suggested future solutions
Reverse power flow	Limited choice to connect solar power plants to power grids	(1) Minimizing the load on the specific feeder (2) Using smart relays
Voltage instability	Using On-Load Tap Changers (OLTCs) and Dynamic Voltage Regulators (DVRs)	(1) Using Flexible Alternating Current Transmission System (FACTS) devices with Fault Ride Through (FRT) capability
Frequency Instability	Geographical smoothing of solar power plants output	(1) Using distributed ESSs with FRT capability based on System Identification (SI) (2) Using advanced control and distribution management systems
Harmonics	(1) Using active harmonic filters (2) Compliance with UL 1741 Standard for inverters, converters, controllers and interconnection system equipment for use with RESs	Using FACTS devices
Protection scheme	Coordination in relays, sectionalizers, fuses, reclosers	(1) SI for real-time monitoring and control of faults (2) Using advanced communication infrastructure and intelligent protection systems
Power factor correction	Using capacitor banks	Dynamic reactive power control based on SI
Power losses	Determining optimal size and location of solar power plants	(1) Robust optimal energy dispatch (2) Using high-performance hybrid grid-connected DGs
Communication systems	(1) Using intelligent electronic devices (2) Using communication and control protocols based on IEEE Standard 2030	Using fast computing and advanced communication infrastructure with compliance of IEEE Standard 2030
Challenges	Existing Solutions	Suggested Future Solutions
Load ability of power grids	Determining optimal size and location of solar power plants	Robust optimal energy dispatch

(continued)

Table 11.1 (continued)

Challenges	Existing solutions	Suggested future solutions
Big Data and Cybersecurity	Using data streaming frameworks and cloud computing	(1) Using advanced data encryption methods and defense mechanisms (2) Using artificial intelligence to mitigate the impacts of cyber-attacks
Power Dispatch and Scheduling	Optimal power dispatch and scheduling based on considering forecasting and economics criteria	(1) Robust optimal energy dispatch (2) Implementing demand response program and demand-side management
Power systems security	Accurate executions of optimal unit commitment and economic dispatch	Robust optimal energy dispatch considering essential aspects of power system planning
Forecasting	Accurate environmental conditions forecasting based on available data	Using hybrid forecasting methods based on knowledge discovery and deep machine learning

are at the early stage and become more complicated with the increased level of penetration of solar power plants. Therefore, further investigations need to be performed to ensure the reliability, security, and resiliency of power systems.

References

1. A. Abdollahi, A.A. Ghadimi, M.R. Miveh, F. Mohammadi, F. Jurado, Optimal power flow incorporating FACTS devices and stochastic wind power generation using krill herd algorithm. Electronics **9**(6), 1043 (2020)
2. F. Mohammadi, G.-A. Nazri, M. Saif, A bidirectional power charging control strategy for plug-in hybrid electric vehicles. Sustainability **11**(16), 4317 (2019)
3. F. Mohammadi, Design and electrification of an electric vehicle using lithium-ion batteries, in *Proceedings of the 3rd International Conference on Electrical Engineering* (Tehran, Iran, 2018)
4. F. Mohammadi, Electric vehicle battery market analysis: lithium-ion, in *Proceedings of the 1st International Conference on Modern Approaches in Engineering Science (ICMAES)* (Tbilisi, Georgia, 2018)
5. F. Mohammadi, Design, analysis, and electrification of a solar-powered electric vehicle. J. Solar Energy Res. **3**(4), 293–299 (2018)
6. F. Mohammadi, Hybridization of an electric vehicle using lithium-ion batteries, in *Proceedings of the 1st International Conference on Modern Approaches in Engineering Science* (Tbilisi, Georgia, 2018)
7. F. Mohammadi, G.-A. Nazri, M. Saif, An improved droop-based control strategy for MT-HVDC systems. Electronics **9**(1), 87 (2020)
8. D. Pitt, G. Michaud, Assessing the value of distributed solar energy generation. Current Sustain Renew Energy Rep **2**, 105–113 (2015)
9. H. Alatrash, R.A. Amarin, C. Lam, Enabling large-scale PV integration into the grid, in *Proceedings of the 2012 IEEE Green Technologies Conference* (Tulsa, OK, USA, 2012)

10. H. Setiadi, A.U. Krismanto, N. Mithulananthan, M.J. Hossain, Modal interaction of power systems with high penetration of renewable energy and bes systems. Int. J. Electr. Power Energy Syst. **97**, 385–395 (2018)
11. A. Anzalchi, A. Sarwat, Overview of technical specifications for grid-connected photovoltaic systems. Energy Convers. Manage. **152**, 312–327 (2017)
12. P. Mohammadi, S. Mehraeen, Challenges of PV integration in low-voltage secondary networks. IEEE Trans. Power Delivery **32**(1), 525–535 (2017)
13. W.E. Ibraheem, C.K. Gan, M.R. Ab. Ghani, Impact of photovoltaic (PV) systems on distribution networks. Int Rev Modell Simul **7**(2), 70–74 (2014)
14. J.V. Appen, M. Braun, T. Stetz, K. Diwold, D. Geibel, Time in the sun: the challenge of high PV penetration in the german electric grid. IEEE Power Energ. Mag. **11**(2), 55–64 (2013)
15. A. Rodriguez-Calvo, R. Cossent, P. Frías, Integration of PV and EVs in unbalanced residential LV networks and implications for the smart grid and advanced metering infrastructure deployment. Int. J. Electr. Power Energy Syst. **91**, 121–134 (2013)
16. L. Mukwekwe, C. Venugopal, I.E. Davidson, A review of the impacts and mitigation strategies of high PV penetration in low voltage networks, in *Proceedings of the 2017 IEEE PES PowerAfrica* (Accra, Ghana, 2017)
17. D. Cheng, B. Mather, R. Seguin, J. Hambrick, R.P. Broadwater, PV impact assessment for very high penetration levels, in*Proceedings of the 2015 IEEE 42nd Photovoltaic Specialist Conference* (New Orleans, LA, USA, 2015)
18. D. McPhail, B. Croker, B. Harvey, A study of solar PV saturation limits for representative low voltage networks, in *Proceedings of the 2016 Australasian Universities Power Engineering Conference* (Brisbane, QLD, Australia, 2016)
19. A. Anzalchi, A. Sundararajan, A. Moghadasi, A. Sarwat, Power quality and voltage profile analyses of high penetration grid-tied photovoltaics: a case study, in *Proceedings of the 2017 IEEE Industry Applications Society Annual Meeting* (Cincinnati, OH, USA, 2017)
20. S. Ghosh, S. Rahman, Global deployment of solar photovoltaics: its opportunities and challenges, in *Proceedings of the 2016 IEEE PES Innovative Smart Grid Technologies Conference Europe* (Ljubljana, Slovenia, 2016)
21. M.D. Baquedano-Aguilar, D.G. Colomé, E. Agüero, M.G. Molina, Impact of increased penetration of large-scale PV generation on short-term stability of power systems, in *Proceedings of the 2016 IEEE 36th Central American and Panama Convention* (San Jose, Costa Rica, 2016)
22. N. Phuangpornpitak, S. Tia, Opportunities and challenges of integrating renewable energy in smart grid system. Energy Procedia **34**, 282–290 (2013)
23. R. Shah, N. Mithulananthan, R.C. Bansal, V.K. Ramachandaramurthy, A review of key power system stability challenges for large-scale PV integration. Renew. Sustain. Energy Rev. **41**, 1423–1436 (2015)
24. M.J. Alam, K.M. Muttaqi, D. Sutanto, A novel approach for ramp-rate control of solar PV using energy storage to mitigate output fluctuations caused by cloud passing. IEEE Trans. Energy Convers. **29**(2), 507–518 (2014)
25. R. Reinhardt, F. Pautzke, M. Schröter, M.A. Wiemers, Case study of sustainable manufacturing strategy: comparative LCA of wheel hub engine for solar car application, in *Proceedings of the 2017 International Conference on Research and Education in Mechatronics* (Wolfenbuttel, Germany, 2017)
26. B. Huang, J. Zhao, J. Chai, B. Xue, F. Zhao, X. Wang, Environmental influence assessment of China's Multi-Crystalline Silicon (Multi-Si) photovoltaic modules considering recycling process. Sol. Energy **143**, 132–141 (2017)
27. T. Tsoutsos, N. Frantzeskaki, V. Gekas, Environmental impacts from the solar energy technologies. Energy Policy **33**(3), 289–296 (2005)
28. A. Mohsenzadeh, S. Pazouki, S. Ardalan, M.R. Haghifam, Optimal placing and sizing of parking lots including different levels of charging stations in electric distribution networks. Int. J. Ambient Energy **39**(7), 743–750 (2018)

29. C. Cao, L. Wang, B. Chen, Mitigation of the impact of high plug-in electric vehicle penetration on residential distribution grid using smart charging strategies. Energies **9**(12), 1024 (2016)
30. E. Alghsoon, A. Harb, M. Hamdan, Power quality and stability impacts of vehicle to grid (V2G) connection, in *Proceedings of the 8th International Renewable Energy Congress* (Amman, Jordan, 2017)
31. G.H. Reddy, A.K. Goswami, N.B.D. Choudhury, Impact of plug-in electric vehicles and distributed generation on reliability of distribution systems. Eng. Sci. Technol. Int. J. **21**(1), 50–59 (2018)
32. M. Akcin, A. Kaygusuz, A. Karabiber, S. Alagoz, B.B. Alagoz, C. Keles, Opportunities for energy efficiency in smart cities, in *Proceedings of the 2016 4th International Istanbul Smart Grid Congress and Fair* (Istanbul, Turkey, 2016)
33. K. Maslo, Impact of photovoltaics on frequency stability of power system during solar eclipse. IEEE Trans. Power Syst. **31**(5), 3648–3655 (2016)
34. Z. Liu, Q. Wu, S. Huang, H. Zhao, Transactive energy: a review of state of the art and implementation, in *Proceedings of the 2017 IEEE Manchester PowerTech* (Manchester, UK, 2017)
35. D.S. Terzi, B. Arslan, S. Sagiroglu, Smart grid security evaluation with a big data use case, in *Proceedings of the 2018 IEEE 12th International Conference on Compatibility, Power Electronics and Power Engineering* (Doha, Qatar, 2018)
36. S.V. Nandury, B.A. Begum, Big data for smart grid operation in smart cities, in *Proceedings of the 2017 International Conference on Wireless Communications, Signal Processing and Networking* (Chennai, India, 2017)
37. W.-L. Chin, W. Li, H.-H. Chen, Energy big data security threats in IoT-based smart grid communications. IEEE Commun. Mag. **55**(10), 70–75 (2017)
38. A.W. Miranda, S. Goldsmith, Cyber-physical risk management for PV photovoltaic plants, in *Proceedings of the 2017 International Carnahan Conference on Security Technology* (Madrid, Spain, 2017)
39. X. Liu, M. Shahidehpour, Y. Cao, L. Wu, W. Wei, X. Liu, Microgrid risk analysis considering the impact of cyber attacks on solar PV and ESS control systems. IEEE Trans Smart Grid **8**(3), 1330–1339 (2017)
40. X. Zhong, I. Jayawardene, G.K. Venayagamoorthy, R. Brooks, Denial of service attack on tie-line bias control in a power system with PV plant. IEEE Trans. Emerg Topics Comput Intell **1**(5), 375–390 (2017)
41. K. Hao, S. Achanta, B. Rowland, A. Kivi, Mitigating the impacts of photovoltaics on the power system, in *Proceedings of the 2016 Saudi Arabia Smart Grid* (Jeddah, Saudi Arabia, 2016)
42. M.M. Aman, G. Bin Jasmon, H. Bin Mokhlis, Q.A. Khan, A.H.B.A. Bakar, M. Karimi, Modeling and simulation of digital frequency relay for generator protection, in *Proceedings of the 2012 IEEE International Conference on Power and Energy* (Kota Kinabalu, Malaysia, 2012)
43. R. Bründlinger, Advanced smart inverter and DER functions requirements in latest european grid codes and future trends, in *Proceedings of the Solar Canada 2015 Conference* (Toronto, ON, Canada, 2015)
44. R. Mack, M. Sakib, S. Succar, Impacts of substation transformer backfeed at high PV penetrations, in *Proceedings of the 2017 IEEE Power and Energy Society General Meeting* (Chicago, IL, USA, 2017)
45. A. Moghadasi, A. Sargolzaei, A. Khalilnejad, M. Moghaddami, A. Sarwat, Model predictive power control approach for three-phase single-stage grid-tied PV module-integrated converter, in *Proceedings of the 2016 IEEE Industry Applications Society Annual Meeting* (Portland, OR, USA, 2016)
46. S.D. Patil, S.G. Kadwane, S.P. Gawande, Current control of grid tied inverter through SHEPWM method. Energy Procedia **117**, 643–650 (2017)
47. H. Li, C. Wen, K.H. Chao, L.L. Li, Research on inverter integrated reactive power control strategy in the grid-connected PV systems. Energies **10**(7), 912 (2017)

48. H. Jafarian, N. Kim, B. Parkhideh, J. Enslin, Analysis of smart inverter functions of decentralized grid-connected AC-stacked PV inverter architecture, in *Proceedings of the IEEE Applied Power Electronics Conference and Exposition* (Tampa, FL, USA, 2017)
49. G. Todeschini, Control and derating of a PV inverter for harmonic compensation in a smart distribution system, in *Proceedings of the 2017 IEEE Power and Energy Society General Meeting* (Chicago, IL, USA, 2017)
50. B. Wang, B. Zhang, Z. Hao, Control of composite energy storage system in wind and PV hybrid microgrid, in *Proceedings of the 2013 IEEE International Conference of IEEE Region 10* (Xi'an, China, 2013)
51. M.C. Kocer, C. Cengiz, M. Gezer, D. Gunes, M.A. Cinar, B. Alboyaci, A. Onen, Assessment of battery storage technologies for a Turkish power network. Sustainability **11**(13), 3669 (2019)
52. A. Al-Hafri, H. Ali, A. Ghias, Q. Nasir, Transformer-less based solid state transformer for intelligent power management, in *Proceedings of the 2016 5th International Conference on Electronic Devices, Systems and Applications* (Ras Al Khaimah, UAE, 2016)
53. F. Vaca-Urbano, M.S. Alvarez-Alvarado, Power quality with solid state transformer integrated smart-grids, in *Proceedings of the 2017 IEEE PES Innovative Smart Grid Technologies Conference* (Quito, Ecuador, 2017)
54. J. Abdella, K. Shuaib, Peer to peer distributed energy trading in smart grids: a survey. Energies **11**(6), 1560 (2018)
55. J.S. Hong, M. Kim, Game-theory-based approach for energy routing in a smart grid network. J. Comput. Netw. Commun. **2016**, 1–8 (2016)
56. Y. Kado, D. Shichijo, K. Wada, K. Iwatsuki, Multiport power router and its impact on future smart grids. Radio Sci. **51**(7), 1234–1246 (2016)
57. K. Hayashi, R. Kato, R. Torii, H. Taoka, R. Abe, Bi-directional power flow through a digital grid router. J. Int. Council Electr. Eng. **5**(1), 42–46 (2015)
58. Y. Liu, Y. Fang, J. Li, Interconnecting microgrids via the energy router with smart energy management. Energies **10**(9), 1297 (2017)
59. J. Wang, Q. Ou, H. Shen, An overview of smart grid routing algorithms. AIP Conf. Proc. **1864**(1), 020098-1–020098-6 (2017)
60. N. Goel, M. Agarwal, Smart grid networks: a state of the art review, in *Proceedings of the 2015 International Conference on Signal Processing and Communication* (Noida, India, 2015)
61. M.S. Hossan, B. Chowdhury, J. Schoene, S. Bahramirad, Advanced distribution management system: implementation, assessment, and challenges, in *Proceedings of the 2018 IEEE Power and Energy Society General Meeting* (Portland, OR, USA, 2018)
62. E. Boardman, Advanced applications in an advanced distribution management system: essentials for implementation and integration. IEEE Power Energ. Mag. **18**(1), 43–54 (2020)
63. S. Voima, H. Laaksonen, K. Kauhaniemi, Adaptive protection scheme for smart grids, in *Proceedings of the 12th IET International Conference on Developments in Power System Protection* (Copenhagen, Denmark, 2014)
64. M.M. Eissa, M.H.A. Awadalla, Centralized protection scheme for smart grid integrated with multiple renewable resources using internet of energy. Global Trans. **1**, 50–60 (2019)
65. V.A. Papaspiliotopoulos, G.N. Korres, N.D. Hatziargyriou, Protection coordination in modern distribution grids integrating optimization techniques with adaptive relay setting, in *Proceedings of the 2015 IEEE Eindhoven PowerTech* (Eindhoven, Netherlands, 2015)
66. Y. Ates, A. Boynuegri, M. Uzunoglu, A. Nadar, R. Yumurtacı, O. Erdinc, N. Paterakis, J. Catalão, Adaptive protection scheme for a distribution system considering grid-connected and islanded modes of operation. Energies **9**(5), 378 (2016)
67. A. Hussain, H.M. Kim, A hybrid framework for adaptive protection of microgrids based on IEC 61850. Int. J. Smart Home **10**(5), 285–296 (2016)
68. A. Arshad, M. Lehtonen, Multi-agent system based distributed voltage control in medium voltage distribution systems, in*Proceedings of the 2016 17th International Scientific Conference on Electric Power Engineering* (Prague, Czech Republic, 2016)

69. G.H. Merabet, M. Essaaidi, H. Talei, M.R. Abid, N. Khalil, M. Madkour, D. Benhaddou, Applications of multi-agent systems in smart grids: a survey, in *Proceedings of the 2014 International Conference on Multimedia Computing and Systems* (Marrakech, Morocco, 2014)
70. J. Xie, C.C. Liu, Multi-agent systems and their applications. J. Int. Council Electr. Eng. **7**(1), 188–197 (2017)
71. A.S. Nair, T. Hossen, M. Campion, D.F. Selvaraj, N. Goveas, N. Kaabouch, P. Ranganathan, Multi-agent systems for resource allocation and scheduling in a smart grid. Technol. Econ. Smart Grids Sustain. Energy **3**(1), 1–15 (2018)
72. R.F. Sampaio, L.S. Melo, R.P. Leão, G.C. Barroso, J.R. Bezerra, Automatic restoration system for power distribution networks based on multi-agent systems. IET Gener. Transm. Distrib. **11**(2), 475–484 (2017)
73. R. Sitharthan, M. Geethanjali, T.K.S. Pandy, Adaptive protection scheme for smart microgrid with electronically coupled distributed generations. Alexandria Eng. **55**(3), 2539–2550 (2016)
74. P. Peidaee, A. Kalam, M.H. Moghaddam, Developing a simulation framework for integrating multi-agent protection system into smart grids, in *Proceedings of the 2017 Australasian Universities Power Engineering Conference* (Melbourne, Australia, 2017)
75. V. Sok, T. Tayjasanant, Determination of optimal siting and sizing of energy storage system in PV-connected distribution systems considering minimum energy losses, in *Proceedings of the 2017 14th International Conference on Electrical Engineering/Electronics, Computer, Telecommunications and Information Technology* (Phuket, Thailand, 2017)
76. S.Y. Reddy, D.S. Reddy, G.K. Rao, Optimal siting and sizing of solar power sources in interconnection grid system. Int. J. Appl. Power Eng. **5**(1), 1–13 (2016)
77. Q. Gong, J. Lei, J. Ye, Optimal siting and sizing of distributed generators in distribution systems considering cost of operation risk. Energies **9**(1), 61 (2016)
78. L. Luo, W. Gu, X.P. Zhang, G. Cao, W. Wang, G. Zhu, D. You, Z. Wu, Optimal siting and sizing of distributed generation in distribution systems with PV solar farm utilized as STATCOM (PV-STATCOM). Appl. Energy **210**, 1092–1100 (2018)
79. S. Sukumar, H. Mokhlis, S. Mekhilef, M. Karimi, S. Raza, Ramp-rate control approach based on dynamic smoothing parameter to mitigate solar PV output fluctuations. Int. J. Electr. Power Energy Syst. **96**, 296–305 (2018)
80. H.M. Al-Masri, M. Ehsani, Feasibility investigation of a hybrid on-grid wind photovoltaic retrofitting system. IEEE Trans. Ind. Appl. **52**(3), 1979–1988 (2016)
81. A.A. Menaem, V. Oboskalov, A. Mahnitko, R. Varfolomejeva, Optimal integration of renewable power into distribution network based probabilistic bus voltage-feeder current uncertainty analysis, in *Proceedings of the 2019 IEEE 60th International Scientific Conference on Power and Electrical Engineering of Riga Technical University* (Riga, Latvia, 2019)
82. B. Mangu, S. Akshatha, D. Suryanarayana, B.G. Fernandes, Grid-connected PV-Wind-Battery-based multi-input transformer-coupled bidirectional DC–DC converter for household applications. IEEE Journal of Emerging and Selected Topics in Power Electronics **4**(3), 1086–1095 (2016)
83. U. Akram, M. Khalid, S. Shafiq, Optimal sizing of a Wind/Solar/Battery hybrid grid-connected microgrid system. IET Renew. Power Gener. **12**(1), 72–80 (2018)
84. A. Askarzadeh, Electrical power generation by an optimised autonomous PV/Wind/Tidal/Battery system. IET Renew. Power Gener. **11**(1), 152–164 (2017)
85. H. Wang, J. Huang, Joint investment and operation of microgrid. IEEE Trans. Smart Grid **8**(2), 833–845 (2017)
86. H. Wang, J. Huang, Cooperative planning of renewable generations for interconnected microgrids. IEEE Trans. Smart Grid **7**(5), 2486–2496 (2016)
87. T.T. Nguyen, L. Pham, F. Mohammadi, L. Kien, Optimal scheduling of large-scale wind-hydro-thermal systems with fixed-head short-term model. Appli. Sci. **10**(8), 2964 (2020)
88. A.H. Shojaei, A.A. Ghadimi, M.R. Miveh, F. Mohammadi, F. Jurado, Multi-objective optimal reactive power planning under load demand and wind power generation uncertainties using ε-constraint method. Appl. Sci. **10**(8), 2859 (2020)

89. I. Kim, J.-A. James, J. Crittenden, The impact of microturbines and PV systems of office buildings in energy-efficient, economical, and environmental aspects, in *Proceedings of the 2015 North American Power Symposium* (Charlotte, NC, USA, 2015)
90. L. Ma, N. Liu, J. Zhang, W. Tushar, C. Yuen, Energy management for joint operation of CHP and PV prosumers inside a grid-connected microgrid: a game theoretic approach. IEEE Trans. Industr. Inf. **12**(5), 1930–1942 (2016)
91. N.K. Kandasamy, K.J. Tseng, S. Boon-Hee, Virtual storage capacity using demand response management to overcome intermittency of solar PV generation. IET Renew. Power Gener. **11**(14), 1741–1748 (2017)
92. H. Mortaji, S.H. Ow, M. Moghavvemi, H.A.F. Almurib, Load shedding and smart-direct load control using internet of things in smart grid demand response management. IEEE Trans. Ind. Appl. **53**(6), 5155–5163 (2017)
93. A. Karapetyan, M. Khonji, C.K. Chau, K. Elbassioni, H.H. Zeineldin, Efficient algorithm for scalable event-based demand response management in microgrids. IEEE Transactions on Smart Grid **9**(4), 2714–2725 (2016)
94. B. Sivaneasan, N.K. Kandasamy, M.L. Lim, K.P. Goh, A new demand response algorithm for solar pv intermittency management. Appl. Energy **218**, 36–45 (2018)
95. E. Rakhshani, K. Rouzbehi, A.J. Sánchez, A.C. Tobar, E. Pouresmaeil, Integration of large scale PV-based generation into power systems: a survey. Energies **12**(8), 1425 (2019)
96. A. Sundararajan, A.I. Sarwat, Roadmap to prepare distribution grid-tied photovoltaic site data for performance monitoring, in *Proceedings of the 2017 International Conference on Big Data, IoT and Data Science* (Pune, India, 2017)
97. S.A. El-Seoud, H.F. El-Sofany, M.A.F. Abdelfattah, R. Mohamed, Big Data and cloud computing: trends and challenges. Int. J. Interactive Mobile Technol. **11**(2), 34–52 (2017)
98. C. Yang, Q. Huang, Z. Li, K. Liu, F. Hu, Big Data and cloud computing: innovation opportunities and challenges. Int. J. Digital Earth **10**(1), 13–53 (2017)
99. C. Yang, J. Yao, W. Lou, S. Xie, On demand response management performance optimization for microgrids under imperfect communication constraints. IEEE Internet Things J. **4**(4), 881–893 (2017)
100. Y. Zhou, D. Zhang, N. Xiong, Post-cloud computing paradigms: a survey and comparison. Tsinghua Science Technol. **22**(6), 714–732 (2017)
101. N. Mohamed, J. Al-Jaroodi, I. Jawhar, S. Lazarova-Molnar, S. Mahmoud, SmartCityWare: a service-oriented middleware for cloud and fog enabled smart city services. IEEE Access **5**, 17576–17588 (2017)
102. A. Alrawais, A. Alhothaily, C. Hu, X. Cheng, Fog computing for the internet of things: security and privacy issues. IEEE Internet Comput. **21**(2), 34–42 (2017)
103. L. Wei, A. Sundararajan, A.I. Sarwat, S. Biswas, E. Ibrahim, A distributed intelligent framework for electricity theft detection using Benford's Law and Stackelberg Game, in *Proceedings of the 2017 Resilience Week* (Wilmington, DE, USA, 2017)
104. X. Lu, H. Shi, R. Biswas, M.H. Javed, D.K. Panda, DLoBD: a comprehensive study of deep learning over big data stacks on HPC clusters. IEEE Transactions on Multi-Scale Computing Systems **4**(4), 635–648 (2018)
105. M. Mohammadi, A. Al-Fuqaha, S. Sorour, M. Guizani, Deep learning for IoT big data and streaming analytics: a survey. IEEE Communications Surveys and Tutorials **20**(4), 2923–2960 (2018)
106. H. Yan, J. Wan, C. Zhang, S. Tang, Q. Hua, Z. Wang, Industrial big data analytics for prediction of remaining useful life based on deep learning. IEEE Access **6**, 17190–17197 (2018)
107. D. Puthal, X. Wu, S. Nepal, R. Ranjan, J. Chen, SEEN: a selective encryption method to ensure confidentiality for big sensing data streams. IEEE Trans. Big Data **5**(3), 392–397 (2018)
108. F. Mohammadi, G.-A. Nazri, M. Saif, A real-time cloud-based intelligent car parking system for smart cities, in *Proceedings of the 2nd IEEE International Conference on Information Communication and Signal Processing* (Weihai, China, 2019)
109. A. Youssef, M. El-Telbany, A. Zekry, The role of artificial intelligence in photo-voltaic systems design and control: a review. Renew. Sustain. Energy Rev. **78**, 72–79 (2017)

110. K.R. Kumar, M.S. Kalavathi, Artificial intelligence based forecast models for predicting solar power generation. Mater. Today Proc. **5**(1), 796–802 (2018)
111. B.G. Burduhos, M. Neagoe, A. Duta, Adaptive stepwise orientation algorithm for non-concentrated dual-axis solar tracking systems, in *Proceedings of 14th World Congress in Mechanism and Machine Science* (Taipei, Taiwan, 2015)
112. B.G. Burduhos, C. Toma, M. Neagoe, M. Moldovan, Pseudo-equatorial tracking optimization for small photovoltaic platforms from Toronto/Canada. Environ. Eng. Manage. J. **10**(8), 1059–1068 (2011)
113. F. Mohammadi, G.-A. Nazri, M. Saif, A fast fault detection and identification approach in power distribution systems, in *Proceedings of the 2019 International Conference on Power Generation Systems and Renewable Energy Technologies, Istanbul* (Turkey, 2019)
114. M. Kermadi, E.M. Berkouk, Artificial intelligence-based maximum power point tracking controllers for photovoltaic systems: Comparative study. Renew. Sustain. Energy Rev. **69**, 369–386 (2017)

Part IV
Sustainable Energy Mixes in Communities

Chapter 12
Decarbonisation of Heating—Towards 2050

Nevem Duić and Goran Krajačić

Having in mind that European Union (EU) has committed to Paris agreement goal of keeping temperature at less than 1.5 °C above preindustrial level, has made decision to reach climate neutrality reached by 2050, and have greenhouse gas (GHG) emissions reduced by 50–55% by 2030, the issue of avoiding stranded costs should be immediately considered. Since natural gas boilers have life of 20+ years, and infrastructure even more, it is time to start preparing the exit strategy, in order to avoid stranded costs being loaded on consumers and market players.

There are three possible pathways for decarbonisation of heating, decarbonisation of gas, replacing gas by decarbonised district heating and replacing gas by decarbonised electricity but the choice is dependent on general energy transition. As it is obvious up to now, primary energy will be mostly based on solar and wind energy, which are cheap and available nearly everywhere. While other renewables will help, they are limited either geographically (hydro, geothermal), or by sustainable potential (biomass), and they are much pricier.

The problem with wind and solar is that they are variable, which means that in some time periods there may be excess of them, and in some lack. In tropical areas this will probably be solved by batteries, due to daily variation of solar energy, but in moderate climates, with it seasonal variations, batteries can only play a marginal role of daily balancing and those batteries could mainly be electric vehicle batteries, since vehicles are parked 96% of time, and they offer good balancing buffer.

When variable renewable energy sources (VRES) will be abundant, electricity price will fall, due to their low variable cost and thus low ask price, resulting in low marginal price and thus low market price. Such low electricity market price will help development of demand response technologies, and the biggest will be power-to-heat and smart charging, since those two demands are huge, each representing

N. Duić (✉) · G. Krajačić
Faculty of Mechanical Engineering and Naval Architecture, University of Zagreb, Ivana Lučića 5, 10000 Zagreb, Croatia
e-mail: neven.duic@fsb.hr

I. Visa and A. Duta (eds.), *Solar Energy Conversion in Communities*,
Springer Proceedings in Energy, https://doi.org/10.1007/978-3-030-55757-7_12

around third of European final energy demand. Competing with those two demand responses will be hydrogen and energy storage.

Thus, decarbonisation of heating by electrification with renewables is one obvious pathway which will be helped by increasing penetration of VRES. The competing option is to decarbonise gas. Currently we use natural gas which is fossil fuel, and which should obviously be soon abandoned. It could be replaced by biomethane (from biogas) but that option is limited to maybe 10% of current natural gas used, if we want to use biomethane sustainably. It could also be replaced by renewable methane produced using renewable hydrogen for hydrogenation of renewable carbon dioxide, either from biomass combustion or from direct air carbon capture, but this is a process with low efficiency and high price, so it will be used only where no other energy carrier is available, mainly in transport and high temperature processes in industry. Electrification will thus be more viable option for decarbonisation of heating than decarbonisation of gas.

The problem is that heating demand peak is 5–20 times higher than electricity demand peak, which would make electrification possible only with enough heat storage, around 4 m^3 per dwelling. That may be acceptable in a rural house, but in urban apartments may be taking too much space. The solution is heat networks, with big storage at the heating plants site.

Chapter 13
Target for 100% Renewable Energy Systems Use in Cyprus for Electricity Production

Rafaela A. Agathokleous and Soteris A. Kalogirou

Abstract Cyprus covers 93% of the energy demand using fossil fuels which are fully covered from imports. This does not only stand as a barrier to the various European Directives and measures which apply for all EU members, but it is also a drawback for the economy of Cyprus. Around 1000 tons of heavy fuel oil is required every year by the Electricity Authority of Cyprus to cover the energy demand. The household electricity prices including taxes paid in Cyprus are the 7th highest in Europe. From 2009, renewable energy systems have started to be used, but due to the price of the systems they were limited. However, price was not the main problem of the systems to explain why they didn't expand more. Many other parameters are still affecting the use of RES in Cyprus; the public acceptance, the education of people, the environmental consciousness, but the most important is the motivation and ease of the applications and licensing from the government parties. Here in this study, an overview of the current energy situation is discussed and suggestions to increase the use of RES are mentioned as well. These include the use of electricity production RES for the domestic needs and measures for the improvement of energy efficiency of the building sector in general, which is the second biggest energy consumer in Cyprus after the commercial sector.

Keywords Renewable energy systems · 100% RES · Cyprus electricity production · Energy efficiency

R. A. Agathokleous (✉) · S. A. Kalogirou
Cyprus University of Technology, Limassol, Cyprus
e-mail: rafaela.agathokleous@cut.ac.cy

S. A. Kalogirou
e-mail: soteris.kalogirou@cut.ac.cy

I. Visa and A. Duta (eds.), *Solar Energy Conversion in Communities*,
Springer Proceedings in Energy, https://doi.org/10.1007/978-3-030-55757-7_13

13.1 Introduction

Climate change is impacting every continent on Earth at increasing intervals. There is an urgent need for the global community to collectively pursue a pathway towards net zero greenhouse gas (GHG) emissions by launching a rapid transition of the energy sector.

The 2012 Energy Efficiency Directive (2012/27/EU) [1] established a set of binding measures to help the EU reach its 20% energy efficiency target by 2020. Under the Directive, all EU countries are required to use energy more efficiently at all stages of the energy chain, including energy generation, transmission, distribution and end-use consumption. The 2030 Climate and Energy Framework includes EU wide targets and policy objectives for the period from 2021 to 2030 [2]. The key targets for 2030 are; at least 40% cuts in greenhouse gas emissions (from 1990 levels), at least 32% share for renewable energy and at least 32.5% improvement in energy efficiency.

During the last years, many countries have set 100% renewable energy production targets. Hansen et al. [3] presented the status and perspectives on 100% renewable energy systems (RES). The same author in [4] studied the Germany's energy system transition towards 100% renewable energy.

According to Climate Council [5], Germany is a world leader in renewable energy and in the first half of 2018 it produced enough electricity to power every household in the country for a year. The country has also set an ambitious target to get 65% of their electricity from renewables by 2030. Thanks to its unique geography and commitment to the environment, small but mighty Costa Rica has produced 95% of its electricity from hydro, geothermal, solar and wind over the past four years. Moreover, according to Climate Council [5], Costa Rica aims to be entirely carbon-neutral by 2021. Uruguay is now almost 100% powered by renewables almost after less than 10 years of concerted effort. Denmark gets over half of its electricity from wind and solar power and in 2017, 43% of its electricity consumption was from wind—a new world record. That's the highest percentage of wind power ever achieved worldwide. The country aims to be 100% fossil-fuel-free by 2050 [5].

Connolly et al. [6] presented the first step towards a 100% renewable energy-system for Ireland. Child et al. [7] studied the role of storage technologies for the transition to a 100% renewable energy system in Europe. A year later Child et al. [8] studied the transition pathways towards a 100% renewable energy power sector of Europe by 2050. It was concluded that a 100% RE energy system for Europe is economically competitive, technologically feasible, and consistent with targets of the Paris Agreement.

Chen et al. [9] examined the pathway for small islands to replace fossil fuels with RES up to 100%, to economically achieve energy security and limit temperature rise as close as possible to 1.5 °C. Another study for 100% renewable island is carried out by Meza et al. [10]. Alves et al. [11] mentions that the integration of RES in islands is crucial to improve their economy allowing them to be energy independent.

However, RES integration in islands presents both opportunities and constraints. Islands are usually very dependent on imported, expensive, fossil fuels to satisfy all their energy needs [12]. Thus, the island's inhabitants depend on outside nations to guarantee their energy security. On the other hand, RES integration in isolated energy systems brings forward some difficulties related to most RES intermittent nature. Grid stability problems and mismatch between demand and supply become more crucial as RES integration share increases.

The aim of this study is the analysis of the current energy situation of Cyprus in order to assess the 100% RES scenario for Cyprus Island. Cyprus has an isolated energy system and has no primary sources of energy. The electricity production of the island depends entirely on imported fuel, mainly heavy fuel oil. In the first part of this study, the current energy situation of Cyprus is assessed, and then the energy consumption through the years is presented. Then Cyprus' renewable energy resources assessment is presented which is then followed by the actions taken for the energy efficiency of the buildings sector in Cyprus, to reach the 100% RES target. At the end, the barriers to this target are discussed and final conclusions were given.

13.2 Current Energy Situation in Cyprus

Since Cyprus has no primary energy sources, the electricity production from the Electricity Authority of Cyprus (EAC) entirely depends on the imported fuel, mainly heavy fuel oil. Currently, the EAC owns and operates three Power Stations with a total installed capacity of 1478 MW as shown in Fig. 13.1 and the net electricity production in Cyprus is about 5000 million kWh every year as shown in Fig. 13.2.

Vasilikos Power Station is the biggest infrastructure project ever carried out in Cyprus with an installed capacity of 868 MW. It is an advanced technology power station consisting of three conventional generation units of a generating capacity of

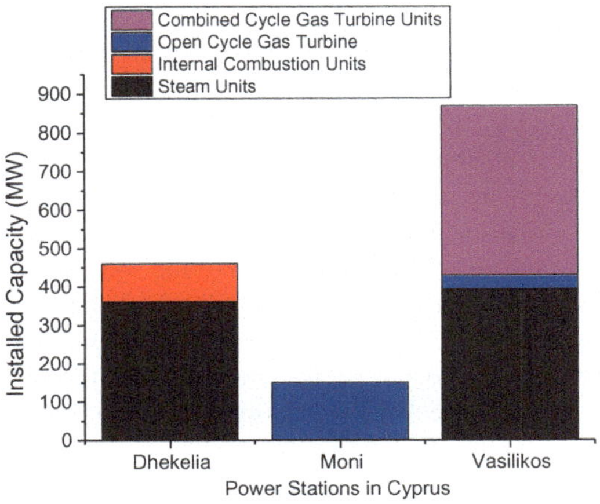

Fig. 13.1 Installed power capacity of the three power stations in Cyprus, from the Electricity Authority of Cyprus

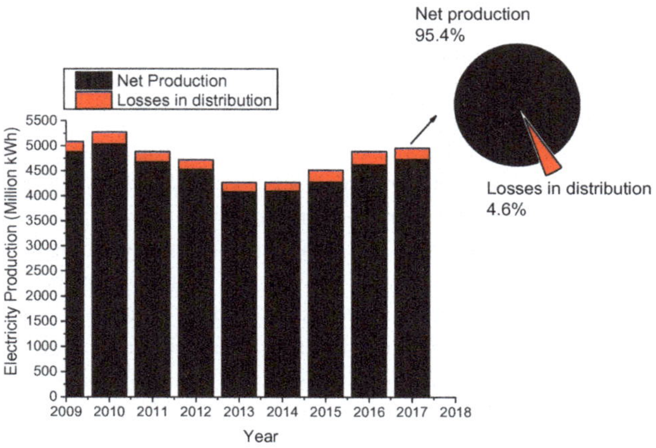

Fig. 13.2 Gross electricity production in Cyprus from 2009 to 2017 including losses in transport and distribution

130 MW using heavy fuel oil, one 38 MW Gas Turbine fueled by diesel oil, and two combined cycle gas turbine units of 220 MW capacity each fueled by diesel oil.

Dhekelia power station uses heavy fuel oil, while Moni and Vasilikos power stations use mainly gasoil. Figure 13.3a shows sales of petroleum products in Cyprus and Fig. 13.3b the heavy fuel oil breakdown showing the large amount of heavy fuel oil being used by the EAC over the years. However, a rapid decrease has been observed after 2010 which can be explained by the fact that renewable energy systems (RES) have started to be used in Cyprus after 2010. However, this could also be explained with the economic crisis during this period.

13.3 Renewable Energy Systems in Cyprus

The binding national target of Cyprus regarding the share of renewable energy in the final energy consumption by 2020 was 13% [13]. There is also another binding target by 2020 that of the use of biofuels in the transport sector by 10%. According to the National RES Action Plan, the overall 13% target can be achieved by the three sectoral targets which are: 23.5% share of RES for heating and cooling to the final energy consumption, 16% RES electricity and 4.9% the use of biofuels for transport. The contribution of the solar thermal systems (solar hot water heating systems) contribute the most to the total share of RES in the Cyprus Energy Balance by 46.84%, followed by biomass utilization for heating 21.33% (mainly wood stoves and fireplaces), electricity generation from wind farms 13.22%, electricity generation from PV systems 8.61%, biofuels due to import obligation 6%, electricity generation from biogas 3% and geothermal heat pumps 1% [13]. However, there is still large

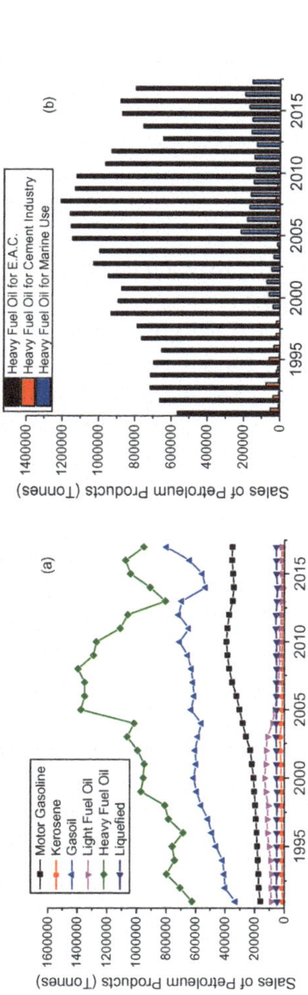

Fig. 13.3 **a** Fuel sales of petroleum products, **b** Heavy fuel oil breakdown including the Electricity Authority of Cyprus usage for electricity production

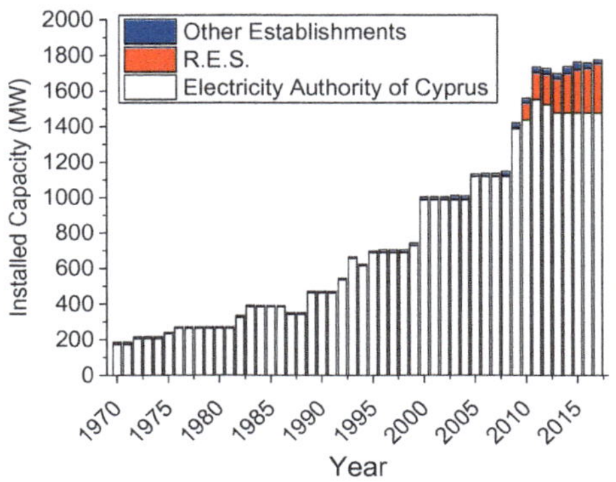

Fig. 13.4 Installed capacity of electric generators in Cyprus from 1970 to 2017

space of improvement in Cyprus in order to reach targets of 2030 for GHG emissions reduction and increase the share of RES.

As previously mentioned, since 2010 the government started motivating people to shift to renewables giving licenses for solar parks, wind parks and building PV installations. As can be seen in Fig. 13.4, the major installed capacity of electric generators is for the EAC while the appearance of RES is obvious from 2010 and later. Regarding the RES usage after 2010 it is important to see that there is an increase in the use of RES every year. However, as Fig. 13.4 shows, most of the electricity gross production comes from the EAC from fossil fuels burning.

Table 13.1 shows the number of RES in Cyprus and their nominal capacity by type of system. It is obvious that the majority of the installed capacity corresponds to wind parks and then to PV parks while roof PV systems and biomass systems are holding a lower percentage although the number of units installed are the bigger. There are currently 1868 PV parks in Cyprus with installed capacity of 86 MW and 12,053 roof PV systems with 47 MW installed capacity. Although there are more PV systems installed than other systems, the highest capacity is hold by the wind systems. There are 6 wind parks in operation with total installed capacity of 157.5 MWe and licenses for the construction of a further 18 MWe capacity wind parks have been given. In Cyprus, there are 14 biomass units in operation with total installed capacity of 9.7 MWe.

The total electricity production by RES in Cyprus is shown in Fig. 13.5. As can be

Table 13.1 Installed R.E.S in Cyprus

Type	Number of systems	MW
PV parks	1868	86
Roof PVs	12,053	47
Wind parks	6	157.5
Biomass systems	14	9.7
Total		*300.2*

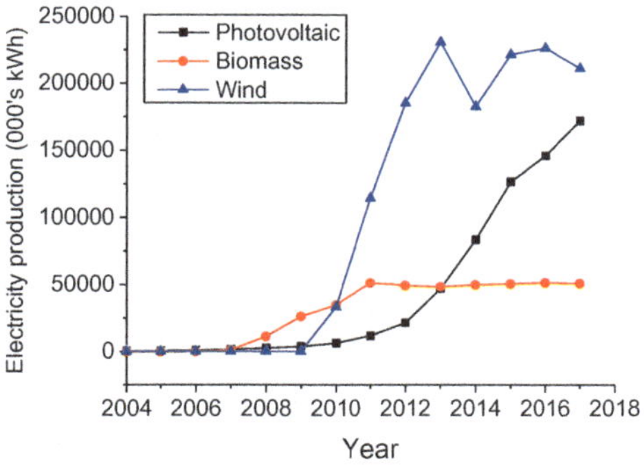

Fig. 13.5 Production of electricity from RES from 2004 to 2017

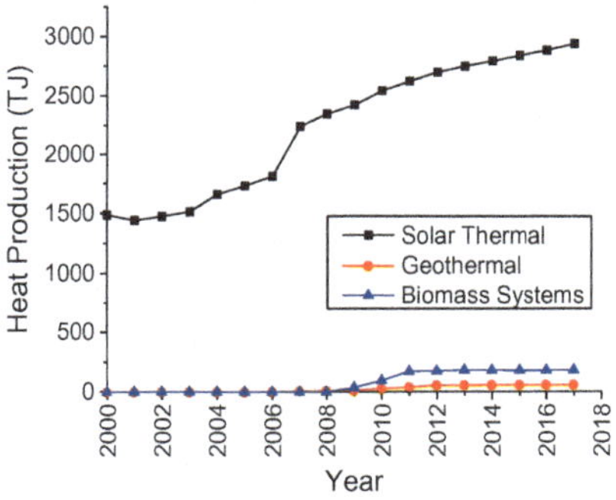

Fig. 13.6 Production of heat from R.E.S in Cyprus from 2000 to 2017

seen, in relation to Table 13.1, the biggest production comes from the wind systems although there are more units of solar PV systems installed.

Besides the electricity production, the amount of thermal energy is also very important. In this type of energy it is known that the use of solar thermal collectors for the production of domestic hot water is widely spread. Figure 13.6 shows the thermal energy production from RES in Cyprus from 2000 to 2017. According to REN21 report [14], Cyprus remained the world solar heating leader on a per capita basis at the end of 2012, with 575 kWt_h per 1000 inhabitants, followed by Israel with 394 kW_{th}.

13.4 Electricity Consumption in Cyprus

In order to set targets to 100% RES energy production in Cyprus, apart from the current energy production situation, the energy consumption should be analyzed. Consequently, Fig. 13.7 shows the electricity consumption by sector from 1960 to 2017. It can be clearly seen that the commercial sector was and still is the biggest consumer, followed by the domestic and industrial sector.

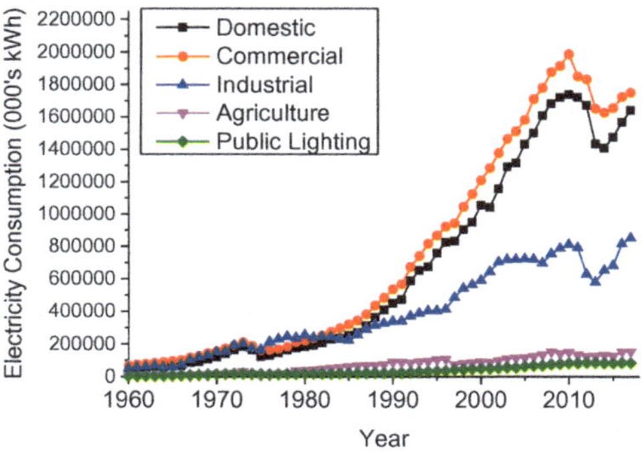

Fig. 13.7 Electricity consumption by sector from 1960 to 2017

According to data from the Statistical Service of the Ministry of Finance the household sectors accounts for 19% of the final energy consumption and another 13% to commerce, hotels and services, i.e., mainly office buildings. The share in final energy demand of the housing sector has been increasing rapidly over the last 10–16 years: From 14.1% in 1995 to 17% in 2000 up to 18% in 2013 [15]. Electricity consumption has gone up by 150% in the same period mainly due to the installation of air conditioners and increasing number of home electric appliances [15]. A deeper examination to the energy consumers extracted from the electricity balance of the island for the year 2017 is shown in Fig. 13.8. Households accounts for 36% of the electricity consumption and requires major attention to reduce their energy consumption.

13.5 Renewable Energy Resource Assessment

Cyprus is the biggest island in the Mediterranean, located at 35° latitude and 33° longitude. Thus Cyprus is located in a hot spot and is projected to face significant temperature rise and rainfall level drop the next years. As a result, important negative effects on climate change should be expected in the next decades in various sectors. Studies assessed the impact of climate change on electricity use based on the assumption that average temperature in the Eastern Mediterranean is expected to rise by about 1 °C by the year 2030 [16].

The climatic conditions of Cyprus are predominantly very sunny with daily average solar radiation of about 5.2 kWh/m^2 on a horizontal surface. The mean daily global solar radiation varies from about 2.3 kWh/m^2 in the cloudiest months of the year December and January, to about 7.2 kWh/m^2 in July. A typical year in Cyprus includes more than 300 sunshine hours. The total annual solar irradiation in horizontal can exceed the 1727 kWh/m^2. The direct normal irradiation map in Cyprus is shown in Fig. 13.9. The island has excellent solar potential throughout the year as the cloudy days do not exceed three continuous days, and the average global radiation exceeds the value of 2000 kWh/m^2. To follow the EU targets of 2020 and

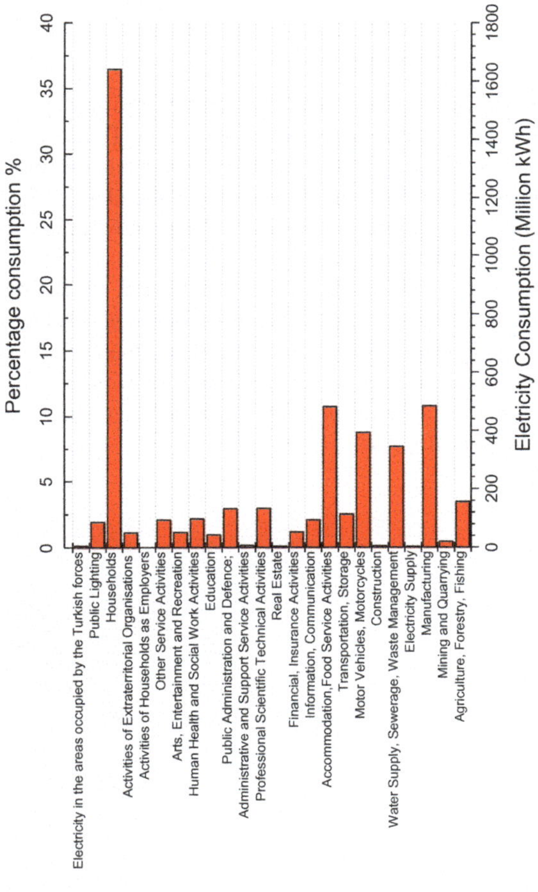

Fig. 13.8 The percentage of energy consumption by sector as shown in the electricity consumption balance for 2017

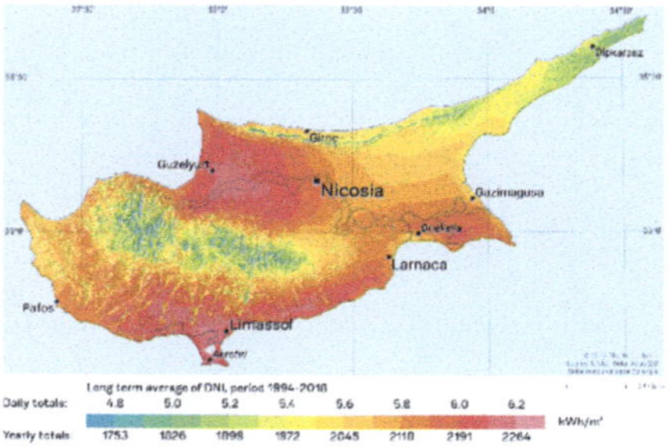

Fig. 13.9 Direct normal irradiation in Cyprus [17]

2030, Cyprus has started to invest in Renewable Energy during the last years, with solar energy being the most valuable source of exploitation. The solar water heating and photovoltaics systems installed in Cyprus have low payback time and cover a significant thermal and electricity demand. Based on the abundance of solar radiation on the island the solar energy potential is very high. The photovoltaic power potential map of the island is shown in Fig. 13.10.

Regarding the other renewable sources, Figs. 13.11 and 13.12 show the mean annual wind speed and the available energy from animal sewage biomass respectively. As can be seen, there is high wind energy potential and biomass energy in most areas of the island. However, regarding the wind speed as shown in Fig. 13.11, most of the area shown with high wind speed is under the Turkish occupation so it cannot be exploited by the government of Cyprus.

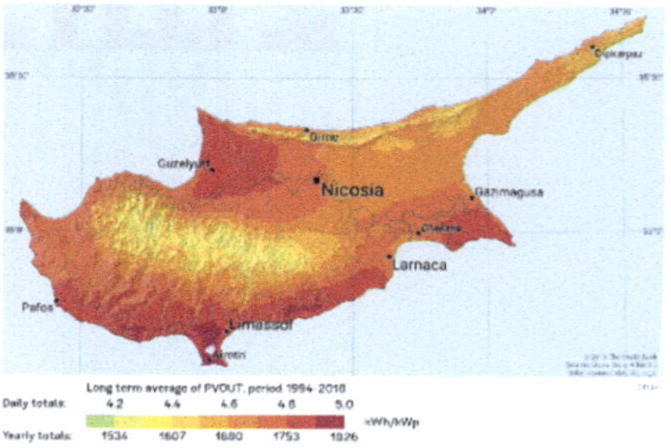

Fig. 13.10 Photovoltaic power potential in Cyprus [17]

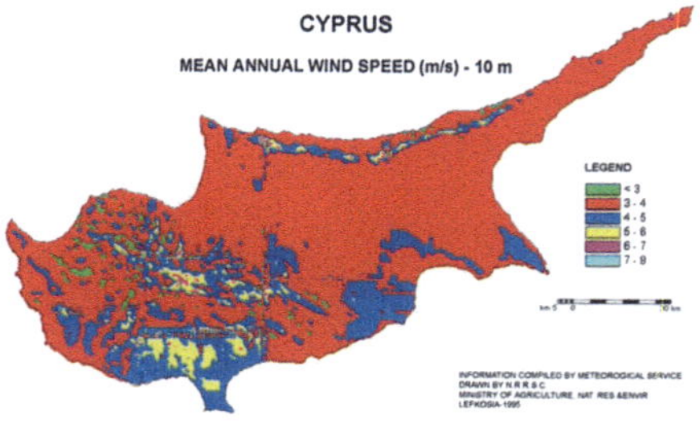

Fig. 13.11 Mean annual wind speed in Cyprus [18]

Fig. 13.12 Available energy from animal sewage biomass in Cyprus [19]

In addition to the renewable energy resource assessment there is another source of energy that can be used in Cyprus. It is believed that waste treatment can support the energy production without too much effort from the people, only attention required is by the government to set measures to proper dispose or use the municipal waste or biodegradable waste to produce heat. Figure 13.13 shows the total waste generated in Cyprus from 2004 to 2017 and per capita waste generation. As can be seen in Fig. 13.14, most of the generated waste is disposed to landfills, very small amount of waste is recycled and less is composted. Although large amount of waste is generated in Cyprus there are still no waste incineration plants or other plants to take advantage of the generated waste to produce energy.

The technical potential for energy efficiency improvements in agriculture seems to be relatively limited e.g. in comparison to the corresponding values in buildings since the available potential for co-generating heat and power from animal waste has largely been exploited in Cyprus, with 14 biomass/biogas plants already operating with a total capacity of around 10 MW biodegradable waste. In Cyprus this predominately consists of the biodegradable fraction of the municipal solid waste, sewage sludge, solid and liquid agricultural residues and solid and liquid wastes from food and drinks industries. Kythreotou [20], estimated that the potential annual electrical energy

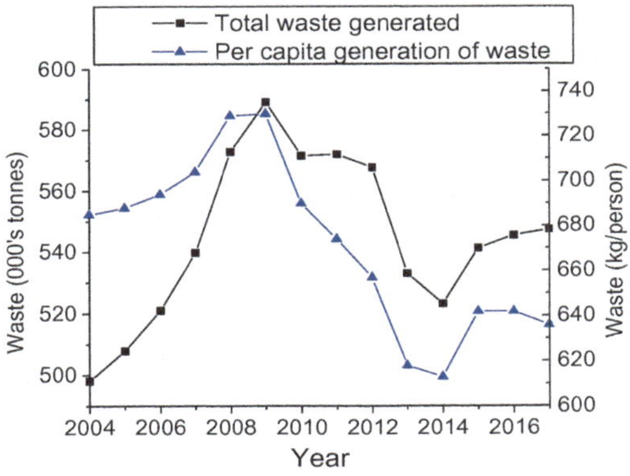

Fig. 13.13 Total waste generated in Cyprus from 2004 to 2017 and per capita waste generation

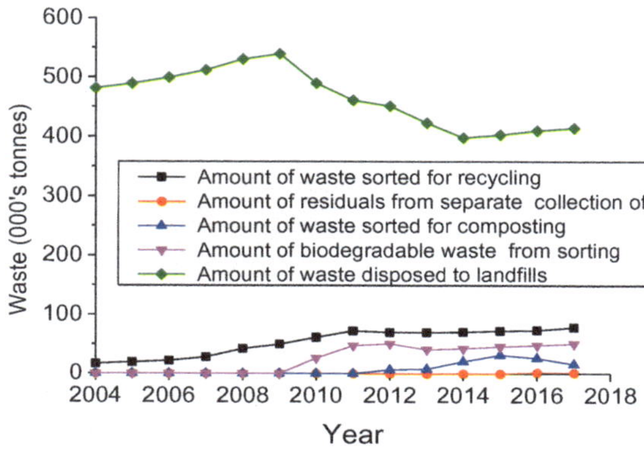

Fig. 13.14 Municipal solid waste in Cyprus from 2004 to 2017 by type

generation in Cyprus is around 1500 GWh and the potential annual thermal energy generation is about 2100 GWh.

13.6 Energy Efficiency in Buildings

As previously mentioned, it is necessary to focus on the most intensive sectors in terms of energy consumption and ease of RES adoption to cover the energy requirements. Thus, since the building's sector is analyzed more. According to Vougiouklakis et al. [21], the building stock in Cyprus comprises of 431.059 residential buildings and 85.198 non-residential buildings. Of the residential buildings almost half are single-family houses and 22% apartments.

The building sector is expected to play a key part in the achievement of the national targets for 2020 covering almost 98% of the energy saving target (BUILD UP Skills Project, 2012). The potential is huge since 91% of all buildings (94% of residential buildings 83% in the service sector) were built before the introduction of mandatory energy performance requirements and 50% do not have any kind of thermal insulation.

Cyprus did not have any mandatory building energy performance requirements until 2007. Since then the requirements have been tightened twice and it is estimated that a new building consumes around 50% less energy than a similar building that was built before the implementation of energy performance requirements [22]. The law on Energy Efficiency of Buildings (No. 210(I)/2012) is the main legal act for transposing the requirements from the EU Performance of Buildings Directive (2010/31/EU) into Cypriot Law. Specific performance requirements have been set by ministerial decree, which have subsequently been tightened in 2009, 2013, and 2016 and are set as shown in Table 13.2 [22]. Directive 2010/31/EU on the energy performance of buildings requires Member States to introduce Energy Performance of Building Certificates

Table 13.2 Evolvements of energy performance requirements in Cyprus

Requirement	Since January 2010	Since December 2013	Since January 2017	Since August 2014 for NZEB
Elergy class according to the EPC	B	B	B	A
U_{value} for walls (W/m^2K)	0.85	0.72	0.4	0.4
U_{value} for floors and roofs (W/m^2K)	0.75	0.63	0.4	0.4
U_{value} for windows (W/m^2K)	3.8	3.23	2.9	2.25

(EPCs) to be issued for buildings at the time of construction, sale or renting or when undergoing major renovation. It has been transposed through a series of laws and regulations with the Law on Energy Efficiency of Buildings (No. 210(I)/2012) as the main legal act.

Requirements and characteristics of nearly-zero energy buildings (NZEB) have as well been set by ministerial decree in 2014 requiring U-values for energy class A, a maximum consumption of primary energy and at least 25% of the demand to be covered by RES. This requirement applies for new and existing renovated buildings.

According to Enerdata [23], energy efficiency in the buildings sector of Cyprus has improved steadily since the adoption of energy performance standards for new buildings in the mid-2000s, and as a result of the implementation of all relevant EU legislation.

Based on the statistical services of Cyprus for the construction and housing statistics [24], a total of 152,487 building permits were given from 2000 to 2016 and a total of 78,551 houses were completed during this period most of them in the district of Limassol. Houses accounts around 70% of total dwellings in the private sector while the rest 30% are apartments.

In 2013 a net metering scheme was introduced for the promotion of small residential photovoltaic systems. This scheme supported the installation of small PV systems with power up to 3 kW in 5000 households. The total allowed installed PV capacity under that scheme was 15 MW. This measure is still due, and upgraded the capacity of the PV systems to 5 kW.

The combination of the NZEB regulations for the energy performance of buildings regarding the U_{value} of the various building elements, and the required use of RES so as each building covers its own energy requirements on site, will help the households sector to minimize its energy consumption from the grid of EAC that comes from the fuels burning and at the same time reduce the carbon footprint of the households sector.

13.7 Barriers to the 100% RES Target and Actions Suggested

The still embryonic state of the energy service market in Cyprus occurs due to the underdeveloped regulatory framework. More emphasis should be put on issues related to standardization of energy services provided, the performance of such services and their procurement and operation in the public sector [21].

The most sever barrier for reaching the planned savings is the limited available budget for such interventions. The private sector has been accustomed to be responsive only when a significant public subsidy is available, while the public sector tends to request full upfront capital coverage. For this reason, the transition to a more market-oriented financial support scheme, will be definitely a challenge and a careful planning along with the mobilization of the appropriate financial and market instruments will be required.

In the various sectors the following actions should be taken:

- In the household/residential sector, any new instrument shall be designed to be cost-attractive as well as implementable in market terms.
- In the service sector emphasis should be given to hospitals and energy efficiency upgrade of street lighting.
- In the agricultural sector the focus should be mainly given to higher penetration of RES systems for heating and cooling, achieving relative significant savings in terms of primary energy use and avoided cost of imported fuel, while also the adoption and support of an energy audit scheme could allow the identification of some significant cost efficient energy saving potential that could be addressed under tailored design national programs either for specific sub-sectors (e.g. wineries) or agricultural process activities (e.g. greenhouses, drying).
- In the transport sector action should focus on modifying the vehicle taxes to accelerate the penetration of higher efficient cars and light commercial vehicles, soft measures to promote a modal shift towards public, e-mobility and other alternative transport modes.

13.8 Conclusions

During this study, the energy situation for the island of Cyprus is analysed as well as the energy consumption showing that the most intensive consumer of energy is the households sector. Accordingly focus should be given in this sector in order to achieve the EU 2030 targets on the first place and then start working on the 100% RES target. Islands with isolated energy systems are facing important difficulties in terms of energy systems. The integration of RES in islands is crucial to improve their economy allowing them to be energy independent.

Although Cyprus has high renewable energy potential in terms of renewable energy resources, the main renewable energy that can be used without limitations is solar energy. From the analysis made on the current status of RES on the island it is observed that in terms of installed units the solar systems are the most, while in terms of installed capacity the biggest producer are the wind energy systems.

Although the solar abundance of the island could provide renewable energy to cover the island's load, there are various barriers to that. These were also discussed and various suggestions were given and national legislative acts that should be taken were suggested. However, even if people accept and use more RES and follow other measures to reach the target, the aim cannot yet be reached because Cyprus has an isolated electricity system. Thus, big changes need to be done by the government and the electricity authority of Cyprus regarding the distribution circuit and consider as well as the most important parameter when considering 100% RES which is the electricity storage.

References

1. European Commission, Energy efficiency directive (2019) [Online]. Available https://ec.europa.eu/energy/topics/energy-efficiency/targets-directive-and-rules/energy-efficiency-directive_en
2. European Commission, 2030 climate & energy framework (2019) [Online]. Available https://ec.europa.eu/clima/policies/strategies/2030_en
3. K. Hansen, C. Breyer, H. Lund, Status and perspectives on 100% renewable energy systems. Energy **175**, 471–480 (2019)
4. K. Hansen, B.V. Mathiesen, I.R. Skov, Full energy system transition towards 100% renewable energy in Germany in 2050. Renew. Sustain. Energy Rev. **102**, 1–13 (2019)
5. Climate council, 11 Countries leading the charge on renewable energy, in *Climate Reality Project* (2019) [Online]. Available https://www.climatecouncil.org.au/11-countries-leading-the-charge-on-renewable-energy/
6. D. Connolly, H. Lund, B.V. Mathiesen, M. Leahy, The first step towards a 100% renewable energy-system for Ireland. Appl. Energy **88**(2), 502–507 (2011)
7. M. Child, D. Bogdanov, C. Breyer, The role of storage technologies for the transition to a 100% renewable energy system in Europe. Energy Procedia **155**, 44–60 (2018)
8. M. Child, C. Kemfert, D. Bogdanov, C. Breyer, Flexible electricity generation, grid exchange and storage for the transition to a 100% renewable energy system in Europe. Renew. Energy **139**, 80–101 (2019)
9. A.A. Chen, A.J. Stephens, R. Koon Koon, M. Ashtine, K.K. Mohammed-Koon, Pathways to climate change mitigation and stable energy by 100% renewable for a small island: Jamaica as an example. Renew. Sustain. Energy Rev. **121**, 109671 (2020)
10. C.G. Meza, C. Zuluaga Rodríguez, C.A. D'Aquino, N.B. Amado, A. Rodrigues, I.L. Sauer, Toward a 100% renewable island: a case study of Ometepe's energy mix". Renew. Energy **132**, 628–648 (2019)
11. M. Alves, R. Segurado, M. Costa, On the road to 100% renewable energy systems in isolated islands. Energy **198**, 117321 (2020)
12. D.F. Cross-Call, in *Matching Energy Storage to Small Island Electricity Systems: A Case Study of the Azores* (2013), pp. 1–141
13. A. Charalambous, in *Cyprus Renewable Energy Targets by 2020: A Nightmare or Not?* (2020)
14. J.L. Sawin et al., Renewables 2012 Global Status Report. REN21 Secretariat, Paris, *Ren21* (2012), pp. 1–172
15. K. Kitsios, M. Kakouris, T. Zachariadis, in *Energy Efficiency Trends and Policies in Cyprus* (2015)

16. T. Zachariadis, Climate change in Cyprus: impacts and adaptation policies. Cyprus Econ. Policy Rev. **6**(1), 21–37 (2012)
17. Solargis: Solar resource maps of Cyprus (2019). Available on https://solargis.com/maps-and-gis-data/download/cyprus
18. WIP GmbH & Co Planungs KG, *Energy Savings Potential in Cyprus*, February (2017)
19. Cyprus Energy Agency, Ανανεώσιμες Πηγες Ενέργειας Στη Κυπρο, *Ενεργειακό Γραφείο Κυπρίων Πολιτών* (2010)
20. N. Kythreotou, S.A. Tassou, G. Florides, An assessment of the biomass potential of Cyprus for energy production. Energy **47**(1), 253–261 (2012)
21. Y. Vougiouklakis, B. Struss, T. Zachariadis, A. Michopoulos, in *An Energy Efficiency Strategy for Cyprus up to 2020, 2030 and 2050* (2017)
22. N. Hadjinicolaou, Implementation of the EPBD in Cyprus, in *Implementing the Energy Performance of Buildings Directive* (EPBD), ed. by E. M. (2015)
23. Enerdata, *Cyprus Profile from Odussee-mure* (2020) [Online]. Available https://www.odyssee-mure.eu/publications/efficiency-trends-policies-profiles/cyprus.html
24. CYSTAT, *Republic of Cyprus Construction and Housing Statistics 2016*, 30 (2016)

Chapter 14
Vertical Axis Wind Turbines: The Behavior of Lift and Drag Airfoils

Bostan Viorel, Bostan Ion, Rabei Ivan, Dulgheru Valeriu, and Ciuperca Rodion

Abstract The blade is the most important element of a wind turbine, as it is the component that largely governs the productivity and it is the main source of efficiency optimization. A blade is defined by the airfoil type. This paper addresses three different airfoil types tested for different wind conditions. The versions considered are: the symmetrical NACA 0018 airfoil, the asymmetrical FX 63-137 airfoil—both of lift type and one drag type of scoop shape. Based on these versions, three sets of blades were obtained and used to form three vertical axis wind turbines which were tested in a wind tunnel. Except for the blade's airfoil, all rotors had the same key parameters which refer to: rotor's swept area, number of blades, height, diameter, aspect ratio, blade length, chord length, solidity and pitch angle. As expected for low wind speeds, the scoop bladed turbine showed significantly higher efficiency than the other two versions. It was also noticed that the performance is not influenced by the curvature position that can be oriented either radially inward or outward in relation to the rotor's axis. For higher wind speeds the FX 63-137 turbine's efficiency increased dramatically while NACA 0018 turbine displayed weak performance in all cases.

Keywords Vertical axis wind turbine · Drag type blades · Lift type blades

14.1 Introduction

The wind turbines cover a significant percentage of the energy needs in many regions of the world and, according to the latest trends, their role is going to be even bigger [1]. This refers especially to the large horizontal axis wind turbines. However recently, vertical axis wind turbines (VAWTs) started to gain more attention from the research centers and companies that look for capitalizing on the advantages they come with,

B. Viorel · B. Ion · R. Ivan · D. Valeriu (✉) · C. Rodion
Technical University of Moldova, Stefan cel Mare Bd. 168, 2004 Chisinau, Republic of Moldova
e-mail: valeriu.dulgheru@bpm.utm.md

B. Viorel
e-mail: viorel.bostan@adm.utm.md

I. Visa and A. Duta (eds.), *Solar Energy Conversion in Communities*,
Springer Proceedings in Energy, https://doi.org/10.1007/978-3-030-55757-7_14

the main ones being the insensitivity to the wind direction (no yaw mechanism needed); placing the heavy components closer to the ground level, etc.

The performance of a wind turbine is almost exclusively dictated by its blades, thus plenty of research is done on this component. The blade is defined by its cross section shape, called airfoil. Depending on the main driving force, there can be two types: lift and drag airfoils. Historically from the lift type category the airfoils of the NACA family were often addressed. Nowadays, many other different airfoils are investigated and adopted [2, 3]. The drag type airfoils are of scoop shape. The blades based on them can be both simpler and cheaper but less efficient for wind energy conversion.

Choosing the right blades for given wind conditions is central. Besides the fact that inappropriate solutions lead to poor efficiency, the environmental impact of the turbine is broader. Life cycle assessment is evaluated for specific wind speeds and a certain productivity of the turbine. A less efficient machine implies a higher environmental footprint [4].

The purpose of this paper is to determine which airfoils are the most appropriate for wind speeds up to 12 m/s. The analysis was experimentally done by testing three small scale vertical axis wind turbines: one having scoop like blades, the second with blades defined by the symmetrical NACA 0018 airfoil and the third involving the asymmetrical FX 63-137 airfoil. The turbines had the same key parameters, thus no other factors except the airfoil induced a difference in performance. It was expected that for low wind speeds the drag type is more productive but the interest was in determining the behavior of the three over a larger wind speed range.

14.2 Methods

14.2.1 The Experimental Setup

The experimental setup consisted of two main components: the wind tunnel and the wind turbines. The tunnel employed in the experiments was Gunt ET 220 with the outer diameter of 600 mm which limits the size of the rotors that can be tested. The maximum wind speeds reached values that varied around 12 m/s.

Three sets of three blades were built: first being defined by the scoop shaped airfoil, the second of NACA 0018 airfoil and the third of FX 63-137 airfoil. The three versions are presented in Fig. 14.1.

The scoop airfoil consisted of a straight line with a semicircle of 24 mm diameter on one end. The airfoils had the chord length c equal to 50 mm.

For the case of asymmetrical airfoils like the scoop or the FX 63-137 ones there can be two orientation modes in relation to the rotor's axis—curvature (camber) inward or outward (Fig. 14.2). The analysis hereby presented focused on both cases.

The blades were developed through 3D printing and taped for increased rigidity. Unlike the NACA 0018 and the FX 63-137 cases, the scoop blades were provided

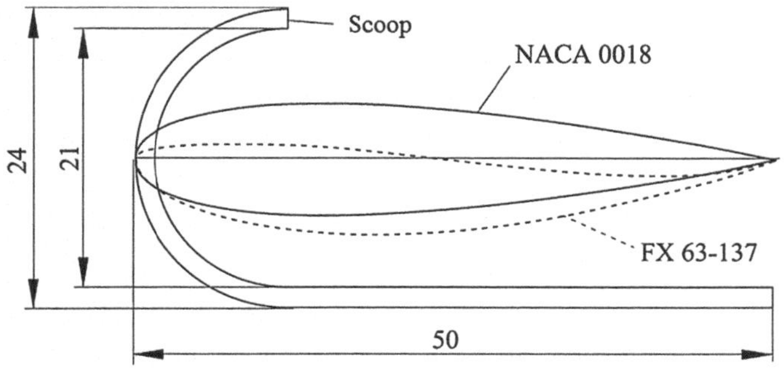

Fig. 14.1 The airfoils of scoop, NACA 0018 and FX 63-137 blades (Dimensions in mm)

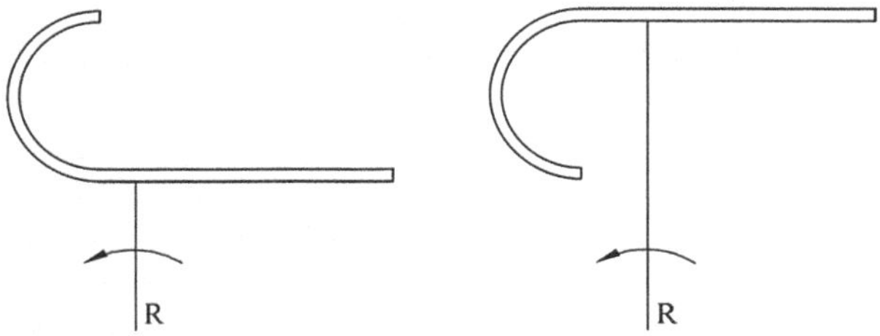

Fig. 14.2 Curvature in and out in relation to the rotor's axis (R—rotor's radius)

with end plates that were larger than the cross section itself by 7 mm (Fig. 14.3). The blades' wall thickness was 1.5 mm.

Fig. 14.3 Printed scoop blades with endplates

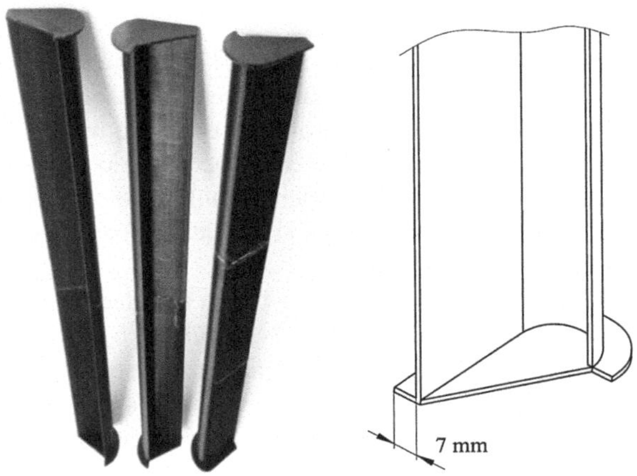

One wind turbine was designed and built up so that the blades could be easily changed. This way the tower, the supporting struts and the generator remained the same while the different sets of blades formed different turbines.

The connection between the turbine's shaft and the generator was mediated by a gear train with the ratio of 1:3.41. The alternating current produced by the three phase generator was transformed into direct current by using a rectifier. The power was then transmitted to a device consisting of NI Elvis II system and the RELab add-on. This module is designed to measure and display the power generated by a small laboratory wind turbine [5]. More tests were done for each wind speed and an average value was calculated. The values obtained for different wind speeds were graphically interpolated and the tendencies were observed. The RELab module and the scoop bladed wind turbine are presented in Fig. 14.4.

14.2.2 Theoretical Considerations

The performance of the VAWT is influenced by its parameters. When optimizing one of them by comparing different variations, the other ones should be kept constant if possible. As for this study the airfoil was optimized while the remaining parameters as the swept area, the rotor's aspect ratio, the chord length, the number of blades, the blade shape (e.g. straight, helical or curved), the pitch angle were kept unchanged in all compared cases. This way, except the airfoil, no other parameter was expected to generate a performance difference between the turbines.

The definitions and effects of the mentioned parameters are expressed below.

The power output P(W) of a wind turbine can be estimated using Eq. (14.1):

$$P = \frac{1}{2} \cdot C_p \cdot \rho \cdot A \cdot U^3 \tag{14.1}$$

where C_p is the power coefficient that is turbine's efficiency index; ρ is the air density (kg/m^3); A is the turbine's swept area (m^2) that is the area crossed by the wind; U is the wind speed (m/s). As of Eq. (14.1), the reasonable way to increase the power is by increasing the swept area or the power coefficient. For a determined value of the area A the turbine's efficiency could be enhanced by optimizing its parameters. The improvement should also consider the prevalent wind conditions under which the turbine will operate.

The solidity σ is the ratio between the blades' area and the turbine's swept area A. The area of a straight blade is calculated by multiplying its chord length c to the height h. For fixed values of A and h, the solidity can be changed with the chord length c, hence the four parameters are strongly related. Generally, high solidities are suggested for turbines that are to operate at low wind speeds and vice versa. The recommended values range between 0.2 and 0.6 [6]. Beside the efficiency, the

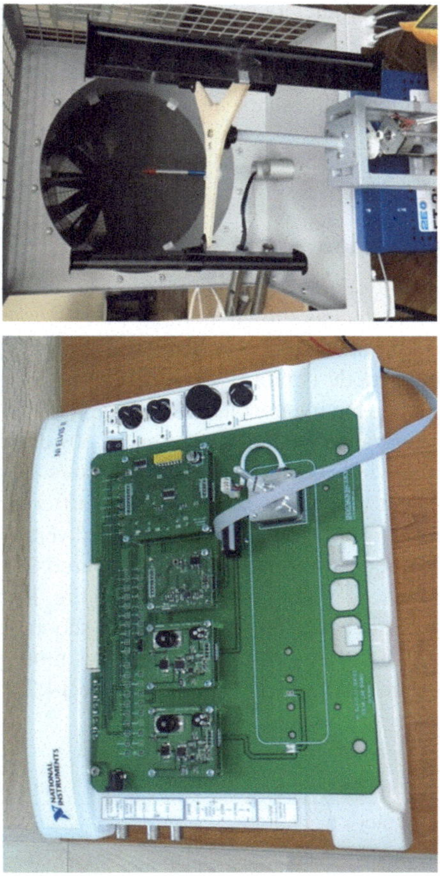

Fig. 14.4 NI Elvis II system with the RELab add-on (left); the scoop bladed wind turbine and the wind tunnel (right)

solidity influences the turbine's tip speed ratio *TSR* (or rapidity λ) which is the ratio between the blade's tip speed and the wind speed.

The angle between the chord line and the tangent line to the blade's circular trajectory is called pitch angle β. It can take negative, null or positive values and can be fixed or variable. Laboratory tests determined that this parameter also influences turbine's productivity thus it was kept constant for all compared cases.

The same swept area can be confined by different values of the diameter D and height h. By dividing the rotor's height h to its diameter value D one can determine its aspect ratio *AR*. Like the other parameters mentioned above, the *AR* also influences the turbine's performance [7].

Blades can be of different shapes, the most common being the straight helical or curved versions. Besides efficiency, the shape also influences the turbine's structural response to the cyclic loads [8].

14.3 Results and Discussion

The three rotors had the parameters as shown in Table 14.1.

The first experiments addressed the scoop bladed turbine. Specifically, it was determined which orientation mode is more efficient in terms of generated power. Two turbines were prepared, one having the blades oriented with curvature *in* and the other with curvature *out*. The graphical results in terms of power coefficients C_p are depicted in Fig. 14.5. For the wind speed ranging up to 12 m/s the power generated for the two modes was almost the same, except a slight difference at the highest points. Therefore, in further tests only the curvature outward version was employed.

Next the scoop bladed turbine's performance was compared to that of the NACA 0018 rotor. The results obtained are graphically presented in Fig. 14.6. The tendencies indicate a significant difference between the two types. The scoop bladed turbine was highly superior to NACA 0018 turbine, generating 39 times more power at 10 m/s. The difference decreases as the wind speed increases, Table 14.2.

Table 14.1 The parameters of the analyzed rotors

Parameter	Unit	Symbol	Value
Rotor's height	m	h	0.4
Rotor's diameter	m	D	0.4
Number of blades	/	N	3
Blade's length	m	L	0.4
Chord's length	m	c	0.05
Swept area	m^2	A	0.16
Solidity	/	σ	0.375
Aspect ratio	/	*AR*	1
Pitch anglex	°	β	0

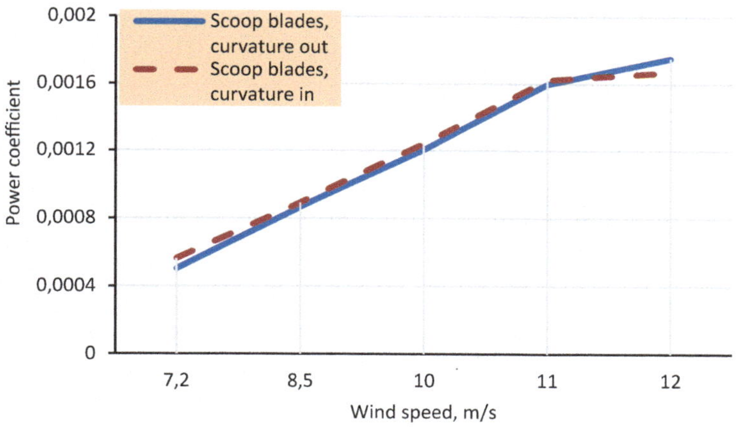

Fig. 14.5 Performance of scoop bladed turbines, curvature *in* and *out*

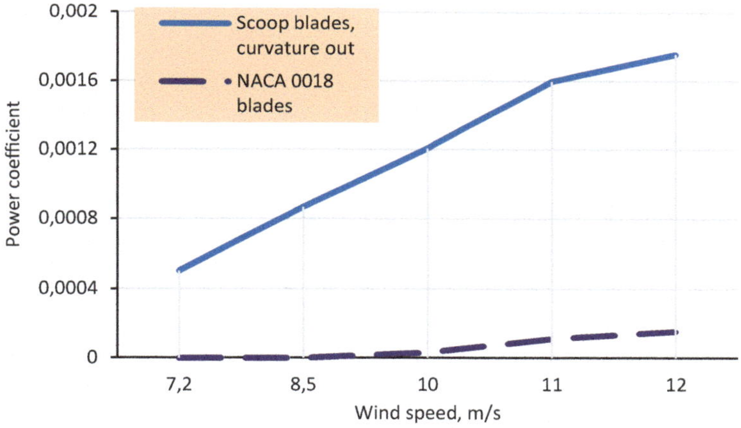

Fig. 14.6 Performance of the scoop bladed turbine versus NACA 0018 turbine

Table 14.2 The power generated by the scoop and NACA 0018 turbines (in mW)

	Airfoil\wind speed	7.2 m/s	8.5 m/s	10 m/s	11 m/s	12 m/s
A1	Scoop airfoil	18	52	118	208	297
A2	NACA 0018	0	0	3	14	26
A1/A2	The ratio	–	–	39	15	11

The scoop bladed turbine started to generate power from wind speeds as low as 5 m/s while exhibiting excellent self-starting abilities. The NACA 0018 turbine started to produce power at 8.5 m/s.

The performance of the scoop bladed turbine was further compared to that of FX 63-137 turbine with the curvature outward. This mode was selected because for the FX 63-137 airfoil the laboratory tests indicated superior efficiency compared to the inward mode. The obtained results are shown in Fig. 14.7.

In this case, the scoop bladed turbine was again superior to its counterpart but the tendencies were different. For wind speeds higher than 10 m/s the difference was small, however at 8.5 m/s the scoop turbine generated 33 times more power, Table 14.3. The FX 63-137 turbine was not tried for speeds higher than 11 m/s as there was the risk of harming the blades due to the high rpm.

Even though the FX 63-137 version was less productive, the number of rotations per minute was higher. Figure 14.8 shows the voltage and current generated by the two turbines for the wind speed equal to 11 m/s. The maximum voltage generated by the FX 63-137 was higher while the current was lower than that of the scoop turbine.

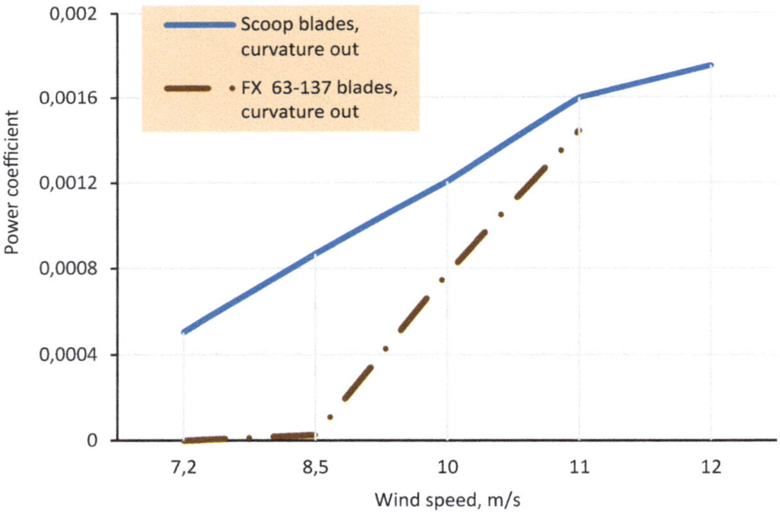

Fig. 14.7 Performance of scoop bladed turbine versus the FX 63-137 curvature outward turbine

Table 14.3 The power generated by the scoop and FX 63-137 turbines (in mW)

	Airfoil\wind speed	7.2 m/s	8.5 m/s	10 m/s	11 m/s	12 m/s
A1	Scoop airfoil	18	52	118	208	297
A2	FX 63-137	0	1.6	76	188	–
A1/A2	The ratio	–	33	1.5	1.1	–

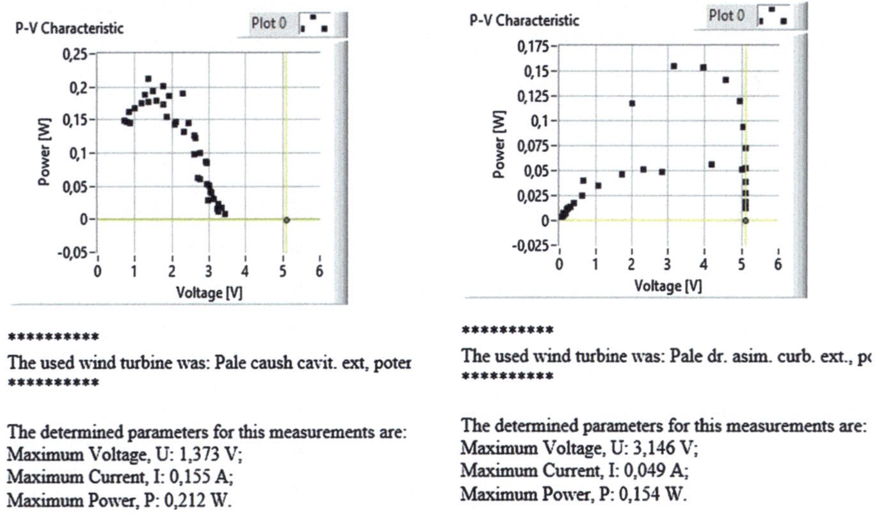

Fig. 14.8 The voltage, current and power generated by the scoop (left) and FX 63-137 turbines (right) at 11 m/s

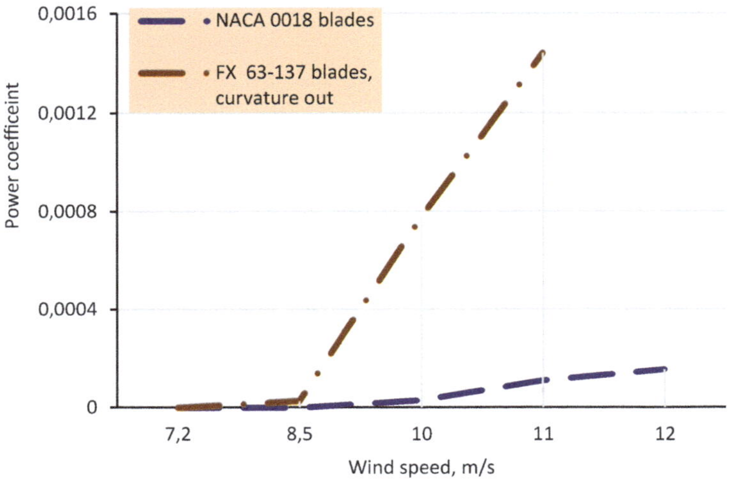

Fig. 14.9 Performance of the NACA 0018 turbine versus the FX 63-137 curvature outward turbine

Hence the superior performance for the last one was due to the superior current and not voltage.

The last comparison was made between the NACA 0018 and the FX 63-137 turbines. The results are depicted in Fig. 14.9. The last version generated throughout more power starting with lower wind speeds.

14.3.1 Results' Interpretation and Application

The common wind turbines operate for Reynolds numbers *Re* higher than 100,000 while having solidities ranging between 0.2 and 0.6 and tip speed ratios between 1 and 6. Maximum power coefficients normally go between 0.1 and 0.4, generally larger wind turbines showing higher values and vice versa.

For the investigated cases the maximum *TSR* exhibited by the fastest FX 63-137 turbine was calculated to be 0.5. Due to the very small scale, the maximum power coefficient was 0.22% of the available energy (Betz limit) and the average Re number was 40,000.

The Reynolds number (Re) is defined as:

$$\mathrm{Re} = \frac{U_f \cdot l}{\nu} \tag{14.2}$$

where U_f is the flow speed (m/s); l is the characteristic linear dimension, (m); ν is the kinematic viscosity of the fluid (m^2/s). For the tested VAWT the Re number was calculated using Eq. (14.3):

$$\mathrm{Re} = \frac{W \cdot c}{\nu} \tag{14.3}$$

where *W* is the blade's relative velocity and *c* is the blades chord length. Due to the specific nature of the vertical axis wind turbines, the relative velocity changes during one rotation, with the Re number changing accordingly. Therefore average values were calculated and used.

Because of the turbines' small scale and low Re numbers, the efficiency peaks could not be reached to be compared. However the information given by the graphs provides insights on tendencies, showing how the different airfoils would perform relative to each other as a general trend.

From the above results, the following statements can be made:

- The scoop blades are the most appropriate for the smallest wind speeds; for this particular airfoil the curvature orientation (either *in* or *out*) in relation to the rotor's axis is of no relevance over the efficiency;
- The tendencies suggest that the FX 63-136 may have a higher efficiency peak than the other two (Fig. 14.7), however this supposition was not proved;
- The NACA 0018 airfoil showed the poorest performance among the three turbines.

Though giving incomplete information, this method is relatively simple to set up and get good ideas about turbines' behavior. Once the appropriate airfoil is selected for the given wind conditions, further optimization could be applied as e.g. determining the proper solidity, pitch angle, aspect ratio etc. eventually by using other methods.

14.4 Conclusions

Three small scale turbines were tested in the wind tunnel. First model had the blades defined by a scoop shaped airfoil, the blades for the second were based on the NACA 0018 airfoil and the blades for the third rotor were based on the asymmetrical FX 63-137 airfoil for which the curvature outward mode was selected. Except the airfoil, the three turbines were characterized by the same key parameters, and undergone the same wind conditions. The parameters refer to: swept area, rotor's aspect ratio, chord length, solidity, number of blades, pitch angle and blade shape (straight blades). Each turbine was tested at: 7.2, 8.5, 10, 11 and 12 m/s wind speeds. The power coefficient values were graphically exhibited, linearly interpolated and then compared and a set of conclusions could be formulated:

1. Having an asymmetrical cross section, the scoop shaped blades can be arranged in two modes in relation to the rotor's axis: curvature *in* and *out*. According to the experiments for the scoop airfoil this aspect has no influence on the generated power, the performance for the two cases being almost the same for all investigated wind speeds;

2. The turbine with scoop blades curvature *out* mode is much more efficient than the NACA 0018 turbine for all investigated wind speeds. For the highest speed of 12 m/s, the scoop version produced 11 times more power with an even higher difference at lower wind speeds;
3. For wind speeds up to 8.5 m/s the scoop bladed turbine is significantly more productive than the FX 63-137 curvature *out* turbine. However, for higher speeds the difference is much lower, at 10 m/s the scoop turbine producing only 1.5 more power and at 11 m/s—1.1. According to the trends, it could be supposed that for even higher wind speeds the FX 63-137 turbine's efficiency would have surpassed that of the scoop turbine, however this supposition could not be proved in this case. Though more efficient, the scoop version had lower number of rotations per minute, generating a lower voltage but a higher current;
4. For wind speeds up to 8.5 m/s both the FX 63-137 curvature *out* turbine and the NACA 0018 turbine performed poorly. For speeds higher than 8.5 m/s, the FX 63-137 was considerably superior to the NACA version, generating 25 times more power at 10 m/s and 13 times more at 11 m/s.

Though for some types of analyzes it provides incomplete information, comparing small scale wind turbines of the same parameters in the same conditions can deliver insights about the influence of specific parameters on the turbine's performance. For this study the trends helped create an idea about which airfoil would be more appropriate to use for specific wind conditions.

References

1. IRENA, Future of wind: Deployment, investment, technology, grid integration and socio-economic aspects (A Global Energy Transformation paper), in *International Renewable Energy Agency*, Abu Dhabi (2019). ISBN 978-92-9260-155-3. https://irena.org/-/media/Files/IRENA/Agency/Publication/2019/Oct/IRENA_Future_of_wind_2019.pdf. Last accessed 28 Feb 2020
2. L. Du, G. Ingram, R.G. Dominy, A review of H-Darrieus wind turbine aerodynamic research. Proc. Inst. Mech. Eng Part C **233**(23–24), 7590–7616 (2019). https://doi.org/10.1177/0954406219885962
3. A. Tummala, R.K. Velamati, D.K. Sinha, V. Indraja, V.H. Krishna, A review on small scale wind turbines. Renew. Sustain. Energy Rev. **56**, 1351–1371 (2016). https://doi.org/10.1016/j.rser.2015.12.027
4. L. Lombardi, B. Mendecka, E. Carnevale, W. Stanek, Environmental impacts of electricity production of micro wind turbines with vertical axis. Renew. Energy (2017). https://doi.org/10.1016/j.renene.2017.07.010
5. P.A. Cotfas, D.T. Cotfas, Design and implementation of RELab system to study the solar and wind energy. Measurement **93**, 94–101 (2016). https://doi.org/10.1016/j.measurement.2016.06.060
6. B.K. Kirke, *Evaluation of self-starting vertical axis wind turbines for stand-alone applications*. Ph.D. thesis, Griffith University (1998)
7. S. Brusca, R. Lanzafame, M. Messina, Design of a vertical-axis wind turbine: how the aspect ratio affects the turbine's performance. Int. J. Energy Environ. Eng. **5**, 333–340 (2014). https://doi.org/10.1007/s40095-014-0129-x
8. Scheurich F, Fletcher TM, The influence of the blade curvature and helical blade twist on the performance of a vertical axis wind turbine, in *48th AIAA Aerospace Science Meeting*, 4–7 January 2010

Chapter 15
Vertical Axis Wind Turbines. Optimal Positioning of the Blades Defined by Asymmetrical Airfoils

Bostan Viorel, Bostan Ion, Rabei Ivan, Gutu Marin, and Dulgheru Valeriu

Abstract Many Vertical Axis Wind Turbines (VAWTs) have blades defined by asymmetrical (cambered) airfoils. There can be two possibilities regarding the orientation of their camber in relation to the rotor's axis: radially inward or outward. The objective of this study was to experimentally determine the relevance of this aspect and the version that comes with higher efficiency. The asymmetrical FX 63-137 airfoil was considered. On the same turbine, the blades were first attached with the camber oriented inward and then oriented outward, for both cases the pitch angle being set to zero. The outward mode proved to be much more efficient. The optimization of the pitch angle was pursued for the camber inward version so besides 0° a few more values were tested: 8°, −8°, −16°. Even though the performance was significantly improved due to this step, the efficiency was still much lower than that for the camber outward mode for which the pitch angle was not optimized at all.

Keywords Vertical axis wind turbine · Asymmetrical airfoil · Camber in · Camber out

15.1 Introduction

The extended use of fossil fuels over the last century facilitated higher living standards but their application is considered as one of the main climate change causes [1]. Renewable energy sources are deemed as replacement solution and extensively implemented with the sun and the wind energy being the top choices [2]. Large horizontal axis wind turbines are indubitable leaders for converting wind energy. Small scale vertical axis wind turbines are recently gaining popularity as research topics and market products. Though less efficient, these offer a series of advantages that make them attractive. The main one is their insensitivity to changes of the wind

B. Viorel · B. Ion · R. Ivan · G. Marin · D. Valeriu (✉)
Technical University of Moldova, Stefan cel Mare Bd. 168, 2004 Chisinau, Republic of Moldova
e-mail: valeriu.dulgheru@bpm.utm.md

B. Viorel
e-mail: viorel.bostan@adm.utm.md

I. Visa and A. Duta (eds.), *Solar Energy Conversion in Communities*,
Springer Proceedings in Energy, https://doi.org/10.1007/978-3-030-55757-7_15

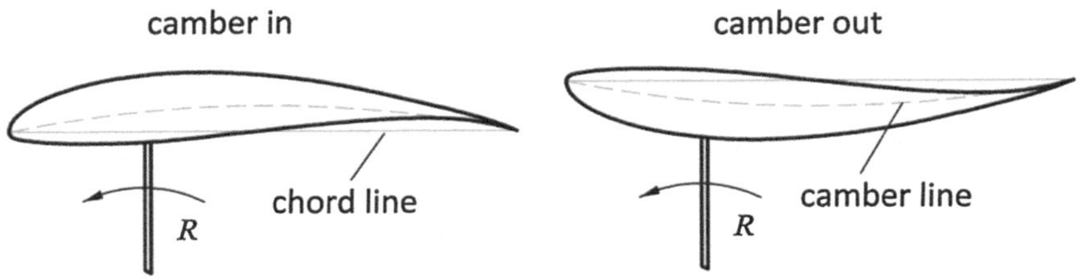

Fig. 15.1 The camber *in* and *out* modes (R—rotor's radius)

direction. They are also less noisy, potentially simpler to build (referred specifically to the small scale versions) and are considered more appropriate for urban implementation. Moreover, some heavier components, as the generator, could be installed near the ground level. Another advantage that can be considered is that the theoretical efficiency limit surpasses that of the horizontal axis wind turbines [3].

The blade is arguably the most important element of a wind turbine and a vast source of efficiency optimization. Improving the blade largely means refining its cross section named airfoil.

The airfoils are divided into two broad categories: of symmetrical and asymmetrical shape. Both versions are extensively used and investigated.

The airfoil's symmetry is considered in relation to the chord line that connects its leading and trailing edges. The camber line is the midline that splits in half the inner and outer sides of the airfoil while having the same starting and ending points (Fig. 15.1). The bigger the difference between the lengths of the two lines the more pronounced the airfoil's asymmetry.

When it comes to the camber positioning in relation the rotor's axis there are two possibilities: radial orientation *in* and *out*. The two modes are depicted in Fig. 15.1.

If doing an internet research over the turbines available on the market, one can notice that for the case of the blades based on asymmetrical airfoils the camber is oriented preponderantly outward but there are also cases where the orientation is inward. There was not found any study on this aspect or any general recommendations related to this.

The goal of this paper is to outline the extent to which the camber orientation influences the turbine's efficiency. One airfoil was adopted, the FX 63-137. The blades were analyzed with the camber *in* and *out* modes. The study also implied the fixed pitch angle optimization.

15.2 Methods

15.2.1 Theoretical Considerations

The wind rotor is defined by a set of parameters, each one influencing its efficiency. The turbines that were analyzed had the same key values that refer to the swept area, rotor's diameter, rotor's height, aspect ratio, blade number, solidity, chord length and airfoil type. This way none of these was expected to create a difference in the generated power except the blade's camber orientation. The definitions and relevant aspects about these parameters are considered below.

The swept area *A* is the area confined by the rotor's height and diameter. This value is referred to in the equation used to calculate the power *P* generated by a wind turbine:

$$P = \frac{1}{2} \cdot C_p \cdot \rho \cdot A \cdot U^3 \tag{15.1}$$

where the C_p is the power coefficient that is a measure of the turbine's efficiency; ρ is the air density and *U* is the wind speed.

The same swept area can be confined by different height and diameter values. This is described by the aspect ratio that is defined as the ratio between the rotor's height and its diameter (*h*/*D*). This parameter also influences the turbine's efficiency [4].

The solidity σ is the percentage of the swept area covered by the blades area. For a straight bladed VAWT the solidity can be calculated using Eq. (15.2):

$$\sigma = (N \cdot c \cdot L)/A = (N \cdot c)/D \tag{15.2}$$

where *N* is the blades number; *c*—the chord length; *L*—the blade's length that for a straight bladed rotor is equal to its height; *D*—the rotor's diameter. The solidity is another important parameter with strong impact on the turbine's efficiency and self-starting capabilities.

The pitch angle is the angle between the chord line and the line tangent to the blade's circular trajectory. It can take positive, null or negative values. The positive value is considered when the blade's leading edge is oriented out of the rotor's axis (Fig. 15.2). Some authors consider variable pitch angles for efficiency boost and for an improved self-starting ability [5].

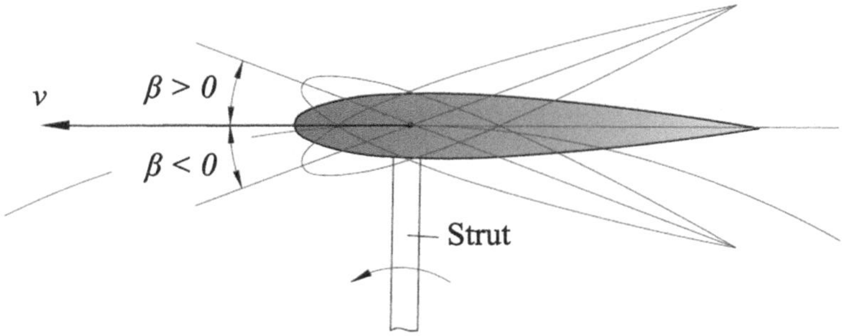

Fig. 15.2 Positive, null and negative pitch angles

15.2.2 Experimental Setup

A small-scale vertical axis wind turbine was built so that the blades could be easily mounted and dismounted, thus different sets of blades can form different wind turbines. This way the same tower (that includes the generator) and blades were used for studying the two modes (Fig. 15.3). The design allows also changing the blades pitch angle as required. A gear train mediates the connection between the rotor's and the generator's shafts. The gear ratio was 1:3.41. The alternating current of the three phase generator was transformed into direct current by using a rectifier. A 200 Ω resistance was used as load in all cases.

The blades were manufactured by 3D printing and taped for better rigidity.

A Gunt ET 220 wind tunnel was used for testing the wind turbines. The machine's outer diameter is of 600 mm with the maximum wind speed of 12.5 m/s.

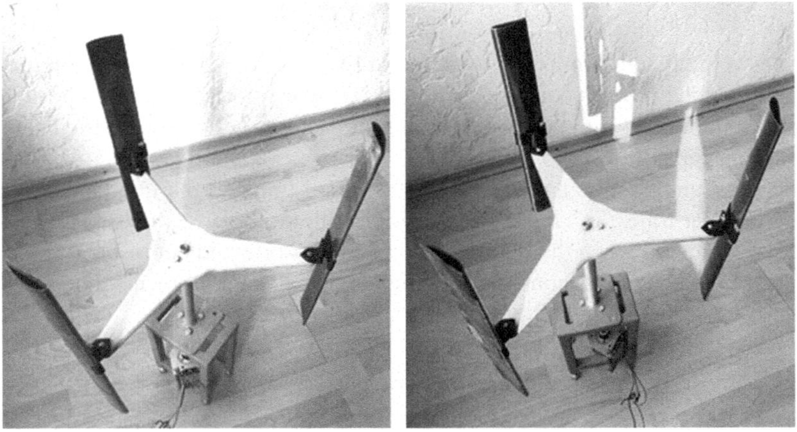

Fig. 15.3 The wind turbines with the blades cambered *in* (left) and *out* (right)

15.3 Results and Discussions

The parameters of the wind turbines are presented in Table 15.1. In all cases the generated power was used to compare the behavior developed by the two modes.

Firstly the camber *in* and *out* modes were tried for a pitch angle equal to zero. The results are presented in Fig. 15.4. For this case the camber *in* turbine was not generating any power, being characterized by vibrations instead of rotations. On the other hand the camber *out* turbine showed good performance generating 106 mW while exhibiting 250–300 rpm at 12 m/s.

Next step was optimizing the pitch angle for the turbine with the lower performance, the camber *in*. Three values were considered: 8°, −8° and −16°. The results are depicted in Fig. 15.5. For the pitch angle equal to +8°, the turbine's behavior

Table 15.1 The parameters of the analyzed rotors

Parameter	Unit	Symbol	Value
Rotor's height	m	h	0.4
Rotor's diameter	m	D	0.4
Number of blades	/	N	3
Blade's length	m	L	0.4
Chord's length	m	c	0.05
Swept area	m^2	A	0.16
Solidity	/	σ	0.375
Aspect ratio	/	RA	1

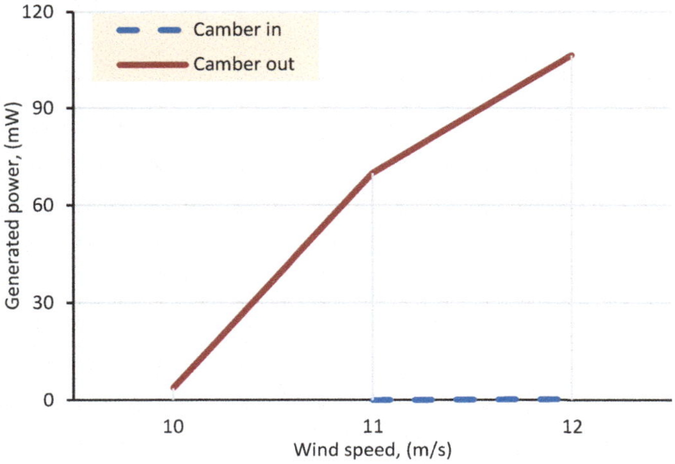

Fig. 15.4 The power generated by two turbines with asymmetrical FX 63-137 airfoil based blades, camber *in* and *out* modes, pitch angle = 0°

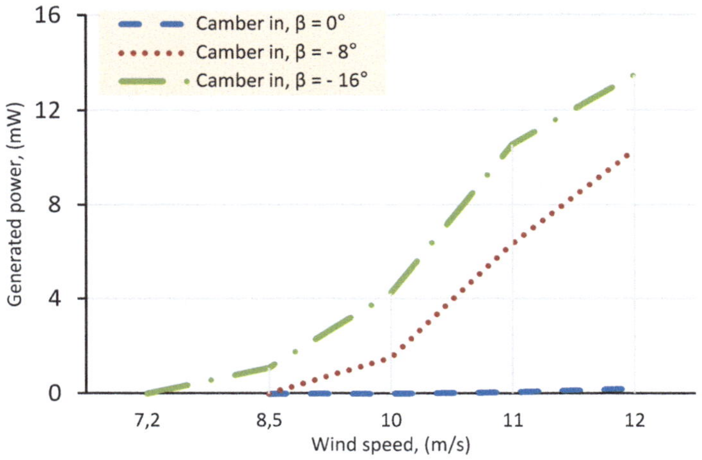

Fig. 15.5 Camber *in* modes with null and negative pitch angles, FX 63-137 airfoil

didn't change, displaying only vibrations thus the results are not included in the graph.

However, for the negative angles the turbine started to generate power. The pitch angle of −16° was the most productive, generating 13 mW followed by the −8° angle for which 10 mW have been produced at 12 m/s.

The graph in Fig. 15.6 shows the trends developed by the camber *out* case at a

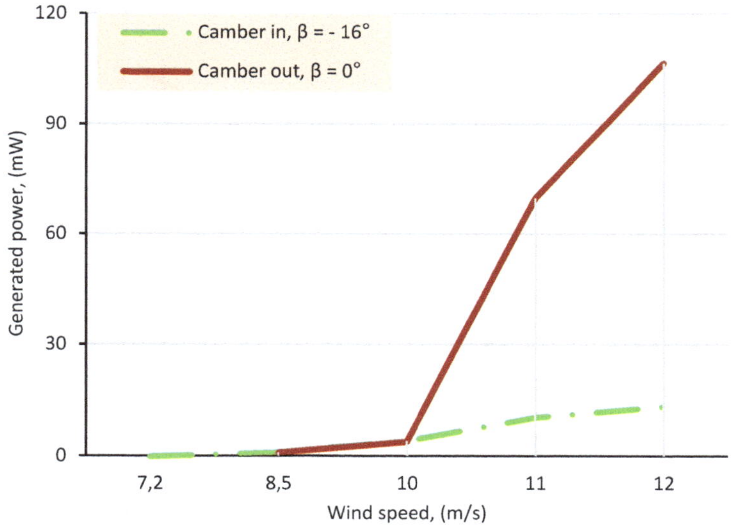

Fig. 15.6 Camber *out* pitch angle 0° and optimized camber *in* pitch angle −16°, FX 63-137 airfoil

pitch angle of 0° and the optimized camber *in* version with a pitch angle of −16°. Even though that the last mode was significantly improved by optimizing the pitch angle, its performance remains far below that of the camber *out* mode. For example at 12 m/s the camber *out* turbine generated 8 times more energy than the camber *in* version.

15.4 Conclusions

The blades based on asymmetrical airfoils can be mounted in two modes in relation to the rotor's axis: camber oriented inward or outward. A series of small scale vertical axis wind turbines were tested with the goal of evaluating the relevance of this aspect over the turbine's performance. The influence of the pitch angle was also considered. The results led to the following conclusions:

1. Camber orientation influences the VAWT's performance. For the case of the FX 63-137 airfoil, the camber *out* mode proved to be much more efficient than the camber *in* mode. As example at 12 m/s the camber *out* turbine generated 106 mW while the camber *in* version did not produce at all, being characterized by vibrations instead of rotation.
2. The fixed pitch angle also influences the turbine's performance, hence selecting an optimum value could bring an important efficiency boost.
3. Even though for the FX 63-137 airfoil the camber *out* mode came with a higher efficiency, it does not necessarily means that this would be the trend for all asymmetrical airfoils. For confirming or infirming the existence of a general trend, more airfoils have to be tested.

References

1. World Meteorological Organization, WMO statement on the state of the global climate in 2019. WMO-No. 1248, ISBN 978-92-63-11248-4.
2. IRENA, Future of wind: Deployment, investment, technology, grid integration and socio-economic aspects (A Global Energy Transformation paper), in *International Renewable Energy Agency*, Abu Dhabi. ISBN 978-92-9260-155-3 (2019) https://irena.org/-/media/Files/IRENA/Agency/Publication/2019/Oct/IRENA_Future_of_wind_2019.pdf. Last accessed 25 Mar 2020
3. I. Paraschivoiu, Wind turbine design with emphasis on Darrieus concept (Ecole Polytechnique de Montréal, 2002). ISBN 2-553-00931-3
4. S. Brusca, R. Lanzafame, M. Messina, Design of a vertical-axis wind turbine: how the aspect ratio affects the turbine's performance. Int. J. Energy Environ. Eng. **5**, 333–340 (2014). https://doi.org/10.1007/s40095-014-0129-x
5. Y.-L. Xu, Y.-X. Peng, S. Zhan, Variable pitch to high-solidity straight-bladed VAWTs for power enhancement. Energy Procedia **158**, 382–387 (2019). https://doi.org/10.1016/j.egypro.2019.01.119

Chapter 16
Hybrid Solar—Biomass System Design for Communities with Collective Houses

Adrian Ilie and Ion Visa

Abstract The use of energy mixes in communities giving value to the renewable energy sources available within or nearby the implementation location became one of the first research and innovation priorities. This trend particularly occurred in the context in which there is a local and regional (European) legislative support for the use of renewable energies, energy performance and energy efficiency that regulates the strategic objectives, the deadlines assumed and the methods of implementation thereof. This paper presents a general method of sizing a hybrid solar-biomass system for a community, starting from the location coordinates, the typical features of the built environment and the specific meteorological profile. The method integrates these data in the sizing process of the systems, based on the existence of district heating networks and available spaces for the individual or districtual installation(s) of hybrid solar-biomass systems and it selects the installation possibilities. In the end, the method is validated using a case study, the community in Taberei District of the Odorheiu Secuiesc City, using meteorological data to create the climatic profile using the Meteonorm software and operation simulations developed using the TRNSYS software.

Keywords Hybrid solar-biomass system · Energy mixes in communities

16.1 Introduction

The Conference of Parties (COP-21), through the ratified document (Paris Agreement), creates the opportunity for all the signatory parts (developed or developing countries) to make an equitable [1] and sustained [2] effort to maintain global warming under the limit of 2 °C [3] and to continue the actions taken for reaching the target of "no greenhouse gas emissions" in developed countries and of reducing the

A. Ilie (✉) · I. Visa
Transilvania University of Brasov, Brasov, Romania
e-mail: adrian.ilie@unitbv.ro

I. Visa
e-mail: visaion@unitbv.ro

I. Visa and A. Duta (eds.), *Solar Energy Conversion in Communities*,
Springer Proceedings in Energy, https://doi.org/10.1007/978-3-030-55757-7_16

level of harmful emissions in the developing countries, bearing in mind the national circumstances [4].

As result of COP-21, a first instrument was formulated to generate examples of technological strategies and related methodologies [5], i.e. the Deep Decarbonisation Pathways Project [6], where 16 countries are involved in order to create personalised visions for meeting the "low-carbon world" objective.

The need for the development of hybrid solar-biomass (HSB) systems has emerged from the need to identify a backup solution for the solar thermal (ST) systems that use the conversion of renewable energies/solar energy [7]. Currently, initiatives have emerged around the world to integrate the HSB systems, most of which are still at pilot stage, by developing the ST systems that are able to cover the thermal energy demand (at least during the warm season), while the remaining (uncovered) demand would be covered by biomass conversion systems [8].

Denmark has built the greatest number of ST stations (using biomass conversion systems as back-up) and HSB stations for heating and DHW (over 50) [9], with capacities of up to 49 MWp [10]. However, there are similar systems in Austria, Germany, France, Italy and Norway, with capacities ranging between 1 and 7 MWp [11].

The projects so far developed provide concrete application examples of the HSB systems [12] that provide thermal energy to the community, by sizing the ST system [13], by creating the district heating system based on the consumption/storage of the solar energy [14] and by implementing a backup system that integrates a biomass conversion system [15] and assessing the two components using solar energy and biomass.

This article proposes a general methodology for the modular sizing of the hybrid solar-biomass systems that can be further applied to any type of community.

16.2 Design Methodology

The proposed methodology is applicable in the design of hybrid solar and wooden biomass systems for communities. The concept is based on the interpretation of the infield data on the existing infrastructure and the estimated or measured energy demand.

The general concept originated from the need to use renewable energies (biomass and solar energy) with a focus on meeting the heating and domestic hot water (DHW) demand throughout the year.

An operation scheme is defined according to the characteristics of the built environment and the possibilities to use solar and biomass energy conversion that can simulate both, the specific energy demand and the energy production. Subsequently, the solar thermal systems and the modular biomass systems are sized according to the possibilities of meeting the demand, by following the steps described hereinafter.

Step 1—Input Data

Sizing the production components and the thermal energy distribution system is performed according to the national standards and it tends to oversize the installed capacities [16]. The collection of data from the implementation location, based on a certain history (at least 1 year) has major benefits, such as the costs optimisation related to the installed capacities [17] and the possibility to personalise the technical solutions depending on the needs [18]. Thus, location-specific data shall be collected and identified as follows:

Input Data on the Built Environment

These data involve the evaluation the physical boundaries and the structure of the community (the type of buildings), the characteristics of the built environment (arrangement, design, useful area), the performance of the built environment (reported and calculated energy consumption), the number of consumers and the behaviour thereof (considering the DHW consumption), the available infrastructure (existing equipment, thermal infrastructure, existing systems).

Considering the community needs, this methodology analyses the possibility to provide energy for the DHW production, using solar thermal collector systems throughout the warm season, i.e. at least 6 months (e.g. April 1–September 30), and partially throughout the cold season when the remaining energy demand would be generated by systems that convert wooden biomass into thermal energy.

The first step would be to provide independent solutions for each block of flats, followed by connection solutions for groups of blocks of flats and the possibility of connecting these blocks of flats to district systems.

Input Data for the Assessment of the Renewable Energy Potential

In order to assess the solar energy potential, meteorological data are required (outdoor temperature, relative humidity, the irradiance value of the solar radiation, reference wind speed). These parameters can be provided by a local weather station (ideally) or can be simulated by specialised software (e.g. Meteonorm) [19].

In order to assess the biomass energy potential, data on the wood areas and the available energy crops where from wood can be annually extracted for thermal energy production, as well as data on other existing biomass sources (e.g. wood waste) are required.

Step 2—Defining the Operational Concept of the System

The TRNSYS software can be used to define the operational concept by using the location-specific input data (defined hereinabove) and a specific simulation scheme (Fig. 16.1). Based on these, various possibilities on use of the resources to meet the energy demand can be predicted and analysed.

The technical operation scheme is based on the interconnection of 4 circuits using the buffer: C1—solar energy conversion circuit, C2—biomass conversion circuit, C3—DHW circuit and C4—thermal energy distribution circuit. The technical solutions are sized to enable the system to maintain a certain balance (consumption *vs.*

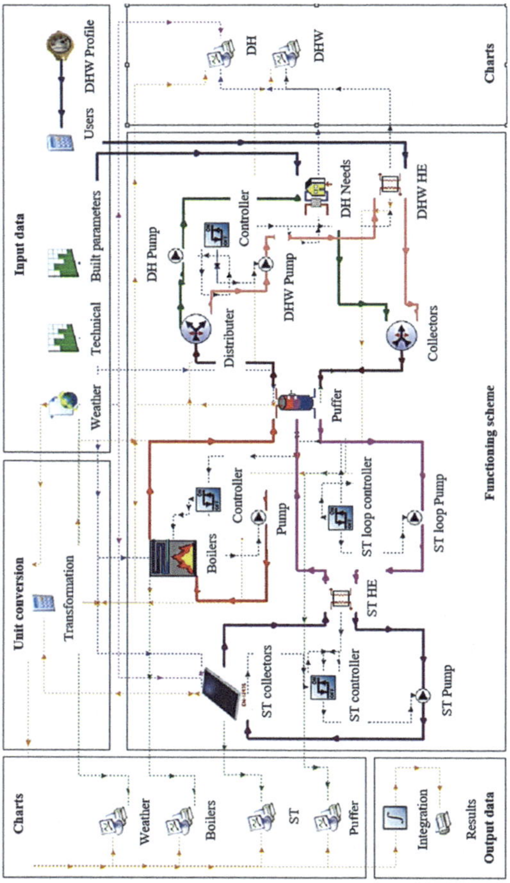

Fig. 16.1 Technical operation scheme (TRNSYS)

production), while the optimisation and final adjustment are carried out using the buffer.

Compared to the classical interconnection schemes of the 4 circuits, in the current system the solar circuit and the DHW circuit are equipped with heat exchangers that optimise the scheme by maximising the energy stored. This increases the maximum storage temperature from 60 to 95 °C and reduces the losses related to the need to cool down the solar energy conversion system during peak production periods.

Input Data for the TRNSYS Simulation

The input data for the simulation(s) created using the TRNSYS software are gradually added, based on each processing stage, as described hereinafter.

Data on the built environment and its performance shall be added during the simulation of the energy demand for heating, i.e. the built areas that serve the community, the specific losses caused by the distribution pipes and the specific consumption according to the recorded data (where available) or to the climate profile. This can be either measured or simulated using specialised software (e.g. Meteonorm).

Data on the number of consumers, the water consumption patterns (daily consumption profile) and the energy losses caused by the distribution system shall be added during the simulation of the energy demand for DHW.

Step 3—Sizing Criteria for the Conversion Systems and the Puffer

When designing the optimal solar-biomass system for a specific community it is important to choose the appropriate components and to size the conversion systems in a way that meets the thermal energy demand in all the operational situations. Thus:

The Solar Thermal Energy Conversion System

According to the analysed input data and location, the sizing of the system must comply with the feasibility and efficiency requirements, while considering:

1. The applicable technical solutions: the type of collector (e.g. flat plate collector or evacuated tube collector) and the type of installation (with individual or centralised storage);
2. The thermal energy produced that depends on how the collectors are laid out according to the specific climate [20]. The simplified calculation algorithm is based on the equations provided in the ISO 9806:2017 standard. This standard only considers the global solar irradiance value in relation with the efficiency factor of the solar thermal collector, the rate at which the assembly angles are modified and the area of the collector. However, this standard does not consider the losses caused by the collector insulation, the outdoor temperature during operation and the wind speed that allows the formulation of Eq. (16.1) [21].

$$\dot{Q} = F'(\tau\alpha)_{en} K_{\theta b} G^* A \tag{16.1}$$

where $\dot{Q}$ = useful energy flow extracted from the solar collector [W]; A = area of the absorber plate [m^2]; F' = efficiency factor of the collector; $(\tau\alpha)_{en}$ = actual transmittance-absorbance gene-rated for the direct irradiance at a normal rate; $K_{\theta b}$ = rate at which the assembly angles are modified for the direct irradiance; G^*= global solar irradiance [W/m^2].

The solar energy conversion systems are sized according to the daily thermal energy consumption or to the peak charge. Generally, solar energy conversion is used to meet the average energy demand in order to balance the size of the ST collector systems and the excess energy generated during peak periods.

For the Biomass Conversion System

It is recommended to use the biomass conversion system only when there is not enough solar energy for DHW and heating (as main energy source). Thus, the biomass conversion boilers must be accordingly sized to meet the energy demand for DHW and heating, particularly for situations of maximum demand, thus providing a back-up solution in the case of malfunction of the solar thermal system.

The wooden biomass conversion systems are sized according to the daily energy consumption, the burning technology used in the selected boiler (with mobile grates, circulating fluidised bed etc.) and, implicitly, the boiler capacity.

For the Buffer

In both cases (individual or centralised storage systems) the capacity of the buffer is limited by the available space and the project infrastructure. In any case, the storage capacity determines the applicable technical solution and is calculated based on 4 vertically overlapping areas, each having a temperature node from entry to the output of the hot water, from the lower part to the upper part. During each stage of calculation, the temperature "i" shall be determined for each point i = {1,..., 4}, according to Eq. (16.2), in relation with the water volume of the buffer between the two nodes (V_{as}), for the selected period of time. Simultaneously, the final temperature variation for the node "i" is calculated according to Eq. (16.3) [22]. The use of the 4 points supports a short-term prediction of the energy consumption allowing to identify the optimal source(s) of energy production, particularly if both systems require long waiting times until they reach full operational capacity.

$$V_i \Delta T_i = V_{as}(T_{i-1} - T_i) \tag{16.2}$$

$$c_{p,a} V_i \frac{dT_i}{dt} = Q_{sol,i} + Q_{aux,i} - Q_{losses,i} \tag{16.3}$$

where V_i is the volume of the calculated area; ΔT_i is the temperature difference between two nodes; T_i is the temperature of the calculated area; $Q_{sol,i}$, $Q_{aux,i}$ are the energies transferred from the solar system and auxiliary systems; $Q_{losses,i}$ ia the environment thermal energy losses; $c_{p,a}$ is the thermal capacity of the water stored in the tank.

Step 4—Analysing the Technical Solutions of Implementation and the Selection of the Best Solution

A comparative analysis of these types of sizing technical solutions is required.

This analysis should include the individual projects (for one block of flats), the districtual projects (2, 5 up to 10 blocks of flats) and the district projects (25, 50 or more blocks of flats) in order for the hybrid solar-biomass systems to be able to use the existing (production and distribution) infrastructure for heating the dwellings and for the DHW production.

For the Solar Thermal System

Sizing the solar energy conversion system has led to the identification of three implementation possibilities at housing community level [23], i.e.:

- The installation of 4–6 solar thermal collectors on each block of flats: the thermal energy generated shall be carried to the thermal unit of each building that shall be equipped with a 1000 L tank and a heat exchanger that will ensure the closed-circuit conditions for the solar thermal system.
- The centralised installation of the solar collectors on one of the blocks of flats in the groups of blocks of flats (consisting of 2, 5 or 10 blocks of flats) and the extrapolation of the system up to the incidence of the total demand. The thermal energy generated shall be transported to a buffer (of 2000, 4000 and respectively 8000 L) and subsequently transported to the blocks of flats in the districtual circuit through a number of underground pipes that shall be laid down.
- The centralised installation of solar collectors on the district heating station and on the blocks of flats and other spaces in their neighbourhood. This technology involves the storage and distribution of the heat carrier for the production of DHW and heat which can be carried out using the existing systems (already installed tanks and pipes), while the heat carrier for production of DHW and heating shall be generated using the existing system (at the existing block staircase modules).

For the Biomass Conversion System

The analysis of the available biomass conversion technologies allowed the identification of three implementation possibilities at housing community level, i.e.:

- Individual boilers: this technical solution involves the installation of individual boilers customised for each housing unit. To this respect, available spaces must be identified within each condominium (building) for installing the boiler, the biomass feed system and the energy storage system.
- Boilers for pairs of condominiums (2 up to 10 blocks of flats): this technical solution represents a viable alternative when considering the conversion of the existing infrastructure. This solution involves the installation of a districtual heating unit that would meet the energy demand and that requires the installation of one or several biomass conversion boilers in a specially designed space, either in the neighbourhood or in other available locations.

- District heating station: this solution involves the conversion of the existing heating systems (thermal stations or district heating stations) into district heating stations for the conversion of wooden biomass. This solution involves minimum interventions on local transport and production infrastructure, but it requires the conversion of the existing thermal energy production systems.

The methodology above described enables the selection of the optimal system and the sizing of the components required to convert solar and biomass energy into thermal energy, by using all the thermal energy obtained from solar energy conversion and using the thermal energy obtained from biomass conversion as a backup source.

16.3 Case Study

The case study considers collective housing units (blocks of flats) in the Taberei District of Odorheiu Secuiesc City, Harghita County, Romania.

The selection of this district is justified by the fact that it has a modernized production and distribution infrastructure for the heat carrier. Moreover, several blocks of flats were here identified, where a significant number of flats are connected to the district heating system (at least 50% of the total number of flats/block).

Step 1—Input Data

The Taberei District is located in the eastern part of the city (46.3° N and 25.3° E, at 385 m altitude above sea level) and it has a population of about 5700 inhabitants in 1496 flats in blocks with the height regim is groundfloor + 4 floors (70 blocks) and groundfloor + 10 floors (16 blocks). The district distribution system of the heat carrier supplies heat and domestic hot water to 506 flats (442 flats at the ground floor + 4 floors blocks and 64 flats in ground floor + 10 floors blocks) for 1950 residents, representing approximately 30% of the total number of potential users.

Input Data Referring to the Built Environment

When analysing the energy consumption that was measured and invoiced to the consumers during 2015–2019, and after comparing the consumption data with the characteristics of the built environment, different patterns could be identified. These depend on the construction materials, the age of the building and especially the building rehabilitation degree that influences the efficiency degree and the energy performance (e.g. for buildings type B1,…, B3 with walls made by briks, and for B4,…, B6 with walls made by concrete), as presented in Fig. 16.2.

The analysis of the distributed energy demand as per months of the year (Fig. 16.3) shows the fluctuations between the energy consumption during the cold season as opposed to that in the warm season, which leads to the need of sizing two different systems with seasonal operation.

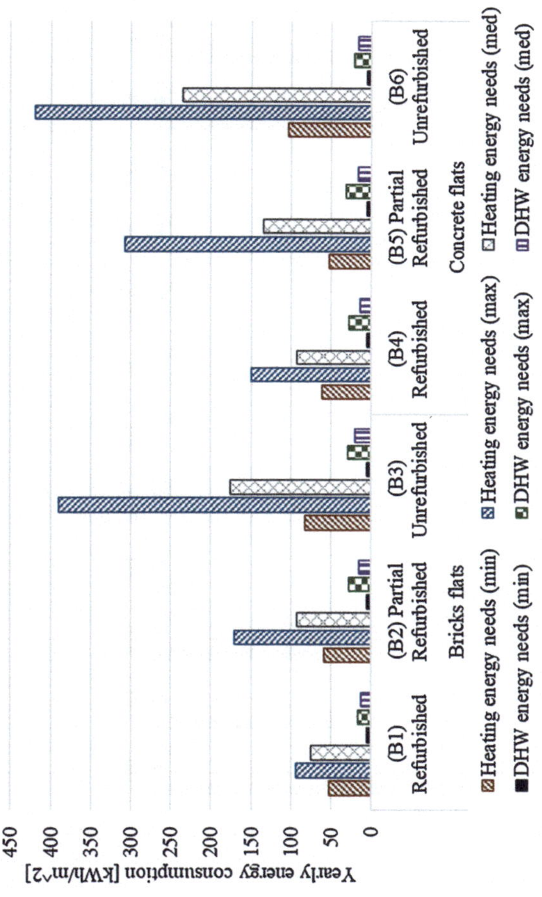

Fig. 16.2 Annual energy consumption for the analysed built environment

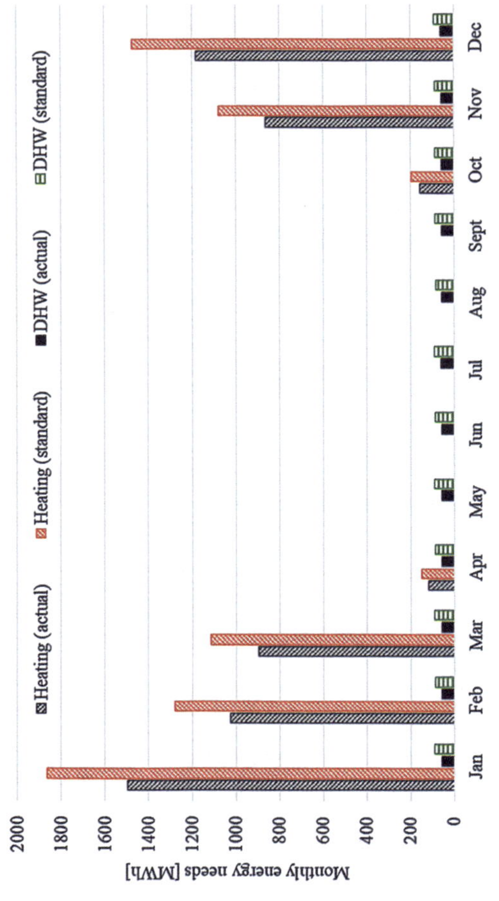

Fig. 16.3 Distribution of the monthly energy demand

The qualitative assessment in Fig. 16.3 shows the differences between the standard way of calculation and the predictions based on onsite data. Considering these results, oversizing the systems may be avoided (with values of up to 25%).

Considering the differences between the energy demand for heating purposes (900,…, 1500 MWh/month) and the energy demand for the production of DHW that is relatively constant (approx. 57 MWh/month) one may notice the need for individualized systems that should operate in different seasons: a system for the warm season (May–September) that should cover the energy demand for the production of DHW and another system for the transition periods and for the cold season that should be able to covert both, the energy demand for heating and that for the DHW production.

Input Data for the Assessment of the Energy Potential

Odorheiu Secuiesc City is located in a mountain area (Harghita Mountains in the Carpathian Mountains) and it is surrounded by a large forest area. The approximately 6500 ha of forest in the proximity of the city might supply about 9500 tons of biomass every year through the valorisation of the wood waste (wood chopping) combined with the waste resulted from the timber processing by the specialized companies.

To evaluate the solar energy potential, several simulations were made using the Meteonorm software; the results show that the location offers the possibility to capture approximately 1320 kWh/m^2 energy during one year.

Step 2—Design of the Functional Concept of the System

Considering the source of energy, the general blueprint of the hybrid system, Fig. 16.4, relies on two circuits for thermal energy production, i.e. the circuit of the solar-thermal collectors and the circuit of the biomass conversion boiler, whereas for the distribution of the energy, it relies on the thermal energy distribution circuit and the DHW circuit. Each of the circuits is equipped with its own pumping system (P1,…, P5), with heat exchangers for isolation (HE1,…, HE4) with collector-distributor systems (DC1,…, DC2).

Thermal energy is supplied through two heat exchangers (HE1 and HE2) and it can be stored in the buffer or pumped, by pump P4, directly through the DC2 distributor to the exchangers for the DHW production and the carrier needed for heating.

The solution adopted is the use of heat exchangers, especially for the energy production circuits so that all the available energy may be extracted, as the exchangers are operational at high temperatures as well.

Step 3—Sizing Criteria for the Conversion Systems and for the Buffer

For the Solar-Thermal Energy Conversion System

The solar energy potential in the selected location was determined using the Meteonorm software by determining the direct, diffuse and global solar irradiance.

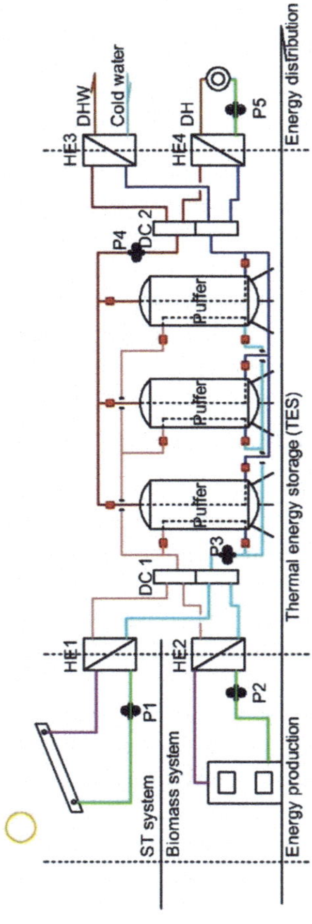

Fig. 16.4 General blueprint of the hybrid system

Based on the prediction on the available solar radiation in the implementation location and considering the technical features of the flat plate solar-thermal collectors installed in favourable locations (azimuth angle $\psi = 0°$ and tilt angle $\chi = 35°$) corroborated with the calculation method, a diagram was generated (Fig. 16.5). The results indicate the useful area of the solar-thermal collectors required to cover the entire DHW demand for 2000 people throughout the year considering a storage capacity of 50,000 m^3.

Considering the large differences of the areas of the solar-thermal collectors required to cover the entire energy consumption for the DHW production throughout the year (areas exceeding 2000 m^2 during the cold season vs. areas of less than 500 m^2 during the warm season), it is recommended to use the solar-thermal system to entirely cover the demand during the summer months and the transition periods, thus avoiding the excessive production of thermal energy during periods of maximum solar radiation.

For the Biomass Conversion System

The current biomass potential that may be used for conversion into thermal energy is of approximately 27,300 MWh/year and may be extended through the development of some energy crops up to approx. 57,000 MWh/year.

When analysing the biomass potential in relation with the total annual energy demand in the studied community (6429 MWh/year for the 509 block of flats community), we find that, currently, the energy potential exceeds the demand by at least 4 times, thus creating the opportunity to develop also other communities in the investigated area.

Step 4—Analysis of the Implementation Solutions and the Selection of the Best Solution

Bearing in mind the multiple technical possibilities to reach the goal, six connection scenarios of the ST systems and the biomass conversion systems were formulated (Table 16.1).

Technical Solution 1

This solution is applicable to Scenario 1 and consists of mounting 6 ST collectors and individual biomass boilers on each of the 50 blocks of flats. The thermal energy thus produced will be carried to the (existing) thermal module(s) of each building that will be supplemented by a 1000 L buffer and a new heat exchanger that should meet the closed circuit condition for the ST system.

Technical Solution 2

This solution is applicable to Scenarios 2, 3 and 4 and consists of centralized mounting the solar collectors on one of the blocks in the residential section (consisting of 2, 5 or 10 blocks of flats). Biomass boilers for pairs of condominiums and the system extrapolation up to the incidence of the 50 buildings that need to be supplied are further considered. The produced thermal energy would be carried to a storage system (with a capacity of 2000, 4000 and respectively 8000 L) and transported to the blocks

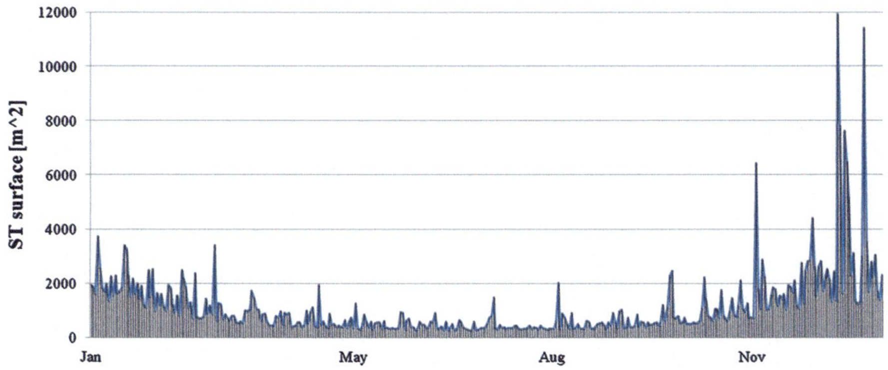

Fig. 16.5 The daily area of the ST system required to cover the DHW demand in the given location for one year

Table 16.1 Definition of the scenarios for the case study

Scenario	Installation type	Distribution	Storage	Supplied persons
1–1 block	Independent	Existing	Additional	38 pers/10 flats
2–2 blocks	Interconnected	Existing	Additional	81 pers/21 flats
3–5 blocks	Interconnected	Existing	Additional	198 pers/52 flats
4–10 blocks	Interconnected	Existing	Additional	396 pers/103 flats
5–25 blocks	District	Existing	Rehabilitated	980 pers/255 flats
6–50 blocks	District	Existing	Rehabilitated	1980 pers/509 flats

of flats in the district circuit through a set of underground pipes that need to be laid down.

Technical Solution 3

This solution is applicable to Scenarios 5 and 6 and consists of centralized mounting of the solar collectors on the district thermal station and, additionally, on the blocks of flats or other constructions or spaces in the proximity thereof and a districtual thermal power plant using biomass. The storage and distribution of the heat carrier for the DHW production can be done with the existing systems (already installed tanks and laid down pipes), whereas the DHW will be produced in an existing system (at the existing block staircase modules).

16.4 Results and Discussions

These technical solutions aim at giving the best use to the solar energy. They adopt systems that may carry and store energy (as heat carrier) at temperatures up to 95 °C, without any limitations related to overheating and the avoidance—to the best extent possible—of the situations when the system should be shut down and cooled during night (Table 16.2).

Table 16.2 Technical solutions applicable to the case study

Scenario	STC area	Number of STC	Type of biomass boiler	Type of storage	Storage capacity
	S [m^2]	[pieces]	[kW]	[°C]	[m^3]
1	10.80	6	80	95–75	1000
2	21.60	12	150	95–75	2000
3	54.00	30	500	95–75	4000
4	117.00	65	1000	95–75	8000
5	291.60	162	2000	95–75	20,000
6	581.40	323	4000	95–75	40,000

Table 16.3 Technical and economic characteristics of the defined scenarios

Scenario	Investment		Energy production			Payback time		
	Solar	Biomass	Solar	Biomass	Surplus	Solar	Biomass	Total
	[10^3 EUR]	[10^3 EUR]	[MWh]	[MWh]	[MWh]	[year]	[year]	[year]
1	11	29	11	5872	11.50	17.56	13.04	14.30
2	20	57	21	5872	11.50	15.96	11.12	12.54
3	47	171	53	5872	12.91	14.25	13.23	12.74
4	85	298	116	5872	13.98	11.88	10.18	10.03
5	193	497	288	5872	17.42	11.39	7.83	8.61
6	358	736	574	5872	17.37	11.35	7.61	8.53

The results in Table 16.3 indicate that the Technical Solution 3 allows the most efficient use of the existing district heating infrastructure. Moreover, the production and storage of the energy at the thermal substations/district thermal station allows the export of the exceeding energy produced during the days of maximum radiance, which obviously shortens the payback time (Table 16.3).

Upon analysis of the amounts of produced energy as compared to the payback time and considering the constant production of energy using biomass conversion systems (produce as needed) and the variable production using the solar thermal system (produces how much radiation it receives), the necessity becomes obvious to select Scenario 6 with the adoption of the Technical Solution 3. This solution allows the construction of a ST system that would secure the thermal energy for the selected community throughout the year, as shown in Fig. 16.6.

16.5 Conclusions

The general design methodology of the hybrid solar-biomass system for the communities of collective housing units consists of four steps and it comprises the specifications of the built environment, the existing energy potential, the drafting of the operational concept of the system, the simulation of the coverage of the energy demand and the implementation of the technical solutions.

The concept allows the complete use of the thermal energy obtained from the conversion of the solar radiation and the use of biomass as a backup source.

The solar-thermal component determines the period of independent operation of the two subsystems.

The proposed concept has a modular structure that allows implementation in communities of different sizes from 1 block of flats up to 50.

By combining the wooden biomass conversion system and the solar-thermal system in a single investment shortens the payback time of the investment as

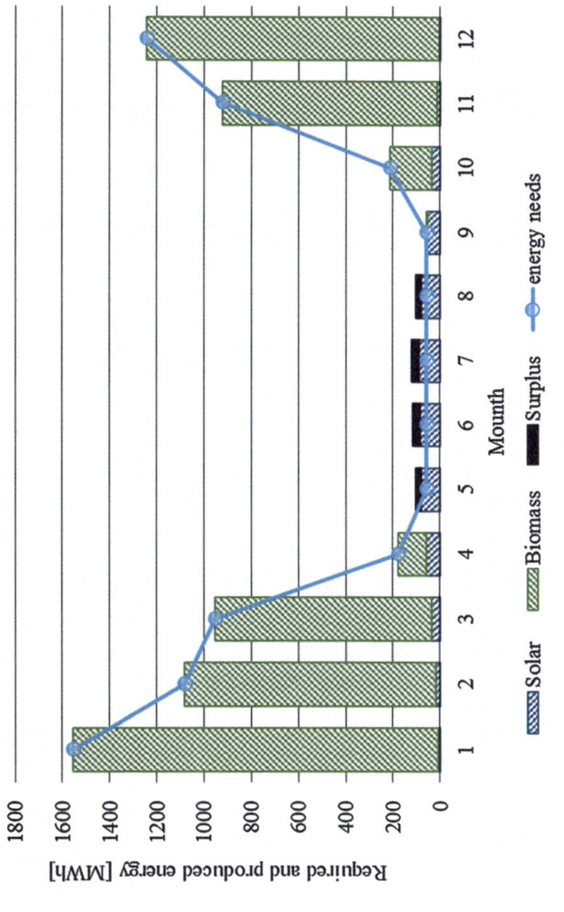

Fig. 16.6 Required and produced energy in Scenario 6

compared to independently built up ST systems, from 11.35 to 8.53 years. Additionally, the condition of an alternative solution in case of break down is also met.

References

1. Climate Action Tracker: Paris Tango, *Climate action so far in 2018: individual countries step forward, others backward, risking stranded coal assets* (2018)
2. A. Izzet, S. Ramazan, Differentiation of developed and developing countries for the Paris Agreement. Energy Strateg. Rev. **18**, 175–182 (2017)
3. R. Stavins, R. Stowe, The Paris Agreement and beyond, in *International Climate Change Policy Post-2020* (2016)
4. United Nations, Conference of the Parties, Twenty-First Session: Adoption of the Paris Agreement (2015)
5. S. Mathy, P. Menanteau, After the Paris Agreement: measuring the global decarbonization wedges from national energy scenarios. Ecol. Econ. **150**, 273–289 (2018)
6. T. Ribera, J. Sachs, Pathways to deep decarbonization, in *Sustainable Development Solutions Network* (2015)
7. A. Ilie, I. Vişa, A. Duţă, Solar-thermal systems for domestic hot water production implemented in communities of collective households. J. Energy Eng. **143**(6) (2017)
8. G. Shaopeng, L. Qibin, S. Jie, J. Hongguang, A review on the utilization of hybrid renewable energy. Renew. Sustain. Energy Rev. **91**, 1121–1147 (2018)
9. J. Nielsen, P.A. Sørensen, Renewable district heating and cooling technologies with and without seasonal storage. *Renewable Heating and Cooling* (2016), pp. 197–220
10. M. Guadalfajara, M.A. Lozano, L.M. Serra, Simple calculation tool for central solar heating plants with seasonal storage. Sol. Energy **120**(1), 72–86 (2015)
11. C. Winterscheid, T. Pauschinger, B.S. Riccardo, SDHp2m... from policy to market, in Advanced Policies and Market Support Measures for Mobilizing Solar District Heating Investments in European Target Regions and Countries (Steinbeis Research Institute Solites, 2019)
12. M.A. Pantaleo, S.M. Camporealed, A. Sorrentino, A. Miliozzi, N. Shah, C.N. Markides, Hybrid solar-biomass combined Brayton/organic Rankine-cycle plants integrated with thermal storage: techno-economic feasibility in selected Mediterranean areas. Renew. Energy **147**(3), 2913–2931 (2020)
13. R.B. Sarkis, V. Zare, Proposal and analysis of two novel integrated configurations for hybrid solar-biomass power generation systems: thermodynamic and economic evaluation. Energy Convers. Manage. **160**, 411–425 (2018)
14. U. Sahoo, R. Kumar, P.C. Pant, Resource assessment for hybrid solar-biomass power plant and its thermodynamic evaluation in India. Sol. Energy **139**, 47–57 (2016)
15. C.M. Iftekhar, H.B. Norton, A. Duffy, Technological assessment of different solar-biomass systems for hybrid power generation in Europe. Renew. Sustain. Energy Rev. **68**, 1115–1129 (2017)
16. A.C. Ferreira, J. Silva, S. Teixeira, Assessment of the Stirling engine performance comparing two renewable energy sources: solar energy and biomass. Renew. Energy **154**(1), 581–597 (2020)
17. A. Ilie, I. Vişa, Energy production in a sustainable community using the local renewable energy sources. Ann. Dunarea de Jos Univ. Galati **1**, 100–107 (2015)
18. R. Strzalka, T. Gabriel, U. Eicker, Analysis and optimization of a cogeneration system based on biomass combustion. Appl. Therm. Eng. **50**(2), 1418–1426 (2013)

19. B.H. Ganthia, S. Sasmita, K. Rout, A. Pradhan, J. Nayak, An economic rural electrification study using combined hybrid solar and biomass-biogas system. Mater. Today **5**(1), 220–225 (2018)
20. N. Aste, M. Beccali, L.C. Tagliabue, Nomograph for rapid technical and economic assessment of solar thermal systems for DHW production. Sol. Energy **86**, 2472–2485 (2012)
21. ISO 9806:2017 Solar energy—Solar Thermal Collectors—Test Methods (2017)
22. S.J.W. Klein, S. Coffey, Building a sustainable energy future, one community at a time. Renew. Sustain. Energy Rev. **60**, 867–880 (2016)
23. A. Ilie, I. Vişa, A. Duta, Solar-thermal systems implemented in communities of collective households for domestic hot water production. Ann. "Dunărea de Jos" Univ. Galaţi **1**, 139–146 (2016)

Chapter 17
Analysis of the Water—Power Nexus of the Balkan Peninsula Power System

Goran Stunjek, Antun Pfeifer, Goran Krajačić, and Neven Duić

Abstract Power generation sector worldwide accounts for high water withdrawal and consumption due to the hydropower generation and cooling of thermal power plants. Hence, the operation of the power generation sector is constrained by the availability of the water resources, as well as the addition of constrains on water resources used for other purposes, such as irrigation, flood control, water supply, agriculture, etc. The optimal utilization of water resources between the water and energy sector is defined under the term water-energy (or water-power) nexus. This study describes the implementation of hydrological LISFLOOD, Medium-Term Hydrothermal Coordination (MTHC) and Unit Commitment and Dispatch (Dispa-SET UCD) models for detailed analysis of impacts on the SEE regional power system for three different hydrological years. Results were validated based on the available ENTSO-E data for the average hydrological (2015) year. Results show increase in hydropower generation from 53.06 TWh for dry year, to 65.24 and 85.13 TWh for average and wet year, respectively, while the average electricity cost falls from 17.79 EUR/MWh for dry year, to 16.36 and 14.05 EUR/MWh for average and wet year, respectively. This analysis successfully replicates the methodology under the WATERFLEX project, with the novelty in run-of-river hydropower generation calculations in MTHC model.

Keywords Energy modelling · Water-energy nexus · Dispa-SET · LISFLOOD

17.1 Introduction

Power generation accounts for high water withdrawal and consumption as a result of hydropower generation and thermal power plant cooling. Besides the water use for power sector, water resources are used for a variety of purposes not related to

G. Stunjek (✉) · A. Pfeifer · G. Krajačić · N. Duić
Faculty of Mechanical Engineering and Naval Architecture, University of Zagreb, 10002 Zagreb, Croatia
e-mail: Goran.Stunjek@fsb.hr

N. Duić
e-mail: neven.duic@fsb.hr

I. Visa and A. Duta (eds.), *Solar Energy Conversion in Communities*,
Springer Proceedings in Energy, https://doi.org/10.1007/978-3-030-55757-7_17

the power sector, such as irrigation, flood control, water supply, agriculture etc. [1]. Several examples of water resource shortages or high river temperatures have been experienced in the last decade. Due to joint effects of bad hydrological conditions and heat waves, several French nuclear power plants in 2013 had to curtail power generation, which generated additional cost of EUR 300 million. In 2006, France, Germany and Spain had to reduce their nuclear power generation due to the high river water temperatures. Poland experienced reduced coal power generation and restricted industrial demand in 2015–2016 due to the same reasons [2]. This forced flexible generation of the inflexible thermal power plants results in demand restrictions, monetary losses and increased wear of generation units. Furthermore, recent examples of unplanned outages in France, Germany and Switzerland have been experienced in 2018 [3]. Mentioned impacts with the forecasts that climate change will cause a number of similar events to rise, raise the questions on how to implement better water management.

The term water-energy nexus is used to refer the interactions between the water and energy sectors for the best utilization of water resources. The hydropower is recognized technology that provides benefits for the total power system operation, such as black start capability, spinning reserve, frequency response, flexibility and reserve with quick start and shutdown capabilities. Mentioned hydropower characteristics identify hydropower as a main cost-competitive resource for integration of variable renewable sources into the European power system [1]. Importance of water-energy nexus is recognized as new challenge for better control of water resources, but present power system models overlook water-related constraints to power system and water resources management. Hydrological related constraints determine hydropower production, which in turn determine the operation of thermal power plants related to its water sources for proper cooling. Thus, the better understanding of the water-energy nexus is needed to enable flexible power generation for the future European power system [2]. With projections of the future extreme droughts in summer and floods in the winter/autumn, adaptation of the hydropower units to a climate change relies on optimal management of water reservoirs [4].

Water-energy nexus has been a popular research topic in the last decade. International Energy Agency started discussion on the energy and water dependence in 2012 in a chapter "Water for Energy: Is energy becoming a thirstier resource?" from the World Energy Outlook 2012 [5]. More thoroughly discussion on the same topic can be found in 2016 World Energy Outlook [6]. US Department of Energy published extensive data and analysis report on water-energy nexus with intention of connecting and encouraging relevant stakeholders in a dialog and joint actions to address the water-energy challenges [7]. Joint cooperation between US Department of Energy, European Commission's Joint Research Centre (JRC) and Directorate-General for Research and Innovation led to organization of a workshop dedicated to understanding the water-energy nexus, with higher emphasis on integrated water and power system modelling [8]. More recent reports published by the JRC introduce integrated analysis of the independencies between energy and agricultural water demand, drinking and urban water provision, and ecosystem flow requirements under the Water Energy Food and Ecosystem (WEFE) Nexus [9, 10]. The report [9] provides a first summary of the WEFE Nexus findings regarding the water and energy usage in Europe, with more emphasis on water availability. Furthermore, report concludes the importance of development of integrated model, and presents coupling water and energy model-based assessments for better understanding of the water-energy nexus. Moreover, policy recommendations are divided into strategic and operational ones, where strategic measures represent long-term actions, while operational measures are based on the existing technological solutions. The Position Paper [10] outlines the importance of the WEFE Nexus as a methodology aiming to an integrated management across water, energy and food security, while ensuring the sustainable usage of ecosystem resources. Moreover, it presents the thoughts and lessons learnt from the experts that attended the 2018 Nexus workshop, adding the recommendations on how to implement the WEFE Nexus approach. The position paper adds on importance of the scientific-technical dimensions as supporting element that provides scientific evidence for evidence-based policy making.

In [2] authors studied the water-energy nexus for a Greek power system. They analyzed the implications of water on the energy system and vice versa for three different historical scenarios. Moreover, the addition of water stress index (WSI) is used to determine the locations and time frames with high possibility of water scarcity under the dry hydrological conditions.

The same approach was used by authors in [11], with addition of vulnerability analysis of cooling-related constraints on allowable water withdrawal for two different power producing units. The analysis included water withdrawal constraints on coal-fired power plants with high marginal cost and moderate installed capacity, and nuclear power plant representing technology with low marginal cost and high installed capacity.

The study [12] includes the water-energy nexus analysis done on the Spanish energy system. Authors reviewed the published work on Spanish water system, with emphasis on separate study of energy-for-water and water-for-energy. The energy-for-water study includes dividing water use stages and calculation of energy cost for water use, with special consideration on irrigation.

On the other hand, water-for-energy study includes evaluation of water needs for power plant cooling. D. Zafirakis et al. studied in [13] the water needs in the Greek electricity sector concluding that promotion of renewable energy sources will ensure conservation of water resources in vulnerable regions. Authors collected the data on operation of thermal power units and renewable technologies to determine the minimum water needs and compared it to existing technologies. Furthermore, authors indicate that water withdrawal coefficient for lignite-fired power plants is as high as expected, but the calculated water consumption coefficient is lower than the ones in available databases. They finished with the conclusion that high RES technologies penetration in water scarce region, such as West Peloponnesus, Crete and West Macedonia, might resolve local water scarcity problems.

The water-energy nexus for Greece region was also studied in [14]. Authors provided calculations on water consumption of several different processes. Calculation included processes of electricity generation in conventional thermal power units, such as lignite, diesel, oil and gas-fired units, production of biodiesel, and extraction and refining processes in the primary energy production sector. To connect the water consumption with energy, they provided the calculation of electricity consumption for purposes of water supply and water treatment. Authors conclude that the most water-intensive sector includes power generation from lignite and oil-fired thermal power plants averaging at water consumption of 1.81 m^3/MWh, followed by CCGT units with water consumption of 1.19 m^3/MWh. The biofuel production accounts for nearly 0.5 m^3/MWh, while the primary fuel production requires the least amount of freshwater. Moreover, the authors conclude that water supply is much more energy-intensive, when compared to the water treatment processes.

Authors in [15, 16] study the Spanish energy sector adaptation to available water resources, as well as integration of water and energy models. The study in [15] is review of available models and recommendations for the future work, including the literature on water-energy nexus. Furthermore, need for water and energy sector integration, as well as barriers in integrated water and energy modelling, with list of recommendations is thoroughly discussed in the study.

In [16] authors took the approach of comparing two different scenarios. Stressed scenario represents integrated water-energy model that takes water constraints into account, while Unconstrained scenario is traditional non-integrated energy model that neglects the importance of water constraints on energy sector. They came with the results that neglecting the water constraints results with unpredicted costs under the climate change scenario. Moreover, authors estimate that the cost of neglecting the water constraints in the future water-restricted scenarios may range from 0.2 and 8% of the total system cost, which is more than double of adaptation costs.

Water stress vulnerability of electricity generation units in the EU region was studied in [17]. Study included 1326 thermal power units and 818 water basins. Authors used year 2014 as reference year, and projection scenarios for 2020 and 2030. Furthermore, study shows energy-water-climate model that integrates power plant, water quantity and water temperature databases. Model also includes the adaptation strategies, such as usage of air cooling for planned and constructed units, additional use of seawater for coastal units cooling, early retirement of older units, and replacement of planned power capacities with renewable energy sources. Results show that regions that experience reduction in power generation due to the waters stress increase from reference 47–57 basins between 2014 and 2030, while including water demand for non-energy related processes. Moreover, authors conclude that highly vulnerable regions are Mediterranean regions, Germany, Bulgaria, Poland and France.

Pereira-Cardenal, S. J. et al. in studies [18–20] focused on interactions between water and energy system to identify methods that could be used to assess spatial-temporal interactions in water-energy nexus. Authors used the approach of including water and energy sector in joint optimisation problem with objective function composed of power production costs, while maximizing the benefits of water allocations. Link between two systems is described using constraints in optimization problem and solved using stochastic dynamic programming. Authors used Iberian Peninsula as case study for method implementation. Authors conclude that climate change may reduce hydropower generation by 24%, increasing thermal power generation and CO_2 emissions. Moreover, authors recommend topics for future research, such as more realistic representation of power market using the hourly values, and further spatial disaggregation of the hydrological system.

Authors in [21] made a comprehensive review of existing optimisation techniques and approaches in planning and design of water supply side in water-energy nexus with objective to identify research gaps. Authors conclude that research on water-energy nexus lacks the holistic approach, and that the problem is mostly addressed from either water or energy side. Authors also add that most of the studies ignore the uncertainties of used parameters in the optimisation models.

The water-energy nexus for the US region was studied in [22, 23]. In [22] the economic implications were studied for shifting from coal to natural gas, and replacement of open-loop with the closed-loop cooling technologies. Results show that on average shift from coal to natural gas saves 32% of water consumption and 37% of water withdrawal. Shift from open-loop to closed-looped system shows the 96% decrease in water withdrawal and 58% increase in water consumption.

In [23] the analysis of 2011 droughts was studied to examine the power plant's vulnerability regarding moderate year 2010. The water consumption in energy-related sectors and the energy consumption in water-related activities were studied, with a discussion on energy and environmental implications for the MEAN region in a study [24].

Authors in [25] developed a model to determine economic impacts, the water consumption and withdrawal, and detailed operation of the power system under different current and future scenarios. Based on a modelling framework developed by the JRC, authors study the water-energy nexus for the West African Power Pool. Results show that future power system operation of the Western African Power Pool regions significantly depends on the water availability, which translates in high volatility of the system cost. Hence, the future policy scenarios should use the technologies that will be most suitable to achieve low volatility, low cost and low emissions.

Implementation of Unit Commitment and Dispatch mode (Dispa-SET UCD) to the six Western Balkan countries is shown in [26]. Authors were using Dispa-SET UCD to prove the hypothesis stating that it is possible to phase out large amount of lignite-powered power units and replace it with renewable energy sources without compromising the flexibility and stability of the power system. The referenced scenario for 2010 including power systems of Albania, Bosnia and Herzegovina, Kosovo, Macedonia, Montenegro and Serbia was developed and validated.

Two additional scenarios that include implementation of national energy strategies for 2020 and 2030 were analysed. Results showed that high RES integration coupled with expansion of cross-border interconnections increases the region's energy independence and security of supply. Authors in [27] used the Dispa-SET UCD model on Western Balkan countries with addition of Croatian and Slovenian power systems. The model was developed for three different years, 2015, 2030 and 2050 used for testing various modelling formulations. The goal of the research was to test usefulness of different types of clustering techniques to lower the computational time. In three different clustering approaches results showed that computation time was 1.4, 2.1 and 19 times lower that the no clustering approach, which on average lasted for about 2.4 h.

Water resources have always been important for the Balkan Peninsula economy with its use for irrigation, drinking water supply, tourism, industry, livestock production and hydropower generation. The hydropower generation accounts for 49% of all electricity generated in the Western Balkan region [4]. Projections are that Balkan Peninsula is getting warmer and that trend will continue with the expected increase in global temperatures due to climate change. Even though precipitation rate changes with terrain, elevation and proximity to the sea, the region is experiencing lower annual precipitation with projections for a further decrease.

Worst case scenario, a 4 °C temperature rise, by the projections in [28] states that Balkan Peninsula Region could encounter reduced water availability with projections of precipitation declining between 20–50%. As most countries in the Balkan region depend on hydropower generation, reduction in water availability would strongly affect the region power system. Moreover, due to the increased possibility of extremely low river flows in summer days, the mean number of days during which electricity production will be reduced by more than 90% is projected to increase.

This study illustrates the implementation of three models for detailed analysis of impacts on the regional power system due to different hydrological conditions. Case study includes Balkan Peninsula region covering analysis of countries: Albania, Bosnia and Herzegovina, Bulgaria, Croatia, Greece, Hungary, Kosovo, Montenegro, North Macedonia, Romania, Serbia and Slovenia. For the water-energy nexus analysis, method includes combining hydrological LISFLOOD model, Medium-Term Hydrothermal Coordination model (MTHC) and Dispa-SET Unit Commitment and Dispatch model (Dispa-SET UCD). Hydrological LISFLOOD model is used as source of water inflow data needed as input for two energy models. The first MTHC energy model, determines reservoir accumulation levels for hydropower units during one-year time period, as well as the hydropower generation from run-of-river units. Results from the MTHC model are used as input data for Dispa-SET UCD model that results in power generation, economical, and commitment and power dispatch values for each power unit included in the model.

The remainder of the paper is as follows: Sect. 17.2 describes the models used to analyse the water-energy nexus. Section 17.3 covers input data regarding the modelled region, as well as the scenario definition. Section 17.4 provides results and discussion, while the Sect. 17.5 presents conclusions of the provided study.

17.2 Methodology

Two energy models, MTHC and Dispa-SET UCD models are both linked to the hydrological LISFLOOD model. LISFLOOD model is used as input data in form of water inflows used for calculation of reservoir levels and run-of-river hydropower generation. MTHC model runs at daily time step to provide results on management

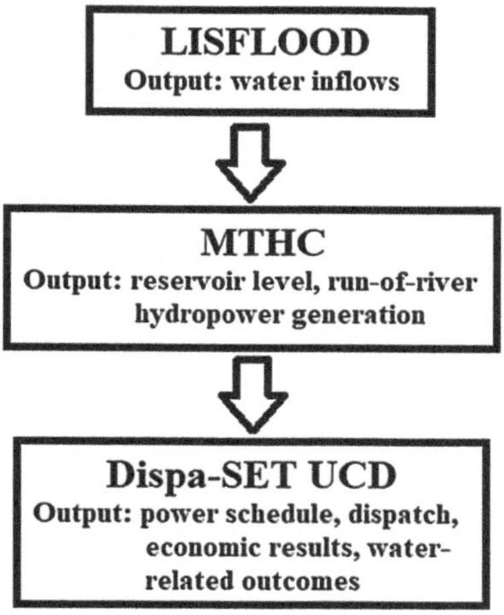

Fig. 17.1 Modelling steps

of the water resources. Reservoir level and run-of-river hydropower generation as results from MTHC model, is used as input data from Unit Commitment and Dispatch model. Dispa-SET UCD model runs at the hourly time step and results in power dispatch and schedule, water-related and economic values (Fig. 17.1).

17.2.1 LISFLOOD Hydrological Model

The LISFLOOD model will be only briefly described as its available output data are used as input values for two energy models. The LISFLOOD model has been developed by the floods group of the Natural Hazards Project of the Joint Research Centre. It is the hydrological rainfall-runoff model that simulates the hydrological processes in a catchment including flood forecasting, effect of the land-use change, assessing the effects of river regulation measures, and effects of climate change [29]. The model is designed to be used across a wide range of spatial and temporal scales. Since it is grid-based, the model can be used on a grid cells ranging from as little as 100 m for the medium-sized catchments, and up to 10 km for global models. The time steps can be daily based for the simulation of the long-term water balance, while the hourly time steps are used for the simulation of the individual flood events. Also, the output of the "water balance" simulation can be used as input data for the "flood" simulations. Even though the primary output is channel discharge, all the internal rate and states variable can be written as the output with the complete user control.

The model is made up of the two-layer soil water balance sub-models, sub-models for the simulation of groundwater and subsurface flow, sub-model for the routing of channel flow, and sub-model for the routing of surface runoff to the nearest river channel. Simulated processes include infiltration, snowmelt, leaf drainage, surface runoff, evaporation, interception of rainfall, water uptake by vegetation, exchange of soil moisture between soil layers, drainage to the groundwater, bypass of the soil layer and flow through the river channel. More on the formulation of the mentioned processes can be seen in [29].

17.2.2 Medium-Term Hydrothermal Coordination Model

The MTHC model is used to determine operation planning of hydropower reservoirs and thermal power plants based on minimization of system cost function composed of the system generation costs over a given planning horizon. The time horizon ranges from one year to several years with daily, weekly or monthly times steps. The degree of detail of hydropower units is greater than in the short-term operation at the expense of clustering the same fuel-powered thermal power plants. That suggests that thermal power units are aggregated by fuel and country, due to the main scope of the MTHC model being results on hydropower generation and reservoir levels. Inclusion of each thermal power unit itself would substantially increase the run time of the model. The MTHC problem can be characterized as large-scale, nonlinear and nonconvex optimization.

The MTHC problem can be solved from two perspectives. The extensive form also knows as deterministic equivalent assumes fixed water inflows, and regarding the formulation of the hydro and thermal related technical features, can be formulated as linear programming, mixed-integer linear programming, and non-linear programming. On the other hand, stochastic form includes uncertainty of hydrological scenarios for each planning stage that consist of the amount of water resources available for the power generation at each stage of the time horizon. Stochastic problem can be solved vertically by stage/time, or horizontal by scenarios, and are mostly used in situations of inherent uncertainty of different variables that could affect real-time operational decisions [30]. The deterministic approach is used for scenario-based analysis and it is used in this paper to define constraint linear programming problem in GAMS.

The model sets are shown in Table 17.1, variables in Table 17.2 and model parameters in Table 17.3.

The objective can be seen in Eq. 17.1. The objective function determines the total electricity generation cost during the simulation period. The objective function includes variable costs of power generation for all units, pumping costs in pumped-storage hydropower units, spillage of excess of water, energy transmission, energy curtailment and load shedding.

Table 17.1 MTHC model sets

Sets	
p	Time periods
ut	Thermal power plants
ur	Renewable power units: SUN, WIN, HROR
uh	Hydropower plants with storage
up	Pumped storage hydropower plant
l	Lines (Transmission lines between neighbouring countries)
n	Nodes (Countries)
t	Technology

Table 17.2 MTHC model variables

Name	Unit	Description
G (*p*,*u*)	GWh	Energy generated in period *p* by power plant *u*
PUMP (*p*,*u*)	GWh	Pumping water at period *p* to storage of plant *u*
RES (*p*,*u*)	Mm^3	Water stored at period *p* in plant *u*
DIS (*p*,*u*)	m^3/s	Water discharge at period *p* by plant *u*
CH (*p*,*u*)	m^3/s	Water charge at period *p* to pumped hydro storage *u*
SPILL (*p*,*u*)	m^3/s	Spillage at period *p* by plant *u*
UPSTREAM (*p*,*u*)	m^3/s	Inflow from upstream hydropower plants at time *p* for plant *u*
FLOW (*p*,*l*)	GWh	Energy transmission at period *p* and line *l*
CURT (*p*,*n*)	GWh	Curtailed RES at time *p* in node *n*
LOSTLOAD (*p*,*n*)	GWh	Unsatisfied demand at time *p* in node *n*

$$
\begin{aligned}
SystemCost = {} & \sum_{p,u} VarCost(u) \cdot G(p,u) + \sum_{p,u} PumpingCost \cdot PUMP(p,u) \\
& + \sum_{p,u} SpillageCost \cdot SPILL(p,u) \\
& + \sum_{p,u} TransmissionCost \cdot FLOW(p,l) \\
& + \sum_{p,u} CurtailmentCost \cdot CURT(p,n) \\
& + \sum_{p,u} LostLoadCost \cdot LOSTLOAD(p,n)
\end{aligned} \tag{17.1}
$$

The objective function (17.1) is constrained by the set of equations from (17.1)–(17.12).

The market clearing Eq. (17.2) state that for each node *n* at period *p* the supply (generation and imports of electricity) must meet the demand:

Table 17.3 MTHC model parameters

Name	Unit	Description
dt	h	Period duration
Gravity	m/s^2	Gravity constant
Density	kg/m^3	Water density
F_1	$GWh/((m^3/s) \cdot m)$	Conversion factor from m^3/s to GWh
F_2	$Mm3/(m^3 \cdot s)$	Conversion factor from m^3/s to Mm^3
Technology (u,t)	/	Power generation technology
Demand (p,n)	GWh	Electricity demand for the node n at period p
Duration (n,t)	day	Minimum number of days a given technology must be producing to match statistics
Location (u,n)	/	Unit location
Pmin (u)	GW	Minimum stable generation of unit
Pmax (u)	GW	Installed capacity
VarCost (u)	k€/GWh	Variable cost of electricity generation
Stmin (u)	Mm^3	Minimum storage level
Stmax (u)	Mm^3	Maximum storage level
Stinit (u)	Mm^3	Initial storage level
eta_pump (u)	%	Pumping efficiency
eta_turb (u)	%	Discharging efficiency
Delay (u, uu)	day	Water transport delay between two unit u
NominalHead (u)	m	Nominal head of hydropower plant
Resources (p,u)	m^3/s	Natural water inflows
Evaporation (p,u)	m^3/s	Evaporation loses from reservoirs
Profiles (p,u)	/	Capacity factor for solar and wind power
Topology (u,uu)	/	Hydropower network (Cascades)
Spillage_max (p,u)	m^3/s	Maximum spillage allowed
Incidence_matrix (n,l)	/	Line-node incidence matrix for power flow
LineCapacity (l)	GW	Transmission line capacity
DemandW (p,u)	m^3/s	Water withdrawal from plant u at period p
Eco_flow (p,u)	m^3/s	Environmental flow
Availability (p,u)	%	Unit availability

$$\sum_{u \in U(n)} G(p,u) + \sum_{l \in L(n)} FLOW(p,l) = Demand(p,n) + \sum_{u \in PUMP(n)} PUMP(p,u) + CURT(p,n) - LOSTLOAD(p,n) \quad (17.2)$$

In Eq. (17.3) minimum and maximum generation bound are set. It determines minimum and maximum power generation capabilities of each unit in every time step:

$$Pmin(u) \cdot dt < G(p,u) < Pmax \cdot dt \quad (17.3)$$

Equation (17.4) represents energy generation by hydropower units.

$$G(p,u) = eta\_turb(u) \cdot DIS(p,u) \cdot NominalHead \cdot F_1 \quad (17.4)$$

Equation (17.5) shows F_1 factor used to calculate hydropower generation in GWh.

$$F_1 = 24(\mathrm{h}) \cdot 60(\mathrm{min/h}) \cdot 60(\mathrm{s/min}) \cdot Gravity \cdot Density \cdot \frac{1}{3600}\left(\frac{\mathrm{Wh}}{\mathrm{J}}\right)\frac{1}{10^9}\left(\frac{\mathrm{GWh}}{\mathrm{Wh}}\right) \quad (17.5)$$

Renewable energy generation is calculated using Eq. (17.6), while Eq. (17.8) sets the line capacities between the modelled countries.

$$G(p,u) = Pmax(u) \cdot Profiles(p,u) \cdot dt \quad (17.6)$$

$$FLOW(p,l) \leq LineCapacity(l) \cdot dt \quad (17.7)$$

Equation (17.8) is the water balance equation for the available water resources for each node n at period p. Factor F_2 is used to convert water resources from m^3/s into Mm^3.

$$RES(p,u) - RES(p-1,u) = F_2 \cdot (Resources(p,u) - Evaporation(p,u) + UPSTREAM(p,u) + CH(p,u) - DIS(p,u) - SPILL(p,u) - DemandW(p,u)) \quad (17.8)$$

Minimum and maximum available reservoir storage is set by Eq. (17.9)

$$Stmax(p, u) \geq RES(p, u) \geq Stmin(p, u) \tag{17.9}$$

Pumped hydropower plant pumping mode is described using Eq. (17.10), while Eq. (17.11) sets the maximum pumping power capacity for the each unit.

$$PUMP(p, u) = CH(p, u) \cdot NominalHead(u) \cdot F_1 \cdot \frac{1}{eta\_pump} \tag{17.10}$$

$$PUMP(p, u) \leq Pmax(u) \cdot dt \tag{17.11}$$

17.2.3 Dispa-SET Unit Commitment and Dispatch Model

Dispa-SET UCD model aims to represent operation of the large-scale power system and it consists of two parts. First part is scheduling the start-up, shut down and operation of available generation units. The problem requires the use of binary variables to be able to represent the start-up and shut down decisions, while also considering constraints connected to the commitment status of the generation units in all time periods. Second part of the problem is allocation of the total power demand to be achieved among the available generation units to achieve minimization of total power system cost. This part of the problem is the economic dispatch problem, which determines the output of all generation units. The problem can be formed as a mixed integer linear problem (MILP) or simplified linear program (LP) depending on the picked level of details for the input data. The implementations of both problems (MILP and LP) exists in both GAMS and PYOMO and can be in more details found in [31].

Continues variables include dispatched power, shed load and curtailed power generation in every time step, while commitment status of each unit represents binary variables. The model features include: minimum and maximum power outputs for the all units, up and down reserves, minimum up and down times, load shedding, ramping limits, curtailment, pumped-hydro storage, non-dispatchable units, outages of all units, constraints on the targets for the renewables and/or CO_2 emissions, schedules for the reservoir storage level, constraints of CHP units and thermal storage, network-related constraints, different clustering methods and costs of start-up, ramping and no load. More on model sets, variables and parameters can be found in [31].

Dispa-SET UCD objective function is composed of all relevant power system costs, such as start-up and shut down costs, fixed, variable, ramping, transmission-related, load shedding and lost load costs. Objective function can be seen in Eq. (17.12).

$$
\begin{aligned}
SystemCost = \sum_{u,n,i} \big[& CostStartUp_{u,i} + CostShutDown_{u,i} \\
& + CostFired_{u} \cdot Committed_{u,i} + CostVariable_{u,i} \cdot Power_{u,i} \\
& + CostRampUp_{u,i} + CostRampDown_{u,i} \\
& + PriceTransmission_{i,l} \cdot Flow_{i,l} \\
& + CostLoadShedding_{i,n} \cdot ShedLoad_{i,n} \\
& + CostHeatSlack_{chp(u),i} \cdot HeatSlack_{chp(u),i} \\
& + CostVariable_{chp(u),i} \cdot CHPPowerLossFactor_{chp(u)} \cdot Heat_{chp(u),i} \\
& + VOLL_{Power} \cdot \left(LostLoadMaxPower_{i,n} + LostLoadMinPower_{i,n}\right) \\
& + VOLL_{Reserve} \cdot \left(LostLoadReserve2U_{i,n} + LostLoadReserve2d_{i,n}\right) \\
& + VOLL_{Ramp} \cdot \left(LostLoadRampUp_{u,i} + LostLoadRampDown_{u,i}\right)\big] \quad (17.12)
\end{aligned}
$$

The main constraint equation is the supply-demand balance in the day-ahead market. In Eq. (17.13), the sum of all power produced by the units in node n, the power imported from neighbouring nodes and the curtailed power must be equal to the sum of the load and power consumed for energy storage, minus the load interrupted and the load shed.

$$
\begin{aligned}
\sum_{p,u} Power_{u,i} \cdot Location_{u,n} + \sum_{p,u} Flow_{l,i} \cdot LineNode_{l,n} = & \, Demand_{DA,n,h} \\
& + \sum_{p,u} StorageInput_{s,h} \cdot Location_{s,n} \\
& - ShedLoad_{n,i} - LL_{MaxPower,n,i} \\
& + LL_{MinPower,n,i} \quad (17.13)
\end{aligned}
$$

Other constraints related to the reserves, ramping, storage, power output, minimum up and down times, heat production, heat storage, network, emissions, curtailment and load shedding can be seen in [31].

17.3 Case Study and Scenario Definition

Balkan Peninsula Region is dependent on energy import, especially the oil and natural gas imports, with the high dependence and use of coal, primily lignite, in power generation. Besides the high carbon density due to the heavy dependence on coal, the excessive use of wood for fuel is a significant environmental concern, as it is the cause of air pollution, deforestation and land degradation [4].

In Fig. 17.2 percentage share of installed capacities in Balkan Region can be seen. Countries Albania, Croatia and Montenegro have more than 50% of installed capacities in form of thermal power units. The highest share of thermal power units is in Hungary and Kosovo, with percentages of 90% and 89%, respectively. In Hungary,

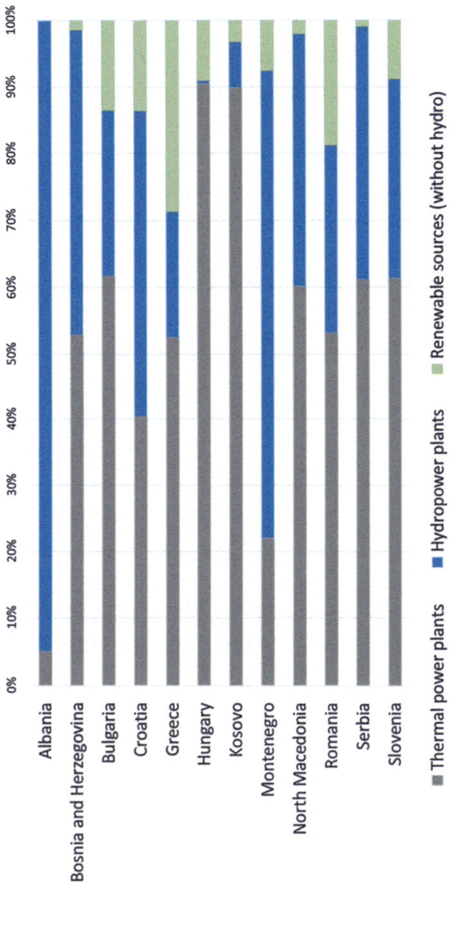

Fig. 17.2 Percentage share of installed capacities in the Balkan Region [32, 33]

nuclear power plant account for 24%, whilst the rest of percentage is related to the fossil-fired thermal power units, mostly gas-fired units. The 89% of thermal power units in Kosovo are lignite-fired thermal power units. Bulgaria, Greece and Romania have the highest share of installed thermal unit capacities, with 7963, 8804 and 12 247 MW, respectively. Excluding nuclear power units, the highest percentages of fossil-fired units are in Hungary, Kosovo and Serbia, with percentages of 67, 89 and 61%, respectively. Countries with highest share of hydropower generation are Albania, Bosnia and Herzegovina and Croatia with shares of 95, 45 and 46%, respectively. Countries with highest installed hydropower capacities are Bulgaria, Greece and Romania with installed capacities of 3204, 3172 and 6490 MW, respectively. Excluding hydropower, countries with high renewable energy sources are Greece, Romania, Bulgaria and Croatia, with shares of 14, 14, 29 and 18%, respectively. Countries with highest installed renewable energy source capacities are Bulgaria, Greece and Romania, with capacities of 4314, 1744 and 4796 MW, respectively.

The study includes scenario-based analysis regarding three different hydrological years. Net water inflows have been provided by the JRC from the rainfall-runoff hydrological LISFLOOD model briefly described in Sect. 2.1. The assumption is that the provided water inflows are the total runoff at studied catchment level.

Figure 17.3 shows total water inflows for the included hydropower plants locations for a period between 1990 and 2016. The yellow highlighted line represents the runoff for the dry (2007), green highlighted for the average (2015), and red highlighted for the wet (2010) year.

17.4 Model Results

17.4.1 Medium-Term Hydrothermal Coordination Model Results

Validation of MTHC model was based on hydropower generation on a country level. Data used to validate hydropower generation was obtained from the European Network of Transmission System Operators for Electricity (ENTSO-E) Transparency platform [32]. Model results, as well as compared values from ENTSO-E can be seen in Table 17.4. Subsequent to model validation, model was solved for the additional wet and dry years with a change in water inflow inputs.

Yearly aggregated hydropower generation for the Balkan Region averaged at 145.91, 175.48 and 232.18 GWh/day, while it peaked at 236.06, 277.96 and 331.86 GWh/day for dry, average and wet year, respectively. Minimum hydropower generation values were 88.66, 89.92 and 135.02 GWh/day for the dry, average and wet year, respectively. Annual Balkan Peninsula region aggregated hydropower generation from MTHC model shows increase from 53 258 GWh for dry year to 64 050 and 84 747 GWh for the average and wet year, respectively.

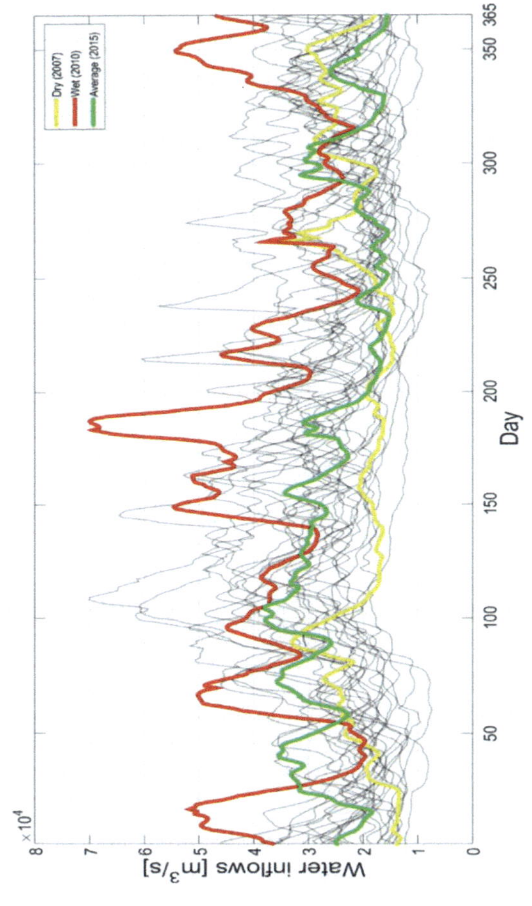

Fig. 17.3 Total water inflows for the studied Balkan Peninsula region. Years between 1990 and 2016

Table 17.4 MTHC hydropower generation for the reference/average (2015) year

Country	MTHC model [GWh]	ENTSO-E [GWh]	Δ/ENTSO-E [%]
Albania	5696	/	/
Bosnia and Herzegovina	5614	5650	−0.64
Bulgaria	5963	6155	−3.12
Croatia	5719	5657	1.10
Greece	6278	6091	3.06
Hungary	237	227	4.51
Kosovo	141	/	/
Montenegro	1442	1415	1.90
North Macedonia	1585	1514	4.71
Romania	16 849	16 545	1.84
Serbia	10 532	10 633	−0.95
Slovenia	3997	4060	−1.56
Sum	64 053	57 947	0.46

Balkan Peninsula region aggregated run-of-river hydropower generation on a daily time scale can be seen in Fig. 17.4. Annual region aggregated run-of-river hydropower generation averaged at values of 59.17, 65.86 and 80 GWh/day for dry, average and wet year, respectively. Run-of-river hydropower generation peaked at values of 86.19, 87.07 and 101.73 GWh/day, while the minimum reached was 40.88, 43.18 and 51.52 GWh/day, for a dry, average and wet year, respectively.

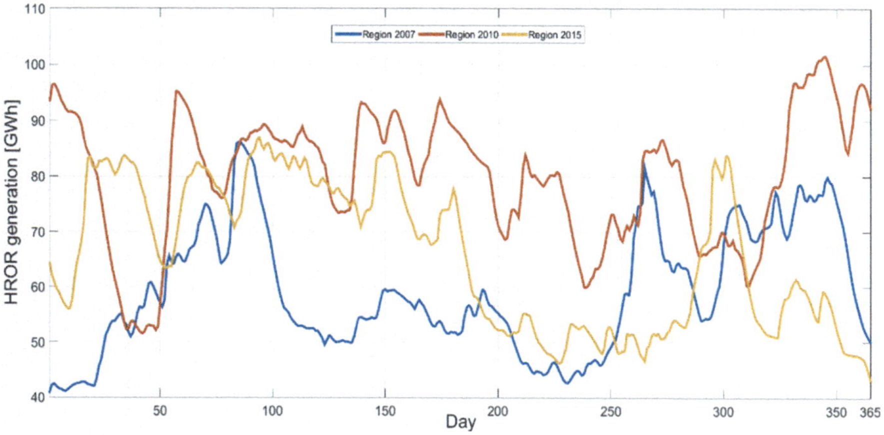

Fig. 17.4 Balkan Peninsula region aggregated run-of-river hydropower generation for dry (2007), average (2015) and wet (2010) year, in GWh

17.4.2 Dispa-SET Unit Commitment and Dispatch Model Results

Dispa-SET UCD model was run for the three different hydrological years using hourly time step. MTHC model results in form of run-of-river hydropower generation and reservoir levels were used as an input for the different scenario models in Dispa-SET UCD.

Difference in average electricity cost between a dry and an average year is due to the lower amount of energy generated from hard coal (decrease from 3.51 TWh during the dry year, to 3.07 TWh for an average year), lignite (decrease from 151.33 TWh during the dry year, to 140.15 TWh for an average year) and gas-powered units (decrease from 6.24 TWh during the dry year, to 4.68 TWh for an average year), that is replaced with hydropower generation (increase from 53.06 TWh for dry year, to 65.24 TWh for average year).

A similar scenario can be observed when comparing average and wet years. One can say that drop in average electricity price from 16.35 to 14.05 €/MWh can be explained by the decrease in electricity generation from hard coal (decrease from 3.07 TWh for average year, to 2.43 TWh for wet year), lignite (decrease from 140.15 TWh for average year, to 125.12 TWh for wet year) and gas-fired units (decrease from 4.68 TWh for average year, to 1.46 TWh for wet year), at the expense of increased hydropower generation (increase from 65.24 TWh for average year, to 85.13 TWh for wet year). Similarly, drop in CO_2 emissions from dry to wet year, due to decrease in fossil-fuel generation, and increase in hydropower generation, can be seen in Table 17.5. Wind and solar generation is the same across the simulated years, because of the same capacity factor being used as an input data.

Compared results on hydropower generation on a country level can be seen in Table 17.6. Table shows that modelled results are closely following statistically obtained data from ENTSO-E, with difference on region level being only 2.12%. Highest difference between modelled results and statistical data are for countries Croatia, Hungary and North Macedonia, with differences of 7.27, 8.62 and 22.55%, respectively. Even though 22.55% difference for North Macedonia is high, modelled results are close to the IEA statistical data, with the difference of only −0.51%. IAE statistical obtained data for Croatian hydropower generation is 6556 GWh, which with ENTSO-E data gives average vale for hydropower generation of 6107 GWh. Comparing that results with modelled hydropower generation for Croatia gives only −0.62% difference. Higher percentage difference for a Hungary can be explained with total smaller amount of hydropower production which in turn gives higher percentages with smaller offsets from statistical data. Pumped storage hydropower units account for 6.88, 6.04 and 9.41 TWh of hydropower generation for average, dry and wet year, respectively. When expressed in percentages, pumped-storage hydropower generation accounts for 10.5, 11.4 and 11% of total hydropower generation for average, dry and wet year, respectively. Hydropower generation as Dispa-SET UCD model result can be seen in Fig. 17.5.

Table 17.5 Balkan Peninsula region aggregated results for a dry (2007), average (2015) and wet (2010) year

Region aggregated statistics	Unit	Dry	Average	Wet
Average electricity cost	€/MWh	17.79	16.35	14.05
Total consumption	TWh	289.22	289.22	289.22
Total system cost	m EUR	4978	4573	3932
Peak load	GW	47.992	47.992	47.992
Net imports	TWh	9.452	9.452	9.452
NUC generation	TWh	48.356	48.356	48.356
LIG generation	TWh	151.33	140.15	125.12
HRD generation	TWh	3.506	3.072	2.430
GAS generation	TWh	6.244	4.682	1.462
WST generation	TWh	0.090	0.090	0.090
SUN generation	TWh	6.919	6.919	6.919
WIN generation	TWh	10.272	10.272	10.272
WAT generation	TWh	53.064	65.237	85.132
CO_2 emissions	$MtCO_2$	164.36	152.67	133.96

NUC—nuclear power generation; LIG—power generation from lignite-fired thermal power plants; HRD—power generation form hard-coal-fired thermal power units; GAS—power generation from gas-fired thermal power plants; WST—power generation from thermal power units using waste as a fuel; SUN—solar power generation; WIN—wind power generation; WAT—hydropower generation

Table 17.6 Dispa-SET UCD hydropower generation for the reference/average (2015) year

Country	UCD model [GWh]	ENTSO-E [GWh]	Δ/ENTSO-E [%]
Albania	5907	/	/
Bosnia and Herzegovina	5664	5650	0.24
Bulgaria	6392	6155	3.85
Croatia	6069	5657	7.27
Greece	6288	6091	3.09
Hungary	247	227	8.62
Kosovo	144	/	/
Montenegro	1515	1415	7.04
North Macedonia	1855	1514	22.55
Romania	16 149	16 545	−2.40
Serbia	10 919	10 633	2.69
Slovenia	4090	4060	0.75
Sum	65 237	57 947	2.12

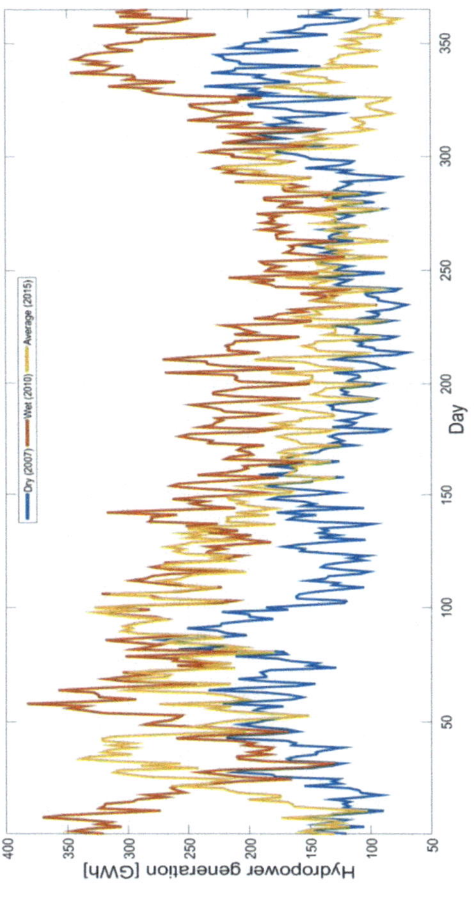

Fig. 17.5 Dispa-SET UCD region aggregated hydropower generation for dry (2007), average (2015) and wet (2010) year

17.5 Conclusion

This study shows successful implementation of two energy models and hydrological LISFLOOD model for detailed analysis of impacts on the regional power system for different hydrological conditions. Countries included in the study are six Western Balkan countries, Albania, Bosnia and Herzegovina, Kosovo, Montenegro, North Macedonia, Serbia, and neighbouring countries Bulgaria, Croatia, Greece, Hungary, Romania and Slovenia. Combining water inflows from hydrological LISFLOOD model with two energy models, three different scenarios for dry, average and wet year were conducted.

Besides power generation, results from UCD model include economical, commitment and power dispatch values for each unit and can be aggregated by country or whole Balkan Peninsula region. Results show an increase of hydropower generation from 53.06 GWh for dry year, to 65.24 and 85.13 GWh for average and wet year, respectively. Increase in hydropower generation is at expense of a decrease in fossil-fuel power generation. Furthermore, average electricity cost decreased from 17.79 €/MWh for the dry year to 16.35 and 14.05 €/MWh for an average and wet year, respectively.

Acknowledgements Contribution by the JRC team, especially Marko Adamović, regarding the LISFLOOD model is greatly acknowledged. Furthermore, we thank Hrvoje Medarac, Matteo De Felice, Ignacio Hidalgo González, Iratxe Gonzalez Aparicio and Sebastian Busch for comments and recommendations regarding the usage of Dispa-SET model. Contributions from European Union's Horizon 2020 programme projects INSULAE and KeepWarm, as well as from the project "Provision of electricity in the event of climate extremes and natural disasters" from European Structural and Investment Fund are greatly acknowledged.

References

1. R. Fernandez-Blanco Carramolino, K. Kavvadias, I. Hidalgo Ganzalez, *Water-related modelling in electric power systems-WATERFLEX Exploratory Research Project: version 1* (2017). https://doi.org/10.2760/482511.
2. R. Fernandez-Blanco Carramolino, K. Kavvadias, A. De Roo, B. Bisselink, I. Hidalgo Gonzalez, *The water-energy nexus and the implications for the flexibility of the Greek power system* (2017). https://doi.org/10.2760/361084
3. Hub, EU Science Service TEC science and knowledge. Freshwater use from the European energy sector (n.d.). https://ec.europa.eu/jrc/en/science-update/freshwater-use-european-energy-sector
4. Mott MacDonald IPF Consortium. REGIONAL STRATEGY FOR SUSTAINABLE HYDROPOWER IN THE WESTERN BALKANS: Background Report No. 2 Hydrology, integrated water resources management and climate change (2017)
5. International Energy Agency (IEA), *World Energy Outlook 2012* (2012)
6. International Energy Agency (IEA), *World Energy Outlook 2016* (2016)
7. Department of Energy U. *The Water-Energy Nexus: Challenges and Opportunities* (2014)
8. Hub, EU Science Service TEC science and knowledge. Exploiting modelling to better address issues related to the water-energy nexus (n.d.). https://ec.europa.eu/jrc/en/news/exploiting-modelling-better-address-issues-related-water-energy-nexus

9. D. Magagna, I. Hidalgo Gonzalez, G. Bidoglio, S. Peteves, M. Adamovic, B. Bisselink et al., *Water-Energy Nexus in Europe* (2019). https://doi.org/10.2760/968197
10. C. Carmona-Moreno, C. Dondeynaz, M. Biedler, *Position Paper on Water, Energy, Food and Ecosystems (WEFE) Nexus and Sustainable Development Goals (SDGs)* (2018). https://doi.org/10.2760/5295
11. R. Fernández-Blanco Carramolino, K. Kavvadias, M. Adamovic, I. Hidalgo Gonzalez, *The Water-Power Nexus of the Iberian Peninsula Power System WATERFLEX Project* (2017). https://doi.org/10.2760/739963
12. L. Hardy, A. Garrido, L. Juana, Evaluation of Spain's water-energy Nexus. Int. J. Water Resour. Dev. **28**, 151–170 (2012). https://doi.org/10.1080/07900627.2012.642240
13. D. Zafirakis, C. Papapostolou, E. Kondili, J.K. Kaldellis, Evaluation of water-use needs in the electricity generation sector of Greece. Int. J. Environ. Resour. **3**, 39 (2014). https://doi.org/10.14355/ijer.2014.0303.01
14. I. Ziogou, T. Zachariadis, Quantifying the water–energy nexus in Greece. Int. J. Sustain. Energy **36**, 972–982 (2017). https://doi.org/10.1080/14786451.2016.1138953
15. Z. Khan, P. Linares, J. García-González, Integrating water and energy models for policy driven applications. A review of contemporary work and recommendations for future developments. Renew. Sustain. Energy Rev. **67**, 1123–1138 (2017). https://doi.org/10.1016/j.rser.2016.08.043
16. Z. Khan, P. Linares, J. García-González, Adaptation to climate-induced regional water constraints in the Spanish energy sector: An integrated assessment. Energy Policy **97**, 123–135 (2016). https://doi.org/10.1016/j.enpol.2016.06.046
17. P. Behrens, M.T.H. van Vliet, T. Nanninga, B. Walsh, J.F.D. Rodrigues, Climate change and the vulnerability of electricity generation to water stress in the European Union. Nat. Energy **2**, 17114 (2017). https://doi.org/10.1038/nenergy.2017.114
18. S.J. Pereira-Cardenal, H. Madsen, K. Arnbjerg-Nielsen, N. Riegels, R. Jensen, B. Mo et al., Assessing climate change impacts on the Iberian power system using a coupled water-power model. Clim. Change **126**, 351–364 (2014). https://doi.org/10.1007/s10584-014-1221-1
19. S.J. Pereira-Cardenal, B. Mo, N.D. Riegels, K. Arnbjerg-Nielsen, P. Bauer-Gottwein, Optimization of multipurpose reservoir systems using power market models. J. Water Resour. Plan. Manag. **141**, 04014100 (2015). https://doi.org/10.1061/(ASCE)WR.1943-5452.0000500
20. S.J. Pereira-Cardenal, B. Mo, A. Gjelsvik, N.D. Riegels, K. Arnbjerg-Nielsen, P. Bauer-Gottwein, Joint optimization of regional water-power systems. Adv. Water Resour. **92**, 200–207 (2016). https://doi.org/10.1016/j.advwatres.2016.04.004
21. N. Vakilifard, M. Anda, P. Bahri, G. Ho, The role of water-energy nexus in optimising water supply systems—Review of techniques and approaches. Renew. Sustain. Energy Rev. **82**, 1424–1432 (2018). https://doi.org/10.1016/j.rser.2017.05.125
22. T.A. DeNooyer, J.M. Peschel, Z. Zhang, A.S. Stillwell, Integrating water resources and power generation: The energy–water nexus in Illinois. Appl. Energy **162**, 363–371 (2016). https://doi.org/10.1016/j.apenergy.2015.10.071
23. B.R. Scanlon, I. Duncan, R.C. Reedy, Drought and the water–energy Nexus in Texas. Environ. Res. Lett. **8**, 045033 (2013). https://doi.org/10.1088/1748-9326/8/4/045033
24. A. Siddiqi, L.D. Anadon, The water–energy Nexus in Middle East and North Africa. Energy Policy **39**, 4529–4540 (2011). https://doi.org/10.1016/j.enpol.2011.04.023
25. M. De Felice, I. Gonzalez Aparicio, T. Huld, S. Busch, I. Hidalgo Ganzalez, *Analysis of the Water-Power Nexus in the West African Power Pool* (2018). https://doi.org/10.2760/362802
26. M. Pavičević, S. Quoilin, A. Zucker, G. Krajačić, T. Pukšec, N. Duić, Applying the Dispa-SET model to the Western Balkans power system. J. Sust. Dev. Energy, Water Environ. Syst. **8**, 184–212 (2020). https://doi.org/10.13044/j.sdewes.d7.0273
27. M. Pavičević, K. Kavvadias, T. Pukšec, S. Quoilin, Comparison of different model formulations for modelling future power systems with high shares of renewables—The Dispa-SET Balkans model. Appl. Energy **252**, 113425 (2019). https://doi.org/10.1016/j.apenergy.2019.113425
28. World Bank 2014, *Turn Down the Heat: Confronting the New Climate Normal* (World Bank, Washington, DC). License: Creative Commons Attribution—NonCommercial—NoDerivatives 3.0 IGO (CC BY-NC-ND 3.0 IGO) (n.d.)

29. P.A. Burek, J. Vand der Knijff, A. De Roo, *LISFLOOD—Distributed Water Balance and Flood Simulation Model* (2013). https://doi.org/10.2788/24719
30. R. Fernandez Blanco Carramolino, K. Kavvadias, I. Hidalgo Gonzalez, *Hydro-related Modelling for the WATERFLEX Exploratory Research Project* (2016). https://doi.org/10.2760/386964
31. S. Quoilin, I. Hidalgo Gonzalez, A. Zucker, *Modelling Future EU Power Systems Under High Shares of Renewables—The Dispa-SET 2.1 Open-Source Model* (2017). https://doi.org/10.2760/25400
32. ENTSO-E Transparency Platform (n.d.) https://transparency.entsoe.eu/dashboard/show
33. International Energy Agency (IEA), IEA Statistics (n.d.) https://www.iea.org/statistics/?country=FYROM&year=2016&category=Electricity&indicator=ElecGenByFuel&mode=chart&dataTable=ELECTRICITYANDHEAT

Chapter 18
An Algorithm for the Design of a Stand-Alone Hybrid System

Ciobanu Daniela, Săulescu Radu, and Jaliu Codruţa

Abstract According to the report provided by REN 21 (Renewable Energy Network for the 21st Century), over 73.8% of the amount of energy produced in Europe comes from non-renewable energy sources. However, the energy produced by fossil sources leads to an increase in the greenhouse gases emissions and to global warming. To reduce these effects, more and more countries have decided to increase the use of renewable energy sources. Sun, wind, water, biomass and geothermal energies are the main sources of renewable energy, but part of them are unpredictable and are not available all the time. Therefore, to meet the required energy, the hybrid systems, based on a combination of renewable energies, are preferred. These systems are often used in remote areas, but, due to the variability of the renewable sources, their design and analysis is difficult to be performed. An algorithm for the selection and design of a hybrid energy system based on solar, wind and/or hydro energy is presented in this paper depending on the renewable resources onsite available. The algorithm is applied to a case study that enables to select the appropriate hybrid system that provides the required energy for a guesthouse located in a remote area. Conclusions and recommendations are formulated on the design of hybrid systems.

Keywords Hybrid system · Wind turbine · PV system · Hydropower system · Stand-alone system

---

C. Daniela (✉) · J. Codruţa
Renewable Energy Systems and Recycling R&D Centre, Transilvania University of Brasov, Brasov, Romania
e-mail: daniela.ciobanu@unitbv.ro

J. Codruţa
e-mail: cjaliu@unitbv.ro

S. Radu
Design of Mechanical Elements and Systems R&D Centre, Transilvania University of Brasov, Brasov, Romania
e-mail: rsaulescu@unitbv.ro

I. Visa and A. Duta (eds.), *Solar Energy Conversion in Communities*,
Springer Proceedings in Energy, https://doi.org/10.1007/978-3-030-55757-7_18

18.1 Introduction

A global increase in energy by 2.3% was registered in 2018 due to the economic growth and the increase in the energy demand for heating and cooling [1]. Worldwide it was proven that the main source of electricity generation is the renewable energy. Renewable energy is increasingly cost-competitive compared to conventional fossil fuel based energy. Thus, at the end of 2018, the price of energy obtained using photovoltaic and wind systems was lower than the price of the energy obtained using fossil fuels. At that moment, the amount of electricity produced from renewable sources represented 26.2% of the total global energy production, Fig. 18.1 [1].

Special attention is devoted to the implementation of renewable energy systems for electricity production in households far from the grid. According to the report provided by REN21, at the end of 2018, 5% of the African population and 2% of the population in Asia had access to electricity through off-grid photovoltaic (PV) systems. Globally, starting from 2017, the population without access to electrical energy felt below 1 billion [1].

The main renewable energies for electricity production are the solar, hydro and wind energies. At the end of 2018, 181 GW of renewable electric power were added: 100 GW were obtained using PV systems, 50 GW using wind systems, 20 GW using hydropower systems and 11 GW geothermal and bio-power systems [1].

However, the main disadvantage of the renewable energy systems is the variability of the energy sources, that depends, among other predictable factors, on the weather conditions. In order to ensure the required energy, hybrid variants, such as: PV-Wind, PV-Hydro or Hydro-Wind are proposed and implemented [2–6]. The design of these

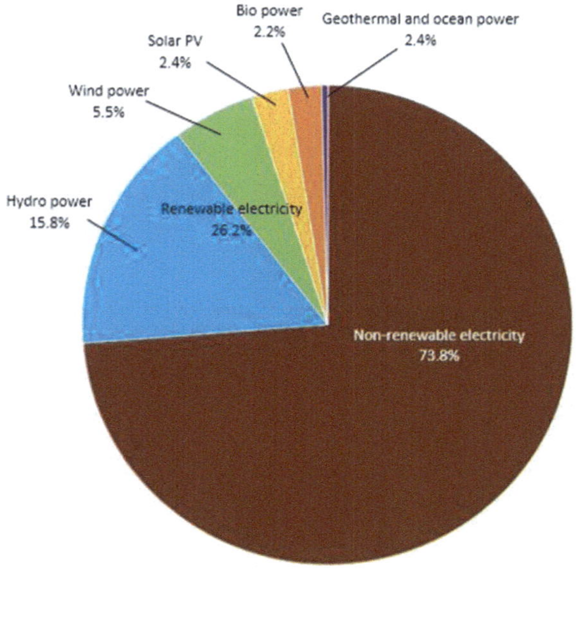

Fig. 18.1 Renewable energy share of global electricity production in 2018 [1]

hybrid systems is still a challenge for the developers as they are site dependent, algorithms being proposed for the design of each type [2–4].

Based on the extended use of PV systems, this paper presents a general algorithm for the design of a stand-alone hybrid system for electrical energy generation, a system where the main source of renewable energy is considered the solar radiation and the secondary source is selected between hydro and wind. In order to obtain the optimal solution, the algorithm takes into account the surface available for implementing the PV systems and the lowest cost of the secondary renewable system. The general algorithm is further applied on a case study, i.e. a guesthouse located in a rural area.

18.2 Algorithm for the Design of the Hybrid System

The proposed algorithm considers the hybrid solution consisting of combinations of photovoltaic, hydro or wind systems, Fig. 18.2. This stand-alone hybrid system can be implemented in locations where all these three renewable sources are available.

The research carried out and published highlights that the photovoltaic systems have the lowest initial cost of the electric energy, i.e. 1.18 €/W [2]. The operation and maintenance costs represent 1% of the installation price in the case of PV systems, between 2.2 and 3% in the case of hydropower systems and 3% in the case of wind turbines [2, 7, 8]. Therefore, in order to choose the optimal hybrid system, the proposed algorithm considers the photovoltaic system as the primary component of the hybrid system, while the secondary one is either a small hydropower plant or a wind turbine.

The initial data necessary for choosing the optimal variant are: the weather data for the implementation location (the solar irradiance, the amount of precipitations, the wind speed, the ambient temperature, etc.), the required amount of electrical energy, the available area for implementing the photovoltaic system and the technical data (database) with renewable energy systems.

The algorithm starts by selecting the type of photovoltaic modules to be used and by determining the energy provided by these obtaining the required number of photovoltaic modules by using Eqs. (18.1) and (18.2):

$$E_{PV} = E \cdot A_{PV} \cdot \eta_{PV} \left[\text{kWh/day}\right] \tag{18.1}$$

$$N_{PV} = \frac{E_{demand}}{E_{PV}} \tag{18.2}$$

where E is the solar energy provided per square meter [kWh/m^2/day], A_{PV} is the area of a PV module [m^2]; η_{PV} is the efficiency of the PV modules and E_{demand} is the required energy [kWh/day].

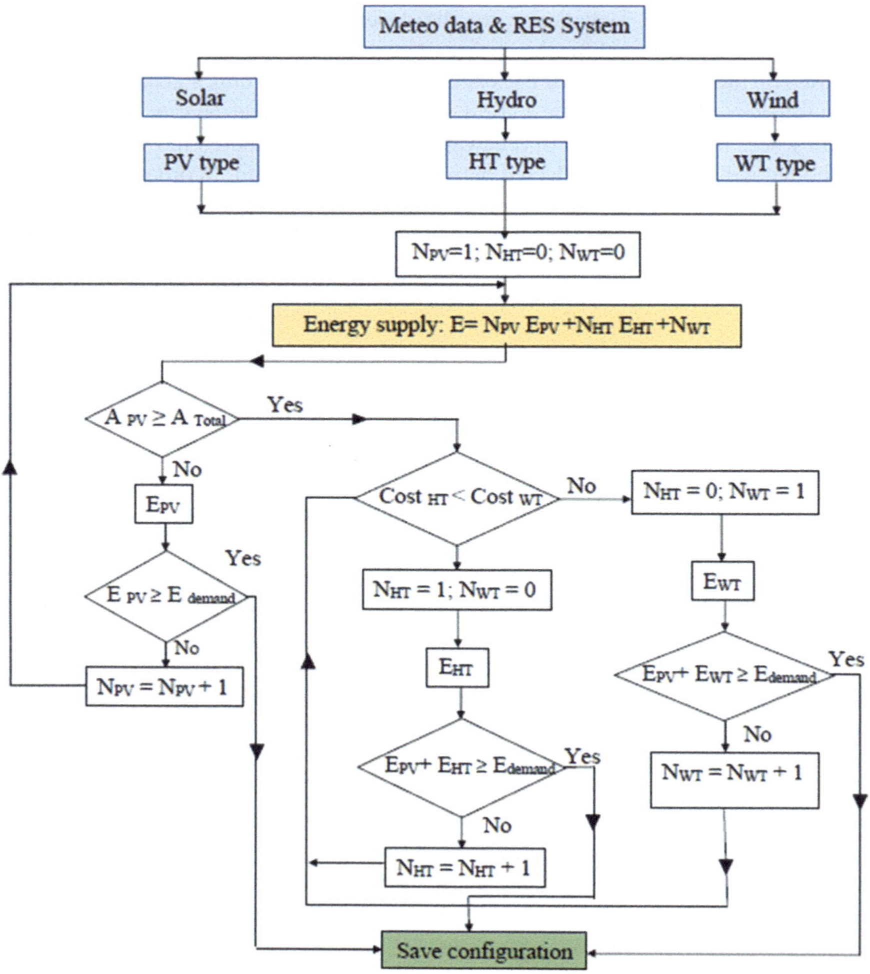

Fig. 18.2 Algorithm for the design of a stand-alone hybrid system

The area A_{PV} required for the installation of the photovoltaic modules is obtained based on the number of photovoltaic modules by using Eq. 18.3 [2]:

$$A_{PV} = L_{PV} \cdot l_{PV} \cdot \sqrt{2} \cdot N_{PV} \left[\mathrm{m}^2\right] \tag{18.3}$$

where L_{PV} represents the length of the PV panel [m]; l_{PV} is the panel width [m]; N_{PV} is the PV modules number.

The area required to install the N_{PV} photovoltaic modules, A_{PV}, is further compared with the available area A_{Total}. If the required area is larger than or equal to the available area, then the number of PV modules that can be installed is calculated using Eq. 18.3. The amount of energy supplied by the PV modules, $E_{PV,}$ is further obtained and the difference between the supplied energy E_{PV} and the required energy E_{demand} is calculated, this amount being covered with one of the other two types of renewable energy systems.

If the occupied area A_{PV} is smaller than the available area A_{Total}, then the algorithm saves the generated configuration.

Because the operation and maintenance costs are similar for hydro and wind systems, the equipment and implementation costs are compared to get the type of system to be used for producing the difference between the required energy and the energy provided by the photovoltaic system. The cheaper system is selected to provide the additional energy.

The energy supplied by the hydropower system can be calculated using Eq. (18.4), and the energy produced by the wind turbine using Eq. (18.5) [4]:

$$E_{HT} = \rho \cdot g \cdot Q \cdot H_n \cdot \eta \cdot h_{operating} \times 10^{-3} \left[\text{kWh/day}\right] \tag{18.4}$$

$$E_{WT} = 0.5 \cdot C_p \cdot \rho_a \cdot A \cdot v^3 \cdot \eta \cdot h_{operating} \left[\text{kWh/day}\right] \tag{18.5}$$

where ρ is the water density [kg/m^3]; g is the gravitational acceleration [m/s^2]; Q is the water flow [m^3/s]; H is the net head [m]; $h_{operating}$ is the number of operating hours [h]; ρ_a is the air density [kg/m^3]; C_p is the power coefficient [-]; A is the rotor swept area [m^2]; v is the average wind speed [m/s] and η is the efficiency of the hydropower/wind unit [-].

Under favourable conditions, η of the hydropower unit is typically around 65–85% and for the wind turbine is about 50–53% [4, 9].

18.3 Case Study

The general algorithm above discussed is further applied for dimensioning a stand-alone hybrid system to be implemented at a guesthouse in the Stupini district in the Brasov city outskirts. The guesthouse, with 10 rooms and a restaurant, is located in an area where all three renewable energy sources can be exploited. The initial data required for selecting and sizing the hybrid system are: the weather data that are recorded by the weather station located on the roof of the L7 Laboratory of the R&D Institute of the Transilvania University of Brasov during 2015–2016 (located at about 500 m from the guesthouse), the electricity demand of the guesthouse is around 100 kWh/day, the available implementation area for the PV modules is of 170 m^2. The river close to the guesthouse is Ghimbasel, at about 80 m from the house. According to the weather station data, the average daily wind speed during April 2015 and July 2016 was 4.1 m/s at a height of 10 m [2]. The precipitations influence the flow of the Ghimbasel river by 10–15% and the flow (Q), varies between

Table 18.1 Technical data of 3 types of PV panels [10]

Type	Monocrystalline Silicon PMS275-300M-60	Polycrystalline Silicon NS-250-280P-60	PERC AE M6-60_280-320W
Power range (Wp)	275–300	250–280	280–320
Efficiency (%)	16.94–18.48	15.4–17.2	17.11–19.24
Dimensions (mm)	1640 × 990 × 40	1640 × 992 × 35	1650 × 992 × 35
Price (€/Wp)	0.212	0.184	0.205

0.15 and 0.65 m^3/s, while the gross head, H_g, at the site is of 12.5 m (12 m net head H). The most unfavorable month was considered in sizing the PV system in terms of solar irradiance, i.e. March. The average amount of solar radiation that can be captured per square meter in March is E = 2897 kWh/m²/day.

Starting from these initial data, according to the algorithm in Fig. 18.2, 3 types of photovoltaic **modules** that exist in the database and are sold on the market are considered [10] (Table 18.1).

In order to reach the lowest possible cost of the hybrid system, the polycrystalline variant is selected to be implemented, that is the cheapest (0.184 €/Wp), even if it has the lowest efficiency. The energy produced by the PV system in March is calculated using Eq. (18.1) where the efficiency is considered to be 16.5%: E_{PV} = 0.777 kWh/day. In order to cover the energy demand, the number of solar modules is calculated using Eq. (18.2): N_{PV} = 129 modules. The area required to ensure the installation of these modules is of 296.8 m^2 (Eq. 18.3). As the surface required for installation is much larger than the available area, the number of modules that can be mounted on the available area is calculated: N_{PV} = 73 PV panels. In this case, the electric energy provided by the PV system consisting of 73 panels is E_{73PV} = 56.721 kWh/day. The difference up to the energy required for the guesthouse is of 43.28 kWh/day and this amount has to be provided by the secondary renewable energy system.

In case of using as secondary system a wind turbine, 3 types wind turbines of 100, 50 and 25 kW are considered. The technical data for the three types of turbines are included in Table 18.2, also considering the recorded weather data (the onsite wind speed and frequency diagram) [2, 11].

The data analysis highlights that the 25 kW wind turbine cannot provide the required daily energy, while in the case of the 100 kW turbine, the cost is higher than that of the 50 kW variant. Therefore, the secondary system can be the 50 kW wind turbine.

Table 18.2 Technical data for the wind turbines [2, 11]

Type (Power) [kW]	100	50	25
Cost [€]	30,000	27,250	18,397
E_{WT} [kWh/day]	142.52	44.84	28.551

If a hydropower system is used as the secondary system, a Crossflow turbine is selected as water turbine, based on the flow and head. These turbines can operate for a flow variation between 0.2 and 8 m^3/s and a net head H between 2 and 200 m [9]. The energy provided by a Crossflow turbine directly connected to the generator is obtained according to Eq. (18.4), where the net head $H = 12$ m is considered, the operation time being 3 h/day, while the most unfavorable variants are considered for the other parameters: the river flow is 0.2 m^3/s and the efficiency $\eta = 65\%$. Under these conditions, the energy provided by the Crossflow system is $E_{HT} = 45.774$ kWh/day. The small hydro system can operate for about 2500 h/year at this flow, the extra amount of energy being stored in batteries.

The Crossflow water turbines can be found on the market at prices ranging between 7359 and 11,038 €, depending on the dimensions [12]. If the turbine cost is 11,038 € and considering that the operation and maintenance costs for 30 years of lifetime are about 3% of the initial costs, it results that the cost of this system is about 11,362 €.

The comparison of the costs of the two types of secondary systems (the 50 kW wind turbine and the run-of-river hydro system) indicates the hydropower system as the lower cost option for providing electricity to the guesthouse.

Thus, the hybrid system proposed to meet the energy demand of the guesthouse located in Stupini—Brasov consists of a photovoltaic system with 73 polycrystalline silicon modules that can be installed on an area of 170 m^2 and a small hydropower system of Crossflow type as the secondary system, operating for 3 h a day.

During cloudy days, when the photovoltaic system cannot provide the calculated amount of energy, this energy is produced by the small hydro system.

18.4 Conclusions

The paper proposes a general algorithm for the design of a stand-alone hybrid system that provides the required energy for a location where the connection to the electrical grid is difficult to be done, either due to the high costs or due to its remote location. The general algorithm considers a primary renewable energy system of PV type and a secondary system that can be either a wind turbine or a small hydropower plant. The algorithm takes into account the following initial data: the potential of the renewable energy sources, the implementation area for the PV system and the lowest cost of the secondary renewable energy system.

This algorithm can be applied for locations with wind and hydro potential, the type of the secondary component of the hybrid system being imposed by the system costs (equipment, implementation, maintenance and operating costs).

For stand-alone applications, e.g. in remote mountain areas with higher values of the wind speed, the wind turbine is recommended to be used instead of the hydro systems. In these locations, at higher altitudes, although the head value increases, the river flow is much smaller; in these cases, another type of water turbine is suitable for the application (Francis or Pelton turbines) with higher initial costs, which increase the total cost of the hydropower system. Moreover, on these sites, the available area for installing PV systems is much smaller, and, therefore, a larger amount of energy has to be supplied by the secondary system.

The stand-alone hybrid system can also contain a back-up source, e.g. a diesel generator, that can supply the required energy when the batteries are discharged and the sun, wind and hydro potentials are very low, situation that may occur during winter time.

The use of stand-alone hybrid systems in remote locations represents the best option in meeting the daily energy demand of the consumers.

References

1. Renewable Energy Network for the 21st Century homepage. https://www.ren21.net/gsr-2019/chapters/chapter_01/chapter_01/#sub_4. Last accessed 6 Apr 2020
2. M. Gheorghe, D. Ciobanu, R. Saulescu, C. Jaliu, Economic analysis algorithm of a PV-wind hybrid system. Acta Tech Napocensis Ser.—Appl. Math. Mech. Eng. J. **61**(3), 107–114 (2018)
3. C. Jaliu, R. Săulescu, D. Ciobanu, Fl. Panainte, PV-wind hybrid system for the energy supply of an off-grid application, nearly zero energy communities, in *Nearly Zero Energy Communities*, Conference for Sustainable Energy 2017, ed. by I. Visa, A. Duta (Springer, Heidelberg, 2017), pp. 443–460
4. C. Jaliu, R. Saulescu, D. Ciobanu, Hybrid system for a stand-alone application, in *International Conference on Production Research—Africa, Europe and the Middle East; 4th International Conference on Quality and Innovation in Engineering and Management 2016* (2016), pp. 125–130
5. F.M. Pimenta, A.T. Assireu, Simulating reservoir storage for a wind-hydro hybrid system. Renew. Energy J. **76**, 757–767 (2015)
6. A. Hina Fathima, K. Palanisamy, Optimization in microgrids with hybrid energy systems—a review. Renew. Sustain. Energy Rev. **45**, 431–446 (2015)
7. A. Kaabeche, R. Ibtiouen, Techno-economic optimization of hybrid photo-voltaic/wind/diesel/battery generation in a stand-alone power system. Sol. Energy **103**, 171–182 (2014)
8. https://cleanleap.com/4-current-cost-hydropower/43-operation-and-maintenance-costs. Last accessed 12 Apr 2020
9. I. Visa, C. Jaliu, A. Duta, M. Neagoe, a.o., *The Role of Mechanisms in Renewable Energy Systems* (Transilvania University Publishing House, Brasov 2015)
10. https://www.enfsolar.com/pv/panel. Last accessed 12 Apr 2020
11. https://www.alibaba.com/product-detail/Wind-turbine-25kw-industrial-use_1682908901.html?bypass=true. Last accessed 14 Apr 2020
12. https://www.alibaba.com/showroom/cross+flow+turbine.html. Last accessed 03 Apr 2020

Chapter 19
Renewable Energy Production in Decommissioned Power Plant Sites for Sustainable Cities—The Case Study of Romania

Adrian Ciocănea, Mircea Scripcariu, Ştefăniţă Pluteanu, and Bogdan Tofan

Abstract Minimizing the energy inputs and decentralized energy production by using renewable sources are high priorities for the future of sustainable cities. Also, rehabilitation and reusing of decommissioned power plant sites are specific objectives of urban design when public decision is concerned. The paper presents an original solution aiming at the revival of five former industrial sites: Baia Mare, Braşov, Piteşti, Targu Jiu and Copşa Mică where tall industrial chimneys and available space is still existing. For each site, a solar park is proposed, where two different solar applications provide electricity: a solar chimney with upward air flow and a solar photovoltaic farm. After technical evaluations, the average installed power for each solar park is 5 MW. An economic assessment is provided based on energy production and other activities such as a charging system for electric vehicles and electrical buses used by the local transportation company and rental services to public/private consumers operating in the park. An analysis of the basic investment of approx. 19.5 M Euro—65% for the solar park and 35% for the architectural conversion works—revealed that the project is financially viable only if the revenue from the proposed activities is over 2.6–2.8 M Euro/year.

Keywords Solar PV energy · Solar chimney · Grid connection · Industrial site reuse · Sustainable city

19.1 Introduction

By 2050 two-thirds of the world's population will be urban [1]. According to recent studies [2] up to 40% of the primary energy consumption in OECD countries are reported in buildings and up to 80% of this occurs in urban agglomerations [3]. At

A. Ciocănea (✉) · M. Scripcariu · Ş. Pluteanu
Universitatea Politehnica Bucharest, 313 Splaiul Independenţei, 060042 Bucharest, Romania
e-mail: adrian.ciocanea@upb.ro

B. Tofan
"Ion Mincu" University of Architecture and Urbanism Bucharest, 18-20 Academiei, 010014 Bucharest, Romania

I. Visa and A. Duta (eds.), *Solar Energy Conversion in Communities*, Springer Proceedings in Energy, https://doi.org/10.1007/978-3-030-55757-7_19

this point, energy strategies at local level are needed but obstacles in energy planning are present at urban scale [4], starting with the lack of common agreement on the exact definitions [5] both for classical approach on supply, conversion, storage and transportation and for the new approach concerning decentralized and renewable energy technologies [6].

One direct approach in the design of energy strategies at city scale is to define a "community" in urban context as a geographic area with no fixed size but composed by similar physical characteristics and connected by a shared identity [7] in order to assess the actual energy demand and future energy needs. Another approach is based on assessing renewable energy generation potential at city scale for identifying the most viable sites and investments [8].

Decentralized energy production, the increasing share of renewable energy installations in the energy mix and the new approach regarding smart grids provide opportunities for locally developed projects. In this context, the concept of "sustainable energy community" known and initially implemented in the UK since 1991 [9, 10] evolved to illustrate the new approaches in underlying the energy policy making process at local level [8, 11–13]. In this context, this paper studies a rehabilitation solution for a former thermoelectric power plant located in the metropolitan area of the Pitesti city by creating a solar plant and considering additional commercial activities enabling project sustainability.

Using the results for the Pitești case study, the authors extended the results to other four sites to assess the viability of this approach.

19.2 Methodology

The case study presents an application that will produce electricity pursuing three objectives: supplying energy to a community next to the site, providing electricity for a number of public urban electric buses (local storage), supplying energy for a number of onsite applications including energy compensation with the National Grid (NG). Five locations available in Romania were selected in order to assess the potential of this idea. The methodology has three steps: the geographic description of the site; evaluation of the onsite solar energy and computing the energy systems output; identifying solutions for the distribution network and connection to the NG. After technical solutions were settled down, the associated financial costs have been analyzed. The methodology was described for one location–Pitești city—and was further extended to the other four.

19.2.1 Site Description and Administrative Characterization

The considered site is located in the southern region of Pitesti city (44°51′38″ N, 24°52′4″ E), an area of approx. 80 ha, where there are: a stack of 280 m height, two cooling towers of 50 m each and a number of structures fit for rehabilitation, Fig. 19.1. Next to the site there are 4 villages covering an area of 183 km^2 and a population of about 20,000 inhabitants.

In terms of solar radiation, the site has a yearly average solar horizontal radiation $H_h = 3860$ Wh/m^2/day [14] as presented in Fig. 19.2. Two solar applications will be considered for the site: a photovoltaic power station (located on the former coal deposit) and a classic solar chimney equipped with a horizontal axis wind turbine (by reusing the old power plant's smokestack). The facilities will be connected to the NG but will be able to operate in an isolated grid, too—Fig. 19.3.

19.2.2 Solar Energy Potential and the Conversion Systems Analysis—Pitești

Sizing the photovoltaic power station. The photovoltaic power station is located on the former coal deposit and has a total area of 130,000 m^2. For this surface and the condition of the solar energy potential (an insolation level estimated at 1.620 kWh/m^2/year) an installed capacity of 5.2 MW and an annual electricity production of 6020 MWh/year results, considering the losses caused by the increased temperature and the reduced solar radiation of about 9.9%, losses caused by reflection of approx. 2.9%, losses in power cables, inverters etc. approx. 18%—Fig. 19.4. The 18,000 selected photovoltaic modules are of monocrystalline silicon type with 250 Wp capacity per unit.

Figure 19.4 outlines the average daily production of electrical energy (E_d, kWh), the average monthly production of electrical energy (E_m, kWh), the average daily global radiation per m^2 received by the system modules (H_d, kWh/m^2) and the average global radiation per m^2 received by the system modules (H_m, kWh/m^2). The calculated data have errors of approx. 5–7% due to some approximations on both the accuracy of incident solar radiation "on site" assessment and on the characteristics of the PV modules.

Solar chimney sizing—Pitești. The former power plant smokestack is transformed to be further used as solar chimney. Due to the solar radiation under a transparent canopy placed at the base, the greenhouse effect is created. Raising the temperature under the transparent canopy, induces an increase in the mean air velocity to the lower section of the tower, where is placed a wind turbine with horizontal axis. Sizing the solar chimney can be achieved by computing the thermal pressure that causes a displacement of the air mass on the height h [15, 16]. According to the calculation diagram from Fig. 19.5, equations may be written between pressures and air velocities in the inlet section of the canopy, the inlet of the turbine and the outlet of the chimney (Fig. 19.6).

The thermal pressure that determines the movement of air masses on the height h is:

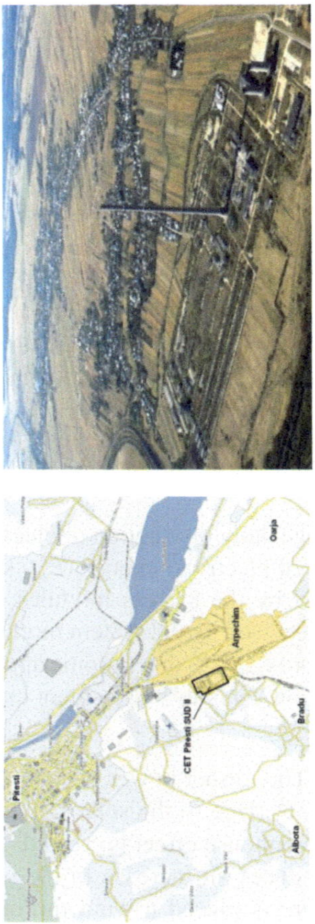

Fig. 19.1 Location of the industrial site CET SUD II Pitesti (wikimapia.org)

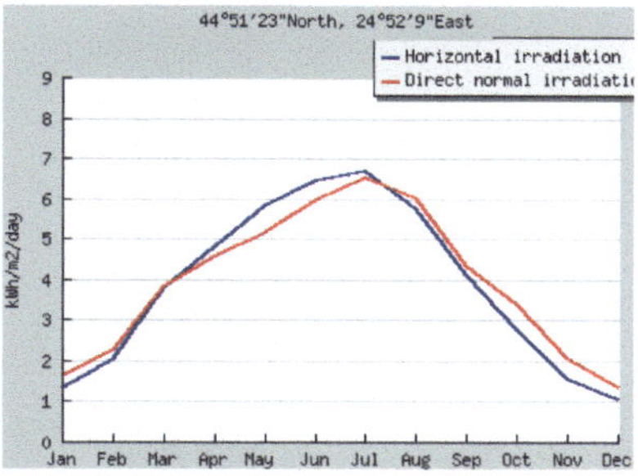

Fig. 19.2 Monthly solar radiation [14]

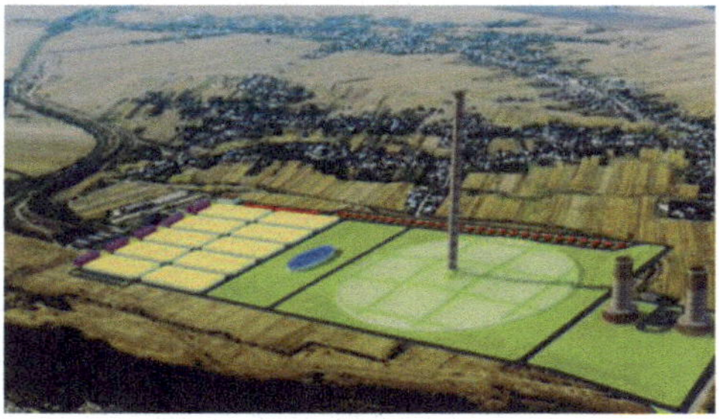

Fig. 19.3 Solar park (www.positive-energy.wix.com/positive-energy)

$$p_t = (\rho_1 - \rho_2)gh, \tag{19.1}$$

the pressure gap which causes the inlet air movement is:

$$\Delta p_1 = p_a - p_1, \tag{19.2}$$

and the air velocity at the entrance in section A1 $= \pi D_1^2/4$ will be:

$$w_1 = \sqrt{\frac{2\Delta p_1}{\rho_1}}. \tag{19.3}$$

One can compute the mass flow rate through the inlet section A1:

$$\dot{m}_1 = \mu_1 A_1 \sqrt{2\Delta p_1 \rho_1}, \tag{19.4}$$

where $\mu_1 = 0.2 \div 0.65$ is the flow coefficient at the inlet section of the chimney. The pressure gap needed to raise the air and to exit into the atmosphere ($p_a = p_{atm}$) and the outlet pressure in the section will be:

$$p_2 = p_1 + p_t = p_1 + (\rho_1 - \rho_2)gh, \tag{19.5}$$

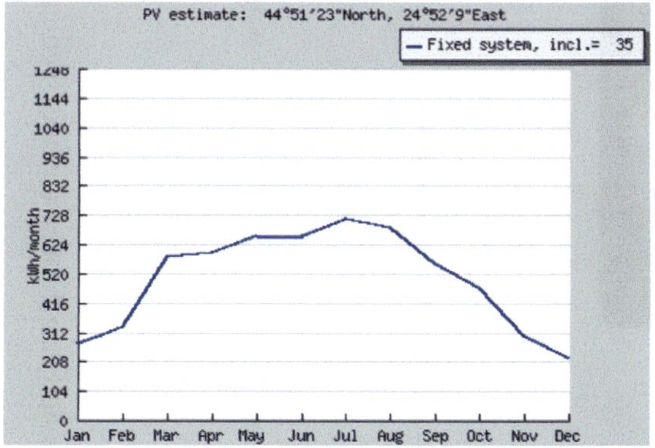

Fixed system: inclination=35°, orientation=0°				
Month	E_d	E_m	H_d	H_m
Jan	8.69	269	2.12	65.8
Feb	11.70	329	2.92	81.8
Mar	18.60	577	4.82	150
Apr	19.90	596	5.34	160
May	21.00	652	5.79	180
Jun	21.90	656	6.12	184
Jul	23.00	713	6.49	201
Aug	22.00	683	6.19	192
Sep	18.40	553	5.02	151
Oct	15.20	470	3.99	124
Nov	10.00	301	2.55	76.5
Dec	7.13	221	1.75	54.3
Yearly average	**16.5**	**502**	**4.43**	**135**
Total for year		**6020**		**1620**

Fig. 19.4 Photovoltaic power plant performance (installed capacity, 5.2 MW) [14]

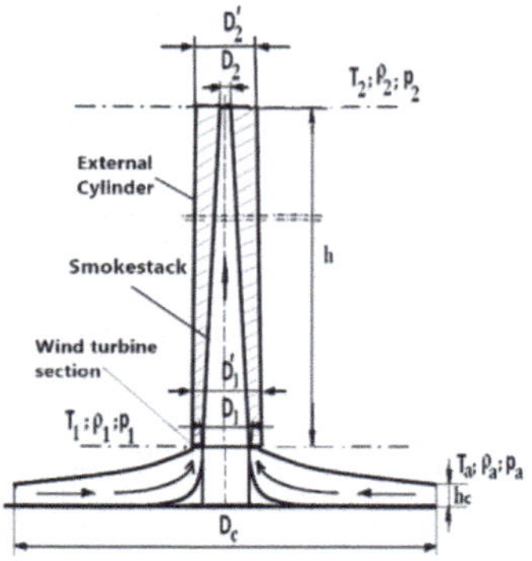

Fig. 19.5 The solar chimney calculation diagram

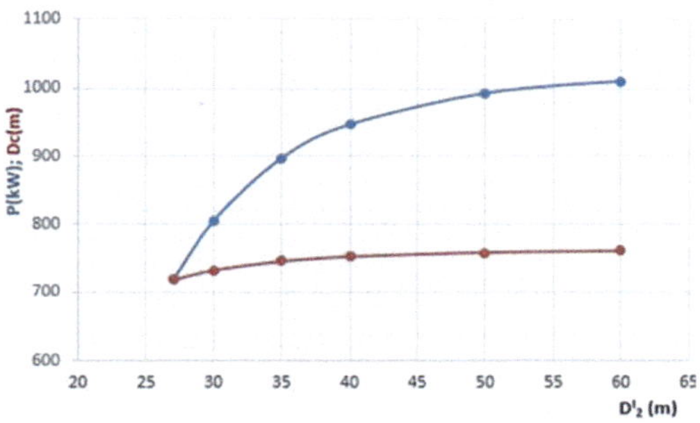

Fig. 19.6 Sensitivity diagram for the Baia Mare solar chimney

and:

$$\Delta p_2 = (\rho_1 - \rho_2)gh - \Delta p_1. \tag{19.6}$$

Using the above equations the velocity is calculated, respectively the mass flow rate in the outlet section:

$$w_2 = \sqrt{\frac{2\Delta p_2}{\rho_2}} = \sqrt{[(\rho_1 - \rho_2)gh - \Delta p_1]\frac{2}{\rho_2}};$$
$$\dot{m}_2 = \mu_2 A_2 \sqrt{2\rho_2[(\rho_1 - \rho_2)gh - \Delta p_1]}. \tag{19.7}$$

Considering continuity in the mass flow rate between Sects. 19.1 and 19.2, the balance equation for the airflow through the installation can be calculated along with Δp_1 if the area ratio is known. For computing the canopy surface it is necessary to estimate the ratio $\beta = A_2/A_1$, the air inlet and outlet characteristic temperatures in the canopy section, and the amount of solar radiation H. By using the air flow rate equation and the equality between the heat flow received by the transparent canopy through radiation from the sun, respectively discharged through the outlet section, it will result the total area required to create an ascending stream with a velocity $w_{air} = w_1$. The equation for the heat output between the canopy and the chimney is:

$$\dot{Q} = HS = \gamma_a c_p \Delta_t \frac{\pi D_1^2}{4} w_a 3600, \tag{19.8}$$

where velocity w_1 is calculated using Eq. (19.3) through the air mass balance:

$$\Delta p_1 = \frac{\mu_2^2 \beta^2 \rho_2 (\rho_1 - \rho_2) g h}{\mu_1^2 \rho_1 + \mu_2^2 \beta^2 \rho_2}, \tag{19.9}$$

resulting the canopy diameter:

$$D_c = \sqrt{\frac{\gamma_a c_p \Delta t D_1^2 w_a 3600}{H}}, \tag{19.10}$$

where γ_{air} is the air specific weight; c_p is the air specific heat; Δt is the difference between the inlet air temperature under the canopy and the chimney base (assimilated entrance point placed at the base of the wind turbine); D_1 is the chimney diameter at the level of the wind turbine; *S* is canopy's surface; $\mu_1 = 0.2 \div 0.75$ and $\mu_2 = 0.5 \div 0.95$ are the flow coefficients for the inlet and the outlet sections. The wind turbine power can be calculated using:

$$P = C_p \left(\frac{\rho_1}{2}\right) \left(\frac{\pi D_t^2}{4}\right) w_1^3, \tag{19.11}$$

where C_p is the power coefficient of the turbine; D_t is the turbine's rotor diameter assimilated with diameter D_1 in the corresponding section of the chimney. Evaluating the operating time t_o of the turbine, depending on the solar radiation period (h/day), the energy production during the considered interval can be calculated using:

$$E = t_o P. \tag{19.12}$$

If the computing algorithm is used for the case of a conrgent-top solar chimney i.e. $D_1 > D_2$ (geometry of all the smokestacks in this study) the value for the installed power of the solar facility is low (i.e. installed power for the wind turbine). In order to enhance the facility performance, one considers the results of recent studies on using divergent-top chimneys and sloping collectors. Some of these studies reported that by using systems with optimized sloping collector (i.e. the optimization of the collector to the chimney transition region [17]) and a divergent-top solar chimney with chimney area ratio A_2/A_1 of 16, one can produce power as much as 400 times higher comparing to the ratio $A_2 = A_1$ [18, 19]. In this paper, convergent-top smokestacks are available and for obtaining a divergent-top section one considers covering these structures with a cylinder made of thin iron plating. A variable circular crown section is created around the smokestack from the base to the top as presented in Fig. 19.5. Therefore, between the external cylinder and the internal convergent-top smokestack one obtains a divergent-top solar chimney. By using this solution, the smokestack becomes the support structure for the external cylinder and the air flow throughout this chimney is blocked for obtaining the divergent-top solar chimney effect. The value for the diameter of the external cylinder (D'_1) is calculated considering the equality between the actual inlet section of the smokestack (D_1) and the circular crown section at the same level: $D'_1 = 20.5 \times D_1$; this value is the lowest one as result of the geometric conditions and represents the reference in this study. For the Pitesti site, the geometric data are as follows: $D_1 = 15$ m; $D_2 = 8$ m; $D'_1 = D'_2 = 20$ m. Table 19.1 presents the results of the calculations for the solar chimney (Pitești) based on the site characteristics and new geometric assumptions while Table 19.2 contains the results for all five locations considering similar input data. The current status of the chimneys and the position of each site are presented in Fig. 19.7.

The total installed capacity for all solar chimneys is of about 1.7 MW because of the poor performance of such facilities. However, considering that the initial investment cost of such a tower would be about 60–70% of the total cost for a solar thermal application, the existing building represents an advantage. Moreover, for enhanced installed power, higher values for the diameter of the external cylinder can be achieved since the solid concrete structure of the smokestack is available. Figure 19.6 presents a sensitivity diagram for the Baia Mare site where a value of 1 MW of installed power can be reached if the external cylinder diameter is $D'_2 =$ 50 m. A special advantage for the Pitești and Targu Jiu sites consists of the fact that

Table 19.1 Computation results for the solar chimney—Pitești

t_a; t_1	$\rho_{1(t1)}$; $\rho_{2(t2)}$	c_p	μ_1; μ_2	H	C_p	Δp_1	w_1	β
[°C]	[kg/m^3]	[kJ/kg°C]	[–]	[W/m^2]	[–]	[N/m^2]	[m/s]	[–]
30; 65	1.0436; 1.165	0.241	0.7; 0.7	750	0.46	260	22.3	1.78

Table 19.2 Sizing results for all solar chimneys in considered sites

Site	h	D_1/D'_1	D_2/D'_2	w_1	D_c	P
	[m]	[m]	[m]	[m/s]	[m]	[kW]
Baia Mare	351.5	20/27	9.0/27	25.1	719	718
Pitești	280	15/20	8.0/20	22.3	508	283
Brașov	280	15/20	8.0/20	22.3	508	283
Targu Jiu	256	14/19	6.0/19	21.8	465	220
Copșa Mică	250	13/17	6.0/17	23.2	436	199
TOTAL	–	–	–	–	–	1703

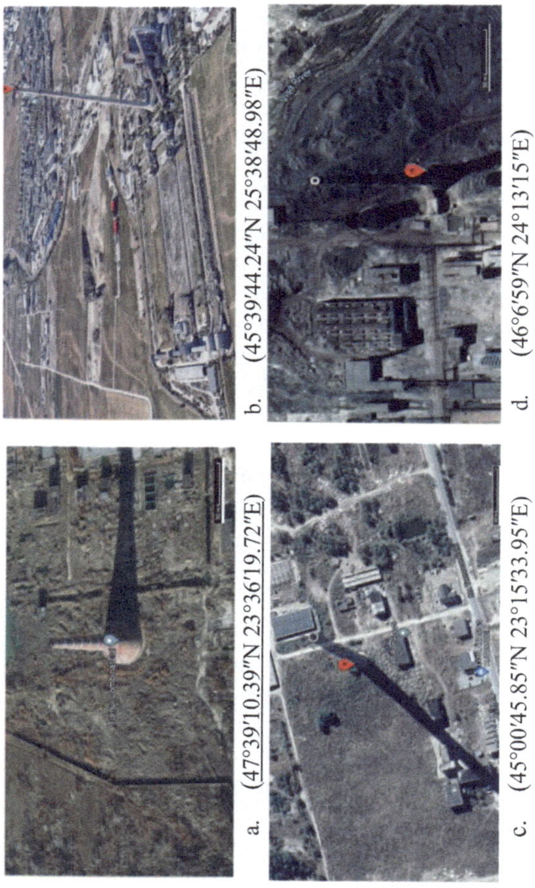

Fig. 19.7 Current status of the smokestacks **a** Baia Mare; **b** Brașov; **c** Targu Jiu; **d** Copșa Mică (Google Maps)

these facilities were never operated and therefore no thermal stresses are expected in the infrastructure.

19.2.3 Power Supply and Solutions for the Connection to the NG—Pitești

The main data and assumptions considered for the electricity network analysis were as follows: the distance from the site to the nearest substation (Bradu 400/220/110/20 kV) is of about 5 km; the maximum produced electrical energy is 6600 MWh/year (for the local weather conditions); a flexible storage solution by charging electrical buses batteries (of the Pitești city transport company) was considered. Based on the electrical energy demand and the on-site energy usage (Table 19.3), the distribution network and the connection to NG were sized. In Fig. 19.8 it is

Table 19.3 Electricity demand (kW)—Pitești site area

Electricity demand	Scenario A		Scenario B	
	Day	Night	Day	Night
Charging electrical plugs	640	640	400	400
Outdoor on-site lighting	0	50	0	50
Indoor demand (for buildings)	120	40	120	40
Special night lighting	0	6	0	6
Entertainment	12	0	12	0
TOTAL	772	736	532	496

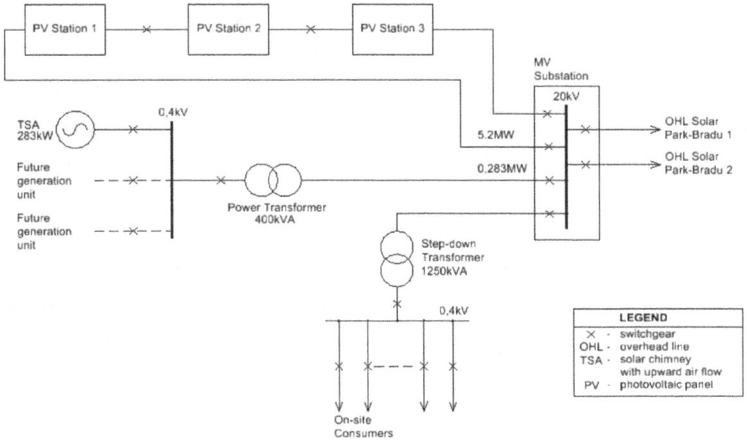

Fig. 19.8 Single-line diagram of the electrical energy distribution network in the solar park

presented the single-line diagram of the medium-voltage network (MV), designed, and sized to supply electricity from the solar park [20]. Each PV station features a step-up transformer, 0.35/20 and 20 kV switchgear for the connection to the on-site distribution grid.

The main investment in the electrical network will involve costs for the photovoltaic modules, for the wind generator implemented in the solar chimney, for the power substation for the connection to NG, for two 20 kV overhead lines, for the power plugs for charging public transportation buses in Pitesti city etc. The charging system will be accessed by other vehicles, considering the proximity of the Bucharest—Pitesti freeway and of the Pitesti—Craiova major highway, leading to important contributions in the national carbon dioxide emissions reduction target—the electricity distribution system was sized for 50 electrical plugs. In the design and sizing the renewable power plant it was considered the total electricity demand on-site, balanced between daytime & nighttime (see Table 19.3 for the power demand scenarios).

Main applications/facilities of the solar park and investment costs for the Pitești site. The solar energy park was designed to meet the following objectives: to demonstrate the reliability and energy efficiency of renewable energy systems by using a new approach; to assess the viability of the public/private investment for industrial sites rehabilitation both for clean energy production and social purposes; to support research and development activities on renewable energy resources, their exploitation and energy efficiency.

To assess the investment effectiveness, the generation installations, the distribution network and the applications were considered in terms of costs and forecasted revenues from: charging electrical vehicles and public transportation buses; renting services to public/private consumers operating in the park (innovative companies, research laboratories, social activities, entertainment etc.); electricity supply in the system through Bradu substation. To provide all the services of the solar park, the investment required for architectural-volumetric conversion of the deprecated industrial site includes demolition and decontamination actions, roads and platforms, architectural conversion of remaining buildings etc. The costs for architectural-volumetric conversion were assessed at about 7 M Euro. These were added to the investment cost in the main electrical installations estimated at about 12.5 M Euro. Architectural-volumetric conversion costs are to be considered even when the site will have another destination.

Electricity balance of the Pitești site. For sizing the electricity demand on-site, two scenarios were considering the European Directive on the infrastructure for alternative fuels [21]. Following this, Romanian authorities are willing to incentivize the installation of 10,000 plugs for charging vehicles/buses by 2020.

- *Scenario A* considers a demand factor of 0.8 for the charging plugs that corresponds to a successful public policy for the stimulation of the electrical transportation.

Table 19.4 Yearly energy balance—Pitești case study, MWh/year

Scenario A	Scenario B
−5	2100

- *Scenario B* considers a demand factor of 0.5 for the charging plugs that corresponds to a less-successful public policy for the stimulation of the electrical transportation.

A successful policy on electrical transportation encourages potential investors in the solar park project to access grants for lowering the commercial founds raising. In Table 19.3, the electricity demand is described for both scenarios.

Considering the day/night electrical energy demand approach, Table 19.4 presents an estimate of the annual energy balance.

Only in the B scenario, the solar park exports electricity to the local community. All the energy flow can be ensured by the link to the local electricity distribution network (through the Bradu substation) with two proposed OHLs, as shown in the Fig. 19.8. With a demand factor for the charging plugs lower than 0.8, the site is exporting green electricity.

19.3 Financial Performances of the Solar Park Project for Two Investment Cases—Pitești

As mentioned, the investment cost for the solar park is related to the energy part, 12.5 M Euro and to the architectural part, 7 M Euro. The financial performances were analyzed considering two investment cases: with and without grants, e.g. grants available in Romania for investment in renewable energy facilities, for purchasing charging plugs for vehicles/buses and for job creation. For the Pitesti-Romania case study, the financial viability was analyzed using the Net Present Value (NPV) and the Internal Rate of Return (IRR) methods. Costs and revenues were estimated for both electricity demand scenarios (see 19.2.3). In the cash-flow analysis, the provision of the Romanian legislation were considered that requires producers to sell all the electricity to the market. For revenues from selling the electricity (the constant part of the revenues), we considered the electricity stimulation system with green certificates available in Romania (three green certificates for each MWh produced by PV modules). The total revenue from the electricity supplied in the grid was estimated using average local market values for electricity generation and green certificates. A value of 0.9 M Euro/year was calculated for the revenue from electricity sales. For estimating the revenues from rental activities and charging electrical vehicles are the variable part of the revenues, authors made assumptions as presented in Fig. 19.9.

The financial performance indicator trend (NPV) for a continuously increasing value of the variable part from the annual revenue, for both A and B electricity demand scenarios, in the without grant case, is shown in Fig. 19.9. The results outline that

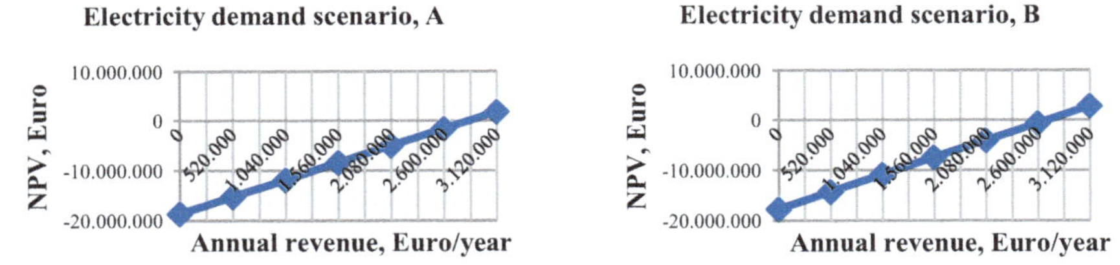

Fig. 19.9 Financial performance indicator trend (NPV) for a continuously increasing variable part of the annual revenue, in A and B electricity demand scenarios, without grant

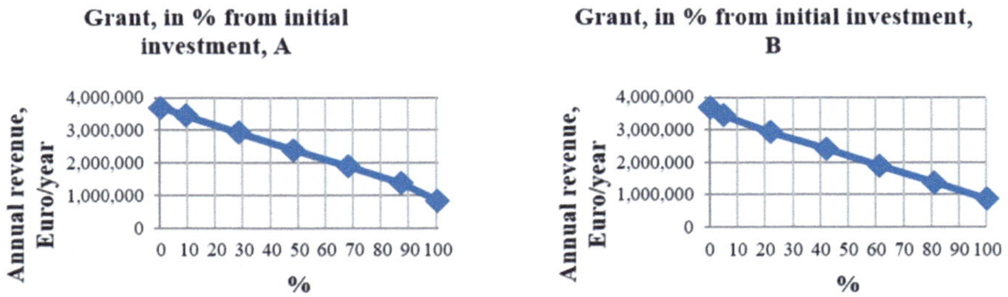

Fig. 19.10 Grant requirement in connection with the annual revenue for the A and B electricity demand scenarios

in the B electricity demand scenario the revenue requirement for a positive NPV is lower. From these technical and economical simulations resulted that the project is financially viable as it meets the NPV positive and IRR $\geq$ a (discount rate) criteria—if the revenue requirement from the proposed activities, other than selling electricity, is over 2.8 M Euro/year in the A electricity demand scenario and is over 2.6 M Euro/year in the B scenario. Otherwise, non-refundable financial incentives are needed for the initial investment, in close connection to the revenues forecast.

In Fig. 19.10, the "with grant" case is analyzed, showing that in the B scenario the grant requirement is lower for the same annual revenue, for favorable financial indicators.

The grant requirement rapidly increases with the annual revenue decrease, meaning that project's financial viability highly depends on its commercial success, mainly from other activities than selling electricity. The external benefits were not considered for implementing such a project, e.g. related to reducing the amount of greenhouse gases released into the atmosphere or to an increased level of awareness regarding the benefits of renewable energy resources or to the reduction in the health care and other social related costs.

19.4 Conclusions

The paper presents a new approach on using renewable energy resources by reconsidering the facilities of decommissioned industrial sites. The electricity generation and distribution installations of a solar park are presented, and financial scenarios are considered to assess the viability of such an investment.

The theoretical approach considers an original solution consisting of covering existing tall smokestacks by cylinders manufactured of thin iron plating to obtain divergent-out solar chimneys with higher efficiency. The viability of this solution is supported by the resilience of the concrete structure used only for guiding the solar chimney.

A case study that illustrates the new approach is applied to the former coal-fired power plant CET Sud II Pitesti and extended to other four available sites in Romania. A photovoltaic power station (5.2 MW) and a solar chimney (283 kW) are proposed. For all these five selected sites, the photovoltaic power station could be similar. Therefore, an overall installed power of about 28 MW could be obtained for the basic scenario; this value could be extended to say 30 MW without substantial supplementary costs.

The opportunity of investments in post-industrial sites *vs.* their demolition is discussed in the context of energy sustainability and social responsibility. The revival of the industrial sites might be a successful project, despite the initial investment cost that, in this case study, may rise to 19.5 M Euro. However, some costs if another solution is selected for using the site are mandatory as demolition, decontamination etc. The project is financially viable if the revenues from rental activities and charging electrical vehicles are higher than 2.6 M Euro/year leading to a simple payback period of roughly 8 years, without any additional grant. For an onsite electricity demand factor less than 0.8, the solar park is able to export clean electricity to the local community through the national grid. For this case study the environmental benefits were not taken into account, but additional positive financial effects can be expected.

References

1. UN, *World Urbanization Prospects: The 2014 Revision, Highlights*. United Nations, Department of Economic and Social Affairs, Population Division (2014)
2. L. Pérez-Lombard, J. Ortiz, C. Pout, A review on buildings energy consumption information. Energy Build. **40**(3), 394–398 (2008). https://dx.doi.org/10.1016/j.enbuild.2007.03.007
3. L. Kamal-Chaoui, A. Robert (eds.), *Competitive cities and climate change*. OECD Regional Development Working Papers N∘2 (OECD Publishing, Paris, France, 2009)
4. M. Cajot, J.-M. Peter, F. Bahu, A. Guignet, F.M. Koch, Obstacles in energy planning at the urban scale. Sustain. Cities Soc. **30**(2017), 223–236 (2017)
5. R.D. Prasad, R.C. Bansal, A. Raturi, Multi-faceted energy planning: a review. Renew. Sustain. Energy Rev. **38**, 686–699 (2014). https://doi.org/10.1016/j.rser.2014.07.021
6. A.M. Adil, Y. Ko, Socio-technical evolution of decentralized energy systems: a critical review and implications for urban planning and policy. Renew. Sustain. Energy Rev. **57**, 1025–1037 (2016). https://doi.org/10.1016/j.rser.2015.12.079

7. J.P. Petersen, Energy concepts for self-supplying communities based on local and renewable energy sources: a case study from northern Germany. Sustain Cities Soc **26**(2016), 1–8 (2016)
8. K. Adam, V. Hoolohan, J. Gooding, T. Knowland, C.S.E. Bale, A.S. Tomlin, Methodologies for city-scale assessment of renewable energy generation potential to inform strategic energy infrastructure investment. Cities **54**(2016), 45–56 (2016)
9. G. Seyfang, J.J. Park, A. Smith, A thousand flowers blooming? An examination of community energy in the UK. Energy Policy **61**(13), 977–989 (2013)
10. G. Walker, P. Devine-Wright, S. Hunter, H. High, B. Evans, Trust and community: exploring the meanings, contexts and dynamics of community renewable energy. Energy Policy **38**(6), 2655–2663 (2010)
11. Y. Noorollahi, R. Itoi, H. Youse, Modeling for diversifying electricity supply by maximizing renewable energy use in Ebino city southern Japan. Sustain. Cities Soc. **34**(2017), 371–384 (2017)
12. C.H. Vermette, E. Cubi, J. Bergerson, Energy performance of a solar mixed-use community. Sustain. Cities Soc **27**(2016), 145–151 (2016)
13. MCEC, *Somerset Power Plants Reuse Study*. Massachusetts Clean Energy Center—Final Report (2015)
14. JRC, *Photovoltaic Geographical Information System (PVGIS)* (2015). https://re.jrc.ec.europa.eu/pvgis
15. M. Grigoriu, *Air Conditioning Installations* (in Romanian Language) (University Politehnica of Bucharest, 1993)
16. A. Ciocanea, Solar tower for electricity generation, in *International Conference of Energy and Environment CIEM 2005*, Bucharest, 22–24 November 2005, vol. II (2005) pp. 5.85–5.92. ISSN 1454-234x.Ed.UPB.
17. P. J. Cottam, P. Duffour, P. Lindstrand, P. Fromme, Effect of canopy profile on solar thermal chimney performance. Solar Energy **129**, 286–296 (2016)
18. D. Hu, Y.C.D. Leung, Mathematical modelling of the performance of a solar chimney power plant with divergent chimneys, in *1st International Conference on Energy and Power, ICEP2016*, 14–16 December 2016, RMIT University, Melbourne, Australia. Energy Procedia **110**, 440–445 (2016)
19. A.B. Kasaeian, Sh. Molana, K. Rahmani, D. Wen, A review on solar chimney systems. Renew. Sustain. Energy Rev. **67**, 954–987 (2017)
20. M. Scripcariu, *Electricity Distribution Equipment* (in Romanian Language) (Politehnica Press, Bucharest, 2008)
21. UECOM, Directive 2014/94/EU of the European Parliament and of the Council, from October 22, 2014, regarding the installation of the infrastructure for alternative fuels (2014)

Chapter 20
Sustainable Biorefinery and Bioenergy Models Based on Seaweeds, as a Potential Feedstock for the Black Sea Coastal Area

Sandor Bartha, Florbela Carvalheiro, Patricia Moniz, Luis C. Duarte, and Santino Di Berardino

Abstract The Romanian coastline of the Black Sea has abundant seaweed resources, which create in summer period discomfort when the coastal area is flowed by thick lever seaweeds and that has a negative impact on the touristic area. That process reaches its maximum during the end of July and August. Usually, these seaweeds are collected when they reach the coast and are then stored in regional landfill waste deposits. Statistical data show that more than 9000 tons of seaweeds were collected during the last year. These were naturally degraded, causing an important environment impact. This paper focuses on the presentation of the results obtained at lab scale for the conversion of the Black Sea coast seaweeds. The first part of this work presents the state of art and the assessment of the chemical and physical composition of this feedstock. In the second part is presented the optimal technological scheme for developing one bioenergy model to produce biogas by the anaerobic digestion process of the pre-treated and untreated biomass. This digester aims to be used by the owners of the fish farms and the owners of the local costal touristic units that are affected by this algae blooming. The digester model also integrates one cogeneration unit designed to use the resulting biogas to produce bioenergy (electricity and heat) for the local use.

Keywords Biorefinery · Biogas · Digester · Macroalgae · Seaweeds

---

S. Bartha (✉)
Bio-C-Silox Research LTD, Sf. Gheorghe, Romania
e-mail: sbarthacv@yahoo.ro

F. Carvalheiro · P. Moniz · L. C. Duarte · S. Di Berardino
Unidade de Bioenergia E Biorrefinarias, LNEG—Laboratório Nacional de Energia e Geologia, Lisboa, Portugal
e-mail: florbela.carvalheiro@lneg.pt

P. Moniz
e-mail: patricia.moniz@lneg.pt

L. C. Duarte
e-mail: luis.duarte@lneg.pt

S. Di Berardino
e-mail: santino.diberardino@lneg.pt

I. Visa and A. Duta (eds.), *Solar Energy Conversion in Communities*,
Springer Proceedings in Energy, https://doi.org/10.1007/978-3-030-55757-7_20

20.1 Introduction

As presented in the European Energy Strategy, the use of renewable energy sources is projected to increase until the end of 2020 to reach a share of 20% in the final energy consumption [1] and 10% renewables in transport. Until 2030 the share of the renewables is expected to increase to at least 27% according to the 2030 Climate and Energy Policy Framework as proposed by the European Commission [2] and included in the proposal for a revised RED II/ (2018).

On December 4, 2018, the European Council of Ministers adopted the EU's entirely revamped Renewable Energy Directive (REDII). The Directive sets a minimum blending mandate for advanced biofuels starting in 2022 something that the Leaders of Sustainable Biofuels (LSB), a coalition of advanced biofuel producers, welcomes as a "decisive step" forward in promoting decarbonisation of EU transportation [3].

To reach these targets the bioenergy production can play an important role.

One sustainable way to produce bioenergy is by anaerobic digestion of different organic materials and wastes; in this way the biogas is produced or the bio-methane.

The resulted biogas is further used in electricity generation, in cogeneration units with different installed capacity or in public transport in the form of biofuel (e.g. bio-methane).

Biogas is a renewable energy fuel produced by the anaerobe degradation of the complex organic matter and the resulted gas consists of 50–75% methane and 25–45% carbon dioxide with small amounts of water vapors, oxygen, nitrogen and H_2S, hydrogen and ammonia [4]. The general composition of the gas is presented in Table 20.1 [5].

Biogas as a fuel comes mainly from its methane content; it exists in gaseous state at normal condition and can be liquefied at pressure (over 20 MPa) and low temperature (112.15 K). The energetic value of biogas depends on the methane content, therefore it is important to use in the anaerobic digestion different organic matter with high energy density.

Table 20.2 [6] shows the heat energy output of the biogas with different methane

Table 20.1 General biogas composition

Biogas composition	
Content of the biogas	Concentration levels
Methane (CH_4)	50–80% by vol
Carbon dioxide (CO_2)	20–50% by vol
Ammonia (NH_3)	0–300 ppm
Hydrogen sulphide (H_2S)	50–5000 ppm
Nitrogen (N_2)	1–4% by vol
Oxygen (O_2)	<1% by vol
Moisture (H_2O)	Saturated 2–5% by mass

Table 20.2 Selected gas properties at standard conditions

Properties/biogas composition	Methane (CH_4)	60% CH_4, 40% CO_2	65% CH_4, 34% CO_2, 1% Others
Heating value [MJ/m^3]	35.46	21.6	24.48
Ignition ratio [% air]	5–15	6–12	7.7–23
Ignition temperature [°C]	650–750	650–750	650–750
Change of state pressure [MPa]	4.7	7.5–8.9	7.5–8.9
Transition temperature [°C]	−82.5	−82.5	−82.5
Density [kg/m^3]	0.72	1.2	1.2
Heat capacity [kJ/m^3/°C]	1.6	1.6	1.6

content in comparison with the natural methane gas fuel used in energy production and transport is presented.

Without biogas production, the organic matter resulted in the economy and in nature is collected and deposited in municipal landfill deposits, and through the natural degradation of the organic matter the resulted gas has as main component methane, that has a 25 times higher impact on climate than carbon dioxide [7].

In the developing countries this organic wastes are usually discarded into a nearby water supplies or direct to the unused lands, where by auto degradation they pollute the environment.

These good thermal proprieties can offer the possibility for the use of the biogas in energy production, instead of fossil fuels, and this brings sustainable benefits as the energy output of 1 m^3 biogas could replace 0.46 kg of LPG, 0.67 L of gasoline, 0.55 L of oil or is enough to produce 1.2–1.4 kWh of electric power [8]. The main thermal properties of the biogas in comparison with same classical fossil fuels are presented in Table 20.3.

That is the reason why the number of biogas plants is growing year by year in the European Union; based on the statistics data in the European Biomass Association Report, at the end of 2018 the number of biogas plants in EU was of 18,202 units, that had 11,082 MW installed electric capacity and produced 63,511 GWh by the

Table 20.3 Selected fuel properties

Energy source	Density STP [kg/m^3]	Fuel heating value [kJ/kg]	Ignition temperature [°C]	Air/fuel ratio [kg/kg]
Methane	0.75	50,000	650	17.2
LPG	540	46,000	400	15.5
Natural gas	0.83	57,000	600	17.0
Propane	2.02	46,300	470	15.6
Biogas (60% CH_4)	1.2	18,000	650	10.2

use of biogas. During the last period the bio-methane sector had a high dynamic development as the total of 610 plants working in EU in 2018 produced 2.28 billion m^3 bio-methane. The statistical data resulted from the quantitative estimation of the potential and recently used biomass and organic wastes in biogas production in the EU, show that the production could be increased from the current level of 14.9 Mtoe towards 28.8–40.2 Mtoe in 2030 [9, 10]. That bio-methane can be directly injected in the national distribution grids or can be used in public transport; one best practice example is in Malmo, Sweden where the bus transport in municipality use bio-methane as fuel [11]. The bio-methane is produced in the city region and one of the biogas plant is in Jordberga where yearly 110 GWh biogas is produced that is equivalent with over 11 million L of oil. The organic wastes resulted at the end of the AD process are used as fertilizer and that plant returned yearly 150 t high grad bio-fertilizers for use in the regions soils. The total biogas produced in this plant contributes to a carbon dioxide reduction equivalent to 30,000 t/year. The used substrates for the AD process, the raw materials, consists mainly of agricultural residues, that can includes beet tops and damaged beets or corn and grain that does not measure up to the commercial standards. There are also used as substrates residues from milling and grain handling industry, carrots, peanuts, etc. [12].

In Europe, up to 70% of the feedstock used in the biogas production comes from the agricultural sector such energy crops, manure and different residues resulting from agriculture. The use of organic wastes like manure is possible by one particularly management system implemented in different co-generation technologies. The distribution of the feedstock used in the European working biogas plants in 2017 is presented in Fig. 20.1 according to the technological data received from the European Biogas Association (www.europeanbiogas.eu). That shows that the important part of the used substrates comes from agriculture, by one well and optimal developed supply and value chain.

To develop this energy production schemes and to reduce CO_2 emissions, is important to find and test new feedstock for biogas production. Today different types of feedstock are estimated, at different research levels (wastes form agriculture, forestry

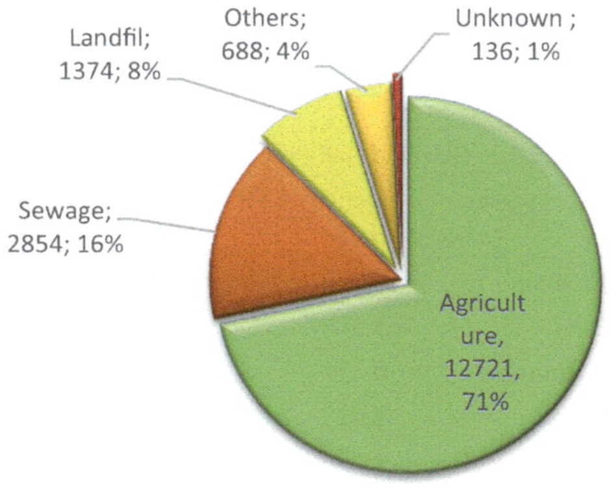

Fig. 20.1 Distribution of the Biogas Plants by feedstock in Europe in 2017 (number of plants in function of the use feedstock

residues, different energy crops and marine biomass, algae, etc.). In the research plans presented by the International Energy Agency (IEA), one important task is to continue to evaluate the potential of novel energy feedstock such as algae and aquatic biomass, with millstones set for the 2017–2040 period [13]. The growing biogas and bio-methane production in the next period can be obtained with the development of new integrated model where one main role is associated with the anaerobic digestion of different feedstock; the Task 42 on Bioenergy developed by IEA is specific on the marine biorefinery concepts, where the macroalgae seaweeds are the main feedstock. The seaweeds cultivation could represent a promising opportunity to reducing the pollution of the seawater and stop the eutrophication process.

Macroalgae contain unique composition of carbohydrates that have different properties than those of terrestrial plants and with higher mineral content than the terrestrial biomass; laminaran and mannitol are energy storage units in algae, like starch in the plants, but alginates are structure compounds like cellulose and lignin in plants [14]. The presence of different polysaccharides in seaweeds make them more attractive also for the pharmaceutical industry. Recently, there were different laboratory and pilot scale biorefinery models developed where the collected or cultivated seaweeds were transformed in biofuels as bioethanol, biodiesel or biogas and, using a complex technology, where added value products could be isolated from that feedstock: alginates, polysaccharides, antioxidants, pigments, protein extracts, oligosaccharides, food additives and bio-fertilizers. Another best suited alternative for the industrial applications is focused on the use of this feedstock in biogas production by AD processes. For the use of this feedstock in AD processes is important to evaluate the chemical composition of the seaweeds. It is know from practice that the optimum level of the C/N ratio for a substrate in the anaerobic process is in range of 20–30/1. The digestion of the nitrogenous substrates with the C/N ratio less than 15 can lead to technical problems caused by the excess levels of ammonia in the digester. The literature data on the state of art of the technological level presents the chemical compositions of different types of seaweeds, collected in West Cork coast in Ireland [15] as presented in Table 20.4.

These data show that the green algae such as *Ulvalactuca*, has a low C/N ratio, that can caused technological problems in the digester due to the high ammonia content of the gas. The methane production potential in the AD process was studied and the

Table 20.4 Characteristics of raw seaweeds collected in Cork in 2013

Seaweeds type/properties	TS % wwt	VS % wwt	Ash % TS	C % TS	H% TS	N% TS	O % TS	C:N ratio
A. nodosum	23.2	19.4	16.1	40.4	5.3	1.6	36.6	26
H. elongate	12.6	8.1	36	30.8	4.1	1.4	2.7	21.4
F. spiralis	19.72	13.92	29.4	36.1	4.7	4.7	27.7	17.3
U. lactuca	18.03	10.88	39.7	30	4.4	4.4	22.4	8.5
S. latissima	15.46	10.9	36.4	29.1	3.8	3.8	32	24

where TS = total solids, VS = volatile solids, wwt = wet weight

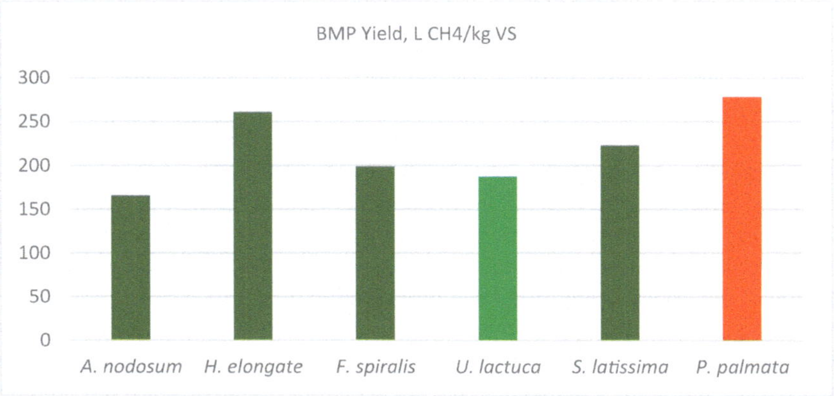

Fig. 20.2 Overview of Bio-methane Potential (BMP) for different species of macroalgae [16]

values reported in literature are summarized in Fig. 20.2. In case of the seaweeds the value of "Bio-methane Potential (BMP)" tends to be between 150 and 350 L_{CH4}/kg VS [16]. That value for different seaweeds types is outlined in Fig. 20.2.

In the case of Black Sea, according to the 2017 annual Report of the Romanian Water Agency, around 14,950 tons of green alga were collected in the coastal area (from Navodari to Vama Veche) and all of it was deposited on the local landfill deposits in this coastal zones [17]. Staring with May till the end of August 2019, in the Black Sea coastal area (Olimp—Vama Veche) there were 6300 t collected and in the area Agigea—Eforie Nord 485 t; all these were stored in the Regional Landfill Deposit [18]). The period of the algae invasion in the Black Sea coast is during middle of June–August.

This paper is focused on the development of different bioeconomy models for this organic feedstock. These models are developed based on the results obtained in one microscale laboratory experiments where the main physical and chemical characteristics of the seaweed harvested in summer were studied.

20.2 Material and Methods

The investigated macroalgae remain to the family of the green seaweeds and were harvested in summer from the coastal area of the Black Sea, at "Golful Pescarilor", near the ICPE- SA Research Centre; the samples were washed with water and naturally dried in well-ventilated places avoiding the direct exposure to the sun. Generally, the seaweeds contain high amounts of carbohydrates (up to 60%), medium/high amounts of proteins (10–47%) and low amounts of lipids (1–2%) and a variable content of mineral ash (7–38%) [18]. In the fresh harvested mass the water content is high (70–80%).

20.2.1 Proximate Analysis Methods

The proximate composition represents the physical properties of the feedstock, as the moisture content, bulk density, particle size, ash content, and the volatile component content of the seaweeds and type of the plant.

The moisture content of the samples was measured using an oven where the samples were dried at 105 °C to constant weight. For the ash content of the seaweeds the samples used for moisture determination were incinerated in a muffle furnace (equipped with a ramping program) to 550 °C. The procedure was according to the NREL/TP/510-42622/2008 protocol and to the CEN-EN 14961 Standards.

20.2.2 Ultimate Analysis and Method

The elemental composition include the data on the elemental composition of the feedstock. That composition is typical for each type of biomass and, based on literature data, the marine biomass was characterized compared with the terrestrial biomass and shows lower carbon content, higher hydrogen content, higher ash content, higher Cl and S content. The seaweed ultimate analysis data used in biorefinery technologies was taken from literature [19]. The data for green seaweeds, that are typical in the Black Sea region, are shown in Table 20.5.

The green seaweeds contain structural polysaccharides that after the optimal treatment the can be transformed in monomeric sugars like, glucose, mannose, galactose, rhamnouse, xylose and also uronic acids [20, 21]. The protein content in the feedstock was estimated by the Kjeldahl method described in the NREL/TP/510-42622 protocol, using the $N \times 4.78$ conversion factor.

The feedstock characterization in the next phase was concentrated on the evaluation of the fermentable sugars content and the acid hydrolysis of the seaweeds was investigated. For this experiment the milled macroalgae were mixed with dilute sulfuric acid solution in a 30 mL glass micro-reactors, respecting the liquid to solid ratio value of 12 (w/w). The hydrolysis was performed in an oil bath heated at 130 °C. Pre-established isothermal periods and sulfuric acid concentrations were tested according to the Doehlert experimental design procedure [22], according to the method generally used in the evaluation process of the biomass hydrolysis. The experiments were carried out using different diluted sulphuric acid concentrations in range of 0–4%, and the influence of the reaction time and acid concentration on the process yield was studied. After the reaction time had elapsed, the micro-reactor was rapidly cooled down and the hydrolysate and solid phase were recovered by filtration

Table 20.5 Typical ultimate analysis of the used algae in biorefinery technologies data

Type of the biomass/element type	C	H	O	N	S	Cl
Macroalgae species, Seaweed—*Ulva rigida* [%]	41.1	7.5	46.2	5.2	6.3	21.7

(Whatman medium filter paper). The liquid part of samples was than filtered using 0.22 μm micro-filter prior to the HPLC analysis that was used in the evaluation process of the sugar content [23]. The solid phase was dried in an oven at 50 °C for 24 h. Subsequently, the recovered biomass was left for minimum 1 h at room temperature, then the recovered mass was measured. The resulted value for the solid part was used in evaluating the global mass balance of the hydrolysis process.

Glucose, xylose, arabinose, acetic acid, formic acid, HMF, and furfural were established by high-performance liquid chromatography (HPLC) using one Aminex HPX-87H (Bio-Rad, Hercules, CA) column operating at 50 °C in combination with a cation H^+ guard column (Bio-Rad). The mobile phase was 5 mM H_2SO_4, and the flow rate was 0.4 mL/min.

20.2.3 Structural Analysis Methods for Seaweeds Evaluation

The structural and storage type carbohydrate was also established by the methods presented in NREL 60957 protocol "Determination of the Total Carbohydrates in Algal Biomass" [24]. In this case, the samples were treated with concentrate sulfuric acid at 30 °C and the resulted mixture was transferred in glass flasks that were diluted with water to reach the final acid concentration, 4%. The samples were treated in autoclave at 121 °C for one hour. The hydrolyzed samples were vortexed, filtered and in this way became ready for use in the HPLC analysis. In case of use the column Aminex HPX-87P, before the evaluation, the sample is required to neutralize. Previous to the analysis, the sample was filtered using a 0.22 μm filter.

20.2.4 Methods for Study of the Biogas Production Kinetics and Digestibility of the Seaweeds

The seaweed pretreatment (autohydrolysis and dilute acid hydrolysis) were carried out under previously [25, 26] optimized conditions using larger volume reactors, of 1000–1500 mL. The resulted liquid and solid phases were used in anaerobic digestion experiments to produce biogas.

The mostly used testing method used in the study of the biogas production kinetics and the process dynamic is the volumetric method, the so called "fluid displacement "that consists of two recipients where one is working as anaerobic reactor to which it is possible to add a suitable stirrer and a thermostatic bath for keeping a stabile temperature and another reactor containing liquid, having a siphon through which the liquid displaced by the gas falls in a graduated cylinder. In this case the displaced volume is equal to the volume of gas resulted in the anaerobic digestion process. Figure 20.3 presents the laboratory system composed of 1000 mL flasks, where the biological activity is carried out (Reactors 4), a thermostatic bath, to maintain a

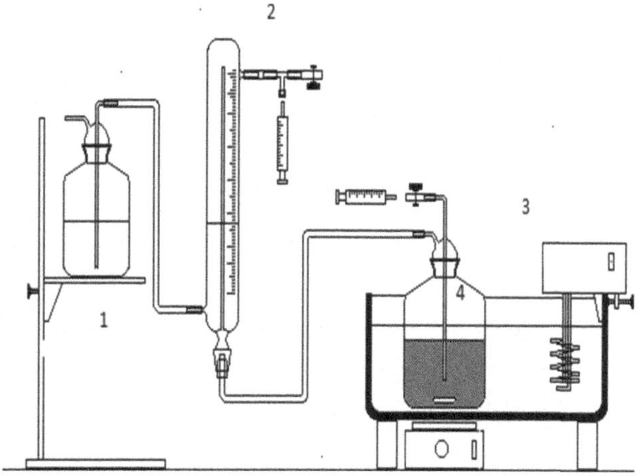

Fig. 20.3 Laboratory scale biogas production unit with volume displacement measurement system

constant temperature in the reactors (3), a syringe to remove the biogas samples and a hydraulic system to measure the biogas production (1, 2).

The biogas metering system consists of graduated tubes connected to the top of the biogas generating reactors, which are filled with a hypersaline aqueous solution composed of H_2O (79%), NaCl (20%) and HCl (1%), to minimize the methane dissolution, and is colored with methyl orange. This solution is kept in a reservoir placed and regulated at the same level of the liquid in the graduated tubes.

The biogas produced during the test inside the flasks generates a pressure that moves the liquid inside the flask to the reservoir in an amount proportional to its volume, and therefore it is possible to record it under hydrostatic conditions based on the volume of the displaced liquid.

The tests were carried out in a batch mode using 2 laboratory reactors in parallel. The reactors were maintained at a temperature of 35 °C (mesophilic conditions), regulated by the thermostatic bath (H_2O). The gas readings produced and accumulated within the graduated collection vessels were periodically carried out, under isobaric conditions at atmospheric pressure. The analytical determinations were performed according to the current methods used in the LNEG laboratory, the gas composition was measured using a portable gas analyzer, type GFM 436.

Table 20.6 Physical properties of the grinded algae

Type of biomass	Moisture content [%]	Ash content [%]	Bulk density [g/L]	Particle average size [mm]
Harvested algae after naturally dried	14.5	20.1	260.5	<4
Algae after drying at 65 °C	9.8	20.2	260.3	<4

20.3 Results of the Seaweeds Characterization Process

20.3.1 Physical Properties of the Grinded Algae

The analysis of the physical parameters of the harvested, dried and grinded (with the average particle size less than 4 mm) macroalgae were evaluated, and the average values for the moisture, ash content and bulk density are presented in Table 20.6.

20.3.2 Structural Composition of the Seaweeds and Composition of the Liquid Phase Resulted from the Acid Hydrolysis

The hydrolysis experiment was done for lower temperature values and with different values for the retention time (8, 60, 112 min.). The experimental design was built to study the effect of the time and of the acid concentration at 130 °C and of the operation temperature in the acid hydrolysis process on the composition of the resulted products in sugars (formed from glucose, xylose and arabinose) and on the values for the resulted inhibitors (acid acetic, HMF, and furfural). It was also established, for each sample, the solid yield. The optimal technological parameter, the highest sugar yield of the hydrolysis process, was observed for the sample where the reaction time was of 60 min and the acid concentration was 4%. The structural and storage carbohydrate type was also established using the methods presented in NREL 60957 Norms [27] and the results and are presented in the Fig. 20.4.

The structural composition of the seaweed shows that this feedstock has an important content of protein. Together with the resulted extractives this value reached 32%. The glucan content is high too; one inconvenient in this case for use of the green algae in different technological biorefinery processes is the high ash content that can raise technological problems. In different agro-technical procedures this can be used as fertilizer in the mixture of the composted organic wastes.

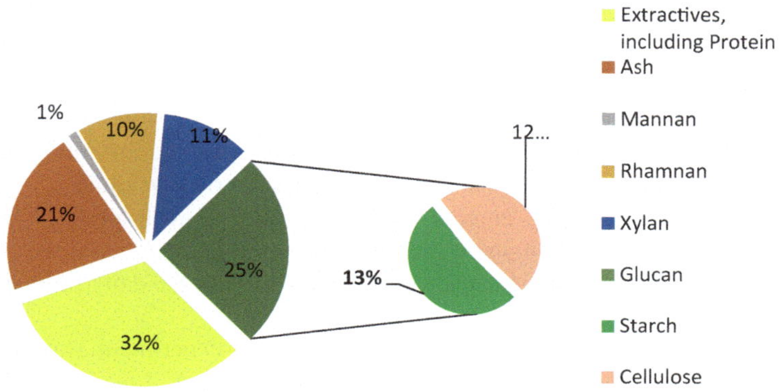

Fig. 20.4 The structural composition of the macroalgae

20.3.3 Biogas Production in the Anaerobic Digestion Process of the Seaweed

During the testing period, the produced biogas volume and composition was periodically recorded and the results are presented in Fig. 20.5, revealing the evolution of the biogas volume generated during the incubation period. The experiment was ended after 30 days, as this was enough time for the resolution of a biodegradability test. In this case, the influence of the pretreatment process on methane generation was studied; in this anaerobic digestion process there were used samples with untreated

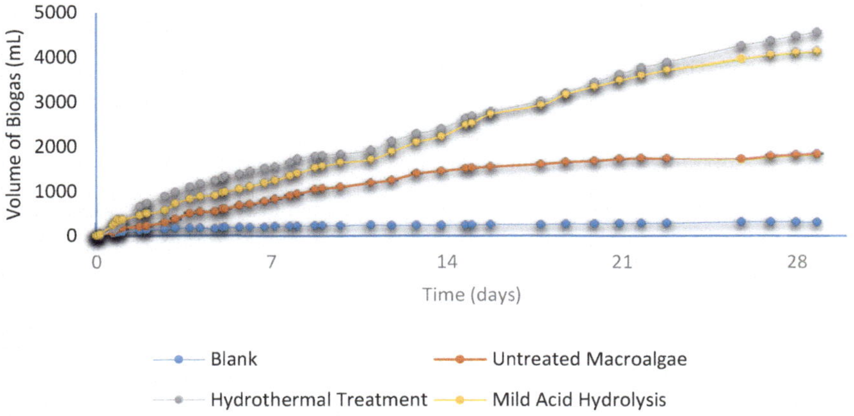

Fig. 20.5 Methan generation by anaerobic digestion of the seaweeds from Black Sea coast (untreated, treated with sulfuric acid and treated in the hydrothermal process)

seaweeds, treated with 3% sulfuric acid and samples resulted by hydrothermal treatment of the feedstock [25, 26]. The values were compared with the data resulted in the digestion, in the same conditions with the sample containing the inoculum.

Observing the curves, we note a first phase with a lower slope and, therefore, a lower evolution, with a relatively low rate, corresponding to an initial phase of bacterial growth and bacteriological adaptation to the nutrient medium.

Subsequently, the system entered the exponential growth phase when the methanogenic bacteria had an abundant substrate, stimulating the cellular growth, as result of a larger degradation capacity and an increase in the daily production of biogas.

The methane content in the biogas produced during the experiment, was measured by gas chromatography. Analyzing the methane composition, it is observed that at the beginning this compound exists in a smaller amount than the carbon dioxide, that is formed in the initial stages of the digestion (hydrolytic and acidogenic phases). When the conditions for methanogenic bacteria progress, the biogas composition evolves increasing the concentration of CH_4, and decreasing that of CO_2. When the methanogenic phase predominates, the methane content increases, reaching the normal and expected values in the biogas composition (60% CH_4). In the discussed case, it can be concluded that the development of the system took place in a normal way for a system not equipped with inoculum. At the end of the test a decrease in the methane concentration occurs because the substrate is less biodegradable, which increases the consumption by cell maintenance on the synthesis and causes a slight increase of the CO_2/CH_4 ratio.

20.4 Conclusions

The investigated biorefinery technologies at lab scale can be integrated in a pilot plant model. The lab scale experiments indicate the structural composition of the seaweeds and indicate the optimal technological parameters useful for the pre-treatments process for obtaining the maximal sugar yield for the treatment.

- The maximum value of the total sugar concentration of the hydrolysate in the experiment was recorded in the case of the *sample* where the used acid concentration was 4% and the reaction time 60 min; in this case the resulted sugar concentration was 36 g/L with the remained value of 14%.
- The macroalgae have an abundant carbohydrate content. This result indicated that this type of plant can be an alternative to the extension of the feedstock base for bioethanol production; based on the theoretical evaluation from one dry tone seaweeds there can be obtained approximately 270 kg bioethanol. This value was obtained from simple material balance data of the fermentation process.
- The experiments with this type of feedstock will be continued to analyze and find different types of lipids and other types of polysaccharides, fibers and colloidal compounds that can be extracted in the first treatment and after that the resulted

organic matter can be used in the co-digestion process for producing biogas/bio-methane [28], that resulted organic matter can be used in the presented AD model as new feedstock.

- Designing a pilot scale biorefinery unit for this type of feedstock that can be implemented in the Black Sea Coastal area, will result in the sustainable use of the collected seaweed from this coast. Based on the literature study, the seaweed has the potential to become a new raw material for producing different added value products for food, pharmaceutical products and cosmetics.
- In the case of biogas production, the pre-treatment process supports the methane production, as it can be observed in Fig. 20.6
- The best technical solution for biogas production is based on the hydrothermal pre-treatment when the resulted methane production is more than six time higher than in case of the AD process with untreated feedstock
- Implementing this biorefinery technologies in Black Sea coast area can reduce the visual and environment pollution of the costal zones. Moreover from the collected seaweeds the experimented biogas yield for green algae is 20 m^3 CH_4/t wwt, (wwt = wet weight) [29]. Thus, from the collected seaweeds from the coastal area there can be produced 180,000 m^3 biogas. That model can be integrated in small scale biogas units working near fish farms and tourist units that administrate the coastal zones.

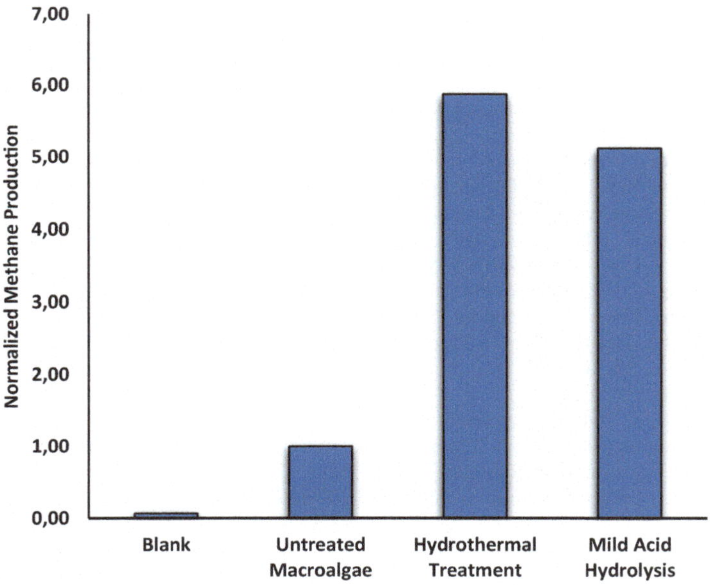

Fig. 20.6 Methane production for different pre-treatment procedure of the seaweeds

Acknowledgements Project Phoenix (MSCA/RISE Contract number 690925) funded by the European Union's Horizon 2020 program is gratefully acknowledged for supporting the work hereby presented.

References

1. European Parliament: Directive 2009/28/EC of the European Parliament and of the Council on the promotion of the use of energy Directives 2001/77/EC and 2003/30/EC. Directive 2009/28/EC. Council of the European Union, Brussels (2009)
2. European Commission: A policy framework for climate and energy in the period from 2020 to 2030. COM (2014) 15 final. Communication from the Commission to the European Parliament, the Council, the European Economic and Social Committee and the Committee of the Regions, Brussels (2014)
3. European Commission: Proposal for a Directive of the European Parliament and of the Council on the promotion of the use of energy from renewable sources (recast). COM (2018) 328/82, Brussels (2018)
4. T. Al Seadi, B. Drosg, W. Fuchs, D. Rutz, R. Janssen, Biogas digestate quality and utilization, in *Biogas Handbook* (2013), pp. 267–270. Sci. Prod. Appl. ISBN 978-0-85709-498-8 (print)
5. S. Koonaphapdeelert et al., *Biomethane Production and Applications* (Springer Nature Singapore Pte Ltd. 2020, Green Energy and Technology (2020), pp. 3–15. ISBN 978-981-13-8306-9
6. J. Walsh, C. Ross, M. Smith, S. Harper, A. Wilkins (1988) *Biogas Utilization Handbook* (Georgia Tech Research Institute, The Environmental, Health and Safety Division, Atlanta)
7. IPCC, *Intergovernmental Panel on Climate Chance, Fourth Assessment Report: Climate Change 2007* (IPCC, Geneva, Switzerland, 2007)
8. A.L. Kohl, R.B. Nielsen, *Gas Purification*, 5th edn. (Gulf Professional Publishing, Houston, 1997)
9. https://www.europeanbiogas.eu/eba-statistical-report-2019-european-overview/. Last accessed 04 May 2020
10. EU—Biofuels in the European Union. ISBN 92-79-01748-9 (2006)
11. EBA Annual Report 2019, European Biogas Association. file:///C:/Users/User/Desktop/EBA-AR-2019-digital-version.pdf. Last accessed 04 May 2020
12. D. Brack et al., *The Green Book: New Directions for Liberals in Government* (Biteback Publishing Library, 2013). ISBN 978-1-84954-561-7
13. B. Jordberga, 1st COASTAL Biogas Conference (Baltic Energy Innovation Energy Centre, 2019). https://www.beic.nu/tag/jordberga-biogas/. Accessed in 04 May 2020
14. IEA—Technology Roadmap Delivering Sustainable Bioenergy (Paris, 2017). www.iea.org
15. G. Roesijadi, S.B. Jones, Y. Zhu, Macroalgae as a biomass feedstock: a preliminary analysis. Analysis, 1–50 (2010)
16. E. Allen, D.M. Wall, C. Herrmann, A. Xia, J.D. Murphy, What is the gross energy yield of third generation gaseous biofuel sourced from seaweed? Energy **81**, 352–360 (2015)
17. https://www.rowater.ro/dadobrogea
18. Agerpress. https://www.radioconstanta.ro/2019/06/29/peste-3-000-tone-de-alge-aduse-pe-plaje-de-furtuni-colectate-in-ultima-luna-pe-litoral/. Last accessed 22 Sept 2019
19. Y.N. Barbot, H. Al-Ghaili, R. Benz, A review on the valorization of macroalgal wastes for biomethane production. Mar Drugs **14**(6), 120. Published online 2016 June 2
20. W. De Jong, R. Van Ommen, *Biomass as a Sustainable Energy Sources for the Future, Fundamental of the Conversion Process* (Wiley, Hoboken, New Jersey, 2015)
21. M. Lahaye, A. Robic, Structure and functional properties of Ulvan, a polysaccharide from green seaweeds. Biomacromolecules
22. D.H. Doehlert, Roy. Stat. Soc. C-App. **19**, 231–240 (1970)

23. H.B. Sluiter, R. Ruiz, C. Scarlata, J. Sluiter, D. Templeton, *Determination of Sugars, Byproducts, and Degradation Products in Liquid Fraction Process Samples*—Laboratory Analytical Procedure (LAP), National Renewable Energy Laboratory—NREL. https://www.nrel.gov/biomass/analytical_procedures.html. Colorado 80401-3393 (2006)
24. Protocol NREL/TP 60957-Determination of the Total Carbohydrates in Algal Biomass (2015)
25. S. Bartha, L.C. Duarte, S.D. Berardino, F. Carvalheiro, P.L. Martins, Anaerobic digestion of coastal seaweeds: A potential solution in the Black Sea Region, COASTAL Biogas conference Malmoe, Sweden, 8–9 May 2019
26. S. Bartha, F. Carvalheiro, P. Moniz, L.C. Duarte, N. Antal, Evaluation of macro algae as potential feedstock for the biorefinery, 4-CIAB-Ibero-American Congress on Biorefineries, Jaén, Spain, 24–26 October 2018
27. Protocol NREL/TP-510-42622-Determination of Structural Carbohydrates and Lignin in Biomass, National Renewable Energy Laboratory (2008)
28. B.M. Smyth et al., What is the energy balance of grass biomethane in Ireland and other temperate northern European climates? Renew. Sustain. Energy Rev. **13**, 2349–2360 (2009)
29. IEA Bioenergy Report, State of Technology Review—Algae Bioenergy, IEA Bioenergy Task 39 (2017), pp. 95–110. ISBN: 978-1-910154-30-4

Part V
New Trends in the Use of Solar Energy in Communities

Chapter 21
Complex Electro-Intra-Conversions Within Rotaxanes' Molecular Machines

Mirela I. Iorga, Nicoleta A. Dudaş, and Mihai V. Putz

Abstract The work advances an original mechanism for the internal frontier (Lowest Unoccupied/LU and Highest Occupied/HO) molecular orbitals (MO), i.e., LUMO > HOMO versus HOMO(i) > HOMO(i + 1) versus LUMO(j) < LUMO(j + 1), for the electronic intra-conversion within molecular machine (MM) systems, here applied on a rotaxane with ring translational movements; the working structure of the present MM system [2]rotaxane $1H^{3+}$ is as follows: dibenzo[24]crown-8 ether macro-cycle, combined with the dumbbell type component containing a center of secondary ammonium ($-NH_2^+-$) and a 4,4′-bipyridine unit (bpy^{2+}), while having the functional group of anthracene at one end, and a 3,5-di-tert-butylphenyl group at the other end, as stoppers. Accordingly, the MM system [2]rotaxane $1H^{3+}$ is considered to work under shuttle movement for which the special activation parameters of the ring shuttle processes for the basic induced process ($A \rightarrow B$) and for the acid-induced one ($AH \leftarrow BH$) are identified. As a result, the quantum-orbitalic mechanism facilitates the Hydrogen activating gates (on/off MM) while driving the machine of the molecular complex within a fundamental acid–base process, i.e., within the "Hard and Soft Acids and Bases" (HSAB) paradigm; it leads to sustainable controlling and eventually to reusing and redirecting in mixed/integrated macromolecular systems the molecular activated energies involved in the individual MM (here I-to-IV) processes.

Keywords Molecular machine · HSAB principle · HOMO–LUMO inter-conversions

M. I. Iorga · M. V. Putz (✉)
Laboratory of Renewable Energies-Photovoltaics, R&D National Institute for Electrochemistry and Condensed Matter—INCEMC—Timisoara, 114 Dr. A. Paunescu Podeanu, 300569 Timişoara, Romania
e-mail: mihai.putz@e-uvt.ro; mv_putz@yahoo.com

M. I. Iorga · N. A. Dudaş · M. V. Putz
Laboratory of Structural and Computational Physical-Chemistry for Nanosciences and QSAR, Biology-Chemistry Department, Faculty of Chemistry, Biology, Geography, West University of Timisoara, 16 Pestalozzi Street, 300115 Timişoara, Romania

I. Visa and A. Duta (eds.), *Solar Energy Conversion in Communities*,
Springer Proceedings in Energy, https://doi.org/10.1007/978-3-030-55757-7_21

21.1 Introduction: Molecular Machines—The New Nanochemistry

The rapid development of many areas in nanotechnology has determined the occurrence of new possibilities in the design of functional materials with special optical, photonic, electronic, electrical, thermal, mechanical, chemical, biochemical, etc. properties [1]. The main objective of nanoscience and nanotechnology is to find solutions to the pressing problems of the new millennium, such as food, health, energy, and the environment [2]. Currently, the trend is to reduce, as much as possible, the size and weight of the components in the devices and machines to perform more complicated operations with increasingly smaller parts [3]. The studies on the possibilities of designing devices, machines, and molecular motors at nanoscale are of great importance for both the fundamental research and the practical applications of nanoscience and nanotechnology. The "bottom-up" molecule-by-molecule approach offers unlimited opportunities for the design and construction of nanoscale supramolecular structures, by combining the high precision of the chemical syntheses with the engineering ingenuity [4]. The main features of the molecular machines [5] can be summarized as follows: type of energy input, operation monitoring, type of component movement (rotation, translation, oscillation), the possibility to repeat cyclic operations, the period required for a complete cycle, the function performed. The vast majority of artificial molecular machines designed so far are based on interconnected molecular species, such as rotaxanes, catenanes, and related species [6].

The rotaxanes are made of a dumbbell molecule (or a thread molecule) with the "axis" surrounded by a macrocyclic compound—the "ring", and terminated at both ends with bulky groups—"stoppers", that have the role of preventing the disassembly of the components [2, 7, 8]. The main interactions [2, 9] present for rotaxanes and catenanes are charge transfer, hydrogen bond formation, hydrophobic-hydrophilic character, π-π arrangement, electrostatic forces, and metal–ligand bonds. In the case of rotaxanes, two main movements with large amplitude can be made: translation—by moving the ring along the axis, or rotation—by moving the ring around the axis (Fig. 21.1). Consequently, rotaxanes are suitable prototypes for the construction of both linear and rotary molecular machines. Indeed, systems of the first type, called molecular shuttles [3], are the most common implementation of rotaxanes in the concept of molecular machines.

Fig. 21.1 Schematic representation of the translational (left) and rotational (right) movement of the ring in the case of rotaxanes (adapted after [3])

Because molecular devices and machines are chemical systems, they work through chemical reactions, and different situations may occur: electronic and nuclear rearrangements, electron transfer without nuclear modifications, or nuclear displacement. Also, since molecular devices and machines must operate in repetitive cycles, an essential requirement is the reset function. Accordingly, the chemical reactions underlying the functioning of molecular devices and machines must be reversible. Such a condition can be fulfilled by processes involving energy transfer, electron transfer (redox), proton transfer (acid–base), photoisomerization, and metal–ligand coordination reactions [2, 3].

An example of chemically driven molecular machines is a rotaxane that acts as a reversible molecular shuttle controlled by acid–base stimulation, namely [2]rotaxane $1H^{3+}$ [10]. It is formed by a dibenzo[24]crown-8 ether (DB24C8), a π-electron donor macro-cycle, and a dumbbell type component containing a center of secondary ammonium (-NH_2^+-) and a 4,4′-bipyridine unit (bpy^{2+}). As stoppers, there are used a functional group of anthracene at one end, and a 3,5-di-tert-butylphenyl group at the other end. The structure of [2]rotaxane $1H^{3+}$ and its molecular components are shown in Fig. 21.2.

However, despite the huge chemical leap in nanosciences made by validation of molecular machines in general and those based on rotaxanes in particular, quantum structural models for such nanochemical compounds in an inter-conversion dynamics was not studied in terms of frontier molecular orbitals and chemical reactivity, according to our best knowledge. Here we report, therefore, first such a quantum structural model through the results and discussion as based on fundamental chemical hardness principles, in Sects. 21.2 and 21.3, respectively.

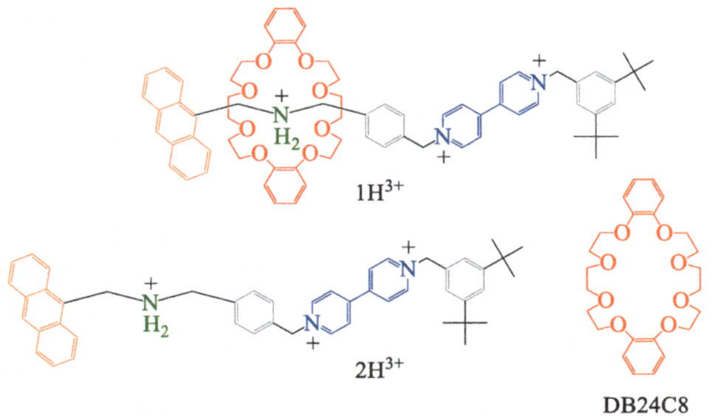

Fig. 21.2 The structure of [2]rotaxane $1H^{3+}$ and its molecular components, dumbbell type component $2H^3$ and the ring DB24C8 (adapted from [10])

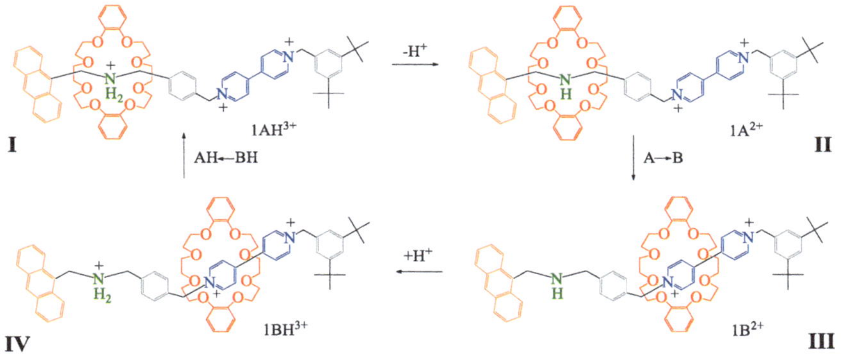

Fig. 21.3 Schematic representation of the shuttling process of the crown ether ring during the deprotonation and re-protonation of the ammonium center (adapted from [11])

21.2 First Quantum Results on Rotaxanes' Molecular Machine

21.2.1 The System

Since the interactions of charge transfer (CT) of the π-electron donor macrocycle with the electron acceptor bpy^{2+} unit are much weaker than the hydrogen bond interactions $[N^+–H\cdots O]$ between the macrocycle DB24C8 and the ammonium center, the [2]rotaxane $1H^{3+}$ will have stable co-conformation when the macrocycle surrounds the ammonium center, as illustrated in Fig. 21.3 in the $1AH^{3+}$ structure.

21.2.2 The Phenomenological Chemistry

If a base is added, the ammonium center is converted to an amine function, and the transient state $1A^{2+}$, which appears after the macrocycle's movement on the bipyridine unit, is transformed into a stable state, $1B^{2+}$. The reverse process can occur if an acid is added and the system passes into the initial state through the transient state $1BH^{3+}$. In the case of a deprotonated rotaxane, the charge transfer interactions are destroyed by the mono-electronic reduction of the bipyridine unit, and the macrocyclic ring can be displaced from the bipyridine station. The schematic representation of the switching process is shown in Fig. 21.4.

The complete chemical reversibility of the acid–base reactions providing energy guarantees the reversibility of the mechanical movement. It should be added that, in the case of a deprotonated rotaxane, it is possible to dislocate the ring from the bipyridine station by destroying the charge transfer interactions by reducing this station. In conclusion, mechanical movements can be induced in the system by two

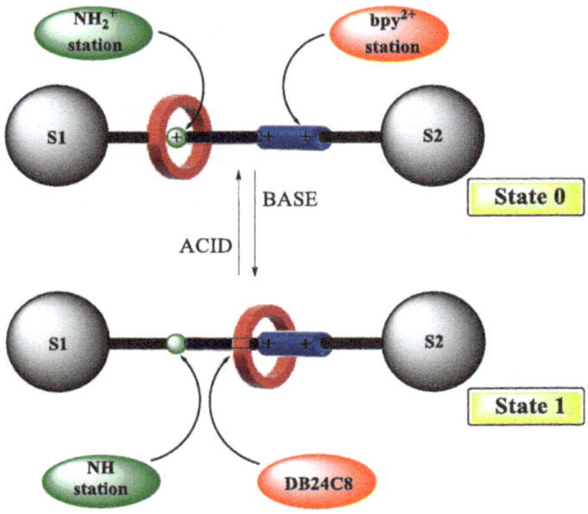

Fig. 21.4 Schematic representation of an acid–base controllable molecular shuttle (adapted from [11])

different types of stimuli, namely acid–base and reduction–oxidation. Besides, this rotaxane may be useful for processing information, as long as it is a *bistable* system and may exhibit binary logical behavior.

21.2.3 The Quantum Chemistry Results

To the best of our knowledge the Molecular Machines systems were not studied so far considering the in-depth quantum structural view or their overall mechanism. Accordingly, this work reports for the first time such an analysis until the 6th degree (!) of spectra of all compounds involved in the machine process of Fig. 21.3, through the results reported on Table 21.1 for the Highest occupied and Lowest unoccupied molecular orbitals, HOMO (1)—HOMO (6) and LUMO (1)—LUMO (6), respectively.

Most interestingly, although not straight, and not at the first side, the searched quantum structural cycle of HOMOs/LUMOs inner (in depth) to trans (in higher virtual) electronic transition successfully find the "closed circle" of linked orbitals covering and explaining therefore the Molecular machine under study, as outlined in Table 21.1 and represented in Fig. 21.5 (spectra and 3D involved HOMO/LUMO orbitals). Moreover, the "closed circle of orbitals" closely follow the experimentally observed free Gibbs energies, also reported in Fig. 21.5, cross-checking the reliability in respect of quantum observable events (compounds spectra).

Table 21.1 The HOMO/LUMO Spectra for the Compounds I–IV of the molecular machine cycle of Fig. 21.3

Spectra: HOMO/LUMO (eV)	Compound I	Compound II	Compound III	Compound IV
LUMO 8	4.990864	7.986239	8.151045	5.505497
LUMO 7	4.837902	7.964234	8.017833	5.225175
LUMO 6	4.664826	7.867704	**7.46166**	4.948896
LUMO 5	4.454133	7.343532	7.219172	4.698437
LUMO 4	3.899544	7.032508	7.115769	4.368488
LUMO 3	3.890334	6.864866	6.571156	4.248714
LUMO 2	3.251346	**6.736673**	6.177625	3.645858
LUMO 1	2.095507	5.88205	5.265115	2.505458
LUMO 0	1.232262	4.83425	4.754953	1.730285
HOMO 0	−2.283457	−0.533813	0.5343327	−1.669896
HOMO 1	−3.411135	−1.423227	−.352209	−3.035007
HOMO 2	**-3.892067**	−1.885268	−2.866274	−3.773967
HOMO 3	−4.359834	−2.315995	−4.049128	**−4.544211**
HOMO 4	−5.518934	−3.417814	−4.278116	−6.35188
HOMO 5	−6.112431	−4.103288	−4.871039	−6.414983
HOMO 6	−6.796926	−4.802192	−5.02403	−7.947465
HOMO 7	−7.026674	−5.036039	−5.815712	−8.257248
HOMO 8	−8.427703	−5.174859	−5.906811	−8.394039

21.3 Quantum Chemical Reactivity Discussion on Rotaxanes' Molecular Machine

21.3.1 HSAB (Hard and Soft Acids and Base) Principle

The HSAB concept was introduced by Pearson [12–14] in the early 1960s in an attempt to unify the chemistry of inorganic and organic reactions [15]. It is widely used in chemistry to explain the stability of compounds and the reaction mechanisms. The theory is used in the context where a qualitative description rather than a quantitative one is required to understand the main factors that govern chemical properties and reactions. To demonstrate this theory, Pearson [15, 16] had been started from the general equation given by Lewis, which has broad applicability in chemistry:

$$A + : B \Leftrightarrow A : B \tag{21.1}$$

while being susceptible to the more detailed chemical exchange such as:

$$A\ : B' + A' : B \ \Leftrightarrow\ A'\ : B' + A : B \tag{21.2}$$

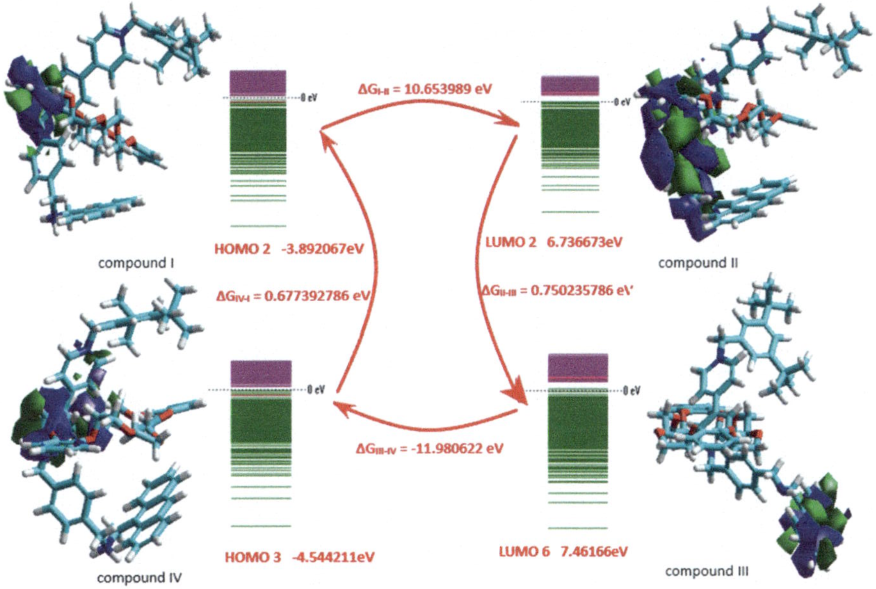

Fig. 21.5 HOMO–LUMO quantum representation of the molecular machine/cycle of Fig. 21.3, with the data in Table 21.1. The selected molecular orbitals among which the electronic transfer promotes the molecular machine cycle; the specific 3D molecular orbital is also emphasized in order to clear highlight the chemical stericity of the quantum reactivity at involved rotaxanes frontiers' orbitals, HOMO and LUMO, respectively

Based on these considerations Pearson [16] classifies Lewis acids as belonging to class (a) or (b), namely the acids of class (a) form more stable complexes with the right donor atoms, and the acids of class (b) prefer the donor atoms on the left side of the electronegativity scale:

$$As < P < Se < S \cong I \cong C < Br < Cl < N < O < F \tag{21.3}$$

and named it: class (a)—hard Lewis acids, and class (b)—soft Lewis acids.

The acceptor atoms in class (a) usually have a high positive charge, are small in size and have no unshared electrons in the valence shell. Acids in class (b) have acceptors with a low positive charge, are large, and usually have unshared electron pairs in the valence shell. All these facts cause the acceptor atoms in class (a) to be not highly polarizable, while the acceptor atoms in class (b) are highly polarizable. Because polarizability means the deformation of the electronic cloud in an electric field, and if they are easy to deform, it means they are soft; this led to the name of the two classes of acids, soft and respectively hard. Looking at the list of donor atoms above presented, it is evident that the polarizability is higher on the left and decreases to the right.

With this terminology, a general statement can be formulated, namely: "hard acids prefer to coordinate hard bases, and soft acids prefer to coordinate soft bases"—this represents the Principle of Hard and Soft Acids and Bases or the HSAB Principle.

The reaction (21.4) can illustrate the HSAB principle in the following sense:

$$hs + sh = hh + ss \quad 0 > \Delta H \tag{21.4}$$

The most important feature is that the hard-hard combinations are held together primarily by ionic bonds, and the soft–soft combinations use the covalent bond.

The problem of expressing chemical hardness was solved by Pearson and Parr since 1983 [17]. They extended the qualitative theory of HSAB with the quantitative definition of chemical hardness (η) as being proportional to the second-order derivative of the chemical system's total energy; they start from the theory of functional density to determine the importance of the electronic chemical potential, μ.

$$\mu = \left(\frac{\partial E}{\partial N}\right)_{\nu} \cong \frac{(I + A)}{2} = -\chi \tag{21.5}$$

$$\eta = \frac{1}{2}\left(\frac{\partial \mu}{\partial N}\right)_{\nu} = \frac{1}{2}\left(\frac{\partial^2 E}{\partial N^2}\right) \cong \frac{(I - A)}{2} \tag{21.6}$$

Pearson believes that Eq. (21.6) gives the perfect definition of chemical hardness, but also the operational definition that was firmly grounded in the density functional theory.

According to the molecular orbital theory and Koopmans' theorem, the term $(I - A)$ represents the energy gap between HOMO and LUMO [18].

Two major principles have been associated with chemical hardness, namely the principle of maximum hardness (PMH) [19] and the principle of HSAB, Eqs. (21.1–21.3), both of which are of utmost importance for understanding the molecular electronic structure and acid–base reactions in general.

From the many circumstantial pieces of evidence since 1987, Pearson [19] concluded that "there seems to be a rule of nature that molecules arrange so as to be as hard as possible". Subsequent studies have supported this principle, which implies the necessary conditions of temperature and constant chemical potential. This fact has been proved by Parr and Chattaraj [20] by a combination of statistical mechanics and the dissipation-fluctuation theorem. Considering the general case:

$$A + B \rightarrow AB \tag{21.7}$$

where A, an atom or a molecule, interacts with B to give a third specie, AB, the interaction energy is given by:

$$\Delta E_{\text{int}} = E[\rho_{AB}] - E\left[\rho_A^0\right] - E\left[\rho_B^0\right] \tag{21.8}$$

Considering that the interaction can be divided into two stages, in the first stage when A and B are placed apart, their chemical potentials are modified to reach a common value [21]. This stage takes place by charge transfer to a constant external potential, and it is assumed that μ_{AB} is equal to the chemical potential of the AB system at equilibrium. The energy exchanged can be written, through modifications of the external potential $v_{AB}(r)$ as:

$$\Delta E_{\nu} = \Delta E_{\nu}^{A} + \Delta E_{\nu}^{B} \tag{21.9}$$

and ultimately in terms of global hardness:

$$\Delta E_{\mu} = -\frac{1}{2} N_{AB}^{2} \left(\eta_{AB} - \eta_{AB}^{*}\right) \tag{21.10}$$

This equation leads to the principle of maximum hardness, because it is assumed that the interaction between A and B leads to a state of equilibrium of lower energy than the total energy of A and B, when they are at a distance from each other. From Eq. (21.10), it is established that under conditions of constant chemical potential, the AB system evolves to a state of maximum hardness. *Moreover, it appears that the exchanged energy in chemical reactivity has a parabolic trend in chemical hardness parameter—just as the free Gibbs energy showcases the parabolic evolution of the internal energy respecting the entropy. From this, another outstanding value of chemical hardness follows, as the ordering parameter of the change transfer among interacting compounds (electronic "open/opening" systems)* [22].

These principles can be used in the context of wave function theory. In this case, the global hardness can be approximated, according to Eq. (21.6):

$$\eta \approx \frac{\varepsilon_{LUMO} - \varepsilon_{HOMO}}{2} \tag{21.11}$$

and thus, the interaction mechanism can be analyzed in terms of the principle of maximum hardness, instead of the principle of minimum energy. This procedure is recommended in the case of semi-empirical methods, where the calculation of the energy of the frontier orbitals is preferred rather than the total energy calculation. Following these results, it was found that the principle of maximum hardness has become a beneficial concept in chemistry, *as it will be used in the case of the rotaxane based molecular machine quantum chemical reactivity description.*

21.3.2 Applied HSAB Principle on Rotaxane Molecular Machine

These general chemical hardness considerations can prove useful for analyzing reaction mechanisms and for understanding the general behavior of a given molecule when it interacts with different chemical species. However, when it is about the *inner-trans electronic transition* that is involving higher degrees of HOMO and LUMO, and among connected molecular compounds in a chain reactivity, the chemical hardness basic formulation (21.11) may be amended as follows:

$$\eta^{\#} \approx \frac{\varepsilon_{LUMO(i)}^{Molecule-A} - \varepsilon_{HOMO(j)}^{Molecule-B}}{2 + \sqrt{i \times j}} \tag{21.12}$$

One notes that the involved "inner/trans" orbital degrees "i & j" indices specifically show that: (1) they appear in the denominator, so that at any non-null contribution decrease the overall chemical hardness,—so sustaining the chemical reactivity within the inner-trans cycle under study (work); (2) they appear under the geometrical contribution being specific to the chemical interaction. Note that the "i & j" indices are orbital indices so behaving likewise the associated wave-functions, from where their multiplication while interacting; moreover, the newly factor $\sqrt{i \times j}$ accounts of the quantum wave modification on the exchanged pair charge "2" among the frontier molecular orbitals involved. This means that the quantum approach of the molecular machines through quantum structural HOMO/LUMO inner-trans spectra method involves the quantum creation/circulation/recirculation of fractional charges along the inner cycle.

Accordingly Tables 21.2 and 21.3 display the chemical hardness results when ordinary HOMO (0)—LUMO (0) *versus* complex inner-trans chemical HOMO (i)-LUMO (j) hardness analysis is performed for the molecules at stake in Figs. 21.3 and 21.5. The results are again spectacular: while with ordinary analysis the machine would be stopped at the conversion into the second compound (II), according to the maximum hardness principle, the complex inner-trans analysis in Table 21.3 suggests

Table 21.2 "*Ordinary*" chemical hardness results based on Eq. (21.11) and the data in Table 21.1 for the molecular cycle of rotaxanes in Fig. 21.5

Compound	HOMO (0) (eV)	LUMO (0) (eV)	η_0 (eV)
I	−2.283457	1.232262	1.7578595
II	−0.533813	4.83425	2.6840315
III	0.5343327	4.754953	2.11031015
IV	−1.669896	1.730285	1.7001175

Table 21.3 "*Complex*" chemical hardness results based on modification of Eq. (21.11) allowing inter-molecular-charge transfer as of Eq. (21.12), and the respective the data in Table 21.1 for the molecular cycle of rotaxanes in Fig. 21.5

Charge transfer	Initial MO (eV)	Final MO (eV)	$\eta^{\#}$ (eV)
I → II	*HOMO (2)* = −3.892067	*LUMO (2)* = 6.736673	(→) 2.65719
II → III	*LUMO (2)* = 6.736673	*LUMO (6)* = 7.46166	Intra-LUMO-growing (↑)
III → IV	*LUMO (6)* = 7.46166	*HOMO (3)* = −4.544211	(←) 1.9232
IV → I	*HOMO (3)* = −4.544211	*HOMO (2)* = −3.892067	Intra-HOMO-growing (↑)

different quantum mechanism: the I → II conversion produces a lower chemical hardness, so the extra-reactivity energy is consumed for activating the trans LUMO (6) virtual orbital in compound (III), from where the inner HOMO (3) level of the compound (IV) is reached, thus driving into even more stable state the molecular machine cycle, while, again the released energy is eventually internally re-used for re-gaining the state HOMO (3) of compound-I, the actually retrieved state as triggering state in compound-I for the entire molecular machine cycle.

21.4 Conclusions and Perspectives

The molecular machine quantum analysis revealed that the HSAB fashioned approach involving frontier (0th levels as of) HOMO and LUMO may be, and perhaps it should be extended to the so called inner-trans HOMO/LUMO (lower HOMO and higher LUMO frontier orbitals) to explain, at the quantum structural level, the complex mechanism of macro cycle inter- and intra-conversion within the molecular machine cycle. The working example is sustained by the experimental observation of the free Gibbs energy slightly gain, Eq. (21.13), across the molecular machine cycle (Figs. 21.3 and 21.5), indeed indicating an non-equilibrium evolution, as it is when "additional charge is rolled" within the intra- and inner- compound conversion.

$$\Delta G(I - II - III - IV - I) \approx 0.100996\,[\text{eV}] \tag{21.13}$$

It is quantum chemically explained by the additional inner-trans orbital interaction factor, here measured by the geometry of involved orbitals' interaction degree $\sqrt{i \times j}$, thus providing a newly chemical hardness based HSAB mechanism. On the other side, the current quantum chemical model is able to identify the correct theoretical-structural complex mechanism at the inter-molecular level assuring the exact zero energy exchanging (losing or gaining) across a machine cycle [23], as one can immediately check with the date in Table 21.3 plugged into chemical inter-rotaxanes selected frontier molecular orbitals involved in the working molecular machine, i.e.:

$$\Delta\varepsilon\left\{\begin{array}{l}\left[\varepsilon^{II}_{LUMO(2)}-\varepsilon^{I}_{HOMO(2)}\right]+\left[\varepsilon^{III}_{LUMO(6)}-\varepsilon^{II}_{LUMO(2)}\right]\\+\left[\varepsilon^{IV}_{HOMO(3)}-\varepsilon^{III}_{LUMO(6)}\right]+\left[\varepsilon^{I}_{HOMO(2)}-\varepsilon^{IV}_{HOMO(3)}\right]\end{array}\right\}=0(!!!)\,[eV] \tag{21.14}$$

Therefore, these remarkable results assure two-folds the present endeavor: (a) The selected frontier molecular orbitals of Fig. 21.5 for the phenomenological rotaxane machine in Fig. 21.3 were correct; (b) The quantum chemical reactivity model in terms of HOMO–LUMO inter-molecular conversion is a viable one, able to sustain the inner equilibrium of the hard and soft acids and bases configuration of the rotaxanes in the given molecular machine/cycle.

Overall, this result is a true confirmation of the quantum model viability. This premiere should be further investigated on a variety of molecular machines [24] to become a recognized quantum chemical structural explanation of the molecular machine phenomenon in itself, eventually leading further synthesis, control, and self-sustained devices (activated by light or by heat).

Acknowledgements This work was supported within the Nucleus Programme under the project PN-19-22-01-02 by the Romanian National Authority for Scientific Research and Innovation (ANCSI) and of its 2020 renewing fund by Ministry of Education and Research.

References

1. M.I. Iorga, M.V. Putz, Quantum Electrochemistry, in *New Frontiers in Nanochemistry: Concepts, Theories and Trends*, vol. 1: General Nanochemistry, ed. by M.V. Putz, pp. 475–492. (Apple Academics & CRC Press, Taylor & Francis Group, Ontario, Canada, 2020)
2. V. Balzani, A. Credi, M. Venturi, *Molecular Devices and Machines: Concepts and Perspectives for the Nanoworld* (Wiley-VCH, Weinheim, 2008)
3. V. Balzani, A. Credi, M. Venturi, Molecular devices and machines. Nanotoday **2**, 18–25 (2007)
4. M. Venturi, M.I. Iorga, M.V. Putz, Molecular devices and machines: hybrid organic-inorganic structures. Curr. Org. Chem. **21**(27), 2731–2759 (2017)
5. A. Credi, M. Venturi, Molecular machines operated by light. Cent. Eur. J. Chem. **6**, 325–339 (2008)
6. M.I. Iorga, N. Mirica, M.V. Putz, Nano-electrochemistry. Part. II. Molecular machines, in *Proceedings of the 23rd International Congress of Chemical and Process Engineering CHISA 2018* (Prague, Czech Republic, 2018), p. 62

7. V. Balzani, A. Credi, F.M. Raymo, J.F. Stoddart, Artificial molecular machines. Angew. Chem. Int. Ed. **39**, 3348–3391 (2000)
8. E.R. Kay, D.A. Leigh, Rise of the molecular machines. Angew. Chem. Int. Ed. **54**, 10080–10088 (2015)
9. M. Venturi, A. Credi, Electroactive [2]catenanes. Electrochim. Acta **140**, 467–475 (2014)
10. P.R. Ashton, R. Ballardini, V. Balzani, I. Baxter, A. Credi, M. Fyfe, Acid-base controllable molecular shuttles. J. Am. Chem. Soc. **120**, 11932–11942 (1998)
11. S. Garaude, S. Silvi, M. Venturi, A. Credi, A.H. Flood, J.F. Stoddart, Shuttling dynamics in an acid–base-switchable [2]Rotaxane. ChemPhysChem **6**, 2145–2152 (2005)
12. R.G. Pearson, Hard and soft acids and bases. J. Am. Chem. Soc. **85**(22), 3533–3539 (1963)
13. R.G. Pearson, Hard and soft acids and bases, HSAB, part I: fundamental principles. J. Chem. Educ. **45**, 581–586 (1968)
14. R.G. Pearson, Hard and soft acids and bases, HSAB, part II: underlying theories. J. Chem. Educ. **45**, 643–648 (1968)
15. R.G. Pearson, *Chemical Hardness—Applications From Molecules to Solids* (Wiley-VCH, Weinheim, 1997)
16. R.G. Pearson, Chemical hardness—a historical introduction, in *Chemical Hardness*, ed. by K.D. Sen (Springer, Heidelberg, 1993), pp. 1–10
17. R.G. Parr, R.G. Pearson, Absolute hardness: companion parameter to absolute electronegativity. J. Am. Chem. Soc. **105**(26), 7512–7516 (1983)
18. R.G. Pearson, Absolute electronegativity and hardness correlated with molecular orbital theory. Proc. Natl. Acad. Sci. U.S.A. **83**, 8440–8441 (1986)
19. R.G. Pearson, Recent advances in the concept of hard and soft acids and bases. J. Chem. Educ. **64**(7), 561 (1987)
20. R.G. Parr, P.K. Chattaraj, Principle of maximum hardness. J. Am. Chem. Soc.**113**(5), 1854–1855 (1991)
21. J.L. Gazquez, Hardness and softness in density functional theory, in *Chemical Hardness*, ed. by K.D. Sen (Springer, Heidelberg, 1993), pp. 27–44
22. M.V. Putz, in *Quantum Nanochemistry. A Fully Integrated Approach* (5 volumes package). (Apple Academic Press & CRC Press, Toronto-New Jersey, Canada-USA, 2016)
23. (a) M.I. Iorga, M.C. Mirica, M.V. Putz, Electrochemical systems and mechanically interlocked molecules. Part 1: Nano-Science. New Front. Chem. **27**(2), 73–87 (2018); (b) M.I. Iorga, P. Svera, D. Buzatu, M.V. Putz, Electrochemical systems and mechanically interlocked molecules. Part 2: Nano-Technology. New Front. Chem. **27**(2), 89–104 (2018)
24. N.A. Dudaș, M.V. Putz, Chemical reactivity driving switchable molecular machines. A case of Bipyridine-Calixarene rotaxane. Fullerenes, Nanotubes Carbon Nanostruct. **27**, 514–524 (2019)

Chapter 22
Quantum Metrological Matrices for Sustainable Graphentronics

Doru L. Buzatu, Paula Ianasi, and Mihai V. Putz

Abstract In this work ZnO/GO (Graphene Oxide) deposited materials in 3Qubit configurations with the arrangement of *matrix-junctions* are discussed, aiming to establish quantum transistor configurations for the quantum renewable energy mixes applications. The graphentronic integrated circuits are developed under applied voltage on the AND, OR, XOR input logical signals of the metal-oxide depositions, resulting in quantum transistors with 3-qubit outputs. Firs results of the output for the logical gates is performed; they can be further processed by using laws of Boolean Algebra and Virial theorem, following the electronic and also quantum tunneling activation resulted data of ZnO/GO depositions influenced by the diversity of metal oxide-graphene oxide placements. By activating quantum Hamiltonian gates it is possible to deliberate the quantum level repulsion effect taking place when valence-conduction graphenic band gap is stretched, including the synergic tunneling transport as a consequence of quantum interferences, resulting in increasing energy over earlier *staking heterojunctions*.

Keywords Graphene oxide · ZnO · 3-qbits · 8-dimensional matrix

22.1 Introduction

The semiconductors industry, since its appearance with the discovery of the transistor in 1947, revolutionized electronics, mainly through the six decades of continuous miniaturization of devices based on complementary metal-oxide semiconductors (*CMOS: complementary metal-oxide semiconductor*) [1]; thus, according

D. L. Buzatu · P. Ianasi · M. V. Putz (✉)
Laboratory of Renewable Energies-Photovoltaics, R&D National Institute for Electrochemistry and Condensed Matter—INCEMC–Timisoara, Dr. A. Paunescu Podeanu 144, 300569 Timișoara, Romania
e-mail: mihai.putz@e-uvt.ro

Laboratory of Structural and Computational Physical-Chemistry for Nanosciences and QSAR, Biology-Chemistry Department, Faculty of Chemistry, Biology, Geography, West University of Timisoara, Pestalozzi 16, 300115 Timișoara, Romania

I. Visa and A. Duta (eds.), *Solar Energy Conversion in Communities*,
Springer Proceedings in Energy, https://doi.org/10.1007/978-3-030-55757-7_22

to the *International Technology Roadmap for Semiconductors* (*ITRS*) [2], the sub-10 nm (*deca-nano*) generation of metal–oxide–semiconductor field-effect transistors (*MOSFET*) is expected to enter mass production immediately after 2020, according to the prediction of doubling the component units in an electronic chip every 12–24 months, from Moore's famous law; particularly are the following:

- Quantum Non-Destructive Information (QND) experiments that have been proposed for the electronic spin by SWAP operations (coherent transfer of electronic state from electron to the nucleus) for Si:P systems; these allow the measurement of the electronic spin related to the donor [3, 4];
- Projective experiments such as Josephson junction, which allows the measurement of Qubit information [5–7];
- Connections between quantum information activation times in donors, coherence times in Qubits transport, as well as the quantum figure of merit for Silicon operating systems [7, 8];
- Quantum information processed by deca-nano-devices via electronic spin [9], and through a junction with extra capture, respectively [10, 11];
- The coherent transfer through the adiabatic passage which was studied for the odd case with three donors A-J-A "Kane" type [12, 13].

It should be noted that the approach of spin information of electrons and donor ions nuclei in nano-electronics with single-atom in Silicon matrix has *certain advantages*—such as [12, 14]:

- The nuclear spin at low-temperature stores the information up to a few seconds;
- The same property is also found in spin systems with electrons in silicon-based semiconductors, due to the ^{28}Si isotope purification;
- Spin systems are relatively stable in the environment (especial at ambient temperature), so they are promising (in principle) for room temperature operations;
- There is a substantial separation of the spin-orbital valley which separates very well the associated Hilbert space from the other system's freedom degrees.

However, the context of electronic and nuclear spin in quantum transport also has *essential disadvantages* precisely in the area where deca-nano-systems should overcome the barrier of Moore's law and current CMOS technology, namely due to [9, 15]:

- A substantial confinement (restriction) of the donors potential, which limits the conduction activation effect and the information quantum transport;
- A large effective mass of Silicon, with the same quantum restrict impact as mentioned above;
- A cumbersome nanofabrication technology (e.g., by single-ion implantation or by bottom-up doping models through scanning tunneling microscopy, STM) to obtain Qubits coupling.

In this context, the present endeavor aims at aggregating the matrix-junction 3oxides-4atoms—graphentronics (Graphene-oxide GO $\cap$ ZnO $\cap$ TiO_2) by smart combinations with quantum effects on enhancing the quantum information transport through the obtained *graphenic gate transistors* (GGT). It will eventually replace the current Silicon based technology of SGT (Silicon Gate Transistors) and the allied quantum effects of *MOSFET* (*metal–oxide–semiconductor field-effect transistors*) by the present version of *GFET* (*graphene-based field effect transistor*).

22.2 The Method

22.2.1 The Design Principle of the Matrix Quantum-Bits Electronic Circuits

Figure 22.1 illustrates the validation of the 3QubitGate concept as based on 4-Atoms (C, Zn, Ti, O) graphentronics. Under applied voltage (V_{DD}) as input of logical signals (here OR, XOR, AND); the structural physical–chemical 4-atoms oxides combinations (by chemical attack, implant, deposition) act as the quantum gate-source-drain connections, so that the obtained quantum transistors (with certain separation such as the quantum tunneling effects being activated, for avoiding the earlier metal contacts

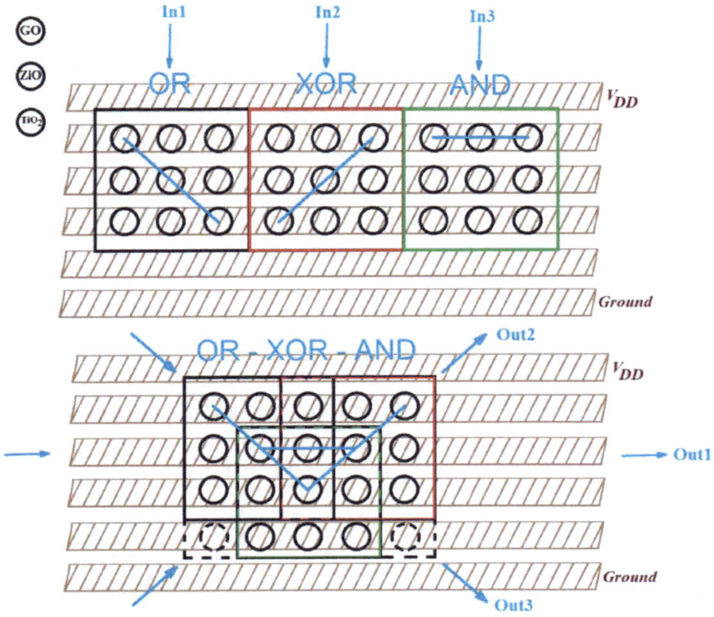

Fig. 22.1 The conceptual design of a 3Qubit4Atoms Transistor: it behaves like the 3Qubit Gate realized by 3oxides (4 atoms) oxides in a graphenic based matrix-junction; this way, either the earlier Silicon based electronics as well as the heterojunctions with their interfaces energy loss by current dissipation (via electronic dispersion)—are to be replaced

which usually slow down the conduction) produce the 3-qubit outputs. They open further quantum computing framework and develop graphentronic integrated circuits.

The experimental validation will be made in two steps: the benchmarking step with single-logical gate through various oxide depositions and 3-quantum gate transistors output; by considering the constructive interference of 3 hetero logical signals as input (Fig. 22.1) in designing the basic (here: the bootstrap charge circuit) to complex integrated circuits (here: pass-transistor and push–pull circuits), observing the positive enhanced difference in recoded 3 quantum gate outputs is accordingly aimed; certain computation rules will emerge for patterning further integrated circuits in quantum graphentronics technology.

22.2.2 The Basic Qubit Electronic Circuits

According to the Horizon 2020 and future Europe 2030 statements, the presented approach unfolds as following:

The basic principles are observed, while practical application are invented and R&D started; the realization of quantum 3oxides-4atoms (Graphene-oxide ∩ ZnO ∩ TiO_2) matrix-junctions is at this stage speculative, Fig. 22.1; It should continue with the effective research and development including analytical and laboratory studies in order to separately validate the *bootstrap charge yield* of the 3oxides-4atoms (Graphene-oxide ∩ ZnO ∩ TiO_2) matrix-junction deposition and buried depletion load contacts within the newly designed *graphene gate transistor (GGT)* by various techniques (i.e. deposition layers, by laser ablation or by spin coating), Fig. 22.2. The basic technological components of the bootstrap charge circuit are integrated to establish that the *pass graphenic gate transistor logic* (Fig. 22.3) and the *push–pull circuit* (Fig. 22.4) types works autonomously. This includes the integration of the components as the hetero-matrix-junctions of 3oxides-4atoms-graphentronics (Fig. 22.1)—in order to test the "low fidelity" of the three-qubits gate response (e.g. the *Toffoli Gate*, Table 22.1) of such integrated systems at small scale in lab

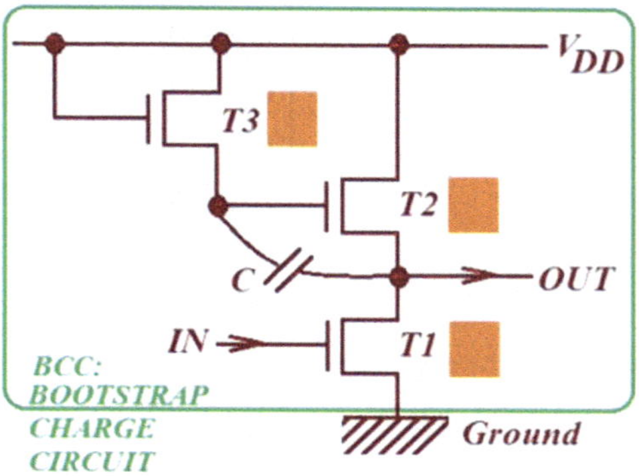

Fig. 22.2 The conceptual design of the bootstrap charge pump circuit using the transistors T1-T3 that can be presently considered via 3oxides-4toms (Graphene-oxide ∩ ZnO ∩ TiO_2) matrix-junctions of Fig. 22.1

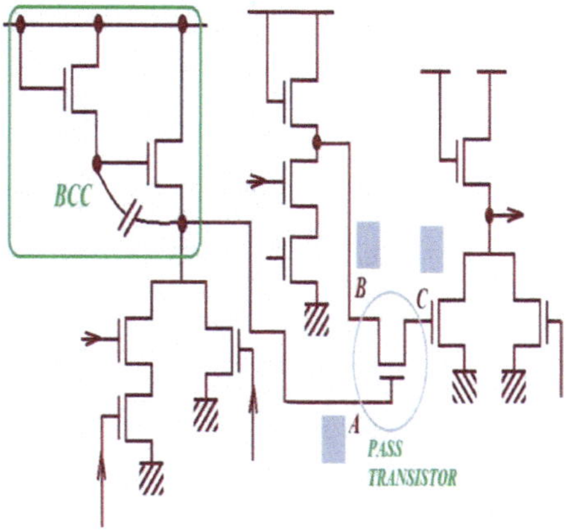

Fig. 22.3 The pass-transistor design, while using the bootstrap charge circuit (BCC) as its working gate (A zone); it may be completed with source (zone B) and drain (zone C) as graphene-based field effect transistors with the currently proposed matrix-junctions technique of deposing 3Qubit-4Atoms graphenic-metallic oxides (Graphene-oxide ∩ ZnO ∩ TiO_2)

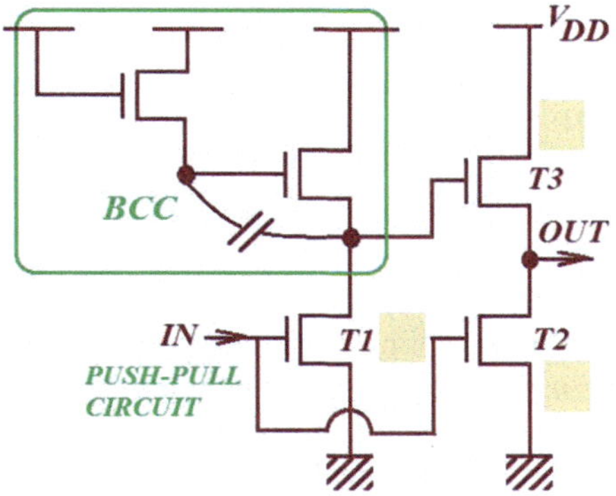

Fig. 22.4 The push–pull circuit also based on bootstrap charge circuit (BCC) of Fig. 22.2; it logically commutes the "0" to "1" charge on BCC node connection complementarily in between (T1 & T2) versus T3 graphenic-based gates

Table 22.1 Toffoli (CCNOT, or Deutsch-π/2) Gate enhancing on 3Qubit gate (its truth table—left and its reversed evolutionary matrix—right); it works as a control flip-gate, which, operating on a basis state $|m, n, l\rangle$ it flips the third bit only if the first two qubits are in the state $|1, 1\rangle$, i.e. it has the operating rule $\hat{U}|m, n, l\rangle = |m, n, l \oplus nm\rangle$ [16]

Input-1	Input-2	Input-3	Output-1	Output-2	Output-3
0	0	0	0	0	0
0	0	1	0	0	1
0	1	0	0	1	0
0	1	1	0	1	1
1	0	0	1	0	0
1	0	1	1	0	1
1	1	0	1	1	**1**
1	1	1	1	1	**0**

$$\hat{U}_{CCNOT} = \begin{bmatrix} 1&0&0&0&0&0&0&0 \\ 0&1&0&0&0&0&0&0 \\ 0&0&1&0&0&0&0&0 \\ 0&0&0&1&0&0&0&0 \\ 0&0&0&0&1&0&0&0 \\ 0&0&0&0&0&1&0&0 \\ 0&0&0&0&0&0&0&1 \\ 0&0&0&0&0&0&1&0 \end{bmatrix}$$

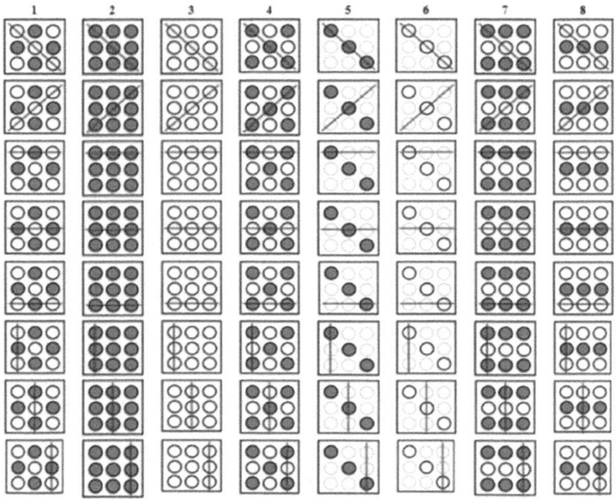

Fig. 22.5 The plates configuration for the 4 atoms oxides depositions (Graphene-oxide ∩ ZnO ∩ TiO2), GO-always as filled circles, in 3Qubit configuration, i.e. 8 × 8 matrix-junction; the empty circles may be either ZnO or TiO_2, in various local metal-oxide quantum transistor configurations; various types of signals are marked by specific directions of tunneling matrix-junctions

environment.

Worth noting that the Toffoli gate maps the state $|m, n, l\rangle$ into $|m, n, l\ XOR\ (m\ AND\ n)\rangle$, Fig. 22.1, meaning that one can use Toffoli gates to build systems that will perform any desired Boolean function computation in a reversible manner. This way, there clearly appears that the 3Qubit (with 8 × 8) quantum evolutionary matrices may be built from OR, XOR and AND logical gates (anticipated in Fig. 22.1, also in Fig. 22.5 in the forthcoming section); they constitute the key of universal cellular automata and reversible quantum computing.

22.3 Preliminary Results

Deposition of GO and ZnO was carried out by drop casting on glass and ITO glass substrate, following a determined matrix pattern as showed in Fig. 22.5. Preparation of the ZnO paste: 0.4 g of ZnO (purchased from Sigma-Aldrich) was weighted and milled, followed by an addition of 12 drops of water. After mixing ZnO and water, 0.08 mL of glacial acetic acid was added. For the following depositions, one type of ZnO paste and a variation of GO were obtained, following two sources of graphene oxide in water dispersion, further reduced as indicated in samples (S1) preparation.

After mixing the compounds, 8 drops of Triton X45 were added. For the preparation of rGO (S1), GO (graphene oxide in water dispersion previously obtained by Cataldo, Putz et al. (2016), [17], was mixed with 1 mL of 1-ascorbic acid (1 M) and

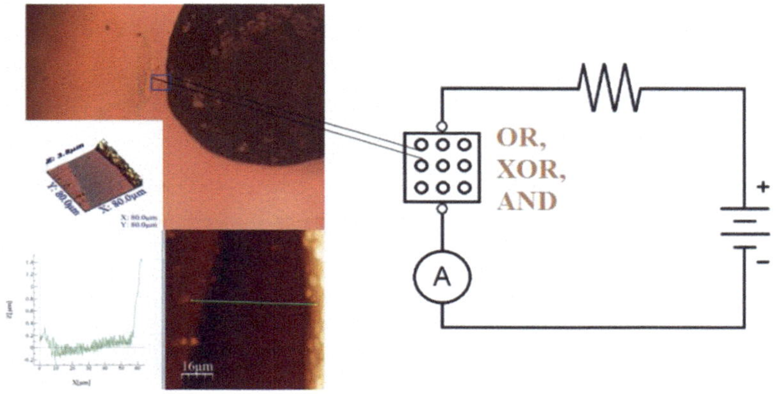

Fig. 22.6 The AFM 3D image and the roughness measurements for the GO-TiO_2 plates' deposition, subject to tunelling conduction under the external applied potential signals

left for one hour in the ultrasonic bath at 60 °C. The mixture was then exposed to a temperature of 90 °C. Then, 0.45 mL of hydrogen peroxide was added and the mixture was left for 30 min at 60 °C in the ultrasonic bath.

The preliminary test employed the imagistic of the graphenic-metallic oxides followed by the first results on the applied signal on the plates of Fig. 22.5, separately for ZnO and TiO_2 depositions. This way, using AFM the close-matrix-junction deposition was checked for the GO/TiO_2 plates (Fig. 22.6) *validating the deposition technique at atomic level*; on the other side, the ohmic resistance and current intensity metrology was initiated on the GO/ZnO paste.

22.4 Discussions

The first electrical resistance measurement was done on the ITO plate. It has been observed that measurements at different distances result in different resistance values. Thereby, measurements were performed at the same distance for all the deposited plates, while indicating 20 Ω for the un-deposited ITO plate. Before the electric measurements of the sample, the resistance of each deposition line was checked, measuring quite high variations in the case of S1 up to 1770 Ω; *this validates the quantum tunneling effects therefore.* Electric measurements were performed with the aid of a multimeter that was connected to a resistance and a battery (Fig. 22.6). In the first place, U, I and R values were recorded in the absence of the deposited plate. With the deposited sample in circuit, no differences in R and U were observed. The applied effects (temperature, reducing the graphene oxide, magnetic field, different ZnO and graphene position depositions) were observed by measuring the I, U and R variations within the samples (Fig. 22.7); *this validates the true output (low resis-*

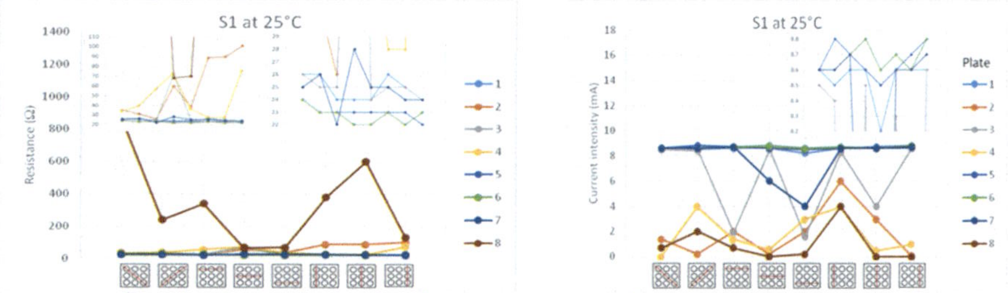

$$\hat{U}_R = \begin{bmatrix} 0 & 0 & 0 & 0 & 0 & 0 & 0 & 0 \\ 0 & 0 & 0 & 0 & 0 & 0 & 0 & 0 \\ 0 & 0 & 0 & 0 & 0 & 0 & 0 & 0 \\ 0 & 0 & 0 & 0 & 0 & 0 & 0 & 0 \\ 0 & 0 & 0 & 0 & 0 & 0 & 0 & 0 \\ 0 & 0 & 0 & 0 & 0 & 0 & 0 & 0 \\ 0 & 0 & 0 & 0 & 0 & 0 & 0 & 0 \\ 1 & \beta & \beta & 0 & 0 & \beta & 1 & 0 \end{bmatrix} \qquad \hat{U}_I = \begin{bmatrix} 1 & 1 & 1 & 1 & 1 & 1 & 1 & 1 \\ 0 & 0 & 0 & 0 & 0 & \beta & \beta & 0 \\ 1 & 1 & 0 & 1 & 0 & 1 & \beta & 1 \\ 0 & \beta & 0 & 0 & \beta & \beta & 0 & 0 \\ 1 & 1 & 1 & 1 & 1 & 1 & 1 & 1 \\ 1 & 1 & 1 & 1 & 1 & 1 & 1 & 1 \\ 1 & 1 & 1 & \beta & \beta & 1 & 1 & 1 \\ 0 & 0 & 0 & 0 & 0 & \beta & 0 & 0 \end{bmatrix}$$

Fig. 22.7 Ohmic resistance and the current intensity measurements for the S1 sample at 25 °C on all 8 deposited plates of Fig. 22.5, with allied 3Qubit graphenic-metal oxide gates; β states for the quantum tunneling parameter driving the current flux through the quantum gate

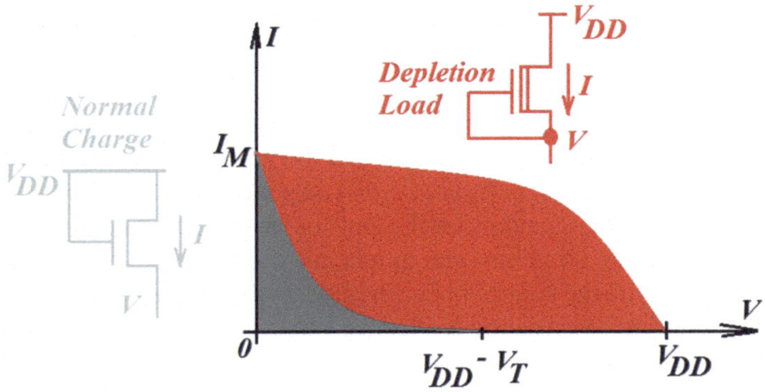

Fig. 22.8 The I-V characteristic of the normal versus depletion load for a transistor configuration; the depletion effect of increasing intensity appears due to the extra free electrons appeared due to the tunneling effects, at triggered by the reduced (electrons' gaining)-graphene-oxide (rGO) used in deposition, as displaying n-type structural defects (Fig. 22.9)

tance—high intensity on selected plates-based quantum transistors) of the matrix-junctions with GO/ZnO deposition. Actually, one can create new working gates even by further staking the plates on the columns of the evolutionary matrices of Fig. 22.7. Further GO/ZnO/TiO_2 combinations are envisaged as deposition on single plates then combined, targeting the bootstrap circuits pump and its further integration within the pass-transistor and push–pull integrated circuits, respectively.

Even more remarkable also at the first results level one may observe the "memory effects" on the sample metrology of Fig. 22.7: the initial idea for using graphene logical gates started from the fact that the atoms-in-oxides have the ability to act as memory unit if the load depletion conditions are ensured (Fig. 22.8). Starting from the idea that a many-electronic state can be transformed from its stable state A to the less stable state B under the influence of external factors (for example an applied potential or by a photon), it may be considered as corresponding to one Bit storage; from its state B, the many-electronic system may return more or less quickly (viz. quantum tunneling activation) to the initial situation A, or may remain as it is, so behaving like the molecular machines, [18]. The first case (fast quantum tunneling) corresponds to an unstable memory (written information that is erased over time), while the second case (low tunneling probability) corresponds to a permanent memory. From here follows another variation of the process, which involves the return from state B to A under the influence of another external stimulus, resulting in the deletion (cancellation) of the previously written information. Accordingly, starting from a simple example of a logic operation, which consists of a lamp, a battery and two switches placed in series, AND, OR and XOR logic can be explained, Fig. 22.6. In a first case, AND logic implies that the lamp (which is part of the logic operation) will light up only if both switches are closed. If the switches are placed in parallel (i.e. as another type of signal applied on GO/Zn/TiO_2 plates), one of the switches (OR logic) must be switched off to turn on the light. Other logical examples may be more complicated, so involving quantum interference and all three involved logical operations, as illustrated in the master Fig. 22.1.

In this context, one concludes that the quantum tunneling activated on the ($GO/ZnO/TiO_2$) depositions follows the laws of Boolean algebra, which is a favorable approach for measuring the logical outcome of the gates. This opens the challenge to employ the Boolean *quantum Hamiltonian computing* (QHC) gate working principle, since based on the quantum level repulsion effect (see the effect of HOMO → LUMO transition in stretching the graphenic valence-conduction band gap in Fig. 22.9) together with the control of constructive and destructive quantum interferences, having as result the synergic (collective coherence) tunneling transport, [19]. In the forthcoming studies we attain to obtain higher tunneling current intensities (larger than few nano-amperes), Fig. 22.9, where more complex QHC logic gates should be accomplished and the graphentronic circuits and quantum computing may enter in a new technological integrative, smart and sustainable stream.

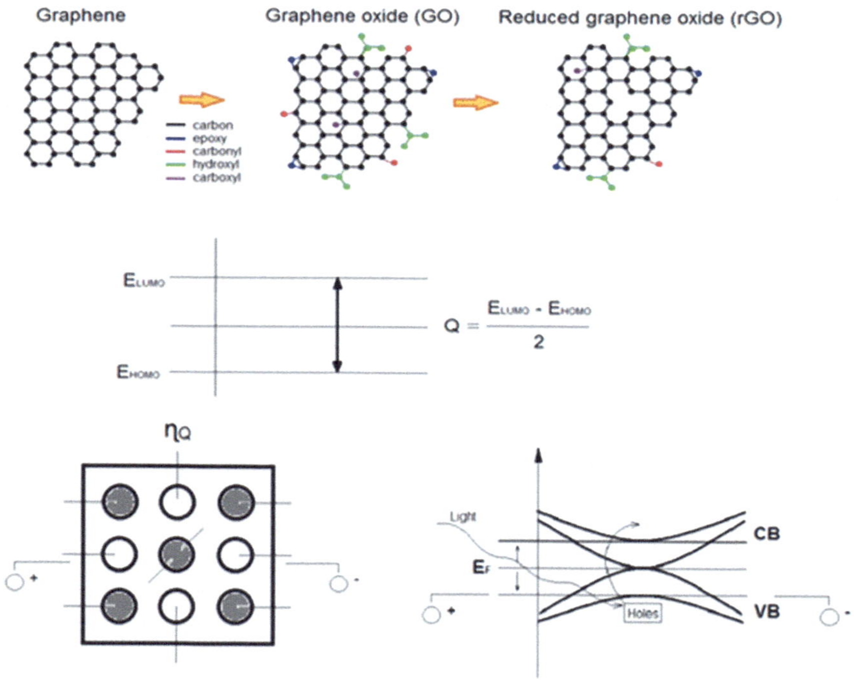

Fig. 22.9 Graphene structural modifications and the representation of its oxides forms in deposition inducing the band gap modification under the external factors, e.g. applied potential, and/or photonic action

22.5 Conclusions

The first integrated 3Qubit gate of graphene-metal oxides (4 Atoms: C, Zn, Ti, O) transistors are designed by the matrix-hetero-junction techniques, with the perspective of enabling the quantum graphenic field technology without the fashioned metallic or ionic or polycrystalline—silicon buried contacts in controlling multi-currents in quantum gate transistors. There are naturally two main objectives, namely:

- Testing and calibrating the I-V responses of the 3oxides-4atoms-graphentronics (Graphene-oxide $\cap$ ZnO $\cap$ TiO_2) embedded layers and buried contacts within the newly designed graphene gate transistor (GGT) as the quantum dot matrix-junctions in bootstrap charge pump circuit;
- Assembling and testing the graphene gate transistor (GGT) quantum 3oxides-4atoms-graphentronics (Graphene-oxide $\cap$ ZnO $\cap$ TiO_2) matrix-junctions for the pass transistor and push–pull circuit types.

The milestones would be marked by establishing the quantum graphenic bootstrap charge pump circuit, followed by its integration in the pass-transistor and push–pull logical circuits, through newly identified quantum evolution operators/gates through the metrological (R-I-V) measurements.

The actual experimental/3qubit circuits' design is supported by the hot preliminary results confirming graphene-oxide GO/TiO_2 atomic deposition as well as by the first measurements of GO/ZnO ohmic resistance versus current intensity output on systematic graphenic based matrix-hetero-junctions. The enhancing current relays on the induced quantum tunneling effect as coming from the GO-to-Reduced GO substrate assuring the depletion load increasing of the registered current due to electronic gain in graphenic deposition.

The success of the experimental design and the so called 3-qubit-4-atoms-Graphenic-based electronics will also indirectly prove the existence of the delocalization quantum particle Bondots [20] responsible for the van der Waals delocalized (tunneling) interaction between layered graphene embedded in matrix-junction—eventually responsible for amplification of the depletion load effect (*smart energy production*), only incremental otherwise. This way, this work specifically responds to the European Union societal challenge of green energy supply by excellent science, so inscribing in the first instance under two of the three pillars of Horizon 2020 vision nowadays extended to Europe 2030. Nevertheless, upon implementation, the acquired know-how will trigger further emerging technology steep in enabling and industrial technologies (EIT), in an inclusive approach of all pillars of *economy of knowledge* in the European Union during 2021–2027.

Acknowledgements This work was supported within the Nucleus Programme under the project PN-19-22-01-02 by the Romanian National Authority for Scientific Research and Innovation (ANCSI) and of its 2020 renewing fund by Ministry of Education and Research.

References

1. H.J.M. Veendrick, *Nanometer CMOS ICs*. From basics to ASICs, 2nd edn. (Springer International Publishing AG, Cham: Switzerland, 2017).
2. ITRS (International Technology Roadmap for Semiconductors) 2.0 (2015) More Moore, https://www.itrs2.net/.
3. M. Sarovar, K.C. Young, T. Schenkel, K.B. Whaley, Quantum nondemolition measurements of single donor spins in semiconductors. Phys. Rev. B **78**, 245302 (2008)
4. J.J.L. Morton et al., Solid-state quantum memory using the 31P nuclear spin. Nature **455**, 1085–1088 (2008)
5. L.E. Ballentine, Realism and quantum flux tunneling. Phys. Rev. Lett. **59**, 1493 (1987)
6. A.J. Leggett, A. Garg, Quantum mechanics versus macroscopic realism: Is the flux there when nobody looks? Phys. Rev. Lett. **54**, 857 (1985)
7. A. Morello et al., Single-shot readout of an electron spin in silicon. Nature **467**, 687–691 (2010)
8. M. Xiao, M.G. House, H.W. Jiang, Measurement of the spin relaxation time of single electrons in a silicon metal-oxide-semiconductor-based quantum dot. Phys. Rev. Lett. **104**, 096801 (2010)
9. L.C.L. Hollenberg, A.D. Greentree, A.G. Fowler, C.J. Wellard, Two-dimensional architectures for donor-based quantum computing. Phys. Rev. B **74**, 045311 (2006)
10. A.L. Danilyuk, A.G. Trafimenko, A.K. Fedotov, I.A. Svito, S.L. Prischepa, Negative differential resistance in n-type noncompensated silicon at low temperature. Appl. Phys. Lett. **109**, 222104 (2016)
11. S. Mahapatra, H. Büch, M.Y. Simmons, Charge sensing of precisely positioned P donors in Si. Nano Lett. **11**(10), 4376–4381 (2011)
12. B.E. Kane, A silicon-based nuclear spin quantum computer. Nature **393**, 133–137 (1998)
13. R. Rahman, S.H. Park, J.H. Cole, A.D. Greentree, R.P. Muller, G. Klimeck, L.C.L. Hollenberg, Atomistic simulations of adiabatic coherent electron transport in triple donor systems. Phys. Rev. B **80**, 035302 (2009)
14. R. Rahman, J. Verduijn, N. Kharche, G.P. Lansbergen, G. Klimeck, L.C.L. Hollenberg, S. Rogge, Engineered valley-orbit splittings in quantum-confined nanostructures in silicon. Phys. Rev. B **83**, 195323 (2011)
15. X. Hu, B. Koiller, S. Das Sarma, Charge qubits in semiconductor quantum computer architecture: Tunnel coupling and decoherence. Phys. Rev. B **71**, 235332 (2005)
16. Y.E. Band, Y. Avishai, *Quantum Mechanics with Applications to Nanotechnology and Information Science* (Elsevier & Academic Press, Amsterdam-New York, 2013)
17. F. Cataldo, M.V. Putz, O. Ursini, G. Angelini, D.A. Garcia-Hernandez, A. Manchado, A new route to graphene starting from heavily ozonized fullerenes: Part 3—an electron spin resonance study. Fullerenes, Nanotubes, Carbon Nanostruct. **24**(3), 195–201 (2016)
18. M. Venturi, M.I. Iorga, M.V. Putz, Molecular devices and machines: hybrid organic-inorganic structures. Curr. Org. Chem. **21**, 2731–2759 (2017)
19. S. Srivastava, H. Kino, C. Joachim, Quantum half-adder Boolean logic gate with a nano-graphene molecule and graphene nano-electrodes. Chem. Phys. Lett. **667**, 301–306 (2017)
20. M.V. Putz, M.A. Tudoran, M.C. Mirica, Quantum dots searching for bondots: towards sustainable sensitized solar cells, in *Sustainable Nanosystems Development, Properties, and Applications*, ed. by M.V. Putz, M.C. Mirica. IGI Global, Hershey Pasadena, USA, Chapter 9, https://doi.org/10.4018/978-1-5225-0492-4.ch009, pp. 261–327

Chapter 23
Influence of the Cd^{2+} Ions on the Crystal Violet Dye Adsorption from Aqueous Solutions by Activated Charcoal Obtained from Pine Cones

Nicoleta Popa, Maria Visa, and Silvia Patachia

Abstract Aiming to obtain low-cost adsorbents for dyes and metallic ions from wastewaters, this study proposed cheap techniques such as carbonization of pine cones, coming from Romanian forests, using low temperature (500 °C) in an auto-generated atmosphere and then, activation by using three types of chemicals: NaOH, KOH and H_3PO_4, respectively and low temperature for thermal treatment (700 °C). The resulted adsorbents were characterized (by AFM, SEM, EDX, XRD, BET specific surface, porosity, FTIR and surface charge) and tested for removal of crystal violet dye (CV) and cadmium ions (Cd^{2+}) from synthetic mono- or di-solute aqueous solutions. The effects of independent variables such as contact time, adsorbent dosage (0.1–0.2 g/L), initial dye concentration (2.5–20.0 mg/L) on the adsorption capacity of the proposed adsorbents were investigated and discussed correlated with the surface properties. The highest adsorption capacity was evidenced by the sorbent activated with NaOH (71.94 mg . g^{-1} CV adsorbed from single solute solution). In solutions with (Cd^{2+}: CV)$_w$ ratio = 1:1, the presence of Cd^{2+} ions strongly diminished the CV adsorption with aprox. 28%. Cd^{2+} seems to be a redoubtable competitor for CV in the adsorption process. In case of adsorption from di-solute systems with (Cd^{2+}: CV)$_w$ ratio = 540:5, the $q_{\max}$ for CV was 24.8 mg . g^{-1}, while for Cd^{2+} cations, was 370.37 mg . g^{-1}, for the same adsorbent, activated with NaOH. The adsorption processes follow the Langmuir model for all the tested species and the process kinetic is well described by the pseudo-second-order reaction model. Considerations on the sorption mechanism were made based on the experimental data.

Keywords Pine cones · Activated carbon · Crystal violet · Cadmium

N. Popa
Department of Forest District Teliu, National Admin. State Forests Romsilva, Brasov, Romania
e-mail: popanicol@yahoo.com

M. Visa (✉) · S. Patachia
Centre of Renewable Energy Systems and Recycling, Transilvania University of Brasov, Eroilor 29, 500036 Brasov, Romania
e-mail: maria.visa@unitbv.ro

S. Patachia
e-mail: st.patachia@unitbv.ro

I. Visa and A. Duta (eds.), *Solar Energy Conversion in Communities*,
Springer Proceedings in Energy, https://doi.org/10.1007/978-3-030-55757-7_23

23.1 Introduction

The wastewaters are loaded with pollutants such as, dyes, heavy metals, surfactants, and contain a variety of harmful organic compounds which also have carcinogenic properties. On the other hand, it was also estimated that about 10,000 of various synthetic dyes and pigments in quantities of over 7×10^5 tons/year are produced worldwide [1]. Their complete biodegradation is slow or even impossible, because all the synthetic dyes and pigments have complex chemical structures contained one, two or more aromatic rings (structure with delocalized of π electrons [2], chromophore groups: antrachinone, azo (–N=N–), thiaxine– [3, 4] and/or heavy metals: Cd, Pb, Cu, Co, Cr (metal-complex dyes) [5]. Moreover, after dying process, approximately 20% of these lost dyes spill into the industrial wastewater [5]. The treatment of the industrial wastewater before its discharge is compulsory, due to exceeding values of the quality indicators (COD, BOD, TOC, turbidity, TDS, color, etc.) in the accepted standards. Crystal violet (CV) is a cationic dye that has been used for various purposes, such as, for biological staining, dermatological agent, veterinary medicine, additive to poultry feed to inhibit propagation of moled, intestinal parasites and fungi, but it is also a mitotic poison and can induce genetic mutations [6].

The presence of the heavy metal ions in wastewater is undesirable. They are soluble in water and thus they can be absorbed by living organisms entering in the food chain and starting to accumulate in the human body, causing serious health disorders. Thus, Cadmium is a common environmental pollutant coming from food, especially from cereals, vegetables, root crops, from sea-food and meat offal that lead to lower intelligence and altered human behavior [7].

Many processes have been applied to remove heavy metals and dyes, including adsorption [6, 8–10], biosorption [11, 12], electrochemical coagulation [13], membrane filtration [6], photodegradation [14], electroflocculation and ozonation [15], photo-Fenton [16], and simultaneous methods (e.g. adsorption and photocatalysis) [17]. The adsorption method is intensely used for removing soluble and insoluble pollutants without the generation of hazardous by-products [18] being inexpensive, while the equipment is simple, easy to operate and highly effective.

Various alternative adsorbents produced by forest, agricultural byproducts have been used such as: rice husk [19], orange peel [20], wheat straw [21, 22], pine tree and coconut shell [23], pine cone powder [24], corncob [25], waste tea [26], sugarecane bagasse [27, 28], ginger waste [29], pine biomass [30], bio-cryogels [8], etc.

Adsorbents prepared from forest and agricultural waste have certain advantages: renewable organic resources are abundant, having low cost, and being characterized by a high selective adsorption. Forest wastes, biomass, such as: small branches, pine cones and leaves are valuable materials for obtaining charcoal or activated carbon. These adsorbent materials proved low adsorption capacity for dyes and heavy metals when comparing with commercial adsorbents such as activated carbon. About 5500 works have been published on the removal of dyes by using activated carbon, a low cost adsorbent obtained from the agriculture wastes [31]. Various papers reported different methods of activated carbon preparation starting from pine cones collected from different areas, which, influences the properties of the charcoal, Table 23.1.

Hereby, for the first time an adsorbent material obtained from pine cones (Pinus Nigra and Pinus Picea Abies) grown/planted on a limestone/acidic soil in the Teliu-Brasov area, Romania, is reported for removing dyes and simultaneously, dyes and heavy metals from complex mixtures.

Many researchers reported on removing from aqueous solutions by using adsorbents obtained from pine cones dyes, but no studies were focused on simultaneous removing of both CV and Cd^{2+} ions. Taking into account that food industry could be a source of cadmium and crystal violet dye simultaneously present in wastewaters; the purpose of this work is to study the possibility to remove them by adsorption.

In this study, charcoal obtained from pine cones was impregnated with NaOH, KOH and H_3PO_4, respectively, to be chemically activated and converted into new activated carbon adsorbents coded: CAC1, CAC2 and CAC3, respectively.

The novelty of this study consists of:

(a) obtaining the adsorbent materials based on specific forest waste biomass, (Romanian pine cones collected from an acidic soil) by using low temperature of carbonization (500 °C) in auto-generated atmosphere, impregnation with alkali or acid and pyrolysis at low temperature (700 °C);
(b) selection of the most effective adsorbent for thet Cd^{2+} ions and for CV;
(c) quantification of the influence of the presence and concentration of Cd^{2+} ions on the CV sorption;
(d) correlation of the sorption process characteristics with the surface properties of the adsorbents.

23.2 Materials and Mehods

Three types of activated carbon were prepared by using raw precursor pine cones and the chemicals reageants such as: phosphoric acid (H_3PO_4), sodium hydroxide (NaOH) (purity $\geqq$ 97%) and potasium hydroxide (KOH) (purity $\geqq$ 97%) were purchased from Sigma-Aldrich). Ultrapure water with resistivity of 18.23 $M\Omega\ cm^{-1}$ was used throughout the whole experiment. Details for the precursor carbonization and then impregnation are given in Fig. 23.1.

The micro-porosity and BET specific surface was measured using an Autosorb-IQ-MP, Quantachrome Instrument. The crystalline structure was recorded by using an X-ray Diffractometer (XRD Bruker D8 Discover Diffractometer, $K\alpha 1 = 1.5406$ Å, 40 kW, 20 mA, step size 0.02, scan speed 2 s/step, 2θ range ranging from 10 to 80°).

The surface morphology (roughness and macro-pores size distribution) of the adsorbents was studied using: scanning electron microscopy (SEM S-3400 N-Hitachi, accelerating voltage of 20 KV) and an atomic force microscope (AFM Ntegra Spectra, NT-MDT model BL222RNTE). Energy dispersive X-ray (EDX, Thermo Scientific Ultra Dry) was used to outline the surface elemental composition.

Table 23.1 Pine cones precursor, activation conditions and sorbent characteristics

Precursor/*collection area*	Activating agent, TC(°C); TP(°C)	Characteristics		Adsorbate		Refs. No.
		S_{BET} (m^2/g)	D_{Pore} (nm)	Type of adsorbate	q_{max} (mg/g)	
Pine cones *China*	−HCl 1 mol/L −TC: 500, −TP: 600 °C	519.69	–	Dimethyl trisulfide	1.098	[32]
Pine cones *Saudi Arabia*	Used without any pretreatment	7.108	1.573	Phenol 2-Cl -Ph	164.5 189.4	[11]
Pine cones *South Korea*	$Zn(NO_3)_2 * 6H_2O$ TC: 500 °C in N_2	11.5	–	As(III)	0.007	[33]
Pine cones *Iran*	H_3PO_4 TP: 500 °C in N_2	1470	1.745	CO_2 CH_4	416.4 70.62	[34]
Pine cones *Iran*	H_3PO_4 (95%); 1:10 TC: 900 °C	869	0.595	Ac.Blue113 Acid Black	458.1 286.0	[35]
Pine eldarica *Iran*	H_3PO_4(1:1–1:4) TP: 750 °C, in N_2	638.4	1.9	Surfactant (SDBS)	93.1	[36]
Pine cones *Australia*	Dried at 70 °C for 24 h., untreated	0.213	–	Methylene Blue	109.8	[37]
Pine cones, *Turkey*	$ZnCl_2$; microwave, N_2	939	1.904	Methylene Blue	60.97	[38]

TC(°C)—carbonization temperature; TP(°C)—pyrolysis temperature

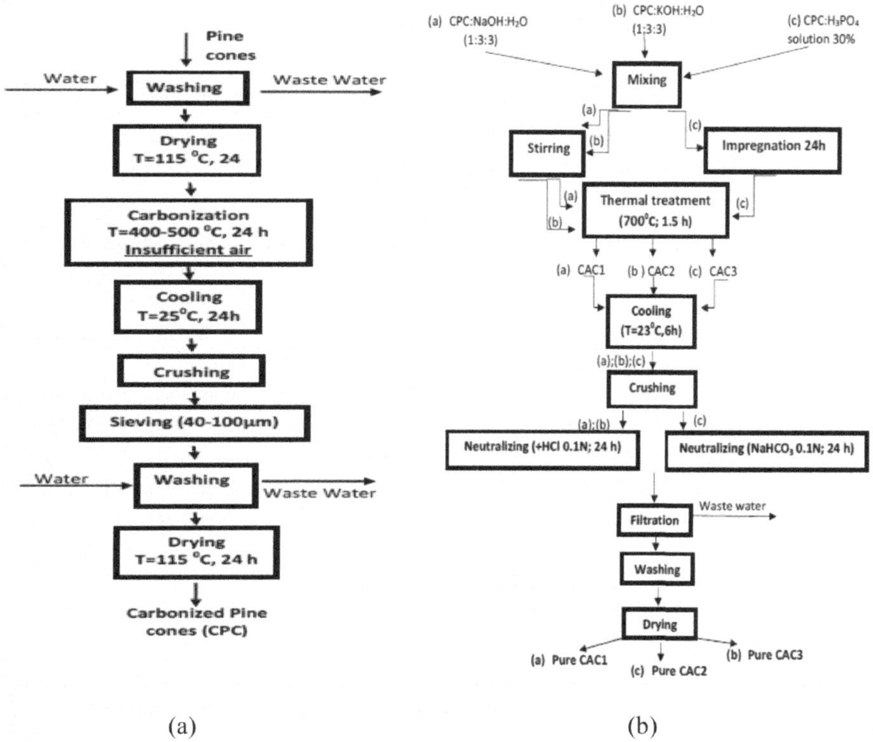

Fig. 23.1 Steps to obtain: **a** carbonized pine cones (CPC), **b** thermal treatment of the charcoal CPC

The size, shape and crystallinity of the samples have been observed by High Resolution. The complementary data were obtained by FTIR spectroscopy (Spectrum BX Perkin Elmer BXII 75,548, $\lambda = 400$–4000 nm). Zero point charge of the surface was determined by using the pH drift method. Plotting the final pH versus initial pH, both for solutions with and without adsorbents and determining the intersection point, allowed us finding the pH_{PZC}.

23.3 Set-Up of the Adsorption Experiments

23.3.1 Adsorption of CV from Single Solute Aqueous Solutions and from Aqueous Solutions Containing Cd^{2+} Ions, by Using Activated Carbonized Pine Cones

The adsorbtion capacity of the adsorbents (CAC1; CAC2 and CAC3), against crystal violet and Cd^{2+} ions, was determined following the steps presented in Scheme 23.1. Batch adsorption experiments were carried out at room temperature (23 °C). All chemicals were analytical grade and used without further purification, Cristal Violet ($C_{25}H_{30}ClN_3$) powder with 407.99 g/mole molecular weight and $CdCl_2 . 2.5H_2O$ (Scharlau Chemie S.A.,c < 98%). The CV solutions were prepared by dissolving an appropriate quantity of CV ($C_{25}H_{30}ClN_3$) in ultrapure water (Direct-Q3 Water Purification System). The CV and Cd^{2+} solutions were prepared by dissolving a certain amount of crystal violet and $CdCl_2 . 2.5H_2O$ in ultrapure water to give an initial dye and cadmium concentration for the experiments. The pH of the initial

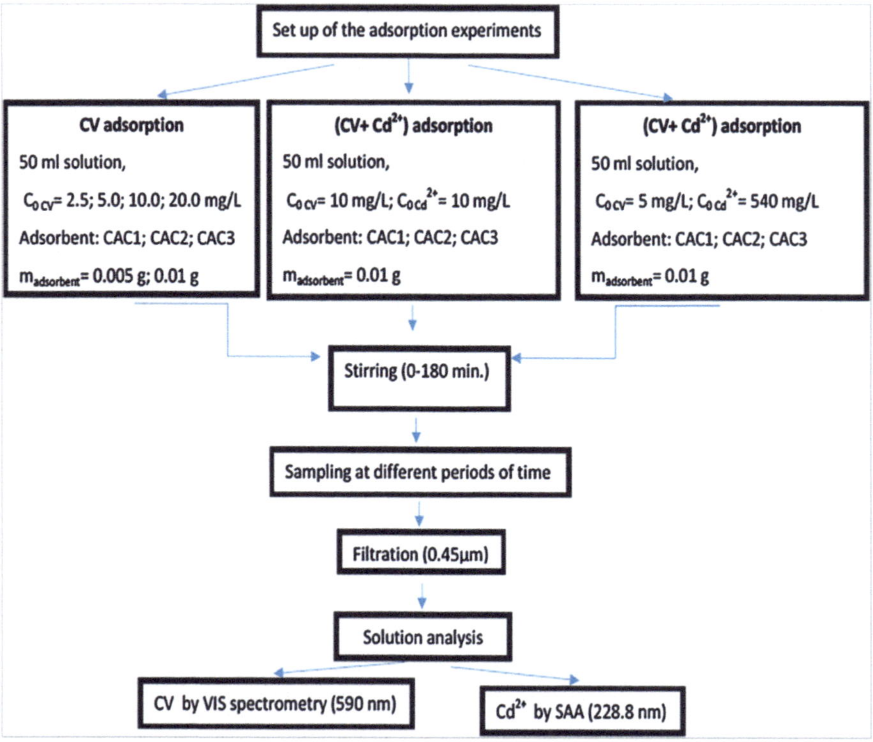

Scheme 23.1 The steps in adsorption experiments

solutions ranges in the 6.5–7 interval. The adsorption experiments were developed in Scheme 23.1. The concentration of CV in the supernatant solution before and after adsorption was analysed by UV–VIS spectrometry (Perkin Elmer Lambda 25), on the calibration curve registered at the maximum absorption peak of CV (λ = 590 nm), in order to evaluate the momentary concentration of dye. The supernatant was further analysed by AAS (Analytic Jena, ZEEnit 700), at: $\lambda_{Cd} = 228.8$ nm.

The amount of the dye or of cadmium cations adsorbed at the equilibrium time reflects the adsorption capacity of the adsorbent under constant operating conditions.

The adsorption capacity, q_t, was calculated based on the initial solution concentration $\mathbf{c}^{i}_{CV/Cd^{2+}}$, momentary equilibrium concentrations of the pollutants, $\mathbf{c}^{t}_{CV/Cd^{2+}}$, volume of crystal violet solution, V_{sol} [L], and the amount of adsorbent, m_{ss} [g], using the following Eq. (23.1).

$$q_t = \frac{(c^{i}_{CV/Cd^{2+}} - c^{t}_{CV/Cd^{2+}}) \cdot V}{m_{ss}} \tag{23.1}$$

The removal efficiency of CV from the single solute aqueous solutions and from di-solute solutions was calculated by Eq. (23.2)

$$\eta = \frac{(c^{i}_{CV/Cd^{2+}} - c^{t}_{CV/Cd^{2+}}) \times 100}{c^{i}_{CV/Cd^{2+}}} \tag{23.2}$$

23.4 Results and Discussions

23.4.1 Adsorption Capacity of the Activated Carbonized Pine Cones

The adsorption capacity of an adsorbent is defined as the amount of adsorbate retained at the surface exposed by 1 g of adsorbent, after reaching the adsorption equilibrium. This measure serves to compare different adsorbents, tested in the same conditions or to compare the effectiveness of different treatments applied on one substrate against adsorption of different molecular or ionic species, it can see in Table 23.1.

23.4.1.1 Influence of the Activation Technique on the Adsorption Capacity of the Carbonized Pine Cones

From Figs. 23.2 and 23.3 it is obviously to note that alkali activated carbonized pine cones (CAC1 and CAC2) are characterized by higher adsorption capacity by comparing to the acid activated carbonized pine cones.

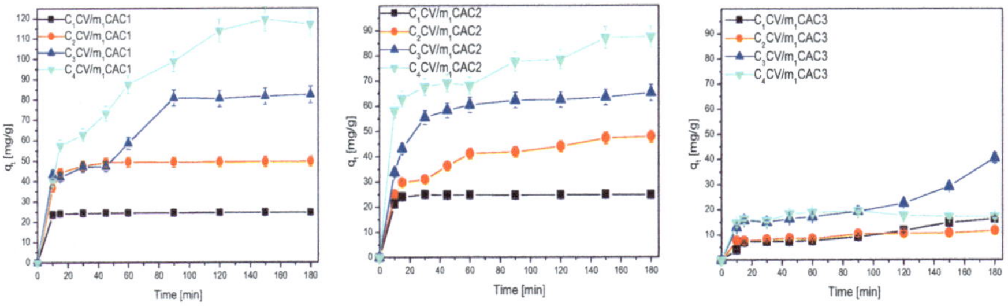

(a) (b) (c)

Fig. 23.2 Adsorption capacity of CAC1, CAC2, CAC3 as a function of time and CV initial solution concentration of $C_1 = 2.5$ mg/L, $C_2 = 5$ mg/L, $C_3 = 10$ mg/L, $C_4 = 20$ mg/L ($m_1 = 0.005$ g adsorbent/50 mL)

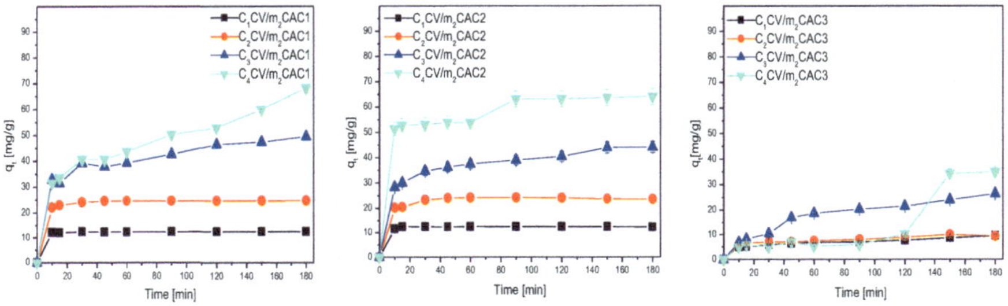

(a) (b) (c)

Fig. 23.3 Adsorption capacity of CAC1, CAC2, CAC3 as a function of time and CV initial solution concentration of $C_1 = 2.5$ mg/L, $C_2 = 5$ mg/L, $C_3 = 10$ mg/L, $C_4 = 20$ mg/L ($m_2 = 0.01$ g adsorbent/50 mL)

The best results were obtained by using CAC1, activated with NaOH. This adsorbent evidenced the highest sorption capacity for all the solution concentrations and all the tested amount of the sorbent used. After 180 min. of contact, the adsorption capacity against CV, for CAC2, activated with KOH, decreased with 25% and for CAC3, activated with H_3PO_4, decreased with 84% by comparing to that of CAC1, when 0.005 g sorbent/50 mL CV solution with the highest tested concentration (20 mg/L), was used. In case of using higher concentration of adsorbent (0.01 g/50 mL solution), keeping constant the other conditions, the situation is similar, but the differences between the adsorption capacity of CAC1 and CAC2 diminished. The adsorption capacity of CAC2 decreased with only 6% for CV and that of CAC3 decreased with 49% by comparing with the CAC1 sorption capacity. The CV sorption profile of CAC3 is completely different by comparing to that of CAC1 and CAC2. It seems that the surface morphology of CAC3 is more compact, with lower and difficult accessible active sites that determines the decrease of the amount of adsorbed CV and an obvious lag-time of the process.

By comparing the results presented in Figs. 23.2 and 23.3 with that from Fig. 23.4,

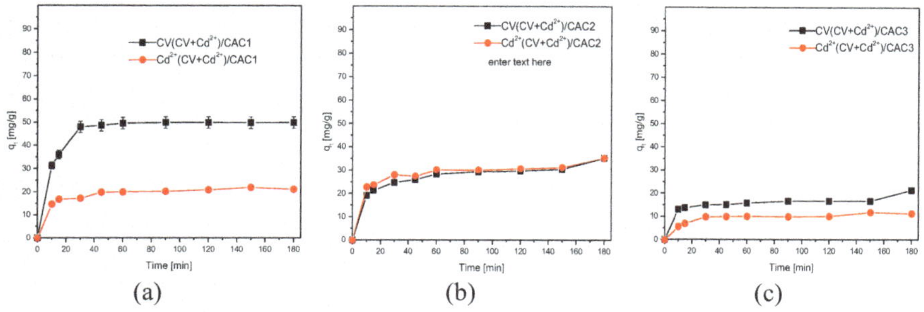

Fig. 23.4 Kinetic curves for the simultaneous CV and Cd^{2+} adsorption on CAC1, CAC2 and CAC3 adsorbents, when the initial solution concentration (for both: CV and Cd^{2+}) was $\mathbf{c}^{i}_{CV/Cd^{2+}}$ = 10 mg/L and the adsorbent mass/solution volume ratio was 0 0.01 g/50 mL

it can be noted that CV adsorption is influenced by the presence of Cd^{2+} ions. The adsorption equilibrium is more rapidly reached in the presence of Cd^{2+} ions, after only 40 min. The differences between the studied substrates maintained the same trend but they are better evidenced than that shown in the absence of Cd^{2+} ions: CAC1 adsorbed 2.5 times higher amount of CV by comparing to Cd^{2+} ions; CAC2 adsorbed approximately the same amount of both species while CAC3 adsorbed only 1.3 times higher amount of CV. But, by comparing the CV adsorption from the single solute and double solute-solutions on the three tested adsorbents, it could be observed that the presence of Cd^{2+} ions strongly decreased the amount of adsorbed CV, while their effectiveness maintained the same trend both for CV and Cd^{2+} ions: $q_{t\,CAC1} > q_{t\,CAC2} > q_{t\,CAC3}$, as it can be noted from Fig. 23.5a. As it can be seen in Fig. 23.5b, expressing q_e in mole of adsorbate/g adsorbent changed the hierarchy of

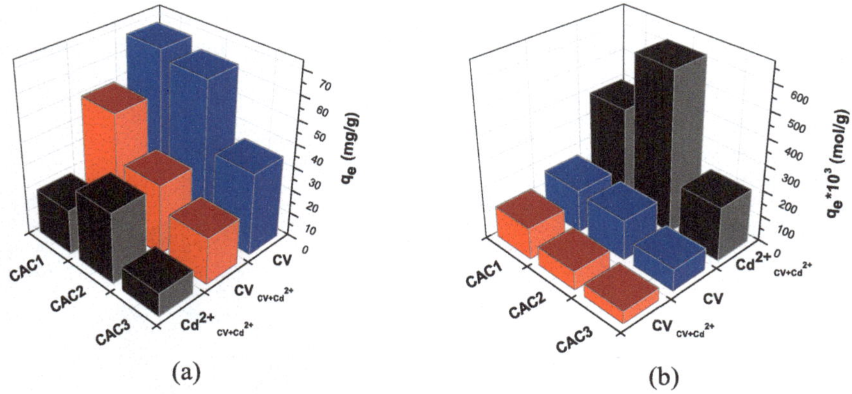

Fig. 23.5 Influence of the substrate type on the q_e of CV adsorbed from the single and di-solute systems and of Cd^{2+} ions from di-solute aqueous solutions (m = 0.01 g adsorbent/50 mL solution; c^{i}_{CV} = 10 mg/L, $c^{i}_{Cd^{2+}}$ = 10 mg/L)

adsorbed species, while that of the adsorbents maintains the same trend. This aspect shows that a higher number of Cd^{2+} ions are adsorbed by comparing to the number of CV cations dye, due to the dimensional reasons. Cd^{2+} ions fill the cracks or deeper and thinner pores, making the adsorbent surface "smother" and changing the CV adsorption mechanism.

CV cations dye will have a limited access, only to a reduced number of the active sites of adsorbent, promoting a simpler adsorption mechanism. From Fig. 23.7 it can be seen that the kinetic curves are better shaped in the presence of Cd^{2+} ions, showing a clear steady state, without inflexion points, evidencing only one mechanism of CV adsorption. A strong decrease of the adsorbed CV amount after reaching the adsorption equilibrium, in the presence of Cd^{2+} ions, was noted: with about 27% from that adsorbed from the single-solute solution, in the case of CAC1; with about 53% for CAC2 and with about 42% for CAC3.

In the case of a very high excess of Cd^{2+} ions (45 times higher than the concentration of CV) in the initial di-solute solutions, it can be noted that the differences between the amount of the adsorbed CV from single and di-solute solution for all the tested adsorbents, is lower and that the capacity of adsorption for both CV and Cd^{2+} ions maintained the same variation trend when using different adsorbents: CAC1 > CAC2 > CAC3, Fig. 23.6.

23.4.1.2 Influence of the Initial Solution Concentration on the Adsorption Capacity of the Activated Carbonized Pine Cones

The following initial concentrations were used: $C_1 = 2.5$ mg/L, $C_2 = 5$ mg/L, $C_3 = 10$ mg/L, $C_4 = 20$ mg/L, to check their influence on the adsorption process of CV, while the other operational parameters were kept constant. From Figs. 23.5 and 23.6 it can be observed that the increase of the initial solution concentration determined the increase of the CV amount removed from the solution, due to the increase of

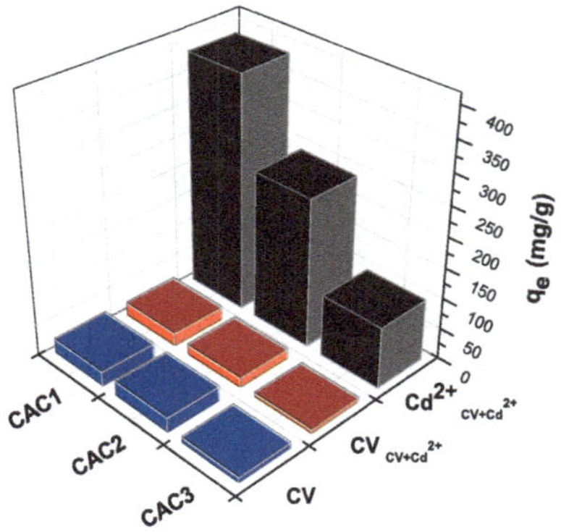

Fig. 23.6 Influence of the substrate type on the q_e of CV adsorbed from the single and di-solute systems and of Cd^{2+} ions from di-solute aqueous solutions (m = 0.01 g sorbent/50 mL solution; $c^i_{CV} = 5$ mg/L, $c^i_{Cd^{2+}} = 540$ mg/L)

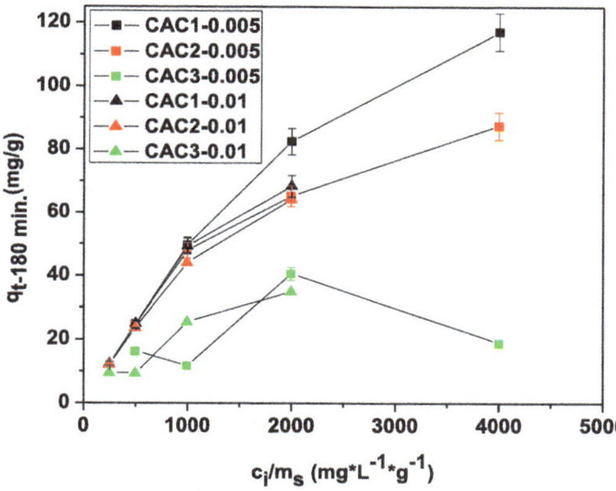

Fig. 23.7 Dependence of the adsorption capacity of the CAC1, CAC2 and CAC3 on the ratio: initial solution concentration/adsorbent mass

the concentration gradient and consequently, the increase of the adsorption driving force.

From experiments, it could be noted that the period necessary to reach the adsorption equilibrium was higher for higher solution concentration.

23.4.1.3 Influence of the Adsorbent Dosage on Its Adsorption Capacity

If the amount of adsorbent was increased, the adsorption capacity of all the tested adsorbents decreased, due to the decrease of the CV/adsorbent mass ratio and consequently the decrease of the available amount of dye per gram of adsorbent (Fig. 23.7).

Figure 23.7 shows that the sorption capacity of all the tested adsorbents is influenced by the ratio: initial solution concentration of CV/sorbent mass and that after 180 min. of contact with solution, the maximum of the adsorption capacity was still not reached. Higher amount of sorbent led to lower their adsorption capacity due to the decrease of the available CV amount. The highest sorption capacity was evidenced by CAC1 when 0.005 g CAC1 was introduced in 50 mL of the highest concentrated solution. Intermediate sorption capacity had CAC2 and the lowest adsorption capacity was noted for CAC3. Irregular dependency of the sorption capacity of CAC3 against solution concentration can be the result of a higher in homogeneity of the adsorbent morphology.

23.4.1.4 Influence of the Contact Time of Adsorbents with CV Aqueous Solution on Their Adsorption Capacity

Figures 23.2 and 23.3 show also that the adsorption mechanism of CV seems to be more complex in the higher concentrated initial solutions. In the case of CAC1

(0.005 g/50 mL solution), the sorption eqilibrium was reached earlier in diluted solutions, after 10 and respectively 15 min. In the case of higher concentrated solutions the sorption process evidenced two steps. The eqilibrium was reached after 90 min. in case of solution with 10 mg CV/L and after 160 min. for solution with 20 mg CV/L. Higher the initial solution concentration, higher the necessary period for reaching the sorption equilibrium. The amount of CV sorbed at sorption equilibrium on CAC1 is higher than that adsorbed by CAC2 and the differences between the adsorbed mass of CV decresed with the initial solution concentration decrease. The lowest amount of CV adsorbed was noted for CAC3. In this case, the increase of the initial solution concentration did not determine the increase of the amount of adsorbed CV. Higher adsorption was obtained by using an intermediary concentrated solution and the equilibrium seems to be not reached after 180 min. These irregular dependencies could be attributed to a higher inhomogeneity of the adsorbent surface, determined by the acid treatment of the cone pines.

23.4.2 Modeling of Batch Adsorption Isotherms

The pollutant uptake capacities for different types of adsorbents can be expressed as batch equilibrium isotherm studies and can be classified by mechanistic models or empirical equations [37]. The adsorption mechanism is able not only to represent but also to explain and predict the experimental behavior [39, 40]. The adsorption isotherm using empirical models for single or di-solute systems indicates how the adsorbate molecules are distributed between the liquid phase and the solid phase at equilibrium, during the process of adsorption. Thus, they play an important role in understanding the adsorbate-adsorbent interactions as well as the mechanism of adsorption.

To model the CV sorption from single and di-solute solutions, on the activated carbonized pine cones, four models were used:

The Langmuir isotherm, Eq. (23.3) (linear form):

$$\frac{c_{eq}}{q_{eq}} = \frac{1}{q_{\max} K_L} + \frac{c_{eq}}{q_{\max}} \tag{23.3}$$

where: $q_{\max}$ represents the maximum monolayer adsorption capacity (mg/g), K_L is Langmuir constant (L . mg^{-1}) related to the adsorption free energy, q_{eq} is the amount of dye adsorbed from the solution at the equilibrium time and C_{eq} is equilibrium concentration of dye in solution (mg/g). From the slope and intercept of the plot C_{eq}/q_{eq} versus C_{eq} the values of K_L and $q_{\max}$ respectively, are calculated. The isotherm model was expressed in terms of the dimensionless constant separation factor $R_{L.}$

$$R_L = \frac{1}{1 + K_L C_0} \tag{23.4}$$

where: C_0 is the initial dye concentration (mg . g^{-1}); R_L—the separation factor that can be determined from Langmuir plot with Eq. (23.4) and that indicates the type of adsorption [41].

The Freundlich isotherm, Eq. (23.5) (linear form)

$$\log q_{eq} = \log K_F + \frac{1}{n} \log c_{eq} \quad (23.5)$$

where: K_F is Freundlich constant indicating the adsorption capacity, and $1/n$ is a dimensionless parameter indicating the adsorption density.

The Temkin isotherm, Eq. (23.6) (linear form) [42]:

$$q_q = \frac{RT}{b_T} \ln A_T + \left(\frac{RT}{b_T}\right) \ln c_e \quad (23.6)$$

$$B = RT/b_T \quad (23.7)$$

where: $A_T$is Temkin isotherm equilibrium binding constant (L/g) b_T is Temkin isotherm constant R = universal gas constant (8.314 J/mole/K) T = temperature at 298 K. B is a constant related to heat of adsorption (J/mole).

Dubinin–Radushkevich isotherm model, Eq. (23.8) (linear form) [42]:

$$\ln q_e = \ln q_s - K_{ad}\varepsilon^2 \quad (23.8)$$

$$E = 1/\sqrt{2B_{DR}} \quad (23.9)$$

$$\varepsilon = RT \ln(1 + 1/c_e) \quad (23.10)$$

where: q_e is the amount of adsorbate on the adsorbent surface at equilibrium (mg/g); q_s is theoretical isotherm saturation capacity (mg/g); K_{ad} is Dubinin–Radushkevich isotherm constant ($mole^2/kJ^2$) and ε is Dubinin–Radushkevich isotherm constant; R, T and C_e represent the gas constant (8.314 J/mole K), absolute temperature (K) and adsorbate equilibrium concentration (mg/L), respectively.

By comparing the results obtained by applying the four above mentioned models for describing the adsorption equilibrium (Table 23.2), it could be concluded that the adsorption of CV from aqueous solution is better described by the Langmuir isotherm (R^2 = 0.950). This means that a monolayer of adsorbate was formed on the outer surface of the adsorbent. No further adsorption occurs after complete coverage of the active surface. The Langmuir isotherm model is valid for monolayer adsorption onto a surface exhibiting a finite number of identical sites. The model assumes uniform energies of adsorption onto the surface and no transmigration of adsorbate in the plane of the surface. K_L (Langmuir isotherm constant) could be correlated

Table 23.2 Constants of adsorption for the checked models Langmuir, Freundlich, Temkin and Dubinin–Radushkevich. Adsorption Isotherms of CV and Cd^{2+}ions onto the adsorbents

Substrate	Langmuir isotherm			Freundlich isotherm				
	q_{max} [mg/g]	$K_L * 10^{-5}$ [L/mg]	R_L	R^2	$1/n$	n	K_F [mg/g]	R^2
CV from mono-solute system								
C1*	71.94	8.0	0.999	0.950	0.1998	5.005	137.35	0.717
C2*	52.63	26.0	0.999	0.977	0.2388	4.187	39.938	0.856
C3*	29.58	86.0	0.983	0.971	0.1981	5.0479	8.6571	0.784
Temkin isotherm				*Dubinin—Radushkevich isotherm*				
	A_T [L/mg]	b_T	B [J/mol]	R^2	q_s [mg/g]	$K_{ed/}$mole^2/kJ^2	E [kJ/mol]	R^2
C1*	1036.4	350.898	7.0607	0.769	64.007	1.00E-08	7.071	0.918
C2*	117.31	374.838	6.6097	0.922	43.0861	3.00E-08	4.082	0.926
C3*	165.02	676.359	3.6631	0.771	25.2241	4.00E-08	3.535	0.872
CV from di-solute system ($CV + Cd^{2+}$)								
Langmuir isotherm				*Freundlich isotherm*				
	q_{max}	$K_L * 10^{-5}$	R_L	R^2	$1/n$	n	K_F	R^2
C1*	23.809	37.0	0.998	0.999	0.3984	2.510	10.762	0.920
C2*	18.727	87.0	0.995	0.955	0.3354	2.982	2.3264	0.672
C3*	7.452	160.0	0.991	0.989	0.4091	2.444	0.3189	0.875
Temkin isotherm				*Dubinin—Radushkevich isotherm*				
	A_T	b_T	B	R^2	q_s	K_{ed}	E	R^2
C1*	70.153	580.336	4.2692	0.974	21.006	3.00E-08	4.082	0.943
C2*	74.153	804.641	3.0791	0.838	14.239	2.00E-08	5.00	0.833
C3*	198.68	1974.63	1.2547	0.881	5.7696	1.00E-08	7.071	0.845

(continued)

Table 23.2 (continued)

Substrate	Langmuir isotherm			Freundlich isotherm				
	$q_{\max}$ [mg/g]	$K_L * 10^{-5}$ [L/mg]	R_L	R^2	*1/n*	*n*	K_F [mg/g]	R^2
Cd^{2+} from di-solute system ($CV + Cd^{2+}$)								
Langmuir isotherm				*Freundlich isotherm*				
	$q_{\max}$ [mg/g]	$K_L * 10^{-5}$	R_L	R^2	*1/n*	*n*	K_F	R^2
C1*	370.37	6.43	0.964	0.973	0.8634	1.1582	0.0776	0.653
C2*	250.00	0.21	0.9999	0.993	0.8709	1.1482	0.0068	0.699
C3*	106.38	210.0	0.4684	0.966	0.7880	1.2690	2.2247E-06	0.873
Temkin isotherm				*Dubinin—Radushkevich isotherm*				
	A_T	b_T	*BJ*	R^2	q_s	$K_{ed\backslash}$	*E*	R^2
C1*	3.276	39.3697	62.93	0.487	406.7098	2.00E-05	0.158	0.645
C2*	6.919	57.8683	42.81	0.303	314.6308	4.00E-05	0.112	0.629
C3*	1.482	123.201	20.11	0.814	108.1804	9.00E-05	0.745	0.979

C1*; C2*; C3* = CAC1; CAC2; CAC3

to the energy of adsorption. Lower the energy of adsorption, higher amount of CV adsorbed. By comparing the data obtained for the three types of the studied adsorbent, the superiority of CAC1 results. The maximum monolayer coverage capacity is the highest for CAC1 (71.942 mg/g), lower with 26.8% for CAC2 and with 58.8% for CAC3, respectively. The calculated adsorption capacity for the tested sorbents is very close to that experimentally obtained. R_L values, situated in the interval $0 < R_L < 1$, indicate the adsorption nature to be favorable. However, R_L values are very close to 1, indicating a close to liner sorption [43].

In the Freundlich model the constant K_F is an approximate indicator of the adsorption capacity. In spite of the lower correlation coefficients of the experimental data with Freundlich model by comparing to the Langmuir model, the same hierarchy of adsorption capacity was obtained: CAC1 $\gg$ CAC2 > CAC3. *1/n* could be considered both a function of the strength of adsorption in the adsorption process [31] and a heterogeneity parameter. For smaller *1/n* values, higher the surface heterogeneity is expected. From the experimental data, it could be noted that adsorption of CV from the single-solute solution is made on a higher heterogeneous surface exposed by the initial adsorbent surface. Also, in all the studied cases, *n* values lie between one and ten, this indicating a favorable adsorption process [27], in agreement with the information obtained by considering the Langmuir model that better fits to experimental data.

In the Temkin model, *B* value is correlated with the heat of adsorption. The Cd^{2+} inons adsorption was found to be an exothermic process [44], while Crystal Violet dye adsorption was found to be endothermic [45, 46]. For the studied systems, low values obtained for *B* indicate a physical adsorption of CV on the used substrates. The highest adsorption heat was obtained for CAC1, meaning a higher interaction sorbate-sorbent and a more stable structure obtaining.

This conclusion is consistent with A_T value, Temkin isotherm equilibrium binding constant, that is the highest for CAC1. Dubinin–Radushkevich isotherm (DR) is generally applied to describe the adsorption mechanism with a Gaussian energy distribution onto a heterogeneous surface [47, 48].

The model was successfully used for high solute activities. The R^2 values obtained by fitting the experimental data with DR model are quite high, the second in the series: Langmuir > DR > Temkin > Freundlich. Analyzing the obtained data (Table 23.2), it can be noted that the value of the free energy of adsorption, *E*, lower than 8 kJ/mole, means that the adsorption process can be considered mainly physical. If *E* values range in the 8–16 kJ/mole interval, the adsorption implies chemical processes [49, 50].

In the present data, the free energy is the highest for CAC1, followed by that for CAC2 and CAC3 respectively. The value for CAC1 is 7.0711 kJ/mole, very close to that for chemical interaction, meaning the strongest interaction between CAC1 and CV molecules, in spite of the dominating role of the physical adsorption of CV onto the studied sorbents.

As it could be seen, by comparing the experimental data fitting with different models and by checking the meaning of the parameters value, the best fitting was obtained with Langmuir model but the R^2 value is not very high (0.950). So, possibly,

CV sorption occurs following a dominating mono-layer adsorption mechanism, but other types of sorption could happen in a minor proportion. The other checked models exhibit lower R^2 value, but for example, DR model fits, with close values of R^2, to Langmuir model. Also, it is important to note that all the isotherm models keep the same hierarchy of the adsorption capacity of the tested sorbents and of the energetic terms describing the adsorption process.

23.4.3 Modeling of Batch Adsorption Kinetics and Mechanism of Adsorption

To determine the uptake kinetic and the adsorption mechanisms of the CV dye from a single solute systems and of CV and Cd^{2+} cations from di-solute systems, three kinetic models were investigated.

23.4.3.1 Pseudo-First Order Kinetics

The pseudo-first order equation of Lagrange is generally expressed as follows [51] Eq. (23.11)

$$\log(q_e - q_t) = \log(q_e) - \frac{K_L}{2.303} \cdot t \tag{23.11}$$

where: K_L (min^{-1}) is the Lagergren constant (the adsorption rate constant), q_t and respectively q_e (mg/g) is the amount of the adsorbate adsorbed at time t (min) and respectively at equilibrium. A plot log *($q_e - q_t$)* versus t gives a straight line from whose slope and intercept, K_L and q_e, can be calculated. The results are presented in Table 23.3. A straight line of log *($q_e - q_t$)* versus *(t)* with high linear regression coefficient (R^2) indicates the validity of this model.

23.4.3.2 Pseudo-Second Order Kinetic

The rate expression is based on the adsorption capacity of the solid phases, which has been investigated using the kinetics of the chemisorption processes. It is expressed by means of Eqs. (23.12, 23.13)[52, 53].

$$\frac{t}{q_t} = \frac{1}{k_2 q_e^2} + \frac{t}{q_e} \tag{23.12}$$

And:

$$h = k_2 q_e^2 \tag{23.13}$$

where: k_2 is the pseudo-second order rate constant (g/mg . min). The adsorption capacity at equilibrium (q_e) and adsorption rate constant (k_2) are determined from the slope and intercept of linear plot of t/q_t against t from Eq. (23.12). A linear relationship between t/q_t against t and close calculated and experimental values of q_e and q_t indicate the applicability of this model. Also, the initial adsorption rate h (mg/g . min) is calculated from Eq. (23.13) when $t \to 0$.

This model is based on the assumption that the rate limiting step may be a chemisorption process involving the valence forces through sharing or exchange of electrons between the adsorbent and the adsorbate.

Intra-particle diffusion model is used to identify the rate controlling step in the adsorption process and the mechanism of adsorption. The mechanism of adsorption involves multiple steps: the CV molecules or cadmium cations migrate from the bulk solution to the surface of adsorbent, diffuse through the boundary layer and then, by intra-particle diffusion, arrive into the interior surface of the adsorbent. The amount of CV or cadmium cations can be calculated by Eq. (23.14).

$$q_t = k_{id} \cdot t^{1/2} + C \tag{23.14}$$

where: k_{id}—is the intra-particle diffusion rate constant (mg/g . min$^{0.5}$) and it can be calculated from the slope of the linear dependency of q_t against $t^{1/2}$ described by Eq. (23.14) The constant C (mg/g) indicates the thickness of the boundary layer [54]. Higher C values indicate higher boundary layer effect describing the inapplicability of pore diffusion as the sole rate-determining step in the dynamics of the adsorption process [43]. The experimental data are presented in Table 23.3.

From Table 23.3, it could be noted that lower the initial concentration of CV is, higher the R^2 for pseudo-second order kinetic fitting is, excepting CAC3. The calculated q_e values are very close to that experimentaly determined for all the tested sorbents. The k_2 and initial adsorption rate h incresed for CAC1 and CAC2 with the initial concentration of CV increase, but an oposite behaviour was noted for CAC3.

It is also interesting to note, that CV adsorption from the solution with the highest tested initial concentration, on the CAC1 substrate, shows very similar regression coefficients R^2, for both pseudo-second order model and intraparticle diffusion model that suggests a complex adsorption mechanism, evidenced also in Figs. 23.5 and 23.6 and sustained by the close regression coefficients for both Langmuir and DR isotherm models. This means that the adsorption rate of CV on CAC1 surface is proportional to the number of active sites occupied onto the adsorbent and the rate-limiting step corresponds to the formation of a chemisorptive bond that involves sharing or exchange of electrons between the adsorbate and the adsorbent [53]. The small value of the constant C indicates a lower boundary layer effect and a low resistance towards mass transfer by the rate of adsorption that leads to a decreased necessary contact time. The C value different of zero, means that intraparticle diffusion is not the only stage

Table 23.3 Kinetic parameters of CV adsorption on the studied substrates for different initial concentration of CV and 0.01 g substrate

k_L (min^{-1})		$q_{e(exp)}$ (mg/g)	R^2	k_2 (g/mg . min)	$q_{e(cal)}$ (mg/g)	h (mg/g . min)	R^2	k_{id} ($mg/gmin^{0.5}$)	C (mg/g)	R^2
Cristal Violet from mono-solute system $c^i = 20$ mg/L										
C1*	0.094	68.38	0.956	0.278	70.92	1395.9	0.961	3.275	20.239	0.964
C2*	0.026	64.19	0.889	0.103	66.23	449.98	0.997	1.433	46.12	0.869
C3*	0.003	20.75	0.934	0.887	11.01	107.38	0.961	0.336	1.432	0.906
Cristal Violet from mono-solute system c^i =10 mg/L										
C1*	0.0150	49.556	0.954	0.177	50.00	441.50	0.992	1.6857	27.10	0.953
C2*	0.0127	44.55	0.959	0.171	44.84	343.86	0.995	1.484	25.187	0.963
C3*	0.0141	26.51	0.963	1.085	29.07	916.37	0.979	1.0727	1.964	0.969
Cristal Violet from mono-solute system $c^i = 5$ mg/L										
C1*	–	24.839	0.649	0.026	24.9	15.858	1.00	–	–	0.60
C2*	–	24.335	0.514	0.022	24.04	12.317	0.999	–	–	0.49
C3*	0.0168	10.143	0.902	1.033	10.0	103.31	0.986	0.365	4.985	0.941
Cristal Violet from mono-solute system $c^i = 2.5$ mg/L										
C1*	–	12.391	0.595	0.0250	12.41	3.848	1.000	–	–	0.675
C2*	–	12.492	0.249	0.0152	12.35	2.317	0.999	–	–	0.309
C3*	0.0097	9.586	0.943	1.5981	9.407	141.42	0.971	0.4115	3.572	0.966
Cristal Violet from ($CV + Cd^{2+}$) di-solute										
C1*	0.0161	24.808	0.894	0.019	24.876	11.757	0.999	–	–	0.268
C2*	0.0173	6.038	0.938	3.884	6.489	163.55	0.966	0.318	1.729	0.912
C3*	0.0041	8.139	0.818	1.146	8.673	86.233	0.995	–	–	0.753

(continued)

Table 23.3 (continued)

	k_L (min^{-1})	$q_{e(exp)}$ (mg/g)	R^2	k_2 (g/mg . min)	$q_{e(cal)}$ (mg/g)	h (mg/g . min)	R^2	k_{id} (mg/gmin$^{0.5}$)	C (mg/g)	R^2
Cd^{2+} from ($CV + Cd^{2+}$) di-solute system										
C1*	–	330.5	0.686	0.004	294.12	787.2	0.952	–	–	0.368
C2*	–	455.5	0.794	0.026	476.19	5963.7	0.983	–	–	0.794
C3*	–	365.0	0.042	0.006	357.14	816.33	0.995	–	–	0.390

limiting of the CV adsorption [36]. A similar situation could be noted in case of CV adsorption on CAC3 surface but the intraparticle transfer rate, k_{id}, is lower, meaning a more compact structure of the substrate by comparing to CAC1. However, the adsorption of CV on CAC3 is not very well fit with the checked models, the *R*2 being lower than that in the case of CAC1. CV adsorption on the CAC2 adsorbent seems to well fit with a pseudo-second order kinetic.

Differences between the CV adsorption kinetics, when using the three tested substrates, suggest differences between the physical and chemical features of adsorbents.

23.4.4 Correlation Between the Sorbents Features and Their Adsorptive Properties

High adsorptive capacity is dictated by: (1) a high porosity of the adsorbent, that leads to a high surface area in contact with the species that will be adsorbed; (2) by the presence at the surface, of functional groups such as –C=C–, >C=O, –C–OH, –COOH, that can interact, physically or chemically, with the adsorbate species and can generate, as a function of pH, a surface charge that acts as driving force of adsorption of opposite charged adsorbate species. Also, the fit of the pores size and form with those of the adsorbed species, as well as the pores inter-connectivity are also important factors affecting the sorption capacity. The technique of obtaining carbonized pine cones, by using a simple raw pine cones pyrolysis in an auto-generated atmosphere, avoiding the use of inert gases and at a relatively low temperature, 700 °C, led to a porous carbonized structure.

The increase of the porosity was achieved by activation with solid NaOH and respectively KOH and by impregnation with H_3PO_4 solution. After pyrolysis, a strong increase of the carbon content was noted along with a decrease of the oxygen and hydrogen content that were released as volatile compounds (EDX and FTIR analysis). Activation process contributed to a deeper degradation of the cellulose, hemicelluloses and lignin, initially present in the wooden structure of the pine cones (FTIR analysis), leading to the release of new formed volatiles, increasing the material porosity (SEM,, BET). The final result was crystalline porous materials (evidenced by XRD), with high carbon content (determined by EDS), containing some organic functional groups and inorganic compounds (determined by FTIR, and XRD) with a relatively high porosity, high surface area (calculated by BET equation) and a high roughness (determined by AFM). All these properties were influenced by the type of the activating agent used and at their turn, these properties influenced the adsorptive features of the obtained materials.

Figure 23.8 evidences the crystalline compounds identified in the activated carbonized pine cones. Carbon and carbon based compounds as well as silicate and aluminate based ones are present in the obtained materials, giving the inorganic content.

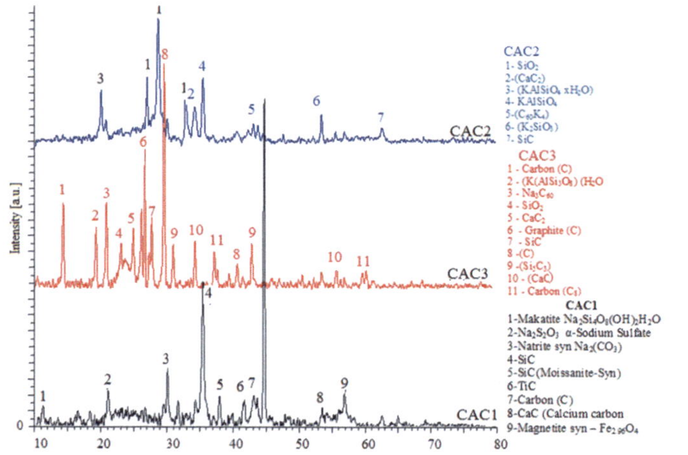

Fig. 23.8 XRD analysis of activated carbonized pine cones by using different activating agents

EDX analysis was performed to characterize the sorbents surface in the initial state and also after sorption process. A high carbon content was evidenced at the samples surface, the highest being that corresponding to CAC1 (72.71%) that presents also, the best adsorption capacity.

CAC2 shows a higher amount of oxygen and a lower content of carbon, by comparing to CAC1. CAC3 presents a more uniform distribution of the elements, with few exceptions, and a lower content of C, N and O elements by comparing to the former samples. This can be attributed to a more compact structure that leads to lower surface exposed to analysis, in agreement with SEM images (Fig. 23.10). The P atoms are also present in CAC3, probably coming from the activation step with H_3PO_4. The presence of N was noted also by other researchers [55], and identified to be a pine cones characteristic. The samples containing high N content (CAC2 and CAC3) seems to have some not completely degraded organic compounds from the initial pine cones, meaning in fact, a lower carbonization and activation degree, and as consequence a lower adsorption capacity, keeping with the adsorption experimental data.

The adsorption of both CV and Cd^{2+} ions changed completely the elements from the samples surface and their distribution. CV adsorbed on the surface can be identified by N appearance in EDX analysis of the samples where the N was not identified in the initial state (CAC1) and by decreasing of the N content from the adsorbent surface in case of N presence in the initial adsorbent structure (CAC2 and CAC3). The last case evidenced the cover of the initial surface by CV. Cd^{2+} ions were clearly identified on the surface of all the studied adsorbents, after adsorption process.

Adsorption process is dependent not only by the type of the elements present on the sorbent surface but also by the types of chemical bonds in which they are involved. Chemical group's polarity and their possibility to dissociate starting with specific values of pH, create the surface charge. This is in fact the driving force that

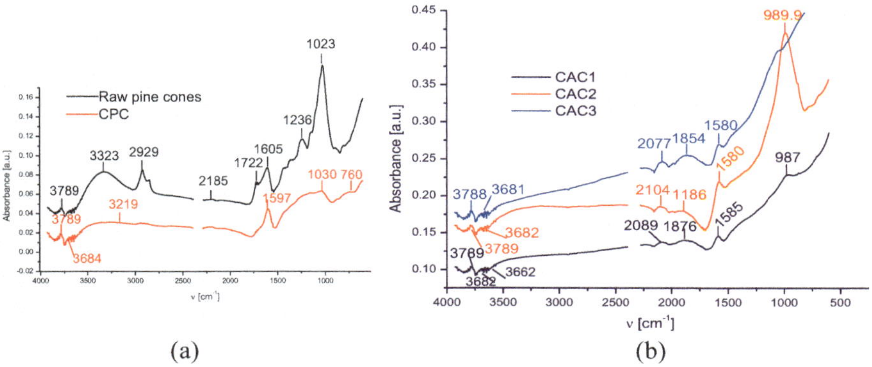

Fig. 23.9 FTIR spectra of the initial pine cones and carbonized pine cones (a); of the activated carbonized pine cones (b)

attracts the opposite charged ions, favoring their sorption by physical or chemical bonds. Polar groups presenting impaired electrons, π bonding, as well as charged groups determines strong interaction with adsorbate molecules, in function of the structural characteristics of the last one.

FTIR analysis allows the identification of the types of polar bonds from the adsorbent surface and determination of pH_{PZC} evidenced the surface charge.

FT-IR spectra presented in Fig. 23.9 show multiple aspects:

- The initial raw pine cones present characteristic groups of cellulose, hemicelluloses and lignin that can be find in all the wooden materials: absorption bands of N–H bonds at 3789 cm^{-1}; a large band centered at 3323 cm^{-1} associated with O–H bonds; C–H bonds at 2929, 2855 and 1365 cm^{-1}; C=O groups at 1722 cm^{-1}; C–O groups at 1023–1030 cm^{-1} as well as C–C bonds from alkenes at 1605 cm^{-1} and 888 cm^{-1} for $C{=}CH_2$.
- Carbonized pine cones show less intense absorption bands, due to the release of volatiles containing these groups. The major part of O–H groups was removed and C–H absorption band disappeared, evidencing the hydrogen removing from the wooden structure. The band from 1722 cm^{-1}, characteristic for C=O from carboxylic acids was shifted to 1585 cm^{-1}, band associated with the anhydride groups, meaning in fact elimination of oxygen and hydrogen from the carboxylic acids.
- Activated carbonized pine cones show simpler FTIR spectra, evidencing less intense absorption bands. However, the FTIR spectra of the materials obtained by different types of activation are different: the spectrum with the smallest number of the absorption bands is that of CAC1, meaning the elimination of the highest amount of C–O, C=O and C–H groups through volatiles formation and removing. Small amount of C=O groups from anhydride can be noted at 1585 cm^{-1} and an intense absorption band can be seen at 987 cm^{-1}, being attributed to $-CH{=}CH_2$.

CAC1, having the smallest number of the polar groups, shows the highest adsorptive properties against the studied species. C=C groups could promote strong bonds with CV molecules through the π electrons. CAC2 and CAC3 present a higher number of absorption bands attributed to C=O, C–O–C and C=C by comparing to CAC1, meaning a lower volatiles formation and release and as consequence, a lower porosity by comparing to CAC1, the best adsorbent obtained in this study. FTIR spectra of the loaded adsorbents, after sorption of Cd^{2+} ions and CV, evidence new absorption bands characteristic for CV and shifted to lower wavenumbers sustaining the interaction between the sorbent and sorbate molecules (1583 cm^{-1} for stretching vibration of C=C bonds from benzene ring and 1168.9 cm^{-1} for stretching vibration of C–N bond from CV [56] to 1541 and 1158.45 cm^{-1} in CAC1; 1560 and 1135.99 cm^{-1} in CAC2 and 1560 cm^{-1} in CAC3). Due to the low amount of CV adsorbed on the CAC3 surface, the band attributed to C-N bond is not evidenced.

The samples morphology was checked by SEM analysis and the results are presented in Fig. 23.10

From Fig. 23.10 it can be noted the effects of the thermal treatment and of the activation process, by using different agents, on the raw pine cone morphology. The obvious wooden structure of the initial pine cones, with clearly delimitated tetrahedral cell walls can be observed in Fig. 23.10a. Carbonization destroyed most of the cell walls, containing cellulose, hemicelluloses and lignin and a low porous structure

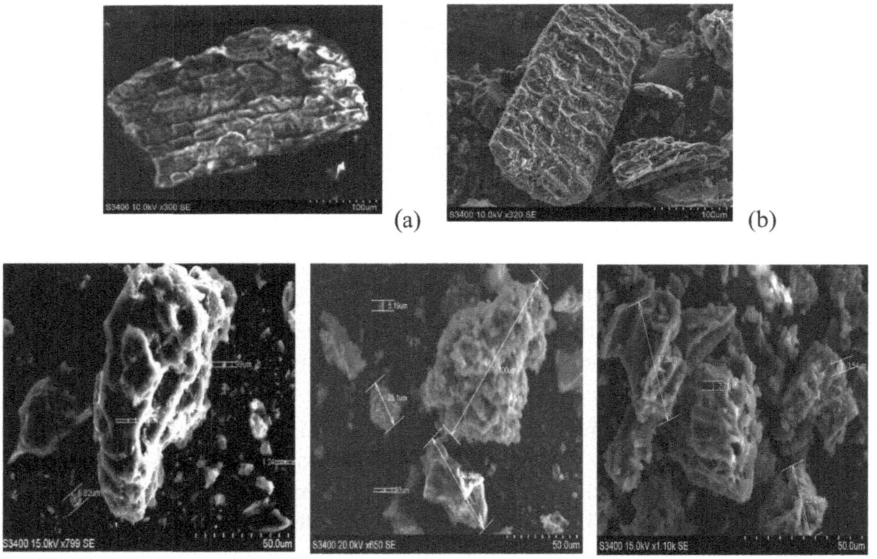

(a) (b) (c) (d) (e)

Fig. 23.10 SEM images of the initial pine cones (**a**), Carbonized pine cones (**b**), activated carbonized pine cones CAC1 (**c**); CAC2 (**d**); CAC3 (**e**)

was obtained (Fig. 23.10b). The highest porous structure, showing large honey-comb—shaped pores, was obtained by treatment with NaOH (CAC1, Fig. 23.10c). Less porous structure was obtained by carbonized pine cones treatment with KOH (Fig. 23.10d) and the lowest porous seems to be CAC3 obtained by activation with H_3PO_4 (Fig. 23.10e). Additionally, all the tested adsorbents present also irregular-shaped pores with different dimensions: 1.24, 1.99, 3.54 and 5.19 μm, as well as large particle agglomerations.

Not only the samples porosity but also their surface roughness is different when the activation was made with different agents: the highest average roughness is shown

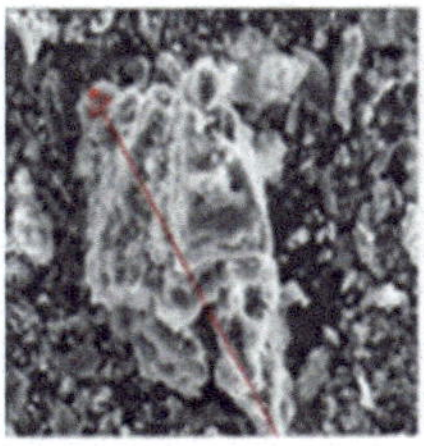

Element	[%]
C	72.7
N	0.0
O	24.35
Na	2.95
Al	0.0
Si	0.0
S	0.0
K	0.0

(a) CAC1 before adsorption

Element	%]
C	17.45
N	19.41
O	20.69
Na	19.21
Si	16.81
Cd	6.43

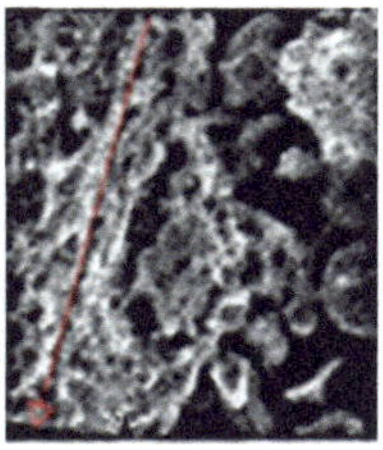

(b) CAC1 after adsorption of CV and Cd^{2+}

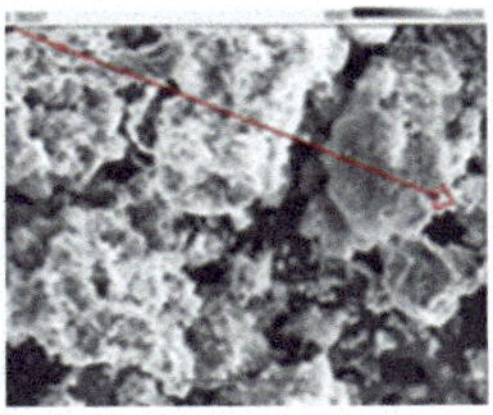

Element	[%]
C	24.79
N	48.23
O	22.73
Na	0.15
Al	0.16
Si	1.92
K	2.01

(c) CAC2 before adsorption

Element	[%]
C	11.49
N	16.54
O	20.09
Na	22.68
Si	20.97
Cd	7.44

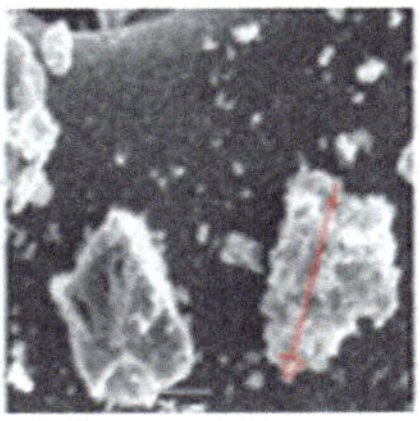

(d) CAC2 after adsorption of CV and Cd^{2+}

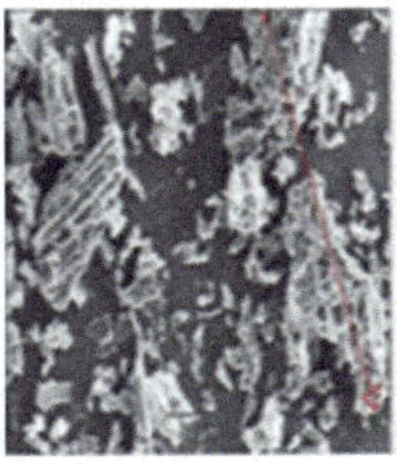

Element	[%]
C	38.12
N	42.62
O	15.76
Na	1.83
Al	0.10
Si	0.32
P	0.76
K	0.48

(e) CAC3 before adsorption

Element	[%]
C	10.31
N	14.13
O	17.41
Na	18.52
Si	17.01
P	16.63
Cd	6.00

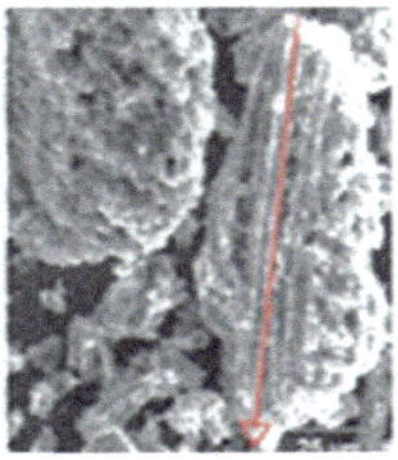

(f) CAC3 after adsorption of CV and Cd^{2+}

Fig. 23.11 EDX results characterizing the adsorbent surface before and after CV and Cd^{2+} ions adsorption

Table 23.4 Surface area and pores characteristics for activated carbonized pine cones

Sample	BET surface (m^2/g)	Pore diameter (nm)	Pore volume (cm^3/g)	Micro pores V (t-plot) (cm^3/g)	V_{micro}/V_t	Micro pores area (t-plot) (cm^2/g)	Average roughness surface (nm)
C1*	353.8	18.69	0.95	0.192	0.200	394.008	447.57
C2*	327.7	19.70	1.91	0.155	0.081	308.136	322.68
C3*	126.9	20.75	0.65	0.048	0.073	97.282	139.46

by CAC1, a lower value was obtained for CAC2 and CAC3 evidenced the smoother surface, far less than that of the CAC1 and CAC2 substrates. The average roughness surface vas calculated from the AFM data and is presented in Table 23.4.

The specific surface area of the tested adsorbents was calculated by BET equation. The pore volume, the pore diameter and pore size distribution were calculated from desorption branch of the isotherm by the Barrett, Joyner and Halenda (BJH) method. The results are presented in Table 23.4.

From Table 23.4 it can be observed that the BET surface decreased in the following order:

$S_{BET\text{-}CAC1} \geq S_{BET\text{-}CAC2} \geq S_{BET\text{-}CAC3}$. The same trend is maintained also by the micro pores volume and micro pores area.

Surface charge of the adsorbent is also an important parameter that influences the type of adsorbed species.

The results of the experiments shows that zero-point charge for the prepared adsorbents falls in the pH range 4.41–4.7. This means that under this pH, the surface is positively charged while at higher values of pH, the surface is negatively charged. The sorption process was performed at the natural pH of the CV and Cd^{2+} ions solutions, ranging in the 6–7.3 pH intervals. These pH values avoid the Cd^{2+} ions precipitation (happening at pH > 8). Below pH = 8, positively charged species such as Cd^{2+} and $Cd(OH)^+$ co-exist in solution and are attracted by negatively charged surface of the adsorbents [44].

From SEM, AFM and BET isotherm studies resulted that CAC1 is characterized by the highest porosity, highest BET surface, highest micro-pores volume and area, highest roughness and all these explain the best adsorptive capacity of the carbonized pine cones activated with NaOH.

Figure 23.11b, d, f show the EDX analysis for the studied adsorbents after CV and Cd^{2+} ions sorption. In a very suggestive way, the highest content of nitrogen, coming from CV, and of Cd^{2+}ions retained at the surface of CAC1, by comparing to the other two studied adsorbents were recorded. The highest uniform distribution of the adsorbed species was noted on the CAC1 surface.

From all these studies it can be noted that adsorbents obtained by alkali activation evidenced the highest efficiency of CV removal by comparing to that of acid activated substrate both from single and di-solute concentration. It can be observed that the process efficiency strongly increases with the initial solution concentration decrease, as expected.

But the level of increase is dependent on the adsorbent type. In case of alkali activated adsorbents (CAC1 and CAC2), the differences between the first two initial concentrations, meaning up to 5 mg/L, are negligible. The differences became more evident for 10 mg/L initial concentration, but in the case of CAC1, after 180 min., the same value of the efficiency was reached as in the case of the lower concentrations. In the case of CAC2, the adsorption efficiency is with approximately 15% lower by comparing to that obtained for diluted solutions. For higher concentrated initial solutions (20 mg/L) the removal efficiency decreased up to 60% in case of alkali activated substrates, but the sorption equilibrium was slower reached by CAC1, probably due to a more complex surface morphology. Figure 23.2c. shows that in the case of CAC3, acid activated adsorbent, the removal efficiency is much lower by comparing to that of the alkali activated substrates. Excepting, the lowest initial solution concentration, when the efficiency reached 75%, all the other initial solutions led to efficiencies lower that 45%. In the case of the highest concentrated solution, an induction/lag time of about 120 min. was evidenced and efficiency about only 20% was obtained. These experiments evidenced the superiority of alkali activated adsorbents and their suitability for adsorption of CV in advanced depollution processes.

23.5 Conclusions

Three types of activated carbon from cones of European pine (*Pinus Nigra* and *Pinus Picea Abies)*, collected from Romania, Brasov (Teliu) county, were obtained by cones carbonization at 400–500 °C, in an auto-generated atmosphere, activated by mixing with (1) NaOH, (2) KOH and water in a ratio of 1:3:3 and respectively (3) by impregnating with aqueous solution of 30% H_3PO_4 in a ratio of 1:3, and then thermally treated for 1.5 h at 700 °C.

The final result [(1)-CAC1; (2)-CAC2; (3)-CAC3)] was crystalline porous materials, with high surface carbon content, containing some organic functional groups and inorganic compounds, with a relatively high porosity, high surface area, a high roughness and the pH_{PZC} values ranging in the 4.41–4.7 interval. The proposed conditions for adsorbents preparation were effective leading to high efficiency of CV adsorption from aqueous solutions in the absence and respectively the presence of Cd^{2+} ions. The best adsorbent was CAC1, NaOH activated, which evidenced the highest BET surface (353.8 m^2/g), the highest micro-pore area (394.0 cm^2/g), and the highest average roughness: (447.57 nm), the lowest number and concentration of the polar groups and the highest >C=C< content. The sorption capacity increased with the increase of the initial solution concentration and the contact period and decreased with the increase of the sorbent dosage.

The presence of Cd^{2+} ions determined the strong decrease of the amount of the adsorbed CV, the first one being a competitive ion in the sorption process that determined the inactivation of a high number of the active sites from the adsorbent surface.

In the presence of Cd^{2+} ions, the adsorption mechanism changed, became simpler and followed the pseudo-second order kinetic model for CAC1 too, as for the other studied adsorbents.

Pollutants removal efficiency from common solution was 99.03% for CV and 57.58% for Cd^{2+} ions when the initial solution concentration was the same (10 mg/L) and 98.94% for CV and 10.04% for Cd^{2+} ions when a very high excess of cadmium was present in the initial solution.

All the above reported results showed that the obtaining of the activated carbonized pine cones in auto-generated atmosphere, avoiding the inert gas use, and at lower temperatures by comparing to other reported methods was effective. Good, low-cost and locally available adsorbents were obtained.

Acknowledgements This work was supported by a grant of the Transylvania University of Brasov, the Doctoral Programme.

References

1. V.K. Gray, R. Kumar, R. Gupta, Removal of malachite green dye from aqueous solution by adsorption using agro industries waste: a case study of Phosopis Ceneraria. Dyes & Pigment **62**, 1–10 (2004)
2. A. Arungagiri, J. Indulekha, P. Kalaichelvi, R. Anantharaj, Removal of crystal violet dye from aqueous solution using triton X-114 surfactants via cloud point extraction. Eng. Sci. Technol. Int. J. **17**, 137–144 (2014)
3. Y. Zhou, L. Zhang, Z. Cheng, Removal of organic pollutants from aqueous solution using agricultural wastes: a review. J. Mol. Liq. **212**,739–762 (2015)
4. M. Visa, F. Pricop, A. Duta, Sustainable treatment of wastewaters resulted in the textile dyeing industry, Clean Techn. Environ. Policy. **13**, 855–861 (2013)
5. M. Malakootian, A. Almasi, H. Hossaini, Sustainable treatment of wastewaters resulted in the textile dyeing industry. Environ. Sci. Techn. **5**, 217–222 (2008)
6. S. Jana, M.K. Purkait, K. Mohanty, Removal of crystal violet by advanced oxidation and microfiltration, Appl. Clay Sci. **50**, 337–341(2010)
7. M. Sulyman, J. Namiesnik, A. Gierak, Utilization of new activated carbon derived from oak leaves for removal of crystal violet from aqueous solution. Pol. J. Environ. Stud. **26**, 2223–2232 (2014)
8. K. Gustina, F. Tofailb, M. Vahtera, M. Kipplera, Cadmium exposure and cognitive abilities and behavior at 10 years of age: a prospective cohort study. Environ. Int. **113**, 259–268 (2018)
9. M. Visa, L. Andronic, D. Lucaci, A. Duta, Concurrent dyes adsorption and photo-degradation on fly ash based substrates. J. Adsorption **17**, 102–108 (2011)
10. R. Dobritoiu, S. Patachia, A study of dyes sorption on biobased cryogels. Apply. Surf. Sci. **285**, 56–64 (2013)
11. A. Papancea, A.J.M. Valente, S. Patachia, PVA cryogel membranes as a promising tool for the retention and separation of metal ions from aqueous solutions. J. Appl. Polym. Sci. **118**(2), 1567–1573 (2010)
12. S. Patachia, R. Dobritoiu, T. Coviello, Determination of the sorption efficiency of Poly (vinly alcohol)/ Scleroglucan cryogels, against Cu^{2+} ions. Environ. Eng. Manag. J. **10**(2), 193–198 (2011)
13. N.S. Kumar, M. Asif, M.I. Al-Hazzaa, Adsorptive removal of phenolic compounds from aqueous solutions using pine cone biomass: kinetics and equilibrium studies. Environ. Sci. Pollut. Res. **25**, 21949–21960 (2018)

14. A. Pholosi, A.E. Ofomaja, E.B. Naidoo, Effect of chemical extract ants on the biosorptive properties of pine cone power: Influence on lead (II) removal mechanism. J. Saudi Soc. **17**, 77–86 (2013)
15. C.L. Yang, J.M. Garrahan, Electrochemical coagulation for textile effluent decolorization. J. Hazard Mater. **B127**, 40–47 (2005)
16. J. Matos, S.M. Cuevas, A. Ruiz-Delgado, I. Oller, S. Malato, Development of TiO_2-C photocatalysts for solar treatment of polluted water. J. Carbon **122**, 361–373 (2017)
17. G. Ciardelli, N. Ranieri, The treatment and reuse of wastewater in the textile industry by means of ozonation and electro-floculation. J. Water Res. **35**, 567–572 (2001)
18. M. Visa, A. Duta, Methyl-orange and cadmium simultaneous removal using fly ash and photo-Fenton systems. J. Hazard Mater **244**, 773–779 (2013)
19. M. Visa, C. Bogatu, A. Duta, Simultaneous adsorption of dyes and heavy metals from multicomponent solutions using fly ash. J. Appl. Surf. Sci. **256**, 5486–549 (2010)
20. N. Popa, M. Visa, The synthesis, activation and characterization of charcoal powder for the removal of methlene blue and cadmium from wastewater. J. Adv. Powder Tech. **28**, 1886–1876 (2017)
21. S. Chakraborty, S. Chowdhury, P.D. Saha, Adsorption of crystal violet from aqueous solution onto NaOH-modified rice husk. Carbohydr. Polym. **86**, 1533–1541 (2011)
22. R. Sivaraj, C. Namasivayam, K. Kadirvelu, Orange peel as an addorbent in the removal of acid violet 17 (acid dye) from aqueous solutions. J Waste Manag. **21**, 105–110 (2001)
23. R. Zhang, J. Zhang, X. Zhang, C. Dou, R. Han, Adsorption of Congo red from aqueous solutions using cationic surfactant modified wheat straw in batch mode: kinetic and equilibrium study. J. Taiwan Inst. Chem. Eng. **45**, 2578–2583 (2014)
24. T. Robinson, B. Chandran, P. Nigam, Effect of pretreatments of three waste residues, wheat straw, corncobs and barley husks on dye adsorption. J. Bioresource Technol. **85**, 119–124 (2002)
25. M.C. Tonucci, L.V.A. Gurgel, S. Francisco de Aquino, Activated carbons from agricultural byproducts (pine tree and coconut shell), coal, and carbon nanotubes as adsorbents for removal of sulfamethoxazole from spiked aqueous solutions: kinetic and thermodynamic studies. Ind.Crop. Prod. **74**, 111–121 (2015)
26. G.Y. Duman, Y. Onal, C. Okutucu, S. Onenc, Y. Yanik, Production of activated carbon from pine cone and evaluation of its physical chemical and adsorption properties. J. Energy Fuels **23**, 2197–2204 (2009)
27. R.L. Tseng, S.K. Tseng, Pore structure and adsorption performance of the KOH-activated carbon prepared from corncob. J. Colloid Interface Sci. **287**, 428–437 (2005)
28. E. Yagmur, M. Ozmak, Z. Aktas, A novel method for production of activated carbon from waste tea by chemical activation with microwav. J Fuel **87**, 3278–3285 (2008)
29. N.K. Amin, Removal of reactive dye from aqueous solutions by adsorption onto activated carbons prepared from sugarcane bagasse pith. J. Desalin. **223**, 152–216 (2008)
30. S.N.d.C. Ramos, A.L.P. Xavier, F.S. Teodoro, L.F. Gil, L.V.A. Gurge, Removal of cobalt (II), copper (II), and nickel (II) ions from aqueous solutions using phthalate functionalized sugarcane bagasse: Mono-and multicomponent adsorption in batch mode. Ind. Crop. Prod. **79**, 116–130 (2016)
31. R. Kumar, R. Ahmad, Biosorption of hazardous crystal violet dye from aqueous solution onto treated ginger waste (TGW). Desalin. J. **265**, 112–118 (2011)
32. A.E. Ofomaja, A. Pholosi, E.B. Naidoo, Application of raw and modified pine biomass material for cesium removal from aqueous solutions. J. Ecol. Eng. **82**, 258–266 (2015)
33. G.Z. Kyzas, M. Kostoglou, Green adsorbents for wastewaters: a critical review. Materials **7**, 333–364 (2014)
34. J. Shang, W. He, C. Fan, Adsorbption of dimethyl trisulfide from aqueous solution on a low-cost adsorbent: thermally activated pinecone. Chin. J. Oceanol. Limno. **33**(1), 169–175 (2014)
35. N. Van Vinh, M. Zafar, S.K. Behera, H.S. Park, Arsenic(III) removal from aqueous solution by raw and zinc-loaded pine cone biochar: equilibrium, kinetics, and thermodynamics studies. Int. J. Environ. Sci. Technol. **12**(4), 1283–1294 (2015)

36. S. Khalili, B. Khoshandam, M. Jahanshahi, A comparative study of CO_2 and CH_4 adsorption using activated carbon prepared from pine cone by phosphoric acid activation. Korean J. Chem. Eng. **33**(10), 2943–2952 (2016)
37. M. Hadi, M.R. Samarghandi, G. McKay, Simplified fixed bed design models for the adsorption of acid dyes on novel pine cone derived activated carbon. Water Air Soil Pollut. J. **218**(1), 197–212 (2011)
38. S. Valizadeh, H. Younesi, N. Bahramifar, Preparation and characterization of activated carbon from the cones of iranian pine trees (Pinus eldarica) by chemical activation with H_3PO_4 and its application for removal of sodium Dodecylbenzene Sulfonate removal from aqueous solution. Water Conserv. Sci. Eng. 1–13. https://doi.org/10.1007/s41101-018-0055-5 (2018)
39. T.K. Sen, S. Afroze, H.M. Ang, Equilibrium, kinetics and mechanism of removal of methylene blue from aqueous solution by adsorption onto pine cone biomass of pinus radiate. Water Air Soil Poll. J. **218**, 499–515 (2011)
40. A. Özhan, Ö. Şahin, M.M. Küçük, C. Saka, Preparation and characterization of activated carbon from pine cone by microwave-induced $ZnCl_2$ activation and its effects on the adsorption of methylene blue. Cellulose, **21**(4). https://doi.org/10.1007/s10570-014-0299, (2014).
41. Y. Balut, H. Aydin, A kinetics and thermodynamics study of methlene blue adsorption on wheat shells. J Desalination **194**, 259–267 (2006)
42. R.K. Gautam, A. Mudhoo, G. Lofrano, M.C. Chattopadhyaya, Biomass-derived biosorbents for metal ions sequestration: Adsorbent modification and activation methods and adsorbent regeneration. J. of Environ. Chem. Engine. **2**, 239–259 (2014)
43. H.N. Bhatti, A. Jabeen, M. Iqbal, S. Noreen, Z. Naseem, Adsorptive behavior of rice bran-based composites for malachite green dye: Isotherm, kinetic and thermodynamic studies. J. of Mol. Liq. **237**, 322–333 (2017)
44. A.O. Dada, A.P. Olalekan, A.M. Olatunya, O. Dada, Langmuir, Freundlich, Temkin and Dubinin-Radushkevich Isotherms Studies of Equilibrium Sorbtion of Zn^{2+} Undo Phosphoric Acid Modified Rice Hush. J. Appl. Chem. **3**, 38–45 (2012)
45. Sewu, D.D., Boakye, P., Woo, S.H., Highly efficient adsorption of cationic dye by biochar produced with Korean cabbage waste. Bioresource Technol. 224, 206–213 92017).
46. A. Bhatnagar, F. Kaczala, W. Hogland, M. Marques, C.A. Paraskeva, V.G. Papadakis, M. Sillanpää, Valorization of solid waste products from olive oil industry as potential adsorbents for water pollution control—a review. Environ Sci Pollut Res Int. **21**(1), 268–298 (2014)
47. N.W. Ingole, S.V. Dharpal, State of art of biosorption technique for treatment of heavy metals bearing wastes. International J. of Advanced Engineering Technology **3**(2), 143–153 (2012)
48. G.Z. Kyzas, J. Fu, K.A. Matis, The change from past to future for adsorbent materials in treatment of dyeing wastewaters. Materials **6**, 5131–5158 (2013)
49. A. Gunay, E. Arslankaya, I. Tosun, Lead removal from aqueous solution by natural and pretreated clinoptilolite: adsorption equilibrium and kinetics. J. Hazard. Mater. **146**, 362–371 (2007)
50. A. Dabrowski, Adsorption—from theory to practice. Adv. Colloid Interface Sci. **93**, 135–224 (2001)
51. K.G. Bhattacharyya, A. Sharma, Azadirachta Indica leaf powder as an effective biosorbent for dyes: a case study with aqueous Congo red solutions. J. Environ. Manag. **71**, 217–229 (2004)
52. M.B. Ibrahim, S. Sani, Comparative isotherms studies on adsorptive removal of Congo red from wastewater by watermelon rinds and Neem-tree leaves. Open J. Phys. Chem. **4**, 139–146 (2014)
53. T. Mahmood, R. Ali, A.N. Naeem, M. Hamayun, M. Aslam, Potential of used Camellia sinensis leaves as precursor for activated carbon preparation by chemical activation with H_3PO_4; optimization using response surface methodology. Process Saf. Environ. Prot. **109**, 548–563 (2017)
54. H.R. Mahmoud, S. Ibrahim, S.A. El-Molla, Textile dye removal from aqueous solutions using cheap MgO nanomaterials: adsorption Kinetics studies and thermodynamics. J. Adv. Powder Technol. 27, 223–231 (2016)

55. Y.S. Ho, G. Mc Kay, The kinetic of sorption of basic dyes from aqueous solution by sphagnum Moss peat. J. Chem. Eng. **76**, 822 (1998)
56. S.J. Allen, G. Mc Kay, Z.H. Khader, Interparticle diffusion of a basic dye during adsorption onto sphangum peat. J. Environ. Pollut. **56**, 39–50 (1999)
57. C. Falco, N. Baccile, M.M. Titirici, Morphological and structural differences between glucose, cellulose and lignocelluloses biomass derived hydrothermal carbon. Green Chem. J. **13** 3273–3281 (2011)
58. M.F.H. Al-Kadhemy, Wafaa Hameed Abaas, Israa Fakher, 20 The Effect of Gamma Radiation on the FTIR Spectrum of Crystal Violet Doped Polystyrene Films, Caspian J. Appl. Sci. Res. **2**(7), 11–17 (2013)

Chapter 24
TiO_2—Graphene Oxide VIS-Active Composites for Advanced Photocatalytic Wastewater Treatment

Anca Duta, Ioana Tismanar, Alexandru Cosmin Obreja, and Octavian Buiu

Abstract Photocatalytic materials are intensively studied in environmental applications as self-cleaning coatings or for advanced oxidation processes for pollutants removal from wastewaters or from air. In the last decade, heterogeneous photocatalysis was largely reported as a viable path for removing pollutants from wastewater, targeting water re-use. For scaling up, the photocatalysts have to be efficient and affordable in usual working conditions, this is why VIS-active thin films gained particular attention. This paper presents the first results obtained in developing VIS-active composite thin-films, aqueously stable, based on graphene oxide dispersed in a TiO_2 matrix. To get a stable film, a stepwise deposition process was employed, involving a first TiO_2 layer developed by spray pyrolysis deposition over an FTO substrate, over which a composite TiO_2–GO layer is deposited. A coherent temperature control over the deposition allows obtaining thin films with good crystallinity and photocatalytic response in the decomposition of methylene blue. The thin film was tested in a demonstrator photo-reactor and the factors influencing the process efficiency are detailed discuss.

Keywords TiO_2–GO thin film · Continuous flow photocatalytic process · Thin film stability

A. Duta (✉) · I. Tismanar
Transilvania University of Brasov, Brasov, Romania
e-mail: a.duta@unitbv.ro

I. Tismanar
e-mail: tismanarioana@yahoo.com

A. C. Obreja · O. Buiu
National R&D Institute R&D for Microtechnologies, Bucharest, Romania
e-mail: cosmin.obreja@imt.ro

O. Buiu
e-mail: octavian.buiu@gmail.com

I. Visa and A. Duta (eds.), *Solar Energy Conversion in Communities*,
Springer Proceedings in Energy, https://doi.org/10.1007/978-3-030-55757-7_24

24.1 Introduction

Large amounts of organics are released in wastewaters raising challenges on the treatment processes for the water discharge. Further on, considering the legal frame that seriously considers the water re-use as a path towards protecting the natural resources, advanced wastewater treatment processes are intensively investigated aiming at removing the pollutants remained in water after the conventional treatment processes (thus at very low concentrations) but not admissible for water re-use.

Advanced oxidation processes (AOP) represent an efficient path for advanced wastewater treatment as these are able to degrade a broad range of organic pollutants at average and low concentrations [1–3]. Heterogeneous photocatalysis is one of the highly investigated AOPs as in the process result oxidation species (e.g. the $HO^{\cdot}$ radical) that will further degrade, up to mineralization, the organic contaminants. The catalyst is a semiconductor that after activation with suitable irradiation allows reactions of the electrons and of the holes to form the active radicals.

The main steps of a heterogeneous photocatalytic process involve [4]:

(1) Adsorption of the pollutant molecules on the photocatalyst surface.
(2) Activate the photocatalyst by irradiating the material with a radiation having the energy equal or higher than the semiconductor bandgap. Under irradiation the excited charges have to be separated and will move to the photocatalyst surface. A side reaction in this step is the electron–hole recombination that significantly diminishes the process efficiency. The electron–hole recombination is more restricted in crystalline structures.
(3) The charges will react on the photocatalytic surface with the hydroxide anion (HO^-) or with the dissolved oxygen to form active hydroxyl ($HO^{\cdot}$) or superoxide ($O_2^{\,-}$) radicals.
(4) The radical species will oxidize the adsorbed pollutant molecules.
(5) The oxidation products are removed from the photocatalyst surface. These products are mainly in gaseous phase (CO_2, NO_x).

Many studies report on various photocatalytic materials. The first one used was TiO_2 that is aqueous stable, has high chemical stability and photosensitivity, is biological inert and has remarkable optical properties. It can be also obtained in up-scalable processes thus being a good candidate for large-scale applications. However, the main polymorphs, anatase and rutile have a broad band gap (close to 3.2 eV) thus the photocatalyst can only be activated by radiation in the UV domain and the use of UV radiation raises the photocatalytic process cost. This is why extending the TiO_2 response towards the VIS spectral range represents a topic investigated but not fully solved yet. Doping TiO_2 with cations (e.g. V, Cr or Ni) or anions (e.g. N or S) showed a narrowing of the band gap as also did the development of composite structures combining n–n semiconductors (TiO_2–ZnO or TiO_2–WO_3) or n-p semiconductors. The last type, mimicking a solar cell, showed a significant reduction in the band gap value, being thus VIS-active, as in the combination of TiO_2–CZTS [5]. These photocatalysts are VIS-active but are of average stability. This is why further combinations, involving more aqueously stable components are investigated.

Graphene is well recognised to allow high electron mobility, thus preventing the recombination when associated with TiO_2 and a high surface area that supports the development of efficient adsorption substrates. However, graphene is non-polar that means that the interfaces developed with the ionic TiO_2 are unlikely to be strong, thus the composite is not stable during long operation durations. This is why composites involving the broad band semiconductor (TiO_2 or ZnO) and graphene oxide (GO) were developed. Because of its structure, GO is dispersible in polar solvents and is able to attach various molecular/ionic structures [6]. Most of these composites are obtained by hydrothermal synthesis; solvothermal and sol–gel syntheses are also mentioned [6–8].

When designing the synthesis of a TiO_2–GO composite care should be taken considering the relative low thermal stability of GO (up to 150 °C) that limits the possibilities for obtaining crystalline structures of the semiconductor matrix.

Further on, while powder composites are obviously more efficient than thin film photocatalysts (because of a larger available surface area), their separation in the end of the process is difficult and occurs with significant losses, contributing to an increased cost. This is why thin film photocatalytic composites of TiO_2–GO type are obtained and tested in environmental applications, as the advanced wastewater treatment or the self-cleaning coatings.

24.2 Materials and Methods

Glass plates covered with a thin film of fluorine doped tin oxide (FTO) were used as substrates for the composite deposition. The substrates were cleaned with water and detergent by ultra-sonication, followed by rinsing in ethanol and drying in air. An overall surface of 20×30 cm^2 was included in the demonstrator photo-reactor. The surface contained five 10×10 cm^2 plates, one 8×10 cm^2 plate and five 2×2 cm^2 plates, used for various analyses before and after the photocatalytic process, as presented in Fig. 24.1. Before use, the photocatalytic plates were conditioned during one hour exposure to UV + VIS radiation ($G_{UV+VIS} = 810$ W/m^2; $G_{UV} = 23$ W/m^2) when the surface became super-hydrophilic (contact angle with water of 1° as compared with 12° before conditioning).

To get structures with an increased crystallinity, that support charges separation, the thin film deposition was designed in two steps:

(1) A first TiO_2 layer was deposited through an up-scalable process, Spray Pyrolysis Deposition (SPD), using an ABB/IRB5400 robot. The SPD deposition used a precursor solution of titanium isopropoxide (TTIP, Sigma Aldrich, 97%), acetylacetone (AcAc, Scharlau, 99%) and ethanol (EtOH, Chemical Company, 99.3%) in a volume ratio TTIP: AcAc: EtOH = 4:4:60 that was sprayed at 400 °C using air as carrier gas. Thirty spraying sequences were employed with a 60 s break between sequences. The titania thin film was further annealed for 3 h, at 450 °C, to increase its crystallinity.

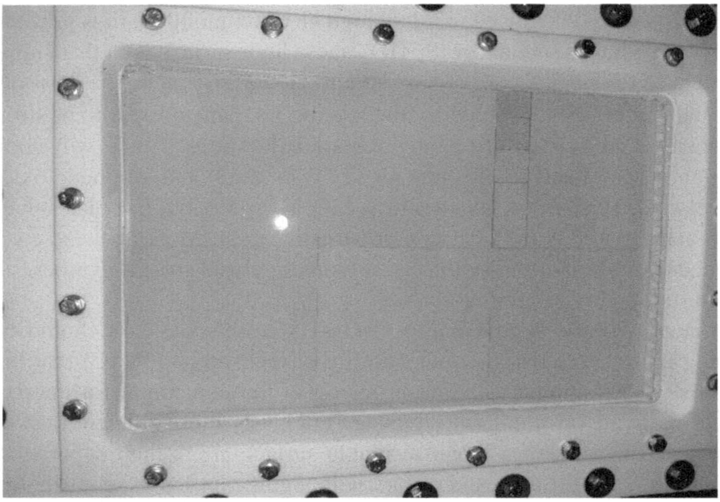

Fig. 24.1 The demonstrator photocatalytic reactor (before use)

(2) The composite TiO_2–GO layer was deposited using 15 spraying sequences on the top of the first layer by spraying a sol–gel dispersion obtained using TTIP, AcAc, EtOH, acetic acid (HAc 99.8%, Scharlau) and an aqueous GO dispersion ($1.4\%_w$) in a volume ratio of TTIP:AcAc:EtOH:HAc:GO(d) = 20:0.89:16:0.18:2.4.

The aqueous GO dispersion was obtained using the Hummers method: 12 g of graphite powder (99.99% purity, Elektrokarbon) were mixed with 247 mL H_2SO_4 (96%, VLSI grade, BASF) and continuously stirred for 24 h. Afterwards, 6 g $NaNO_3$ (99%, Sigma Aldrich) were added and the dispersion was cooled to 5 °C, then 36 g of $KMnO_4$ (99%, Merck) were slowly added. The graphite powder was kept for four days in the oxidation environment; afterwards, the unreacted $KMnO_4$ and the manganese oxide precipitate were removed using 200 mL H_2O_2 5% and 150 mL HCl 5%, respectively. The dispersion was further filtered and the graphite oxide powder was washed several times with deionized water, then it was dispersed in deionized water and was ultrasound treated for 1.5 h to support the exfoliation of the graphene oxide sheets. After sonication, the sol was diluted with EtOH in a 1:5 ratio and sprayed at 100 °C over the first layer, followed by one hour of thermal treatment at 150 °C.

The photocatalytic experiments used 5L of methylene blue (MB) solution with a 10 ppm concentration (according to the ISO 10678:2010 standard, [9]). The wastewater flow in the photo-reactor was of 1L/min and the process was run initially for one hour in dark (to reach the adsorption—desorption equilibrium), followed by 5 h of irradiation at a global irradiance value of 810 W/m^2 out of which 23 W/m^2 UV radiation and the rest VIS, using a solar simulator. After the photocatalytic process is stopped, the photocatalytic plates are regenerated by circulating distilled water in the photo-reactor in two cycles of 30 min, each.

Parallel experiments were developed on small plates (1.5×1.5 cm^2) covered with the same photocatalytic coating, at low irradiance values (55 W/m^2 out of which 3 W/m^2 UV) to investigate the effect of the photons flux on the process. These experiments were done in static regime, using 20 mL MB solution, in a photo-reactor equipped with 2 UV light tubes (Philips, TL-D BLB 18 W/108) and 5 Vis light tubes (Philips, TL-D Super 80 18 W/865). This mix simulates the solar radiation conditions, at much lower irradiance value.

The MB removal efficiency, η was calculated using Eq. (24.1), based on the initial absorbance of the MB solution (A_0) and the absorbance after *t* hours of operation (A_t), measured at the maximum absorbance wavelength for MB ($\lambda = 664$ nm), using a UV–Vis–NIR spectrophotometer (Perkin Elmer Lambda 950):

$$\eta_t = \frac{A_0 - A_t}{A_0} \cdot 100 \tag{24.1}$$

The crystallinity of the composite films was investigated by X-ray Diffraction (XRD, Bruker D8 Discover, step size 0.024, scan speed 1.5 s/step, 2θ range from 5 to 70°). Energy Dispersive X-ray spectrometry (EDX, Thermo) was used for the surface elemental analysis of the thin films while the surface morphology and roughness (RMS) were investigated using Scanning Electron Microscopy (SEM, Hitachi model S-3400 N type II) and Atomic Force Microscopy (AFM, NT-MDT model BL222RNTE).

24.3 Results and Discussions

The photocatalytic degradation of MB under high (810 W/m^2) and low (55 W/m^2) irradiance values run for one hour in dark and five hours under irradiation and the results are presented in Fig. 24.2:

As the results show, adsorption took longer time in the dynamic regime (under the MB solution flow). A fast adsorption was observed in the beginning and adsorption took place about one hour longer, after irradiation was started (when no increase in the efficiency could be observed). This may suggest that the irradiated substrate needed a certain period for developing oxidation species in a useful concentration and this period is significantly higher than in the static regime. Thus the wastewater flow may have a wash-out effect on the photocatalytic surface, when the oxidation species produced through the reaction of the holes (and electrons) with the hydroxide anions are faster removed or deactivated and cannot interact with the adsorbed MB molecules.

However, after 135 min. the process in dynamic regime (under high irradiance) runs faster as compared to the process in static regime (at low irradiance value) as the slope of the efficiency graph is higher for the dynamic flow regime.

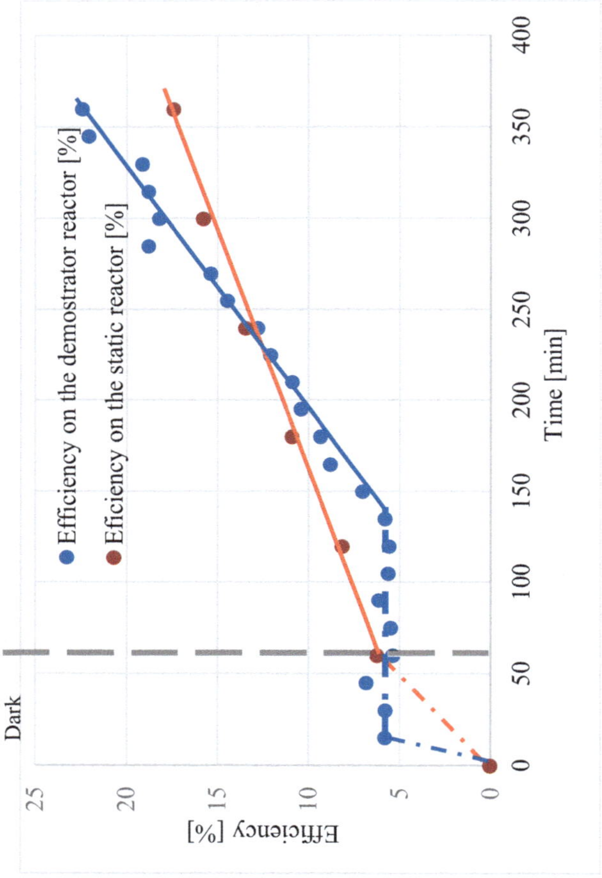

Fig. 24.2 Methylene blue removal efficiency

This was an expected behavior due to the fact that a larger amount of photons is expected to be mirrored by a larger amount of decomposed MB. What is still under a question mark is the effect of the higher irradiance; for an irradiance value about 15 times larger during the dynamic regime as compared with the static regime (810/55 W/m^2) the increase in efficiency is of only 30%. This can be the consequence of several factors, among which the following:

(1) An increased recombination rate of the photo-generated holes and electrons that may interact and deactivate before being involved in chemical reactions in a much broader extend. One way to mitigate this effect is to allow a fast charge separation, e.g. due to an increased crystallinity.
The crystallinity degree of the TiO_2 layer was 37% and the composite, double layer reached a value of 42.5%, proving that the deposition strategy fulfilled its objective and that recombination was limited as much as possible, Fig. 24.3;
(2) A limited amount of pollutant molecules available to be decomposed by the generated oxidation species. The amount of adsorbed MB molecules can be limited by a low surface roughness (a lower surface roughness reduces the amount of adsorption sites) and by the surface charge (that can be tuned considering the point of zero charge of the substrate and the solution pH);
(3) The surface clogging with colored molecules that limits the radiation access to the photocatalyst.

As these results outline, the aqueous flow has various side effects on the process.

Therefore, it is important to assess the stability of the photocatalytic layers during the process by observing the surface roughness before and after the process. The results are included in Fig. 24.4. The SEM images outline a rough surface, with TiO_2

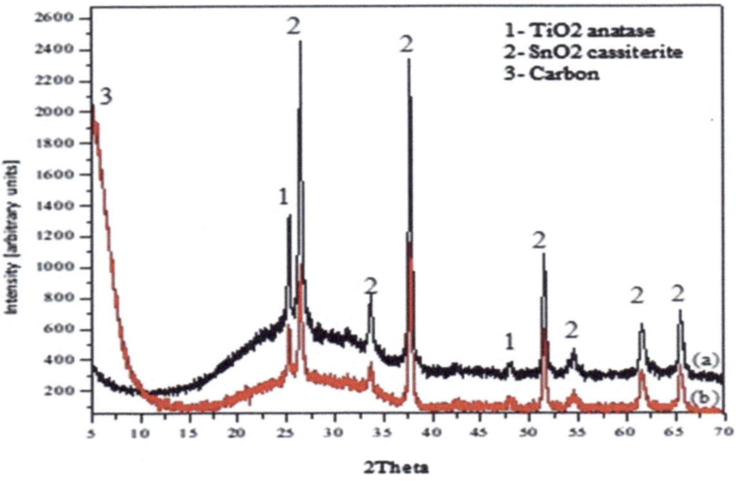

Fig. 24.3 XRD results for **a** the FTO/TiO_2 thin film and **b** for the FTO/TiO_2 / TiO_2-GO composite photocatalytic thin film

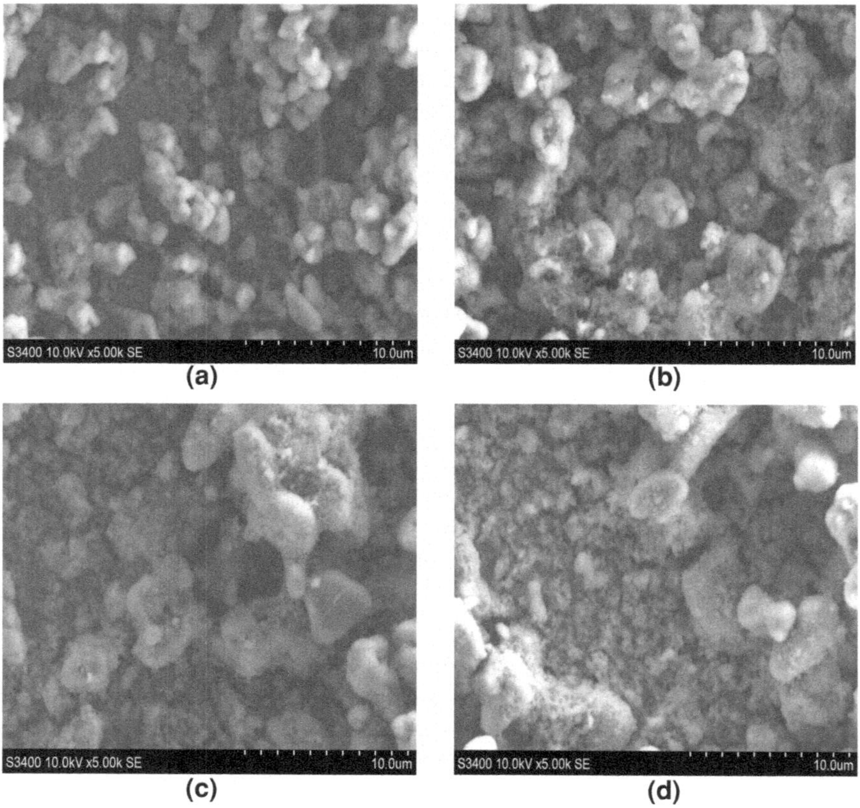

Fig. 24.4 SEM image of the FTO/TiO_2 (SPD)/TiO_2—GO layers: **a** before photocatalysis; **b** after photocatalysis; **c** after photocatalysis and regeneration and **d** in dark

consisting of grains agglomerated in structures adherent to the surface. It is also to be noticed that the layers immersed in water, in dark or under irradiation, are supporting larger agglomerations that are still adherent to the surface.

To quantify the changes in the surface aspect, roughness measurements were done based on AFM images, Fig. 24.5 and the values are inserted in Table 24.1.

As the results show, there is a slight decrease in the roughness values during the five hours of operation under irradiation (of about 13%); a more significant roughness decrease (16%) is observed when the thin films are kept in dark under a water flow for six hours. Basically, in dark or under irradiation, the thin films seem to be more affected when a distilled water flow is employed; the RMS value after the films were exposed to water under irradiation was of 257.2 nm, very close to the value recorded in dark. This proves that the surface re-organization is not likely due to irradiation but can be influenced by the residual photocatalytic by-products adosrbed on the surface.

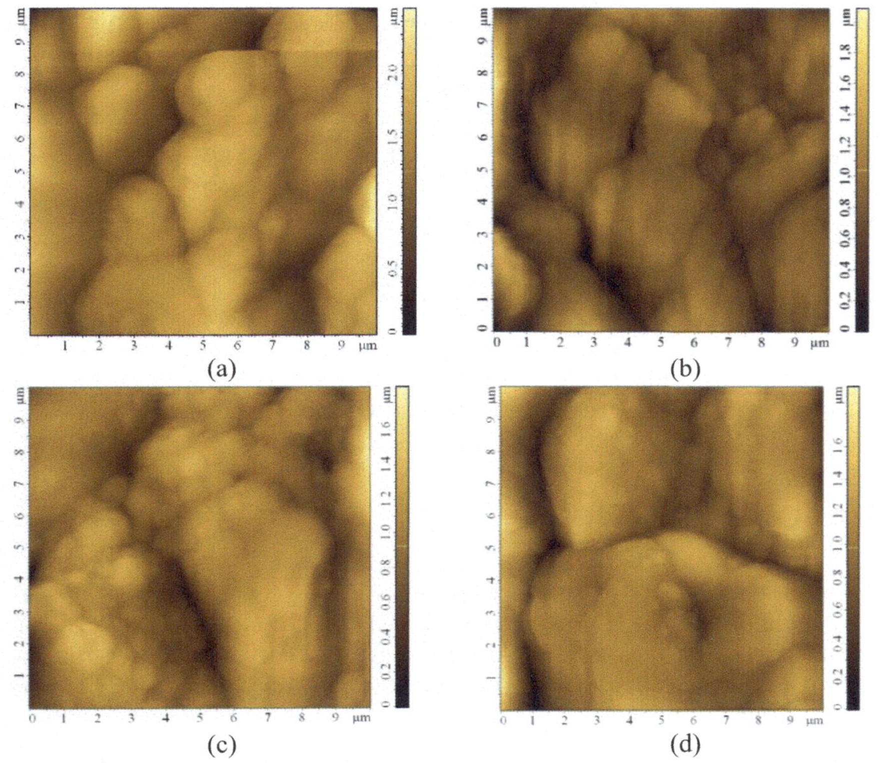

Fig. 24.5 AFM image of the FTO/TiO_2 (SPD)/TiO_2—GO layers: **a** before photocatalysis; **b** after photocatalysis; **c** after photocatalysis and regeneration and **d** in dark

Table 24.1 RMS values of the FTO/TiO_2 (SPD)/TiO_2—GO layers

Sample	Before photocatalysis	After photocatalysis	After regeneration	In dark
RMS [nm]	313.1	272.5	248.0	261.8

The roughness variation may be the results of surface re-organizations as the EDX results show, Table 24.2.

The rather significant decrease in the average carbon content can be the result of the TiO_2 grains that may more densely stuck over the GO plates during operation under the liquid flow. This effect is obvious in the last sample (in dark), where the FTO layer is no longer recorded but, as it cannot be removed from the substrate, one may conclude that the denser coverage remains the only likely option.

The transmittance of the layers can be also of interest when thinking at extended applications as self-cleaning coatings. The transmittance for a radiation with the wavelength of 600 nm was, for the TiO_2 photocatalytic layer of 78 and of 35% for the composite thin film.

Table 24.2 Elemental composition of the photocatalytic surface of the FTO/TiO_2 (SPD)/TiO_2–GO layers

Sample	Elemental composition				
	C [%]	Ti [%]	O [%]	F [%]	Sn, Si, S [%]
Before photocatalysis	6.48	31.58	57.43	0.51	3.98
After photocatalysis	2.00	34.14	61.3	0.23	1.53
After regeneration	1.52	34.1	61.01	0.61	2.75
In dark	1.70	33.15	62.76	0.00	2.01

It can thus be observed that, despite the low GO content, the transmittance is significantly decreased while the photocatalytic response in UV–VIS is much improved and all the other investigated properties have an observable variation.

24.4 Conclusions

Composite thin layers of FTO/TiO_2/TiO_2–GO type were obtained by using up scalable deposition techniques. The thin films proved to be VIS-active and stable during operation. The films had a better decomposition efficiency of methylene blue as compared to the mono-layered TiO_2 thin films but the results did not linearly follow the increase in the irradiance value as a potential result of the increased recombination. Moreover, as the results show, the photocatalytic mechanism is controlled by the adsorption step.

Acknowledgements This work was supported by a grant of the Romanian Ministry of Research and Innovation, CCCDI-UEFISCDI, project number PN-III-P1-1.2-PCCDI-2017-0619, contract no. 42 PCCDI/2018, within PNCDI.

References

1. F. Tisa, A.A. Raman, W.M. Daud, Applicability of fluidized bed reactor in recalcitrant compound degradation through advanced oxidation processes. A review. J. Environ. Manag. **146**, 260–275 (2014)
2. M. Gong, Y. Li, H. Wang, Y. Liang, J.Z. Wu, J. Zhou, J. Wang, T. Regier, F. Wei, H. Dai, An advanced Ni–Fe layered double hydroxide electrocatalyst for water oxidation. J. Am. Chem. Soc. **135**, 8452–8484 (2013)
3. C. Prasse, D. Stalter, U. Schulte-Oehlmann, J. Oehlmann, T.A. Ternes, Spoilt for choice: a critical review on the chemical and biological assessment of current waste water treatment technologies. Water Res. **87**, 237–270 (2015)
4. S. Zhu, D. Wang, Photocatalysis: basic principles diverse forms of implementations and emerging scientific opportunities. Adv. Energy Mater. **7**, 1700841 (2017)

5. M. Covei, C. Bogatu, D. Perniu, I. Tismanar, A. Duta, Comparative study on the photodegradation efficiency of organic pollutants using n–p multi-junction thin films. Catal. Today **328**, 57–64 (2019)
6. A. Tayel, A.R. Ramadan, O. El Seoud, Titanium dioxide/graphene and titanium dioxide/graphene oxide nanocomposites: synthesis. Charact. Photocatalytic Appl. Water Decontamination, Catal. **8**, 491 (2018)
7. R. Wang, K. Shi, D. Huang, S. Zhang, S. An, Synthesis and degradation kinetics of TiO_2/GO composites with highly efficient activity for adsorption and photocatalytic degradation of MB. Sci. Rep. **9**, 18744 (2019)
8. S. Sheshmani, M. Nay, Modification of TiO_2 with graphene oxide and reduced graphene oxide; enhancing photocatalytic activity of TiO_2 for removal of Reamazol black B. Polym. Compos. DOIP1O0L.Y10M0E2R/pcC (2019)
9. ***INTERNATIONAL STANDARD ISO 10678:2010, Fine ceramics (advanced ceramics, advanced technical ceramics)—Determination of photocatalytic activity of surfaces in an aqueous medium by degradation of methylene blue

Chapter 25
Challenges in the Up-Scaled Deposition of Self-Cleaning Thin Films on PV Modules

Maria Covei, Anca Duta, and Ion Visa

Abstract Self-cleaning thin films are applied as top coatings (usually deposited during the manufacturing stage) on the solar glass of photovoltaic (PV) modules. This prevents dust and grime accumulation on the glazing and maintains a good solar radiation transmittance to the PV cells. Due to weathering, the self-cleaning coating can be partially destroyed over time and in-situ re-application on the already installed PV modules may be a viable option. The requirements for the self-cleaning coating are: low-cost materials, easy deposition at ambient temperature, high transparency in the UV–VIS range, high photocatalytic efficiency to degrade organic pollutants using the solar radiation and low contact angle to wash off possible by-products of the photocatalytic process and/or various other pollutants stacked on the surface. These requirements were reached at laboratory level on 1.5×1.5 cm^2 solar glass substrates, using a TiO_2–WO_3–rGO composite thin film obtained using a stable dispersion of sol–gel powder. When scaling up to 20×30 cm^2 surface, thin film homogeneity becomes difficult to maintain. The deposition conditions (spraying area, duration) were re-optimized to support the uniformity of the chemical, morphological, optical, photocatalytic and wetting properties.

Keywords Self-cleaning thin films · TiO_2–WO_3–rGO composite · Spray deposition

---

M. Covei (✉) · A. Duta · I. Visa
Transilvania University of Brasov, R&D Centre: Renewable Energy Systems and Recycling Romania, Brasov, Romania
e-mail: maria.covei@unitbv.ro

A. Duta
e-mail: a.duta@unitbv.ro

I. Visa
e-mail: visaion@unitbv.ro

I. Visa and A. Duta (eds.), *Solar Energy Conversion in Communities*,
Springer Proceedings in Energy, https://doi.org/10.1007/978-3-030-55757-7_25

25.1 Introduction

Self-cleaning thin films were successfully deposited for various applications ranging from clothing to cars, pavements, built environment or the energy sector. The last application targets the use of self-cleaning thin films deposited on the glazing of the PV modules or of the solar-thermal collectors so that dust and dirt will not be collected during in-field functioning. The grime removal process significantly depends on the layer's wetting properties. There are two main self-cleaning mechanisms:

(a) Wash-off the pollutant, supported by super-hydrophobic coatings (water contact angle larger than 150°) that promote the water droplets roll off on the surface, carrying the dust particles with them; the pollutant is removed but not degraded. Polymer and silicon coatings act in this way [1];
(b) Pollutant's degradation, supported by super-hydrophilic coatings (water contact angle lower than 10°) that promote the formation of a water layer on top of the glazing. Soluble, inorganic pollutants are dissolved and removed, while organic pollutants are degraded through photocatalysis. In this case, water acts both as removal medium and as source of reactive OH· radicals. Semiconductor materials with UV or, ideally, Vis-activation are required for these coatings.

Out of the two mechanisms, the latter has the advantage of complete (mineralization) or partial decomposition of the organic pollutants and is therefore more attractive. However, in order to be applicable on PV glazing, the coatings also require special optical properties, mainly high transmittance and low reflectance in order to allow a large share of the solar spectrum (UV and Vis) to pass through the glazing and reach the PV cells. The anti-reflective properties can be optimized by tailoring the morphology and the thickness of the thin film coating.

Another important property of the PV module coatings is their stability. As these films are subjected to weathering, their lifetime is often less than that of the module itself (usually 20 years or more). In order not to decrease the power output due to fouling, the re-application of the self-cleaning, anti-reflective thin film may be an option. This requires a low-cost, ambient temperature, facile deposition method that will also allow large-scale coverage. Considering that most commercial modules contain glazing that was treated at high temperature [2], it is unrealistic to expect the same method to be used for re-application. The use of deposition methods such as spraying is a much more attractive alternative as it meets all the criteria above mentioned.

Some of the materials reported for self-cleaning and anti-reflective applications are metal oxides, such as TiO_2, ZnO, WO_3, etc.[3–5]. These are wide bandgap semiconductors with proved photocatalytic activity, which can be obtained as thin films through a wide variety of deposition techniques that allow tailoring the structural and morphological properties to get the required wetting and optical properties. However, one of the main drawbacks when using wide bandgap semiconductors is their UV-activation, considering that UV represents only 5–10% of the solar spectrum, [6].

Different methods for extending the activation range towards Vis are reported: anion or cation doping, metal loading or developing heterostructures with narrow bandgap materials [7, 8]. Lately, coupling with carbon 2D derivatives such as graphene, graphene oxide or reduced graphene oxide was also reported, [9], as these materials may act as electron scavengers next to the metal oxides (thus decreasing the electron–hole recombination) and as adsorption sites for the pollutant molecules.

Titanium dioxide is one of the mostly reported photocatalysts, being highly hydrophilic under UV-irradiation, [10], and with great aqueous stability. Tungsten oxide is a versatile material that can be obtained with different nanostructures, such as spheres, dendrites, dumbbells, etc., [11]. It is also a Vis-active photocatalyst, with a bandgap of 2.7 eV, [12]. The composite material obtained by coupling TiO_2, WO_3 and rGO, was reported to have a higher photocatalytic efficiency in the degradation of 2,4-dichlorphenoxyacetic acid compared to the individual oxides, under UV radiation, [13]. A ternary compound obtained from TiO_2 nanocrystals and WO_3 nanorods on rGO nanosheets, synthesized using the hydrothermal method, was confirmed as a better photocatalyst than the binary TiO_2-WO_3 composite due to the interfacial electron transfer promoted by rGO that suppresses the electron–hole recombination and boosts the O_2 reduction reactions during the photocatalytic process, [14]. Another hydrothermal TiO_2/WO_3/GO photocatalyst was reported to display excellent photocatalytic activity for the degradation of bisphenol A when irradiated using Vis and solar radiation, as well as for the photodegradation of methylorange under UV radiation, [15]. Still, results on these composites as a self-cleaning layer, where both the photocatalytic activity and the wetting and the optical properties are important, are scarce.

Moreover, the majority of the reports are on small-scale films that are tested in laboratory conditions, simply to validate the deposition method. It is well known that up-scaling to larger thin film deposition area requires a re-optimization of the deposition parameters. Ideally, this should be done with as little changes of the original recipe as possible. Some reports on either hydrophobic or hydrophilic self-cleaning coatings for PV-modules mention screen printing or chemical vapor deposition as viable deposition methods [16–18].

25.2 Materials and Methods

The coatings were obtained following a powder—dispersion—thin film route, as this ensures a good control of the structural and morphological properties of the material without requiring high temperature in the deposition step. This makes the method attractive for the (re)application of the films on already installed PV modules.

25.2.1 Powder Preparation

Two metal oxide powders were obtained using the sol–gel method, as follows:

- The TiO_2 powder was synthesized by mixing titanium tetraisopropoxide (TTIP, Sigma Aldrich, 97%), Ethanol (Chemical, 99.3%), Acetylacetone (AcAc, Scharlau, 99%), Acetic acid (Scharlau, 99%) and water in the volume ratio TTIP:EtOH:AcAc:HAc:H_2O = 10:8:0.445:0.09:0.6. Aging the sol for 24 h in ambient atmosphere was followed by drying it for 1 h at 100 °C. To obtain a crystalline powder, annealing was performed at 550 °C for 3 h.
- The WO_3 powder was obtained usingWCl_6 (Acros Organics, 99%) and Ethanol (Chemical, 99.3%) to get a dispersion of 39.7 g/L. The same aging, drying and annealing steps were applied to the WO_3 material, as discussed above for TiO_2.

To obtain the TiO_2-WO_3-rGO composite powder, the dried and annealed metal oxide powders were mixed together and then the rGO commercial dispersion (Sigma Aldrich, 10 mg/mL) was added to obtain a one-batch mixed powder with the gravimetric ratio TiO_2:WO_3:rGO = 50:50:1, using magnetic stirring, followed by ultrasonication under heating at 100 °C, until completely dried. The powder was milled (MLW, KM 1), for 24 h to obtain the fine particles required for the dispersion preparation.

25.2.2 Dispersion Preparation

The dispersion was obtained by mixing the composite powder with a water: ethanol mixture (3:1) to get a 5% concentration. Preliminary studies showed that a higher concentration leads to a fast destabilization of the dispersion, whereas a lower concentration will not support the thin film homogeneity in the next step (deposition on solar glass). Furthermore, to be in-field sprayed, the dispersion should be stable for at least 30 min. Therefore, sodium dodecylsulphate (SDS, synthesis grade, Alfa Aesar) was used as stabilizing agent in a100 ppm concentration in the continuous medium.

The dispersion was magnetically stirred for 30 min, followed by ultrasonication (ultrasound bath SCIENTZ SB5200) for 3 h. The dispersion stability was checked by recording the UV–Vis transmittance spectra (UV–Vis-NIR spectrophotometer Perkin Elmer) in the Vis range (400–800 nm) immediately after ultrasonication and after 30 min. A transmittance increase suggests a destabilization of the dispersion with the dispersed phase segregating at the bottom of the flask.

25.2.3 Thin Film Deposition

(a) Deposition on small-scale substrates
Small-scale substrates (1.5 × 1.5 cm^2) were used to preliminary test the self-cleaning, anti-reflecting properties of the thin films at laboratory level. The deposition conditions were optimized considering the structure, morphology, optical, wetting and photocatalytic properties.
The stable dispersion was sprayed onto the substrate heated at 80 °C using 10 spraying sequences with 60 s break between them to allow the evaporation of the continuous medium. The carrier gas (air) pressure was 1.2 bars and the distance between the nozzle and the heated substrate was set at 20 cm.

(b) Deposition on relevantly-sized substrates
A larger scale, relevant surface was considered with an area of 20 × 30 cm^2 that corresponds to a commercial PV polycrystalline Si module of about 10 W that can be used to power small electronics such as a mobile phone, a radio, some light bulbs, etc.

The control property in this case was the homogeneity of the deposited layer.

The ABB Foundry Plus industrial robot covers the substrate surface by spraying sequential lines, as illustrated in Fig. 25.1. The spraying cone dimensions are influenced by the dispersion concentration, the spraying height and the carrier gas pressure. The homogeneity of the thin film is correlated with the distance between the spraying lines (s) and the break (t) between two consecutive spraying sequences. To check the homogeneity of the film, two areas were investigated and compared to the small-scale thin film: an area in the center of the cone and an area at the edge of the cone.

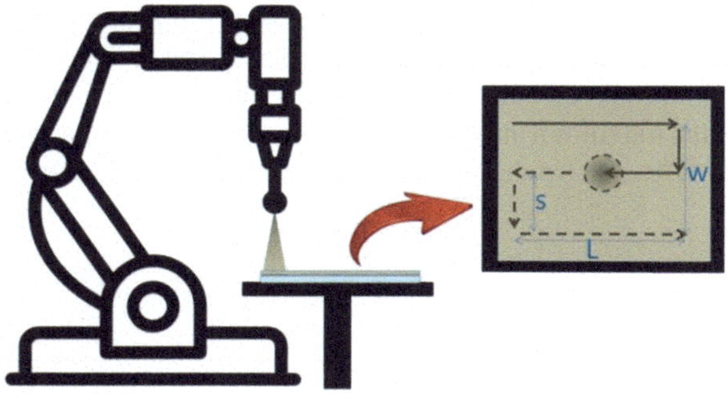

Fig. 25.1 A schematic representation of the spraying equipment and the deposition method of multifunctional, composite thin films on relevantly sized substrates

25.2.4 Thin Film Characterization and Testing

The crystallinity of the powders and of the thin films was investigated using a Bruker D8 Discover X-ray Diffractometer ($CuK_{\alpha 1}$ = 1.5406 Å, step size 0.025, scan speed 1.5 s/step, 2θ = 5–70°). Scanning electron microscopy (SEM) used a Hitachi SEM S-3400 N type 121 II device coupled with an Energy Dispersive X-ray (EDX) unit (Utltra Dry, Noran System7, NSS Model). Surface topology was analyzed through atomic force microscopy (AFM) using an NT-MDT microscope, model NTGRA PRIMA EC, working in semi-contact mode with Si-tips, NSG10. Roughness was calculated using the AFM software over a 5 μm × 5 μm surface.

The optical properties of the thin layers, as well as the dispersions' stability, were tested using UV–Vis spectroscopy (Perkin Elmer Lambda 950). The contact angle measurements used an OCA20 instrument and water as the testing liquid.

The composite films were considered to have good self-cleaning properties if they possess good wetting behavior and promising photocatalytic efficiency in the degradation of methylene blue under simulated solar radiation. For the photocatalytic tests, two UV radiation sources (UVA, 340–400 nm, $\lambda_{UV,max}$ = 365 nm, Philips) and five Vis light sources (TL-D Super 80 18 W/865, 400-700 nm, λ_{Vis}, max = 565 nm, Philips), with a total average irradiance of G = 34 W/m^2 were used. The UV share in the radiation is of 10% to resemble solar radiation profile at a significantly lower irradiance value.

The photocatalytic experiments used a methylene blue (MB, 99.8%, Merck) solution with the initial concentration of 10 ppm. This specific pollutant was selected, following the ISO 10678:2010 standard (*Determination of photocatalytic activity of surfaces in an aqueous medium by degradation of methylene blue*). The thin films (with a surface of 1.5 × 1.5 cm^2) were immersed in 20 mL aqueous pollutant solutions. Before irradiation, the pollutant solution with the immersed thin film photocatalyst was kept 1 h in dark to reach the adsorption/desorption equilibrium. The photodegradation efficiency, η, was calculated based on the initial absorbance of the pollutant solution (A_0) and the absorbance after 1, 2, 4, 6 and 8 h of irradiation (A), recorded at the maximum absorbance wavelength of the pollutant (λ = 664 nm), using a UV–Vis-NIR spectrophotometer (Perkin Elmer Lambda 950) and applying Eq. (25.1):

$$\eta = \frac{A_0 - A}{A_0} \times 100 \tag{25.1}$$

25.3 Results and Discussion

The TiO_2–WO_3–rGO powder was investigated using both XRD and EDX analyses to confirm the rGO incorporation in the composite. The XRD results outline a small peak at 2θ ~ 5° that can be allocated to rGO. The main peak of rGO, at 2θ ~ 25° overlaps with the more pronounced peaks of both anatase and WO_3. The main XRD peaks corresponding to the powder are inserted in Table 25.1.

The EDX analysis (Table 25.2) confirms that rGO is part of the composite as a C content of 3–22% was identified. It is obvious from the large amount of carbon (over 20%) that some residual organic traces from the precursor system remained in the powder, in agreement with the ~30% amorphous phase identified through the XRD analysis. It is also important to notice that the morphology of the composite powders preserves aspects of the individual TiO_2 and WO_3 powders. The TiO_2 powder is rougher and, even after milling, some larger crystallites can be observed, whereas the WO_3 powder is much finer, composed of spherical grains that mix together well with the TiO_2 ones. The EDX results in Table 25.2 show that rGO is likely in all the investigated points, with the rGO platelets covered with metal oxides (e.g. points 2 or 3).

The good mixing of the three components is important for preventing any segregation in the dispersion step. The stability of the dispersion to be used for the sprayed film was investigated using transmittance measurements in the Vis range. Two spectra were considered:

Table 25.1 Main peaks allocation in the X-ray diffractogram of the TiO_2-WO_3-rGO powder

2θ [°]			
TiO_2-anatase	TiO_2-rutile	WO_3-monoclinic	rGO
34.0, 37.7, 48.2, 53.8, 54.9, 62.7	36.7, 48.3	31.2, 31.6, 32.6, 35.4, 38.2, 44.0, 44.9, 46.7, 49.5, 54.7, 59.2, 61.4, 62.5, 64.8, 65.6, 69.4, 70.1, 70.9, 72.2, 77.7, 80.2	5.0, 24.6, 45.2

Table 25.2 Elemental composition in the composite powders [$\%_{at}$]

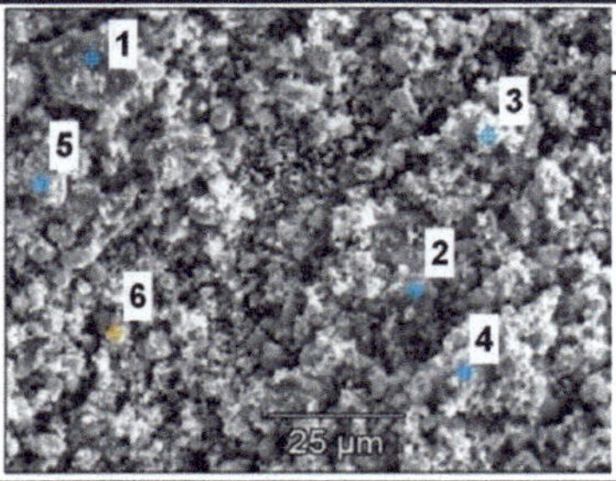

Point	Ti	W	O	C
1	48.24	0.24	48.08	3.43
2	24.38	20.60	41.76	13.27
3	8.36	9.72	59.82	22.09
4	28.08	0.67	65.29	5.96
5	5.90	17.76	59.81	16.53
6	3.71	13.10	65.36	17.83

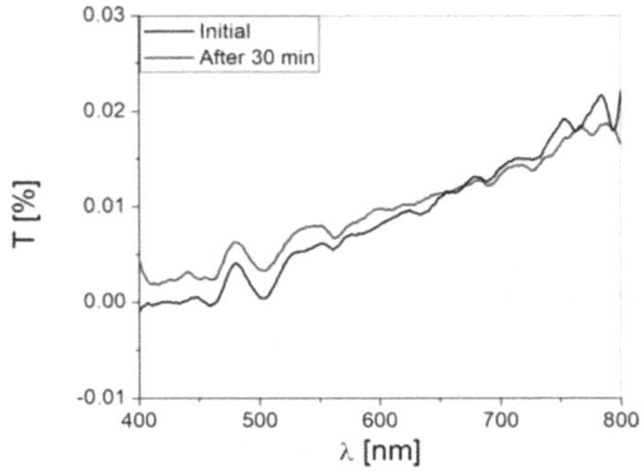

Fig. 25.2 Dispersion transmittance measurement in the Vis spectral range, at the initial moment and after 30 min rest

(1) the initial transmittance values, that should be as low as possible if there is a good distribution of the powder particles in the continuous medium.
(2) the transmittance values after 30 min rest, that should also be as low as possible to indicate the stability of the dispersion during the spraying duration.

As the results in Fig. 25.2 show, both the initial and final transmittance values are very low outlining that the powders were well-dispersed in the water–ethanol mixture. The use of SDS as stabilizing agent was a successful choice, as the system opacity remained almost unchanged for the duration of the investigated rest time (30 min).

The dispersion was further used to deposit homogeneous thin films with anti-reflective and self-cleaning properties on solar glass substrates with small and with relevant sizes. The optimized deposition conditions are inserted in Table 25.3. The step size (s), the spraying line length (L) and width (w) are illustrated in Fig. 25.1.

The following deposition parameters were kept constant: the number of spraying sequences (n = 10), the substrate temperature (T = 80 °C), the distance between the nozzle and the substrate (d = 20 cm) and the carrier gas pressure (P = 1.2 bar).

Up to 20 small-size samples could be arranged on one spraying line, with full coverage of the surface in a single step. Due to the small area of the substrate, the 60 s break between the consecutive spraying sequences was enough to allow the evaporation of the residual continuous medium of the sprayed dispersion. However, when the larger substrate was used, the spraying area needed to be extended (see Table 25.3). The step size was varied between 30 and 60 mm. It was found that at a smaller step size (30 or 40 mm), the sprayed lines significantly overlap and the thin film becomes too thick/opaque and uneven. However, if the step size is 60 mm, the

Table 25.3 Optimized deposition conditions for the small sized and for the relevantly sized thin films

Thin film area	Step size [mm]	Spraying line length [mm]	Spraying line width [mm]	Break between sequences [s]
1.5×1.5 cm^2	–	320	–	60
20×30 cm^2	50	320	220	300

substrate is unevenly covered, with a "zebra-shaped" aspect. Therefore, the step size was fixed at 50 mm, which led to 6 steps required to cover the entire surface.

The large substrate area requires a longer time to fully remove/evaporate the residual continuous medium left after each spraying sequence. Therefore, the break time was increased up to 300 s. If smaller breaks are used, the ongoing spraying sequence will wash off the remaining "wet patches" deposited during the previous sequence, leading to "waves" with different thickness in the thin film.

The optimized parameters included in Table 25.3 led to thin films that appeared to be homogeneous; morphological, wetting and photocatalytic investigations were performed on two different areas of the film (at the edge of the spraying cone and in its center) and the results were compared with the values recorded on the small substrates.

The morphology of the films was investigated through AFM and SEM analyses. The results in Fig. 25.3 indicate the complete coverage of the substrate in all three

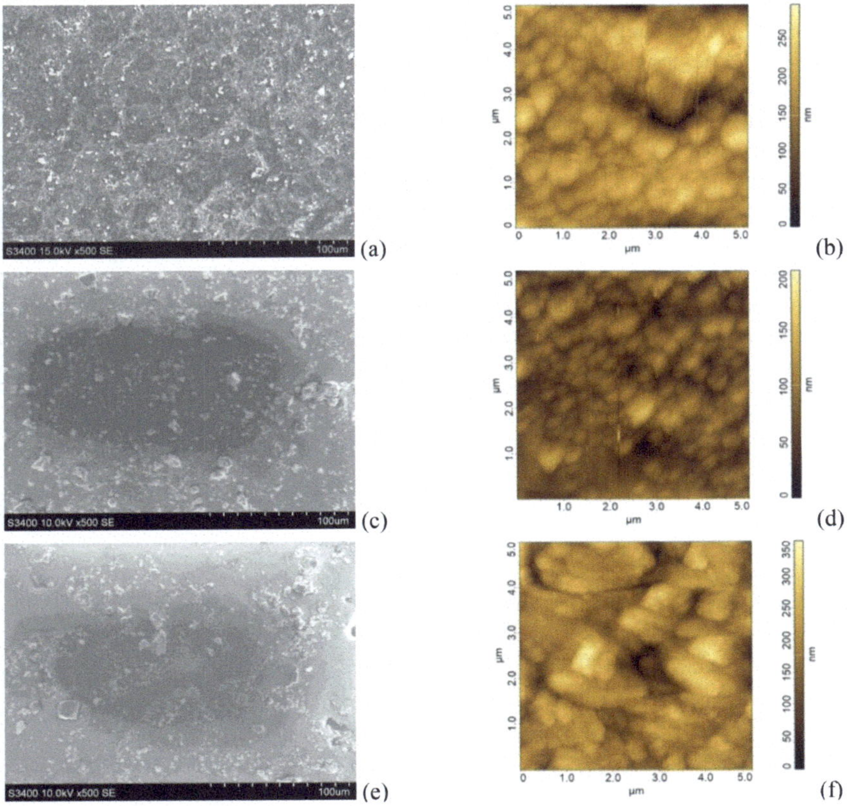

Fig. 25.3 **a**, **c**, **e** SEM and **b**, **d**, **f** AFM images of the **a**, **b** small-sized thin films as well as the **c**, **d** center-cone and **e**, **f** edge-cone areas of the relevantly-sized thin film

cases, with a thin continuous film that exhibits small crystallites that adhered to the surface. These can be correlated with the larger aggregates identified through SEM analyses in the TiO_2–WO_3–rGO powders, before being dispersed.

These structures support the photocatalytic process, as they increase the surface area and therefore the available area for the pollutant molecules adsorption. It is also interesting to note, that the aggregates were still present on the surface after 9 h of immersion, during the photocatalytic process, thus suggesting their relatively good adherence. There is a good similarity between the small-sized and the center-cone surfaces, while the cone-edge surface seems to be slightly less peppered with aggregates, thus suggesting a slightly thinner film.

The cone-edge AFM image also outlines the rGO platelets covered with metal oxide, as result of the variation in the weight of the composite grains containing rGO, significantly lighter than the grains containing only metal oxide, thus being able to be deposited at the edge of the deposition cone. The roughness values for the three areas, as resulted from the AFM measurement are inserted in Table 25.4, outlining that all these are similar and typical for the thin films obtained from powder dispersions. The surface aggregates and pores that contribute to the roughness support the wetting behavior of the coatings. An extended contact between water and the composite is essential in the photocatalytic process, thus contact angles close to (or smaller than) 10° are required. As the results in Table 25.4 show, the contact angles in all three cases are close to 11°, proving the material as a promising candidate for self-cleaning applications. The very small difference between the cone-edge and cone-center surfaces is also an indicator of the film homogeneity. This aspect has also been confirmed by the similar chemical composition of the two areas on the relevantly sized films. The elemental composition results also outline that the films are very thin, as the substrate elements are the components with the highest percentage. Moreover, the carbon percentage is higher than that of the metal (Ti and/or W) oxide, thus the rGO platelets are part of the composite, close to the surface and covered with a very thin metal oxide layer.

Table 25.4 Mean square roughness (RMS), water contact angle and surface elemental composition for the small-size and relevantly-sized thin films

Area	RMS[nm]	θ [°]	Elemental composition [%]				
			Ti	W	O	C	Substrate elements[a]
Small-sized	42.2	11.3	0.58	0.48	66.41	3.42	29.11
Cone-center	40.5	11.0	0.64	0.45	67.41	3.26	28.24
Cone-edge	44.1	11.8	0.46	0.33	62.66	2.98	33.57

[a]Subtrate elements: Si, Na, Ca

Fig. 25.4 Methylene blue photodegradation efficiency of the small-sized and relevantly-sized thin films

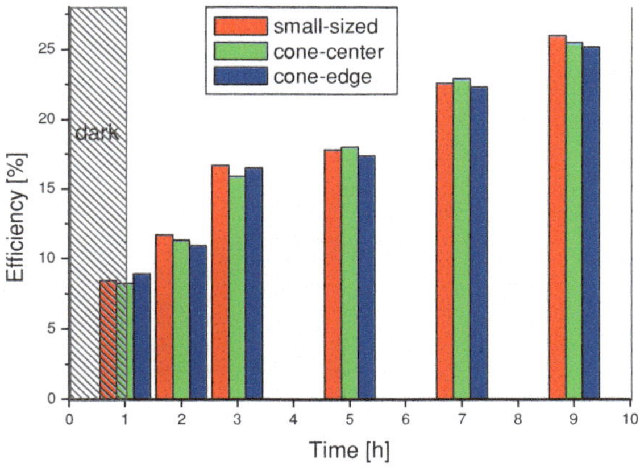

The other requirement for self-cleaning thin films is their ability to degrade organic pollutants under solar irradiation. Thus, the three samples were tested in the photocatalytic degradation of the methylene blue dye, under solar-simulated radiation (Fig. 25.4). The results show that a good adsorption of the dye occurs during the first hour of contact (in dark) between the photocatalyst and the dye solution. The photocatalytic process runs similarly on the small-sized film and on the relevantly-sized sample. Small variations in the photocatalytic efficiency can be expected due to the slight differences in the films surface roughness and contact angle. However, the general trend of the degradation process is similar and confirms that the films maintained their self-cleaning properties.

In the experimental conditions, the overall efficiency values are rather low due to the low irradiance value (34 W/m^2). Once solar radiation ($G = 800 - 1000\ W/m^2$) will be used, the number of generated electron–hole pairs increases and more OH· radicals are formed at the photocatalyst surface, so the overall photodegradation efficiency is expected to increase.

Another control property is the optical performance of the films. The samples need to be sufficiently transparent and anti-reflective in the UV–Vis range to allow the radiation to penetrate the glass casing to the PV cells bellow. As the results in Fig. 25.5 show, high transmittance values (over 80%) were recorded over the Vis and UV range ($\lambda > 320$ nm), making these thin films acceptable as PV casing coatings. As the composite material contains two (rather) broad bandgap semiconductor oxides, low transmittance values are expectable in the far UV range ($\lambda < 320$ nm). Reflectance values lower than 20% were recorded for all samples and this is in line with the requirements for anti-reflective coatings for PV casings. Considering the differences in the surface roughness and the overall morphology, some slight differences in the optical properties between the cone-center and cone-edge samples are expected. However, as the transmittance values are nearly overlapping, it can be concluded that no significant thickness variation between the two areas occurs.

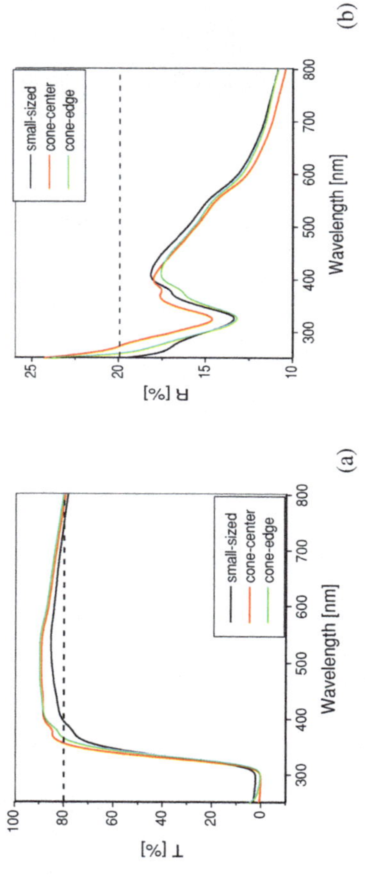

Fig. 25.5 **a** Transmittance and **b** reflectance spectra of the small size and relevantly sized thin films in the UV–Vis range

The values for all the control properties are similar for the relevantly-sized and the small size films and this outlines that, in spite of the changes in the deposition conditions, the re-optimized up-scaling procedure did not affect the significant properties of the thin films. Good self-cleaning and anti-reflective coatings were obtained on the $20 \times 30\ cm^2$ substrate and are recommended for further advancement to in-field deposition on commercial PV modules.

25.4 Conclusions

Composite TiO_2-WO_3-rGO thin films were deposited using a powder—dispersion –spraying route, on small ($1.5 \times 1.5\ cm^2$) and relevantly-sized ($20 \times 30\ cm^2$) solar glass substrates. The morphological, chemical, optical, photocatalytic and wetting properties of the thin films were investigated to find out whether they can be used as self-cleaning, anti-reflective coatings to be applied on in-field PV-modules. The deposition conditions were re-optimized for the larger-scale films, aiming at preserving the high transmittance and low reflectance in the UV–Vis range, a low contact angle and high photocatalytic efficiency in methylene blue decomposition, under simulated solar radiation. To test the homogeneity of the larger size films, two areas were compared: at the center and at the edge of the spraying cone while the effectiveness of the up-scaling process was discussed by comparing the control properties of these two areas with that of the small-sized film. Similar morphologies and chemical composition in the three samples led to similar and very good optical properties ($T > 80\%$, $R < 20\%$), wetting properties ($\theta \sim 11°$) and photocatalytic properties ($\eta > 20\%$ at an irradiance value of $35\ W/m^2$).

It can be thus concluded that the films obtained on relevantly sized substrates preserve the self-cleaning and anti-reflective properties, are homogeneous and can be considered for further deposition on PV modules in field installed.

Acknowledgements This work was supported by the grant of the Ministry of Research and Innovation, CNCS-UEFISCDI, project number PN-III-P1-1.1.-PD-2016-0289, contract no. 78/2018.

References

1. A. Syafiqa, A.K. Pandey, N.N. Adzman, N.A. Rahim, Advances in approaches and methods for self-cleaning of solar photovoltaic panels. Sol. Energy **162**, 597–619 (2018)
2. K. Midtal, B.P. Jelle, Self-cleaning glazing products: A state-of-the-art review and future reserch pathways. Sol. En. Mat. Sol. Cells **109**, 126–141 (2013)
3. S. Banerjee, D.D. Dionysiou, S.C. Pillai, Self-cleaning applications of TiO_2 by photo-induced hydrophilicity and photocatalysis. Appl. Cat. B: Env. **176–177**, 396–428 (2015)
4. B.J. Li, L.J. Huang, N.F. Ren, X. Kong, Y.L. Cai, J.I. Zhang, Superhydrophobic and anti-reflective ZnOnanorod-coated FTO transparent conductive thin films prepared by a three-step method. J. Alloys Compd. **674**, 368–375 (2016)
5. A. Pan, H. Lu, L.Z. Zhang, Experimental investigation of dust deposition reduction on solar cell covering glass by different self-cleaning coatings. Energy **181**, 645–653 (2019)
6. R. Vinu, G. Madras, Environmental remediation by photocatalysis. J. Indian Inst. Sci. **90**, 189–230 (2010)
7. B. Ohtani, Photocatalysis A to Z—What we know and what we do not know in a scientific sense. J. Photochem. Photobiolog. C: Photochem. Rev. **11**, 157–178 (2010)
8. A. Duta, L. Andronic, A. Enesca, The influence of low irradiance and electrolytes on the mineralization efficiency of organic pollutants using the Vis-active photocatalytic tandem $CuInS_2/TiO_2/SnO_2$. Catal. Tod. **300**, 18–27 (2018)
9. F.W. Low, C.W. Lai, Recent developments of graphene-TiO_2 composite nanomaterials as efficient photoelectrodes in dye-sensitized solar cells: a review. Renew. Sust. Energ. Rev. **82**, 103–125 (2018)
10. M. de Jesus, G. Timò, C. Agustín-Sáenz, I. Braceras, M. Cornelli, F.A. de Mello, Anti-soiling coatings for solar cell cover glass: climate and surface properties influence. Sol. Energy Mater. Sol. Cells **185**, 517–523 (2018)
11. M. Govender, L. Shikhwambana, B.W. Mwakikunga, E. Sideras-Haddad, R.M. Erasmus, A. Forbes, Formation of tungsten oxide nanostructures by laser pyrolysis: stars, fibres and spheres. Nanoscale Res. Lett. **6**, 1–8 (2011)
12. M. Parthibavarman, M. Karthik, S. Prabhakaran, Facile and one step synthesis of WO_3 nanorods and nanosheets as an efficient photocatalyst and humidity sensing material. Vacuum **155**, 224–232 (2018)
13. V. Iliev, D. Tomova, L. Bilyarska, Promoting the oxidative removal rate of 2,4-dichlorophenoxyacetic acid on gold-doped WO_3/TiO_2/reduced graphene oxide photocatalysts under UV light irradiation. J. Photochem. Photobiolog. A: Chem. **351**, 69–77 (2018)
14. X. Zeng, Z. Wang, G. Wang, T.R. Gengenbach, D.T. McCarthy, A. Deletic, J. Yu, X. Zhang, Highly dispersed TiO_2 nanocrystals and WO_3 nanorods on reduced graphene oxide: Z-scheme photocatalysis system for accelerated photocatalytic water disinfection. Appl. Cat. B: Environ. **218**, 163–173 (2017)
15. X. Hao, M. Li, L. Zhang, K. Wang, C. Liu, Photocatalyst TiO_2/WO_3/GO nano-composite with high efficient photocatalytic performance for BPA degradation under visible light and solar light illumination. J. Ind. Eng. Chem. **55**, 140–148 (2017)
16. Y. Liu, A. Das, Z. Lin, I.B. Cooper, A. Rohatgi, C.P. Wong, Hierarchical robust textured structures for large scale self-cleaning black silicon solar cells. Nano Energy **3**, 127–133 (2014)
17. S. Maharjan, K.S. Liao, A.J. Wang, K. Barton, A. Haldar, N.J. Alley, H.J. Byrne, S.A. Curran, Self-cleaning hydrophobic nanocoating on glass: a scalable manufacturing process. Mat. Chem. Phys. **239**, 122000 (2020)

18. J.S. Appasamy, J.C. Kurnia, M.K. Assadi, Synthesis and evaluation of nitrogen-doped titanium dioxide/single walled carbon nanotube-based hydrophilic self-cleaning coating layer for solar photovoltaic panel surface. Sol. En. **196**, 80–91 (2020)

Chapter 26
Hydrophilicity Variation of TiO_2—Graphene Oxide Composite Thin Films for Photocatalytic Applications

Ioana Tismanar, Alexandru Cosmin Obreja, Octavian Buiu, and Anca Duta

Abstract Novel photocatalytic materials are currently required for advanced wastewater treatment or for self-cleaning applications. For these applications, (super)hydrophilic surfaces must be developed and UV photo-induced hydrophilicity can be a viable path to obtain such materials. This paper reports on the hydrophilicity variation of the novel TiO_2/TiO_2 GO composite thin films obtained by spray pyrolysis deposition (SPD) followed by the deposition of a diluted sol by spraying at 100°C. The thin films were kept in dark (in normal atmospheric conditions) and under UV irradiation, at different exposure periods (up to 72 h) and the variation in the contact angle of water were analysed along with the morphology, surface composition and roughness (RMS) to correlate the hydrophilicity with the materials properties. The results show an increase in the contact angle values over time for the samples kept in dark and a decrease of it when the films were UV irradiated mainly as results of the roughness variation, due to slight changes in the thin film (surface) composition. The thin films were further tested in methylene blue decomposition and the effect of the hydrophilicity on the photocatalytic efficiency was investigated. The results show that, in the experimental conditions, a pre-conditioning treatment of the thin films can lead to more efficient processes.

Keywords TiO_2—GO thin film · Hydrophilicity variation · Roughness effect

I. Tismanar (✉) · A. Duta
Transilvania University of Brasov, Brasov, Romania
e-mail: tismanarioana@yahoo.com; ioana.tismanar@unitbv.ro

A. Duta
e-mail: a.duta@unitbv.ro

A. C. Obreja · O. Buiu
National R&D Institute R&D for Microtechnologies, Bucharest, Romania
e-mail: cosmin.obreja@imt.ro

O. Buiu
e-mail: octavian.buiu@gmail.com

I. Visa and A. Duta (eds.), *Solar Energy Conversion in Communities*,
Springer Proceedings in Energy, https://doi.org/10.1007/978-3-030-55757-7_26

26.1 Introduction

The photocatalytic processes represent potential routes for the removal, from wastewater, of organic pollutants at low concentrations, targeting water re-use [1]. For up scalable processes the photocatalytic materials and the equipment need to be affordable. In terms of materials, one of the most widely used photocatalyst is titanium dioxide, TiO_2, because it has a high chemical stability in aqueous media, it is non-toxic, has high capacity to produce oxidative species when irradiated with the adequate radiation and has not too high production costs. However, as a wide band gap semiconductor, TiO_2 can be activated only by UV radiation ($\lambda < 380$ nm) and this raises the costs of the entire process. This is why, VIS-active or solar-active photocatalysts are required to mitigate the process cost [2]. To reduce the TiO_2 band gap, various techniques are used as doping or coupling the n-type semiconductor with a p-type one in a diode—type structure. Recently, graphene derivatives as graphene oxide (GO) or reduced graphene oxide (rGO) were used as fillers embedded in a metal oxide matrix [3]. These types of composites are reported to be Vis(solar)-active because of the n-p diode structure developed when using these materials [4], but the development of continuous, stable interfaces between the matrix and the graphene derivatives remains an important challenge. This is the reason why not TiO_2—graphene composites are the targeted materials, as the non-polar graphene has little compatibility with the ionic TiO_2.

The Vis-activation problem is solved when using TiO_2-graphene derivatives composites and further on the photocatalytic properties of this type of samples have to be optimized to reach high performance materials. One important property of any photocatalysts is its surface hydrophilicity. This property is important particularly for applications as self-cleaning coatings, but it is also important in the pollutant adsorption step in the wastewater treatment mechanisms. It is reported that the (super)hydrophilicity of the metal oxide surfaces can be photo-induced. Wang et al. [5], were the first ones to analyse the behaviour of TiO_2 when exposed to UV radiation and they concluded that the surface wettability can be controlled by irradiation to obtain surfaces with antifogging and self-cleaning properties. Different mechanisms are involved in the photo-induced hydrophilicity of the TiO_2 surface, such as photo-induced oxygen vacancy generation [6] along with the adsorption of water molecules at the defects' sites, forming hydrophilic domains [7], along with the possible photo-oxidation of the traces of hydrophobic carbonaceous species from the TiO_2 structure [8]. Most of these mechanisms are reversible and the atmospheric oxygen can replace the adsorbed hydroxyl groups changing the surface wettability of TiO_2 back, from hydrophilic to hydrophobic [9].

It was reported [6], that the hydrophilicity of the TiO_2-rGO composites was increased by exposure to solar radiation and the water contact angle (WCA) decreased from 50° to <5° after 30 min of exposure under simulated solar irradiation. After that, the photo-induced self-cleaning properties were evaluated by measuring the degradation of methylene blue (MB) under simulated solar conditions and the material proved to be solar-active concluding that the graphene derivative is not affected by the irradiation. A similar conclusion is outlined by another study that also proved that the concentration of the filler in the composite can significantly change the wettability properties: a higher concentration of the filler can lead to a super-hydrophilic surface when irradiated only with VIS radiation [10]. Most of the hydrophilicity studies followed the development of materials for self-cleaning coatings and not for advanced wastewater treatment while studies on the behavior of these composites kept in dark (without any irradiation) are scarce.

This paper reports on the hydrophilicity variations of the novel TiO_2/TiO_2—grapehene oxide (GO) two-layered composite thin films obtained by spray pyrolysis deposition (SPD) coupled with spraying (at lower temperature) a sol of the composite material. The surface morphology, composition and roughness (RMS) were discussed to correlate the hydrophilicity variation with the material's properties. The methylene blue (MB) decomposition, under simulated solar radiation at low irradiance values, was investigated to outline the influence of the WCA values on the photocatalytic efficiencies for advanced wastewater treatment applications.

26.2 Materials and Methods

Glass plates (1.5 cm × 1.5 cm) covered with a thin film of fluorine doped tin oxide (FTO) were used as substrates. These substrates were cleaned with water and detergent by ultra-sonication, followed by rinsing in ethanol and drying in air.

The composite thin films were deposited in a two-layered system in order to get a performance material in photocatalytic processes with controlled hydrophilicity. The first TiO_2 layer was obtained using a precursor system containing titanium tetra-isopropoxide (TTIP, Sigma Aldrich, 97%) mixed with acetylacetone (AcAc, Scharlau, 99%) and ethanol (EtOH, Chemical Company, 99.3%) in a 1:1:15 volume ratio. This layer was deposited by Spray Pyrolysis Deposition (SPD) using an ABB/IRB5400 robot. The deposition temperature was 400°C and 30 spraying sequences were used, with 60 s break between consecutive pulses. After deposition, this first layer was annealed for 3 h, at 450°C, to increase its crystallinity degree.

The second TiO_2-GO layer was sprayed on the top of the first one using a diluted sol obtained based on TTIP as precursor, ethanol as continuous medium, acetylacetone as complexing agent and acetic acid (HAc 99.8%, Scharlau) to control the pH (pH ~5). These were mixed, under stirring, in an overall volume ratio: TTIP:EtOH:AcAc:HAc = 1:0.8:0.04:0.009.

An aqueous GO dispersion (30 mg/mL) was obtained using the Hummers method: 12 g of graphite powder (99.99% purity, Elektrokarbon) were mixed with 247 mL H_2SO_4 (96%, VLSI grade, BASF) and continuously stirred for 24 h. Afterwards, 6 g $NaNO_3$ (99%, Sigma Aldrich) were added and the dispersion was cooled down to 5 °C, then 36 g of $KMnO_4$ (99%, Merck) were slowly added for a good control of the exothermic reaction. Further on, the graphite powder was kept in the oxidation environment for 4 days; after that the excess $KMnO_4$ and the manganese oxide precipitate were removed using 200 mL H_2O_2 5% and 150 mL HCl 5% respectively.

The dispersion was then filtered and the graphite oxide powder was washed several times with deionized water. This powder was further dispersed in deionized water and placed in an ultrasonic bath to support the exfoliation of the graphene oxide sheets.

The GO dispersion, prepared as previously discussed, was added in the TiO_2 precursor system, to obtain the composite sol with stable interfaces between TiO_2 and the graphene derivative. Further on, the sol was sonicated for 1.5 h and after sonication, the sol was diluted with ethanol in a 1:5 sol: EtOH volume ratio. The GO weight ratio in the precursor system was 1.4%$_w$.

This diluted sol was deposited, by robotic spraying, over the first TiO_2 layer, using 15 spraying sequences, with 60 s break between pulses. The deposition temperature was selected at 100°C to support the ethanol evaporation while preventing the GO oxidation. The composite thin film was further thermally treated for one hour at 150°C to decompose/remove the carbon containing by-products, without degrading the GO filler [11].

This two-layered composite structure was selected to increase the crystallinity degree of the composite upper layer by using a first layer that has a predominant crystalline structure, considering the low temperature of the thermal treatment applied to the final thin films. Further on, previous results outlined that the adherence between the composite and the substrate is increased by using an intermediate matrix layer.

The thin film composites obtained as above discussed were kept under various conditions: in dark (in regular atmospheric conditions) and under UV irradiation (irradiance value, $G = 8.3$ W/m^2, measured using a Delta-T, type BF3 pyranometer), at different exposure periods (up to 72 h) and the WCA variations were analyzed using a DataPhysics OCA20 EC instrument. The volume of the water drop was 10 μL. The surface elemental composition was analyzed by Energy Dispersive X-ray spectrometry (EDS, Thermo) while the surface morphology and roughness (RMS) were investigated using Scanning Electron Microscopy (SEM, Hitachi model S-3400 N type II) and Atomic Force Microscopy (AFM, NT-MDT model BL222RNTE). The results were correlated with the photocatalytic efficiencies recorded in the photo-degradation of the methylene blue (MB) pollutant, as indicated by the ISO 10678:2010 [12], standard.

The photo-degradation experiments were developed in static regime using a 10 ppm (3.125×10^{-5} M) MB solution and a photo-reactor equipped with 2 UV light tubes (Philips, TL-D BLB 18 W/108) and 5 Vis light tubes (Philips, TL-D Super 80 18 W/865). This mix simulates solar radiation conditions but at much lower irradiance value; the total irradiance was $G = 55$ W/m^2, out of which 3 W/m^2 corresponds to the UV radiation. Before the photocatalytic tests, the films were kept in dark for a shorter time (24 h) and for a longer duration (7 days) and were subsequently pre-conditioned under UV radiation for 24 h. These films were immersed in 20 mL MB solution and left one hour in dark to reach the adsorption/desorption equilibrium, as previously optimized [13]. After one hour in dark the thin film samples were continuously irradiated for 8 h. The MB removal efficiency, η was calculated using Eq. (26.1), based on the initial absorbance of the MB solution (A_0) and the absorbance after up to 8 h of irradiation (A_8), measured at the maximum absorbance wavelength for MB (λ = 664 nm), using a UV–Vis–NIR spectrophotometer (Perkin Elmer Lambda 950):

$$\eta = \frac{A_0 - A_8}{A_0} \cdot 100 \quad (26.1)$$

26.3 Results and Discussions

The investigations on the hydrophilic character show an increase in the WCA values over time, for the samples stored in dark (Fig. 26.1) and, respectively, the decrease of the WCA value when exposing the films to UV radiation (Fig. 26.2). The optimum irradiation duration was found to be of 24 h. After this period of time the values of the contact angles remained constant.

It can be also observed that for the sample containing only TiO_2 SPD kept in dark the WCA variation is rather low, indicating that this type of coatings can be stored without any significant consequences on the hydrophilic properties. For the composite sample, the storage in dark leads to a significant increase in the WCA values, thus a reduction in the hydrophilic character.

To understand these changes in hydrophilicity in different conditions, characterization analyzes were performed. According to the results in Tables 26.1 and 26.2, it can be noticed that the hydrophilicity increase over time when irradiated with UV

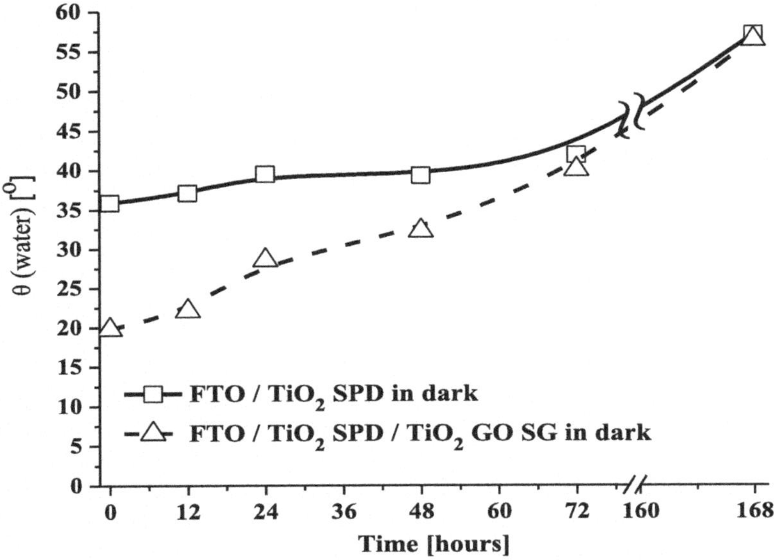

Fig. 26.1 Variation of the WCA in time, for the films kept in dark

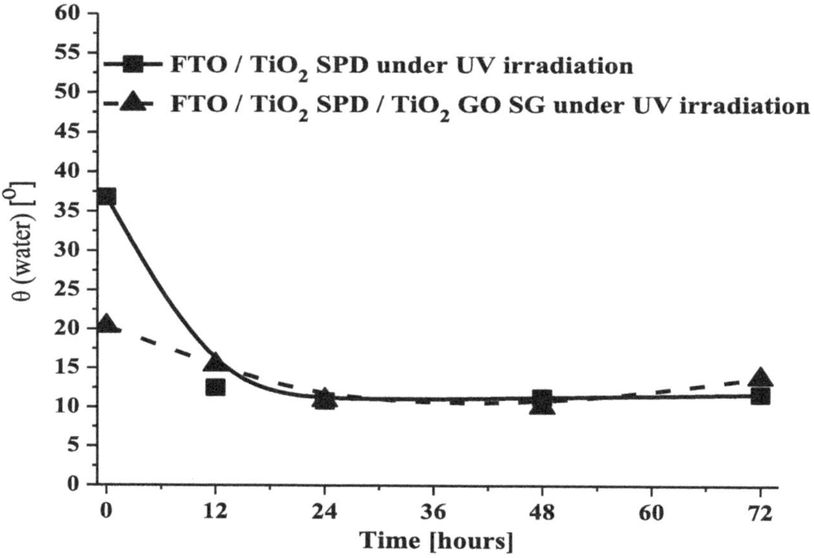

Fig. 26.2 Variation of the WCA in time, for the UV irradiated thin films

Table 26.1 RMS values at different exposure periods under UV irradiation and in dark condition

Sample	Under UV irradiation		In dark	
	Time	RMS [nm]	Time	RMS [nm]
FTO/TiO_2 SPD	Initial	11.5	Initial	11.8
	12 h	9.1	12 h	12.2
	24 h	9.0	24 h	14.5
	48 h	8.9	48 h	17.3
	72 h	8.6	72 h	19.0
FTO/TiO_2 SPD/TiO_2 GO SG	Initial	332.5	Initial	327.9
	12 h	312.3	12 h	348.6
	24 h	304.5	24 h	362.3
	48 h	292.2	48 h	375.4
	72 h	290.0	72 h	387.6

radiation is due to the variation of the film roughness correlated with the decrease in the carbon content from the surface of the samples. For the films kept in the dark, the hydrophilicity decreases and the RMS values have a constant increase over time while under UV irradiation the RMS values decrease in time. These results are attributed to the possible densification of the thin films and less to the oxidation of the carbon existing in the sample from the sol precursors, under UV irradiation. It is also possible a continuation of the reactions in the composite for the samples kept in dark

Table 26.2 The surface elemental composition of the samples innitially measured, after 72 h of UV irradiation and after 72 h of dark condition

Sample	EDX Map [% at.]		
	Initial	After 72 h of UV irradiation	After 72 h in dark
FTO/TiO2 SPD	C: 2.29	C: 1.51	C: 3.38
	O: 64.78	O: 64.98	O: 64.24
	Ti: 23.50	Ti: 18.92	Ti: 23.98
	Sn: 8.34	Sn: 12.17	Sn: 7.10
	F: 0.74	F: 1.09	F: 0.65
	Si: 0.35	Si: 1.33	Si: 0.65
FTO/TiO_2 SPD/TiO_2 GO SG	C: 9.55	C: 6.52	C: 11.30
	O: 59.23	O: 60.87	O: 56.00
	Ti: 30.08	Ti: 29.01	Ti: 32.05
	Sn: 1.01	Sn: 2.38	Sn: 0.61
	F: 0.10	F: 0.19	F: 0.00
	Si: 0.03	Si: 1.03	Si: 0.04

and for the ones irradiated during the first 24 h, when UV radiation can accelerate the processes. Previous results of the photocatalysis tests performed under UV + VIS irradiation compared with those under UV irradiation, prove the Vis-activation of the composites and also that the graphene oxide is not affected by the exposure to UV irradiation (under the experimental conditions).

The SEM images of the composite films kept in dark, Fig. 26.3, outline large aggregates on the surface of the composite films, compared to the initial structure concluding that the increase in the RMS values can be due to these aggregates and hence the decrease of the sample hydrophilicity.

In good agreement with the SEM images, the AFM results, Fig. 26.4, outline surface changes of the composite thin film under UV irradiation and in dark.

To correlate the variation of the hydrophilic character with the performances of the films during the photocatalytic processes, photo-degradation tests of the MB pollutant were performed, under UV + VIS irradiation, in static regime, on the multilayered composite films and, for comparison, on the TiO_2 SPD films. The films were kept, as previously discussed, in dark for 24 h and for a long duration (7 days) and were further pre-conditioned under UV irradiation for 24 h.

The photocatalytic results included in Table 26.3 show that the samples stored for long-term in dark lead to lower photo-degradation efficiencies compared to the more recently deposited and UV pre-conditioned films. The results can be correlated with the higher WCA values for the samples kept 7 days in dark, in agreement with the results already discussed. From Table 26.3 it can also be observed that the adsorption (1 h dark) is lower for the samples kept in dark, and the results show that this step depends on the hydrophilicity of the films. It may be noticed that the composite thin

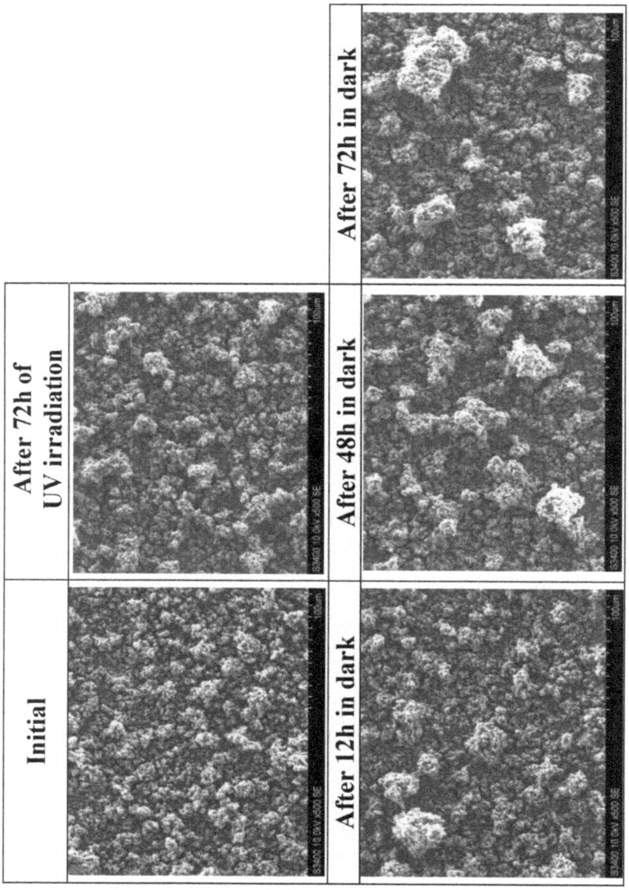

Fig. 26.3 SEM images of the FTO/TiO_2 SPD/TiO_2 GO SG composite at different exposure periods under UV irradiation and in dark condition

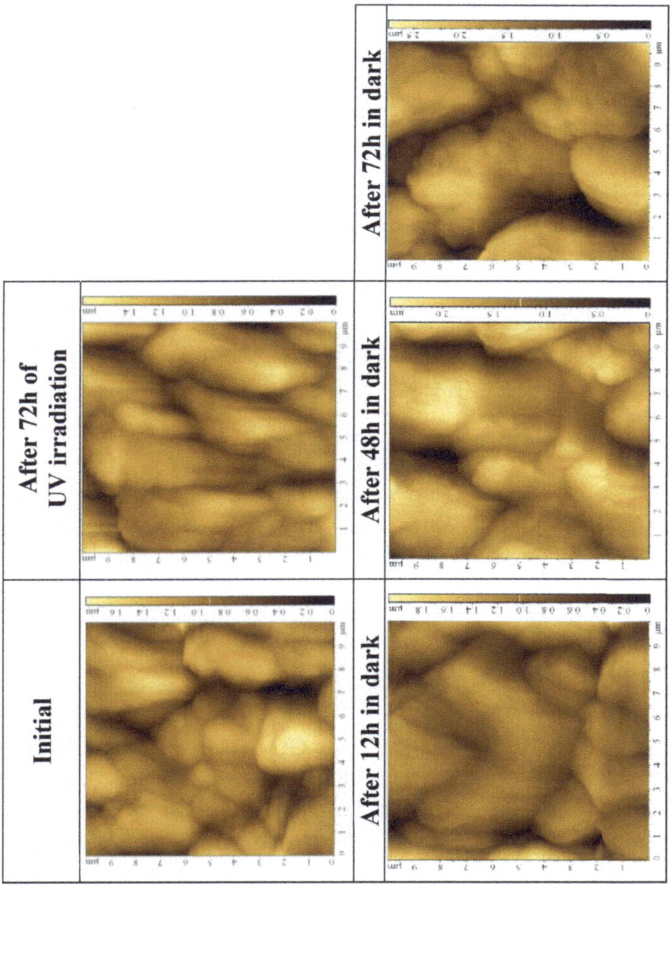

Fig. 26.4 AFM images of the FTO/TiO_2 SPD/TiO_2 GO SG composite at different exposure periods under UV irradiation and in dark

Table 26.3 Photodegradation efficiencies of the MB pollutant and the contact angle for the sample exposed at UV irradiation and kept in the dark

Sample	Condition	WCA [°]	Photocatalytic tests η [%]	
			1 h dark	8 h UV + VIS
FTO/TiO_2 SPD	Initial	31.9	6.36	15.57
	24 h in dark	38.3	4.96	14.62
	24 h in dark + 24 h UV irradiation	10.1	6.50	15.72
	7 days in dark	57.0	4.82	13.06
	7 days in dark + 24 h UV irradiation	10.8	6.26	15.75
FTO/TiO_2 SPD/TiO_2 GO SG	Initial	18.3	8.48	28.26
	24 h in dark	24.2	8.03	26.57
	24 h in dark + 24 h UV irradiation	11.9	8.69	27.27
	7 days in dark	56.3	6.53	23.84
	7 days in dark + 24 h UV irradiation	20.1	8.54	27.17

films have a small variation in the MB removal efficiency (smaller than 4.5%) and that UV conditioning is important.

26.4 Conclusions

The hydrophilicity of the FTO/TiO_2 SPD/TiO_2—GO SG composite thin films varies depending on the conditions under which the samples are exposed, namely: the WCA increases over time for the samples kept in dark and decreases after irradiation of these samples with UV radiation. This is mainly the consequence of the variation in the roughness values of the films, in correlation with the morphology and the surface composition. The variation of the hydrophilicity mainly affects the pollutant adsorption step and hence the photocatalytic efficiencies are lower for the samples with a higher contact angle. However conditioning the thin film through UV irradiation allows reaching similar efficiencies as in the case of recently deposited samples and for the improvement of the photocatalytic efficiency, a first step of pre-conditioning the films by UV irradiation is required followed by the photocatalysis process, especially when the films were stored for longer periods before use.

Acknowledgements This work was supported by the grant of the Romanian Ministry of Research and Innovation, CCCDI-UEFISCDI, project number PN-III-P1-1.2-PCCDI-2017-0619, contract no. 42 PCCDI/2018 within PNCDI.

References

1. M.N. Chong, B. Jin, C.W.K. Chow, C. Saint, Recent developments in photocatalytic water treatment technology: a review. Water Res. **44**(10), 2997–3027 (2010)
2. J. Schneider, M. Matsuoka, M. Takeuchi, J. Zhang, Y. Horiuchi, M. Anpo, D.W. Bahnemann, Understanding TiO_2 photocatalysis: mechanisms and materials. Chem. Rev. **114**(19), 9919–9986 (2014)
3. S.K. Fanourakis, J. Peña-Bahamonde, P.C. Bandara, D.F. Rodrigues, Nano-based adsorbent and photocatalyst use for pharmaceutical contaminant removal during indirect potable water reuse. Npj Clean Water **3**(1) (2020)
4. C. Chen, W. Cai, M. Long, B. Zhou, Y. Wu, D. Wu, Y. Feng, Synthesis of visible-light responsive graphene oxide/TiO_2 composites with p/n heterojunction. ACS Nano **4**(11), 6425–6432 (2010)
5. R. Wang, K. Hashimoto, A. Fujishima, M. Chikuni, E. Kojima, A. Kitamura, M. Shimohigoshi, T. Watanabe, Light-induced amphiphilic surfaces. Nature **388**, 431–432 (1997)
6. S. Prabhu, L. Cindrella, O. Joong Kwon, K. Mohanraju, Superhydrophilic and self-cleaning rGO-TiO_2 composite coatings for indoor and outdoor photovoltaic applications. Sol. Energy Mater. Sol. Cells **169**, 304–312 (2017)
7. M. Janczarek, J. Hupka, H. Kisch, Hidrophilicity of TiO_2 exposed to UV and VIS radiation. Physicochemical Probl. Mineral Process. **40**, 287–292 (2006)
8. T. Zubkov, D. Stahl, T.L. Thompson, D. Panayotov, O. Diwald, J.T. Yates, Ultraviolet light-induced hydrophilicity effect on $TiO_2(110)(1\times1)$. Dominant role of the photooxidation of adsorbed hydrocarbons causing wetting by water droplets. J. Phys. Chem. B **109**(32), 15454–15462 (2005)
9. M. Miyauchi, N. Kieda, S. Hishita, T. Mitsuhashi, A. Nakajima, T. Watanabe, K. Hashimoto, Reversible wettability control of TiO_2 surface by light irradiation. Surf. Sci. **511**(1–3), 401–407 (2002)
10. M.K. Kavitha, L. Rolland, L. Johnson, H. John, M.K. Jayaraj, Visible light responsive super-hydrophilic TiO_2/reduced graphene oxide coating by vacuum-assisted filtration and transfer method for self-cleaning application. Mater. Sci. Semicond. Process. **113**, 105011 (2020)
11. X.Z. Tang, W. Li, Z.-Z. Yu, M.A. Raffie, J. Raffie, F. Yavari, N. Koratkar, Enhanced thermal stability in graphene oxide covalently functionalized with 2-amino-4,6-didodecylamino-1,3,5-triazine. Carbon **49**, 1258–1265 (2011)
12. * * * INTERNATIONAL STANDARD ISO 10678:2010, Fine ceramics (advanced ceramics, advanced technical ceramics)—Determination of photocatalytic activity of surfaces in an aqueous medium by degradation of methylene blue
13. A. Enesca, M. Baneto, D. Perniu, L. Isac, C. Bogatu, A. Duta, Solar-activated tandem thin films based on $CuInS_2$, TiO_2 and SnO_2 in optimized wastewater treatment processes. Appl. Catal. B **186**, 69–76 (2016)

Chapter 27
Correlation Between the Growth Process and Film Properties of RF-PECVD Grown Nanocrystalline Graphite/Graphene

Octavian-Gabriel Simionescu, Elena Anghel, Oana Tutunaru, Cristina Pachiu, Raluca Gavrilă, Andrei Avram, Octavian Buiu, and Gheorghe Dinescu

Abstract This study aims to better correlate the properties of nanocrystalline graphite/graphene thin films with the individual deposition parameters, such as RF power, substrate temperature, chamber pressure and other influential factors. The carbonic films were grown using methane as gas precursor by capacitively coupled radiofrequency plasma enhanced chemical vapor deposition. The morphology, structure and electrical properties of the thin films deposited on both Si and SiO_2 substrates at different process parameters were investigated through scanning electron microscopy, Raman spectroscopy, atomic force microscopy and four-point probing.

Keywords Nanocrystalline structure · PECVD · Morphology · Properties

---

O.-G. Simionescu (✉) · O. Tutunaru · C. Pachiu · R. Gavrilă · A. Avram · O. Buiu
National Institute for Research and Development in Microtechnologies—IMT Bucharest, 126A Erou Iancu Nicolae Street, 077190 Voluntari City, Ilfov County, Romania
e-mail: octavian.simionescu@imt.ro

O. Buiu
e-mail: octavian.buiu@gmail.com

O.-G. Simionescu · G. Dinescu
Faculty of Physics, University of Bucharest, 405 Atomistilor Street, 077125 Bucharest, Romania

E. Anghel
Faculty of Applied Chemistry and Materials Science, University Politehnica of Bucharest, Bucharest 011061, Romania

G. Dinescu
National Institute for Laser, Plasma and Radiation Physics, 409 Atomistilor Street, 077125 Magurele City, Ilfov County, Romania

I. Visa and A. Duta (eds.), *Solar Energy Conversion in Communities*,
Springer Proceedings in Energy, https://doi.org/10.1007/978-3-030-55757-7_27

27.1 Introduction

Graphene is already a well-known material in the scientific community, due to its extraordinary physical and chemical properties, such as: excellent electrical conductivity, reliable mechanical robustness and chemical stability. While graphene exceeds other materials by far with respect to its properties, it fails to surpass them in production costs, large scale fabrication and ease of integration in current thin films industry processes [1–3].

Nanocrystalline graphite/graphene (NCG) demonstrates that it can be a valid substitute, as it solves the challenges of production while also providing enhanced performance when compared to the current materials on the market. This paper treats NCG as a broader class of carbonic materials and recognizes its subcategories as stated in [3]: nanocrystalline graphene (nc-graphene), nanocrystalline graphite (nc-graphite), bulk-NCG and graphene/graphite nanowalls (GNW), also known as carbon nanowalls (CNW) or vertical graphene (VG). Plasma enhanced chemical vapor deposition (PECVD) is one of the most advantageous techniques when attempting to synthesize this type of carbonaceous material directly on an insulating substrate, allowing for improved control on the morphology of the grown film. The diverse possibilities of final morphology and structure, makes NCG thin films suitable candidates for various applications: piezoresistive or electrochemical sensors [1, 2, 4–8], protective coatings [9], NEM/MEM switches and resonators [10, 11], active material for supercapacitors, electrodes [12] and others [3].

In this work, we explore the structural, morphological and electrical properties of NCG thin films grown by capacitively coupled radiofrequency (RF) PECVD at different deposition parameters, and discuss the benefits brought by this material in several fields of application. Well known characterization techniques, such as scanning electron microscopy (SEM), Raman spectroscopy, atomic force microscopy (AFM) and four-point probing, were employed for the analysis of these films.

27.2 Materials and Methods

27.2.1 Substrate Preparation

For our experiments we used *p*-type 4" Si wafers with a <100> crystallographic orientation, a thickness of $525 \pm 25\ \mu$ m and a resistivity of $1 - 10\Omega$ cm. The growth processes were carried out on thermally grown SiO_2 in the case of bulk-NCG [1, 2] and on bare Si wafers for the GNW thin films [13].

Table 27.1 Parameters during the growth step of bulk-NCG and GNW by capacitively coupled RF-PECVD

NCG subclass	Bulk-NCG	GNW
Substrate temperature [°C]	890	700–850
Process pressure [Pa]	200	40
CH_4 flow rate [sccm]	60	10
H_2 flow rate [sccm]	75	–
Ar flow rate [sccm]	–	190
RF power [W]	100	300

27.2.2 Thin Film Growth

The NCG thin films discussed here were synthesized by capacitively coupled RF-PECVD with the Nanofab 1000 system (Oxford Instruments, UK). Both bulk-NCG and GNW can be obtained with the same reactor configuration by adjusting the process parameters accordingly. The differences between the growth steps for each category of NCG discussed in this work can be found in Table 27.1.

27.2.3 Characterization

The SEM micrographs were acquired using the Nova NanoSEM 630 Scanning Electron Microscope (FEI Company, Hillsboro, OR, USA) fitted with UHR detector (Through-Lens-Detector-TLD) at an acceleration voltage of 10 kV. The Raman analysis was performed with a high resolution Scanning Near-Field Optical Microscope fitted with the Raman Module Witec Alpha 300S (Witec, Germany). All the data for the Raman spectra presented here was gathered with a green laser of wavelength $\lambda = 532$ nm. The AFM investigations were carried out with the NTEGRA Aura (NT-MDT Co., Russia) on a 1 μm × 1 μm surface area. The sheet resistance was directly measured with the 4-point probe technique, while the film electrical resistivity and conductivity were calculated according to Eqs. (27.1) and (27.2), respectively [1, 13].

$$\rho = R_S \cdot h \tag{27.1}$$

$$\sigma = \frac{1}{\rho} \tag{27.2}$$

where: ρ is the films resistivity, measured in Ω m; R_S is the sheet resistance of the thin film, measured in Ω/sq.; h is the film thickness in nm (approximated from the growth rate for the films thicker than 200 nm) and σ is the electrical conductivity of the measured film, in S/m.

27.3 Results and Discussion

27.3.1 Bulk Nanocrystalline Graphite

The growth process of bulk-NCG has been well documented over the years and so, this NCG subclass has been synthesized through a large number of techniques. A review on the subject of NCG synthesis methods, as well as applications, can be found in [3]. From this variety of deposition techniques, a large amount of them are dedicated to the PECVD method, resulting in a comprehensive description of the growth mechanism of NCG by this method: the process entails in a nucleation step at first, where the hydrocarbon radicals diffuse to the substrate and, after dissociation, form sp^2 hybridized carbon aromatic rings that begin to grow laterally until they overlap; after the first few layers are formed on the substrate, the growth rate stabilizes and the film begins to grow in thickness. Bulk-NCG has a turbostratic relation between individual nanodomains and layers, the films internal structure and morphology depending mainly on the adjustment of parameters and on whether or not a metallic catalytic layer is used.

From previously published papers [1, 2] it can be observed that the growth rate for bulk-NCG is ~3.16 nm/min and when plotting the growth rate with respect to growth time, the nucleation stage of the process can be clearly observed (Fig. 27.1). The sudden drop in growth rate seen in Fig. 27.1 can be attributed to the transition from the nucleation step of the process to the actual growth step. However, the figure in question is to be treated as a qualitative display of the transition between the two steps, and not as a quantitative one, because we already know from [1] that at 15 min of growth time the NCG thin film reaches a thickness of ~46 nm. For a more precise investigation, measurements should be taken at smaller time intervals, as we predict the curve representing the growth rate has a sharper drop than that presented in Fig. 27.1b. A more suited time interval for this type of investigations would be of 1 min, or even 30 s.

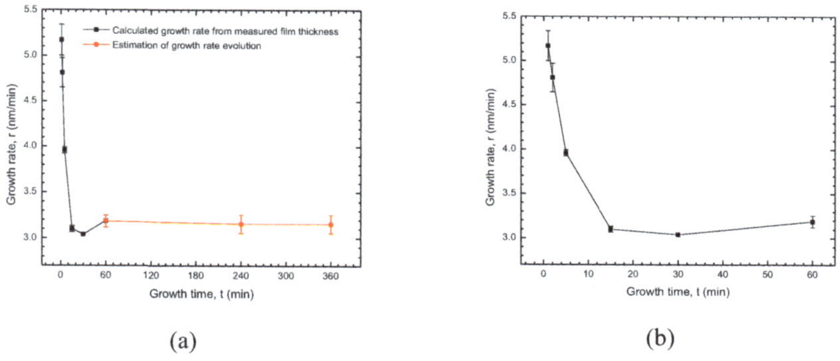

Fig. 27.1 **a** Growth rate of bulk-NCG synthesized by RF-PECVD as a function of growth time with **b** emphasis on the nucleation step of the process. Reproduced from [2]

From the various applications enumerated in the introduction section of this work, our group has investigated the use of bulk-NCG as the active material in piezoresistive sensors. Although in [1] we first started with low temperature oxide (LTO) that presented an RMS roughness of ~1.4 nm when deposited on a Si wafer with 0.15 nm RMS roughness (Fig. 27.2) as the insulating layer, the oxide film presented a high compressive stress of −200 MPa unless thermally annealed at 900 °C [1]. Due to this requirement of the LTO film, we chose to proceed with thermally grown SiO_2 in order to reduce the number of steps required and to simplify the whole technological workflow [2].

The substrate on which the deposition of the NCG thin film is carried out have been known to influence the growing film [14]. The depositions are performed on a differently obtained SiO_2, and although this should not cause changes in the carbonic material as was proven by Jerng et al. with Al_2O_3 substrates of different cutting planes [15], Raman spectra were acquired to confirm the bulk-NCG structure (Fig. 27.3).

The appearance of the well known G, D, 2D and D + D' peaks at ~1347 cm^{-1}, 1586 cm^{-1}, 2680 cm^{-1} and 2934 cm^{-1}, respectively, together with the I_D/I_G ratio of ~1.15 and FWHM of ~50 cm^{-1} and ~74 cm^{-1} for peaks D and G, respectively, confirms the high disorder type bulk-NCG [2, 3]. Seeing how the Raman spectra for the film deposited on 100 nm SiO_2 and for the one deposited on 300 nm SiO_2 are fairly similar, the latter was chosen for the next steps in order to provide better protection for the carbonic material in the plasma processes that are to come.

The testing of the piezoresistive potential of bulk-NCG, as reported in [2], was performed after the carbonic thin film was transferred onto a flexible substrate (polydimethylsiloxane—PDMS) through a series of processes: after the growth of the bulk-NCG thin film by PECVD, the surface was treated in a low power oxygen plasma to improve its adherence and PDMS composite was added on the surface of the film in a 3D printed PLA mold. After the polymerization of the PDMS layer, the SiO_2 layers and the entire Si wafer are etched in CHF_3:Ar plasma and SF_6 plasma, respectively. The samples are diced in multiple strips, which are contacted by Cu foils and encapsulated in PDMS [2].

In order to test the piezoresitive capabilities of the samples, their electrical resistance was monitored during mechanical deformations [2]. Due to its internal structure of nanodomains, charge carriers travel through this type of carbonic material by tunneling through the defects located at the boundaries of these domains. When applying a mechanical strain on the film, the total tunneling distances modifies and so does the electrical resistance, hence the piezoresistive effect [2, 4]. The efficiency of this type of sensors is measured by the gauge factor (GF), which is the proportionality factor between the relative variation of the electrical resistance and the applied mechanical strain and is calculated with Eq. (27.3). The values obtained for the GF are ~50–250 and they show the promise of this material in the field of piezoresistive sensors [2].

$$\mathrm{GF} = \frac{\Delta R}{R(l_0)} \cdot \frac{1}{\varepsilon} = \frac{R(l) - R(l_0)}{R(l_0)} \cdot \frac{1}{\varepsilon} \tag{27.3}$$

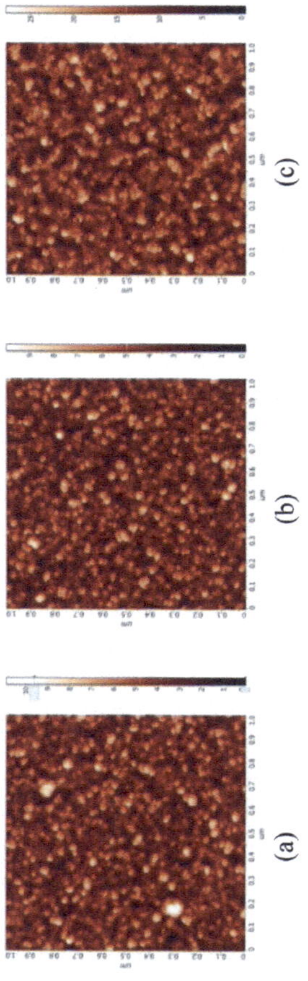

Fig. 27.2 AFM measurements of **a** an as-deposited 300 nm LTO/Si with ~1.4 RMS roughness, **b** thermally annealed LTO/Si with ~1.3 RMS roughness and **c** bulk-NCG/thermally annealed LTO/Si with ~3.7 RMS roughness. Subfigures **b** and **c** are ©IEEE and are reproduced with permission [1]

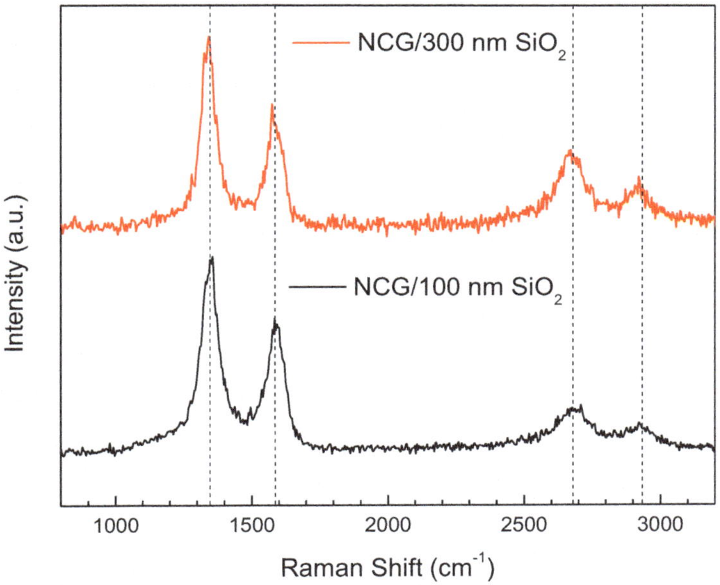

Fig. 27.3 Raman spectra of ~380 nm bulk-NCG deposited by RF-PECVD on 100 nm (black) and 300 nm (red) thermally grown SiO_2. The vertical dash lines are at 1347.45 cm^{-1}, 1585.76 cm^{-1}, 2679.53 cm^{-1} and 2933.58 cm^{-1}, respectively

where ε is the applied strain and $R(l)$ and $R(l_0)$ are the electrical resistances in a deformed and relaxed state, respectively.

When investigating the films electrical conductivity with respect to the growth time, a saturation is observed after 60 min, which is equivalent to a ~200 nm film thickness (Fig. 27.4).

By comparing this graphical representation of the conductivity with the growth rate function described in Fig. 27.1, one can clearly observe a direct correlation between the two, as they both seem to undergo a transition phase and begin to saturate after ~200 nm (60 min growth time). This can be attributed to the fact that after a certain thickness either the hydrogen plasma etching of the amorphous carbon intensifies or that the new forming layers of carbonic material have a better crystallinity as they bond to an already sp^2 hybridized carbon underlayer. It can also be a combination of the above mentioned theories or due to the fact that at a certain distance from the substrate its influence over the growing layer diminishes, but either way it is clear that the number of defects decreases and better conductive paths for the charge carriers appear in the film at higher thicknesses.

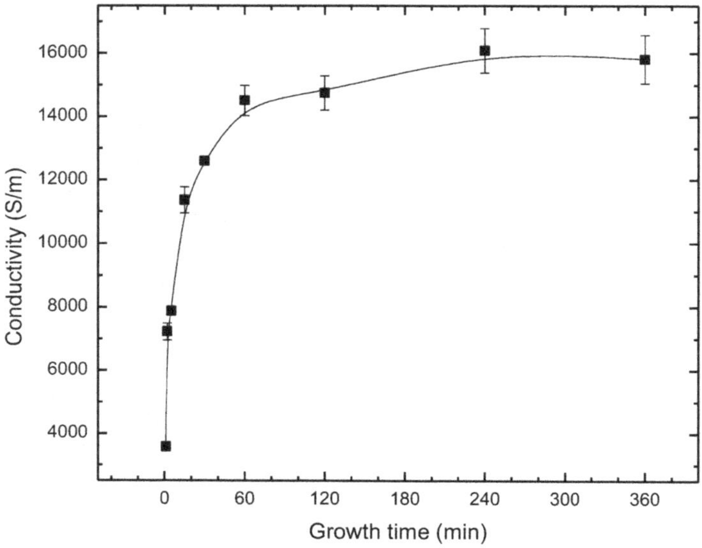

Fig. 27.4 Electrical conductivity of bulk-NCG thin films deposited by RF-PECVD with respect to the growth time

27.3.2 *Graphene/Graphite Nanowalls*

GNW thin films are another form of carbonic materials from the NCG class. With a similar structure of graphitic nanodomains, they distinguish themselves from the other NCG types by their unique morphology of vertical planes with respect to the substrate [3]. The influence of the process parameters on this type of growth can be clearly seen when observed side-by-side with the ones used for bulk-NCG growth, as they are presented in Table 27.1.

In this process, argon plasma is used to dissociate the methane gas precursor. Molecular hydrogen has been removed from the precursors to lower the etching process, as the film consists of vertical thin graphitic planes and a high etching rate could prevent the growth altogether. The CH_4 flow rate has been significantly reduced from 60 to 10 sccm in order to prevent the formation of amorphous carbon between the vertical planes. The process pressure has been lowered to allow for a higher mean free path of the incident particles and the RF discharge power has been increased to obtain more energetically charged particles at the surface of the growing film. These adjustments of pressure and discharge power permits fast bonding between particles at the extremities of the graphitic planes and promotes edge growth of existing planes over the formation of new ones.

The fast edge growth of the vertical graphitic planes can be observed in the growth rate as well. As they have a much smaller density than bulk-NCG, their growth rate is higher, reaching ~8–13 nm/min depending on the substrate temperature. In order

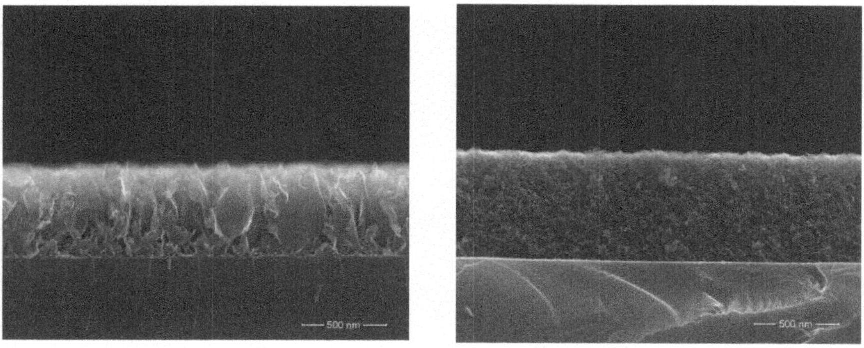

(a) (b)

Fig. 27.5 Cross-section SEM micrographs of GNW thin films deposited by RF-PECVD at a substrate temperature of **a** 750 °C and **b** 850 °C

to obtain denser films of the GNW type, one can increase the substrate temperature of the deposition. By setting a higher temperature at the substrate surface, especially during the nucleation phase of the process, the adsorbed CH_x radicals are more easily dissociated and the resulting carbon adatoms have a higher diffusion energy. Hence, the nucleation density will increase at higher substrate temperatures and the GNW density will implicitly increase as well. This phenomenon can be clearly observed in Fig. 27.5, where SEM micrographs of GNW thin films grown at 750 and 850 °C respectively, are presented [13].

The substrate temperature clearly influences the density of the growing film and the growth rate, but do not appear to modify the internal structure in anyway as can be seen in Fig. 27.6 from the Raman spectra of GNW thin films grown at different substrate temperatures. The individual spectra are relatively similar to each other, presenting the specific D, G, 2D and D + D' peaks at ~1346 cm^{-1}, 1577 cm^{-1}, 2684 cm^{-1} and 2931 cm^{-1}, respectively. The spectra are also similar to that of bulk-NCG with a slight shift to lower regions in the signal, both being subclasses of the NCG type.

GNW thin films have a very high surface area and so, they are very well suited for sensors and have shown remarkable results as the active material for micro-supercapacitors when coated with a metal oxide to induce a pseudo-capacitance [12].

27.4 Conclusions

Over the course of the presented experiments, both bulk-NCG and GNW thin films have been investigated. The growth mechanisms for both types of carbonic materials have been explained in detail and the differences between the two have been pointed

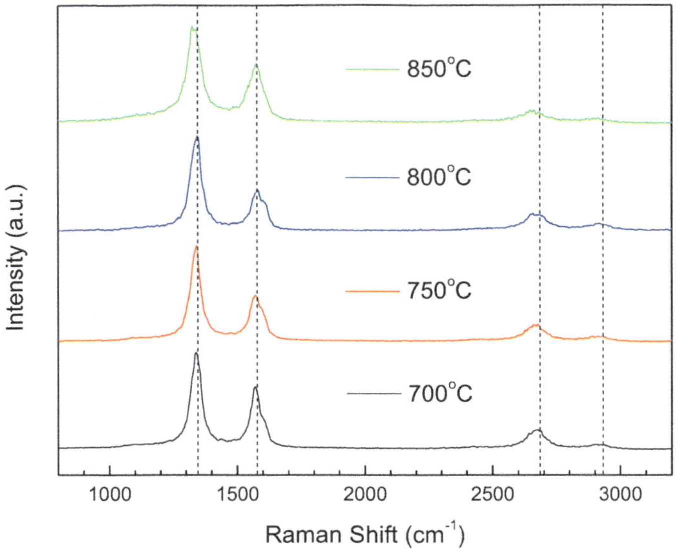

Fig. 27.6 Raman spectra of GNW thin films grown by RF-PECVD at different substrate temperatures. The vertical dash lines are at 1346 cm^{-1}, 1577 cm^{-1}, 2684 cm^{-1} and 2931 cm^{-1}, respectively

out. The correlation between electrical properties, internal structure, morphology and process parameters have been investigated through well know characterization techniques.

The electrical properties for bulk-NCG thin films primarily depend on the lateral crystallite size of the nanodomains, which can be controlled through the nucleation density by the substrate temperature. Hence, bulk-NCG films show a great potential as piezoresistive sensors as they have a tunable GF, by controlling the total length of conductive paths and tunneling distances in the thin film. Also, better crystallinity, with improved electrical conductivity, is observed for thicker films.

The GNW thin films have an unique morphology that sets them apart from the rest of the NCG subclasses. Their vertical arrangement on the substrate surface results in a thin film with a very high specific surface area. This aspect of GNW thin films prescribes them for applications such as electrochemical sensors, micro-supercapacitor electrodes and can even prove useful for piezoresistive sensors.

Acknowledgements The authors would like to acknowledge the financial support from the Romanian Ministry of Education and Research through grant no. PN-III-P1-1.2-PCCDI-2017-0619, project “Nanostructured carbonic materials for advanced industrial applications”.

References

1. O.G. Simionescu, E. Anghel, R. Popa, C. Pachiu, R. Gavrilă, F. Comănescu, A. Avram, G. Dinescu, Bulk nanocrystalline graphite thin films for piezorezistive sensing applications. In: International Semiconductor Conference (CAS) (IEEE, Sinaia, Romania, 2019), pp. 229–232. https://doi.org/10.1109/smicnd.2019.8923644
2. O.G. Simionescu, C. Pachiu, O. Ionescu, N. Dumbrăvescu, O. Buiu, R.C. Popa, A. Avram, G. Dinescu, Nanocrystalline graphite thin layers for low-strain, high-sensitivity piezoresistive sensing. Rev. Adv. Mater. Sci. **59**(1), 306–313 (2020). https://doi.org/10.1515/rams-2020-0031
3. O.G. Simionescu, R.C. Popa, A. Avram, G. Dinescu, Thin films of nanocrystalline graphene/graphite: an overview of synthesis and applications. Plasma Process. Polym. **17**(7), e1900246 (2020). https://doi.org/10.1002/ppap.201900246
4. J. Zhao, C. He, R. Yang, Z. Shi, M. Cheng, W. Yang, G. Xie, D. Wang, D. Shi, G. Zhang, Ultra-sensitive strain sensors based on piezoresistive nanographene films. Appl. Phys. Lett. **101**(6), 063112 (2012). https://doi.org/10.1063/1.4742331
5. J. Zhao, G. Wang, R. Yang, X. Lu, M. Cheng, C. He, G. Xie, J. Meng, D. Shi, G. Zhang, Tunable piezoresistivity of nanographene films for strain sensing. ACS Nano **9**(2), 1622–1629 (2015). https://doi.org/10.1021/nn506341u
6. R.C. Mani, M.K. Sunkara, R.P. Baldwin, J. Gullapalli, J.A. Chaney, G. Bhimarasetti, J.M. Cowley, A.M. Rao, R. Rao, Nanocrystalline graphite for electrochemical sensing. J. Electrochem. Soc. **152**(4), E154 (2005). https://doi.org/10.1149/1.1870772
7. D. Noll, P. Hönicke, Y. Kayser, S. Wagner, B. Beckhoff, U. Schwalke, Transfer-free in situ CCVD grown nanocrystalline graphene for sub-PPMV ammonia detection. ECS J. Solid State Sci. Technol. **7**(7), Q3108–Q3113 (2018). https://doi.org/10.1149/2.0171807jss
8. C. Albu, S.A. Eremia, M.L. Veca, A. Avram, R.C. Popa, C. Pachiu, C. Romanitan, M. Kusko, R. Gavrilă, A. Rădoi, Nano-crystalline graphite film on SiO_2: electrochemistry and electroanalytical application. Electrochim. Acta **303**, 284–292 (2019). https://doi.org/10.1016/j.electacta.2019.02.093
9. S. Rana, J.D. Reynolds, T.Y. Ling, M.S. Shamsudin, S.H. Pu, H.M. Chong, D. Pamunuwa, Nanocrystalline graphite for reliability improvement in MEM relay contacts. Carbon **133**, 193–199 (2018). https://doi.org/10.1016/j.carbon.2018.03.011
10. J. Sun, M.E. Schmidt, M. Muruganathan, H.M.H. Chong, H. Mizuta, Large-scale nanoelectromechanical switches based on directly deposited nanocrystalline graphene on insulating substrates. Nanoscale **8**, 6659–6665 (2016). https://doi.org/10.1039/C6NR00253F
11. S.J. Fishlock, S.J. O'Shea, J.W. McBride, H.M.H. Chong, S.H. Pu, Fabrication and characterisation of nanocrystalline graphite MEMS resonators using a geometric design to control buckling. J. Micromech. Microeng. **27**(9), 095015 (2017). https://doi.org/10.1088/1361-6439/aa7ebb
12. T.M. Dinh, A. Achour, S. Vizireanu, G. Dinescu, L. Nistor, K. Armstrong, D. Guay, D. Pech, Hydrous RuO_2/carbon nanowalls hierarchical structures for all-solid-state ultrahigh-energy-density micro-supercapacitors. Nano Energy **10**, 288–294 (2014). https://doi.org/10.1016/j.nanoen.2014.10.003
13. O.G. Simionescu, A. Avram, R.C. Popa, O. Tutunaru, C. Pachiu, G. Dinescu, Morphology and structural investigations of graphene/graphite nanowalls grown by capacitively coupled RF-PECVD at various substrate temperatures. Poster presentation at International Conference on Plasma Physics and Application (CPPA), Iasi, Romania (2019). https://doi.org/10.13140/rg.2.2.15154.56009
14. T. Schumann, M. Dubslaff, M.H. Oliveira Jr., M. Hanke, F. Fromm, T. Seyller, L. Nemec, V. Blum, M. Scheffler, J.M.J. Lopes, H. Riechert, Structural investigation of nanocrystalline graphene grown on $(6\sqrt{3} \times 6\sqrt{3})R30^{\circ}$-reconstructed SiC surfaces by molecular beam epitaxy. New J. Phys. **15**, 123034 (2013). https://doi.org/10.1088/1367-2630/15/12/123034

15. S.K. Jerng, D.S. Yu, Y.S. Kim, J. Ryou, S. Hong, C. Kim, S. Yoon, D.K. Efetov, P. Kim, S.H. Chun, Nanocrystalline Graphite Growth on Sapphire by Carbon Molecular Beam Epitaxy. J. Phys. Chem. C **115**(11), 4491–4494 (2011). https://doi.org/10.1021/jp110650d

Chapter 28
Multi-Temperature Model for Ultrafast Laser Experiments on Single Layered Graphene

Mihai Oane, Bogdan Alexandru Sava, Lucica Boroica, Rareş Medianu, Ileana Cristina Vasiliu, Mihail Elisa, Ana Violeta Filip, and Catalin Marius Dinca

Abstract The integral transform technique, merging the Anisimov-Nolte Two Temperature Model with the Cattaneo—Vernotte equation, is used in the present paper. Using the integral transform technique, relatively simple expressions for the electrons and phonon temperatures can be obtained, which can provide great help for the experimentalists especially for laser flash experiment on graphene. The obtained expressions are easy to handle in the software package MATHEMATICA 9.0.

Keywords Single layer graphene · Two temperature models · Multi temperature model

28.1 Introduction

The Two Temperature Model (TTM) by laser interaction with solids was proposed almost 45 years ago [1]. Many developments followed since then in specialized literature [2–17]. The solution to TTM can be inferred by solving two coupled differential equations. A simplified TTM was introduced by Nolte in 1997 [2]. In 2012 a TTM with an infinite speed of heat propagation [3] was proposed by our team.

TTM developments can also stand for an essential theoretical tool to be used in connection with the new strongly emerging technologies based on laser additive manufacturing (LAM), also known as solid freeform fabrication, digital manufacturing, or e-manufacturing [18–20].

Recently it was pointed out a more sophisticated model that replaces and is beyond the standard TTM [21–29]. The new model named: Multiple Temperature Model (MTM) is more sophisticated from the theoretical point of view and gives a much better fitting of the experimental data as, for example, in predicting the thermal field

M. Oane · B. A. Sava (✉) · L. Boroica · R. Medianu · A. V. Filip · C. M. Dinca
National Institute for Lasers, Plasma and Radiation Physics, Bucharest, Romania
e-mail: savabogdanalexandru@yahoo.com

I. C. Vasiliu · M. Elisa
National Institute of R & D for Optoelectronics INOE 2000, Măgurele, Romania

I. Visa and A. Duta (eds.), *Solar Energy Conversion in Communities*,
Springer Proceedings in Energy, https://doi.org/10.1007/978-3-030-55757-7_28

during laser-graphene interaction. The main novelty of the MTM in comparison with TTM is that the new model takes into account different electron–phonon couplings for different phonon branches. Thus, instead one electron–phonon interaction, we have multiple interactions with respect to the different types of phonons. We have like example for single layer graphene: longitudinal optical phonons (LO), transverse optical phonons (TO), out-of-plane acoustic phonons (ZA), longitudinal acoustic phonons (LA), transverse acoustic phonons (TA) and out of plane optical phonons (ZO). The TTM, as far as we know, use the QUANTUM-EXPRESSO CODE. It is not very easy to handle because it makes appeal to very advanced knowledge in the area of solid state physics and collision theory. Our main goal is to present a less sophisticated MTM using the integral transform technique as we have succeeded with TTM [2]. It would be a great advantage for experimentalists taking into account that MTM [30, 31] is much more accurate in predictions.

28.2 TTM and MTM

The TTM has the following two coupled equations:

$$A \cdot T_e\left(\frac{dT_e}{dt}\right) = K\left(\frac{\partial^2 T_e}{\partial x^2} + \frac{\partial^2 T_e}{\partial y^2} + \frac{\partial^2 T_e}{\partial z^2}\right) - G\,(T_e - T_i) + P_a(\vec{r}, t) \quad (28.1.a)$$

$$C_i\left(\frac{dT_i}{dt}\right) = G\,(T_e - T_i) \quad (28.1.b)$$

Here T_e and T_i are the electron, respective phonon temperatures. G is the coupling factor between electrons and phonons. $P_a(\vec{r}, t)$ is the heat source, which is induced through laser-metal interaction. The interaction could be classical or steady state quantum mechanical type. A is electron heat capacity and K is the thermal conductivity of the metal.

G can be determined from the equation:

$$G = \frac{\pi^2 m N v^2}{6\tau\, T_i} \cdot \left(\frac{T_e}{T_i}\right)^4 \times \int_0^{T_e/T_d} \left[x^4/\left(e^x - 1\right)\right] \cdot dx \quad (28.2)$$

where: m is the electron mass, N is the conduction electron density, v is the velocity of sound in the metal, τ is the electron–phonon collision time, T_D is the Debye temperature.

Because: $C_i >> A$ [3], we can make the first approximation and write:

$$K\left(\frac{\partial^2 T_e}{\partial x^2} + \frac{\partial^2 T_e}{\partial y^2} + \frac{\partial^2 T_e}{\partial z^2}\right) - C_i\frac{\partial T_i}{\partial t} = -\,P_a(\vec{r}, t) \quad (28.3)$$

Now we can use the formula from Nolte model:

$$T_i = \kappa T_e \tag{28.4}$$

where:

$$\kappa = \frac{\tau_L}{\tau_L + \tau_i} \tag{28.5}$$

(τ_i—is the lattice cooling time and τ_L—is the pulse time)
Consequently, we have:

$$\frac{\partial T_i}{\partial t} = \kappa \frac{\partial T_e}{\partial t} \tag{28.6}$$

Equation (28.3) can be thus written as:

$$K\left(\frac{\partial^2 T_e}{\partial x^2} + \frac{\partial^2 T_e}{\partial y^2} + \frac{\partial^2 T_e}{\partial z^2}\right) - C_i \kappa \frac{\partial T_e}{\partial t} = -P_a(\vec{r}, t) \tag{28.7}$$

or:

$$\left(\frac{\partial^2 T_e}{\partial x^2} + \frac{\partial^2 T_e}{\partial y^2} + \frac{\partial^2 T_e}{\partial z^2}\right) - \frac{1}{\gamma}\frac{\partial T_e}{\partial t} = -\frac{P_a(\vec{r}, t)}{K} \tag{28.8}$$

with:

$$\gamma = \frac{K}{C_i \kappa} \tag{28.9}$$

The MTM in our formalism has the analytical form (for the phonon branch i, and e-p coupling):

$$A \cdot T_e\left(\frac{dT_e}{dt}\right) = K_e\left(\frac{\partial^2 T_e}{\partial x^2} + \frac{\partial^2 T_e}{\partial y^2} + \frac{\partial^2 T_e}{\partial z^2}\right) - \sum G_{ep,i}\left(T_e - T_{p,i}\right) + P_a(\vec{r}, t) \tag{28.10}$$

and:

$$C_{p,i}\left(\frac{dT_{p,i}}{dt}\right) = K_{p,i}\left(\frac{\partial^2 T_{p,i}}{\partial x^2} + \frac{\partial^2 T_{p,i}}{\partial y^2} + \frac{\partial^2 T_{p,i}}{\partial z^2}\right) + G_{ep,i}\left(T_e - T_{p,i}\right) + G_{pp,i}\left(T_{lat} - T_{p,i}\right) \tag{28.11}$$

The links between the two models are represented by the next equations:

$$G = G_{e,p} = \sum G_{ep,i} \tag{28.12}$$

and

$$\sum G_{pp,i}\left(T_{lat} - T_{p,i}\right) = 0 \tag{28.13}$$

where: $G_{pp,i}$ is the *p-p* coupling factor between phonon branch *i* and the scattering lattice reservoir; and T_{lat} is lattice temperature; defined to ensure the energy transfer among phonon branches conservation. Until now we have presented the general formalism of thermal effects in laser-graphene interaction. Our main goal is to solve analytically, the general formalism described above in order to predict the thermal field at laser-graphene interaction as close as possible to reality. The solutions can be found in the monographs [27, 28] or in articles [3, 21].

The solutions are of the following form [Eigen-function and Eigen-values], in polar coordinates:

$$T(r,\varphi,z,t) = \sum_{m,n}\sum_{i=1}^{\infty}\sum_{l=0}^{\infty}\sum_{j=1}^{\infty} \begin{bmatrix} \hat{f}_{2l}(\mu_{il}\lambda_j l)\cdot g(\mu_{il}\lambda_j t) K_r(\mu_{il},r)\times K_\varphi(2l,\varphi)\times K_z(\lambda_j,z) \\ +\hat{f}_{2l-1}(\mu_{il}\lambda_j l)\cdot g(\mu_{il}\lambda_j t) K_r(\mu_{il},r) K_\varphi(2l-1,\varphi)\times K_z(\lambda_j,z) \end{bmatrix} \tag{28.14}$$

here:

$$\hat{f}_{2l}\left(\mu_{ij}\lambda_j,l\right) = \frac{1}{k\pi C_{il}C_j}\int_0^a \alpha e^{-\alpha x}\times K_z\left(\lambda_j,z\right)dz \times \int_0^b\int_0^{2\pi} I_{mn}(r,\varphi) r\times K_r(\mu_{il},r)\times K_\varphi(2l,\varphi)drd\varphi \tag{28.15}$$

and

$$\hat{f}_{2l-1}\left(\mu_{ij}\lambda_j,l\right) = \frac{1}{k\pi C_{il}C_j}\int_0^a \alpha e^{-\alpha x}\times K_z\left(\lambda_j,z\right)dz \times \int_0^b\int_0^{2\pi} I_{mn}(r,\varphi)\cdot r\times K_r(\mu_{il},r)\times K_\varphi(2l-1,\varphi)drd\varphi \tag{28.16}$$

with:

$$I_{mn}(x,y) = I_{max,mn}\left[H_m\frac{\sqrt{2}x}{w}H_n\frac{\sqrt{2}y}{w}\times exp\left(-\frac{x^2+y^2}{w^2}\right)\right]^2, \tag{28.17}$$

$$g\left(\mu_{il},\lambda_j,t\right) = \frac{1}{\left(\mu_{il}^2+\lambda_j^2\right)\left[1-e^{-\beta_{ijl}^2 t}-\left(1-e^{\beta_{ijl}^2(t-t_0)}\right)u(t-t_0)\right]} \tag{28.18}$$

Here: $K_r(\mu_{il}, r)$, $K_\phi(2\,l, \phi)$, $K_\phi(2\text{–}1, \phi)$ și $K_z(\lambda_j, z)$ are the Eigen-functions of the Eigen-values: μ_{il}, $2\,l$, $2\,l\text{-}1$ și λ_j.(see references [3, 21, 27, 28] for full details). We have $K_r(\mu_{il}, r) = J_l(\mu_{il} \cdot r)$,$K_\phi(2l - 1, \phi) = \sin(l\phi)$.

and $K_z(\lambda_j, z) = \cos(\lambda_j \cdot z) + (h/k\lambda_j) \cdot \sin(\lambda_j z)$. Here *i, l, j* are usual integer index parameters which goes for *l* from 0 to infinity, and for *i* and *j* from 1 to infinity.

The big advantage thus is that the solution of Eq. (28.14) is a series which converges very rapidly.

For typical calculations, after 10 iterations the precision of the value for the thermal field has a value of 10^{-2} K, acceptable for many experiments.

Although we can limit the series to 10 iterations we obtain in fact an analytical solution rather than a semi-analytical one. In conclusion Eq. (28.14) can be written as follows:

$$T(r, \phi, z, t) = \sum_{m,n}\sum_{i=1}^{10}\sum_{l=0}^{10}\sum_{j=1}^{10} \hat{f}_{2l}(\mu_{il}, \lambda_j, l) \cdot g(\mu_{il}, \lambda_j, t) \times K_r(\mu_{il}, r) \times K_\phi(2l, \phi) \times K_z(\lambda_j, z)$$
$$+ \sum_{m,n}\sum_{i=1}^{10}\sum_{l=0}^{10}\sum_{j=1}^{10} \hat{f}_{2l-1}(\mu_{il}, \lambda_j, l) \cdot g(\mu_{il}, \lambda_j, t) \times K_r(\mu_{il}, r) \times K_\phi(2l - 1, \phi) \times K_z(\lambda_j, z) \tag{28.19}$$

For a typical single transversal mode {m,n}, the temperature is a sum of 2200 functions. In consequence we can calculate easily the thermal fields using a regular desktop PC.

28.3 Simulations

In this chapter we present our simulations according to integral transform techniques. For great details one has to study references [27, 28]. We used the MATHEMATHICA software package. From the very beginning we have to say that we made simulations using the experimental frame and input data from references [30, 31]. We consider a laser beam with radial focus of about 5 μm and TEM_{00}. The target is one layer graphene. In Fig. 28.1 is presented the electron beam temperature versus space and time, during 100 fs laser irradition on single layer graphene. The arbitrary units are self-consistent in Figs. 28.1, 28.2, 28.3 and 28.4.

The TO and LA phonons temperature versus space and time, during 100 fs laser irradition on single layer graphene, are presented in Fig. 28.2, respectively 3. In Fig. 28.4 the TO phonons thermal field after the end of laser irradiation is shown. We may comment that our model is in good agreement with the model presented in reference [30, 31]. We may conclude that the higer the coupling constant between electrons and phonons of kind i, the better is the concordance between our model and the model presented in reference [30, 31]. When the coupling constant is zero our model give a zero temperature variation which is not the case in reference [30, 31].

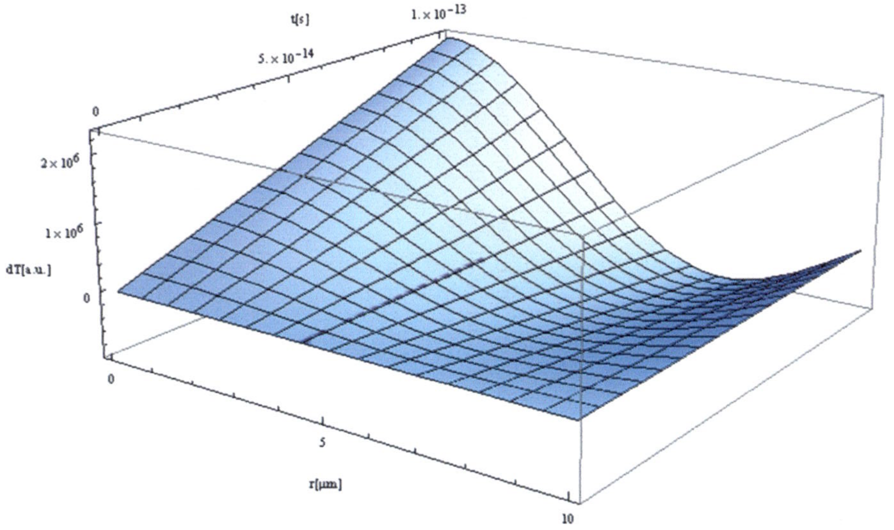

Fig. 28.1 The electron beam temperature versus space and time, during 100 fs laser irradition on single layer graphene

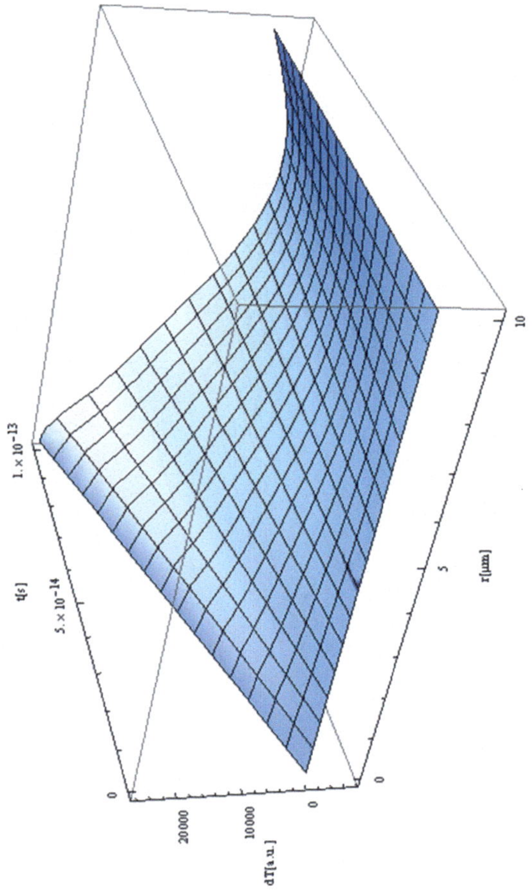

Fig. 28.2 The TO phonons temperature versus space and time, during 100 fs laser irradition on single layer graphene

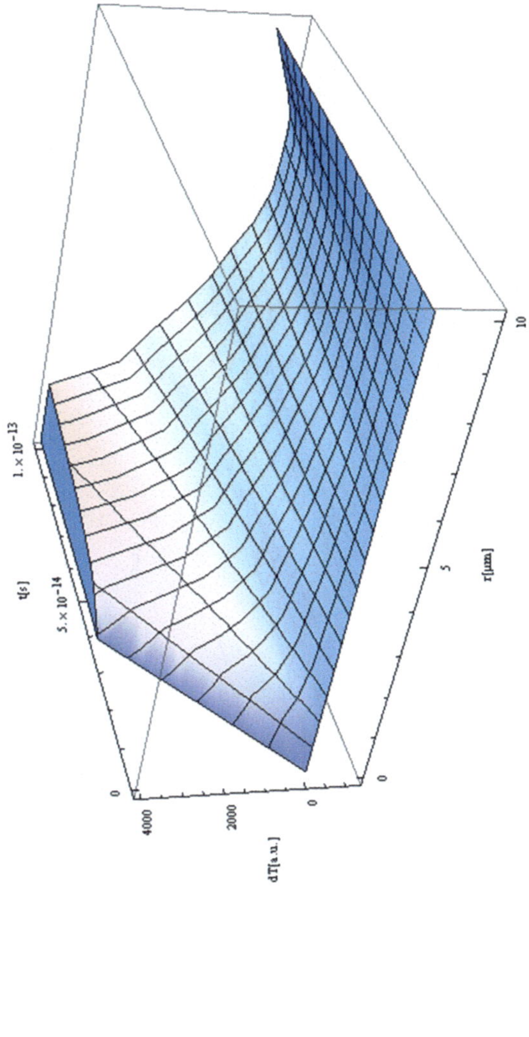

Fig. 28.3 The LA phonons temperature versus space and time, during 100 fs laser irradition on single layer graphene

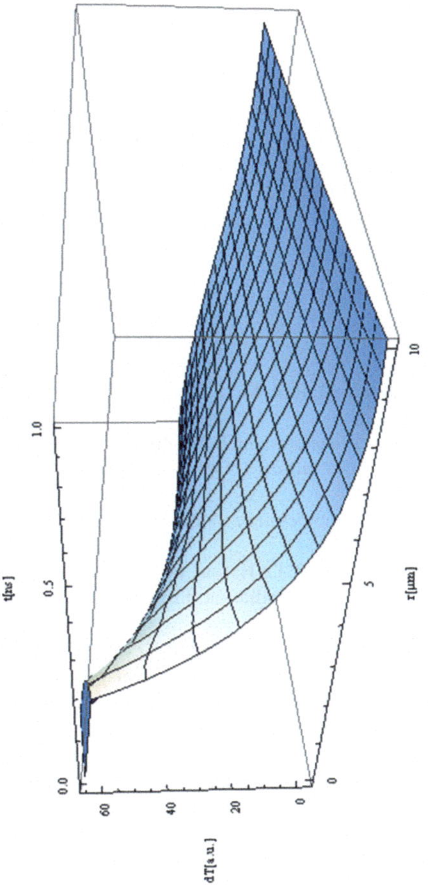

Fig. 28.4 The temperature field decrease for the TO phonons

28.4 Conclusions and Outlook

In this paper, we propose a new model for MTM applied to flash laser experiments on single-layered graphene. Our result are in good agreement with the litertaure data [30, 31]. Despite the model from [30, 31], our model is much simplified, like final results, and is able to reproduce 3D thermal fields instead of 2D thermal fields. Our model could be an excellent tool for the people from experimental area. The crux of mystery of our model is that we have applied a very powerful thermal model, namely the integral transform tehnique [22–24]. We do believe that our model could also be applied to very intense laser beams.

Acknowledgements the financial support in the frame of the project: PN-III-P1-1.2-PCCDI-2017-0871-contract 47PCCDI/2018, PN-III-P1-1.2-PCCDI-2017-0619-contract 42PCCDI/2018, PN-III-P1-1.2-PCCDI-2017-0387-Contract 80PCCDI/2018, and Core Programs: PN 16N/2019 LAPLAS VI, PN 18N/2019 OPTRONICA VI and 1IN/2020

References

1. D.D. Joseph, L. Preziosi, Heat waves. Rev. Mod. Phys. **61**, 41–73 (1989). https://doi.org/10.1103/RevModPhys.61.41
2. S. Nolte, C. Momma, H. Jacobs, A. Tünnermann, B.N. Chichkov, B. Wellegehausen, H. Welling, Ablation of metals by ultrashort laser pulses. J. Opt. Soc. Am. B. **14**, 2716 (1997). https://doi.org/10.1364/JOSAB.14.002716
3. M. Oane, M. Taca, S.L. Tsao, Two temperature models for metals: a new, "radical" approach. Lasers Eng **24**(1–2), 105–113 (2012)
4. J.P. Colombier, P. Combis, F. Bonneau, R. Le Harzic, E. Audouard, Hydrodynamic simulations of metal ablation by femtosecond laser irradiation. Phys. Rev. B. **71**, 165406 (2005). https://doi.org/10.1103/PhysRevB.71.165406
5. I.J. Kim, K.H. Pae, C.M. Kim, H.T. Kim, J.H. Sung, S.K. Lee, T.J. Yu, I.W. Choi, C.-L. Lee, K.H. Nam, P.V. Nickles, T.M. Jeong, J. Lee, Transition of proton energy scaling using an ultrathin target irradiated by linearly polarized femtosecond laser pulses. Phys. Rev. Lett. **111**, 165003 (2013). https://doi.org/10.1103/PhysRevLett.111.165003
6. Y. Yamashita, T. Yokomine, S. Ebara, A. Shimizu, Heat transport analysis for femtosecond laser ablation with molecular dynamics-two temperature model method. Fusion Eng. Des. **81**, 1695–1700 (2006). https://doi.org/10.1016/j.fusengdes.2005.09.011
7. G.L. Eesley, Generation of nonequilibrium electron and lattice temperatures in copper by picosecond laser pulses. Phys. Rev. B. **33**, 2144–2151 (1986). https://doi.org/10.1103/PhysRevB.33.2144
8. T. Balasubramni, S.H. Kim, S.H. Jeong, Enhanced two temperature modeling of ultrashort laser ablation for the investigation of thermionic emission characteristics. Appl. Surf. Sci. **255**, 9601–9604 (2009). https://doi.org/10.1016/j.apsusc.2009.04.080
9. M.E. Povarnitsyn, T.E. Itina, K.V. Khishchenko, P.R. Levashov, Multi-material two-temperature model for simulation of ultra-short laser ablation. Appl. Surf. Sci. **253**, 6343–6346 (2007). https://doi.org/10.1016/j.apsusc.2007.01.103
10. Z. Lin, L.V. Zhigilei, Temperature dependences of the electron–phonon coupling, electron heat capacity and thermal conductivity in Ni under femtosecond laser irradiation. Appl. Surf. Sci. **253**, 6295–6300 (2007). https://doi.org/10.1016/j.apsusc.2007.01.032

11. M.L. da Silva, V. Guerra, J. Loureiro, Two-temperature models for nitrogen dissociation. Chem. Phys. **342**, 275–287 (2007). https://doi.org/10.1016/j.chemphys.2007.10.010
12. N.A. Inogamov, V.V. Zhakhovskii, S.I. Ashitkov, V.A. Khokhlov, Y.V. Petrov, P.S. Komarov, M.B. Agranat, S.I. Anisimov, K. Nishihara, Two-temperature relaxation and melting after absorption of femtosecond laser pulse. Appl. Surf. Sci. **255**, 9712–9716 (2009). https://doi.org/10.1016/j.apsusc.2009.04.139
13. S. Valette, R. Le Harzic, N. Huot, E. Audouard, R. Fortunier, 2D calculations of the thermal effects due to femtosecond laser-metal interaction. Appl. Surf. Sci. **247**, 238–242 (2005). https://doi.org/10.1016/j.apsusc.2005.01.080
14. V. Linss, H. Kupfer, S. Peter, F. Richter, Determination of the neutral gas temperature of nitrogen-containing low-pressure plasmas using a two-temperature model. Surf. Coat. Technol. **200**, 1696–1701 (2005). https://doi.org/10.1016/j.surfcoat.2005.08.035
15. R.W. Schoenlein, W.Z. Lin, J.G. Fujimoto, G.L. Eesley, Femtosecond studies of nonequilibrium electronic processes in metals. Phys. Rev. Lett. **58**, 1680–1683 (1987). https://doi.org/10.1103/PhysRevLett.58.1680
16. B.H. Christensen, K. Vestentoft, P. Balling, Short-pulse ablation rates and the two-temperature model. Appl. Surf. Sci. **253**, 6347–6352 (2007). https://doi.org/10.1016/j.apsusc.2007.01.045
17. V. Damian, M. Oane, A. Bucă, The Fourier approach of the two temperature model for laser beam-metal interaction: experiment versus theory. Lasers Eng **1–3**, 181–186 (2016)
18. D.D. Gu, W. Meiners, K. Wissenbach, R. Poprawe, Laser additive manufacturing of metallic components: materials, processes and mechanisms. Int. Mater. Rev. **57**, 133–164 (2012). https://doi.org/10.1179/1743280411Y.0000000014
19. D. Gu, *Laser Additive Manufacturing of High-Performance Materials* (Springer, Berlin Heidelberg, 2015)
20. A. Gasser, G. Backes, I. Kelbassa, A. Weisheit, K. Wissenbach, Laser additive manufacturing: laser metal deposition (LMD) and selective laser melting (SLM) in turbo-engine applications. LTJ **7**, 58–63 (2010). https://doi.org/10.1002/latj.201090029
21. M. Oane, D. Sporea, Temperature profiles modeling in IR optical components during high power laser irradiation. Infrared Phys. Technol. **42**, 31–40 (2001). https://doi.org/10.1016/S1350-4495(00)00065-7
22. M. Oane, A. Peled, F. Scarlat, I.N. Mihailescu, A. Scarisoreanu, G. Georgescu, Multi-photon temperature profile modeling in solids during powerful pulse laser irradiation. Infrared Phys. Technol. **51**, 242–245 (2008). https://doi.org/10.1016/j.infrared.2007.07.003
23. M. Oane, A. Peled, F. Scarlat, I.N. Mihailescu, G. Georgescu, Quantum Fourier models for semiconductors under multiple laser irradiations. Infrared Phys. Technol. **51**, 348–350 (2008). https://doi.org/10.1016/j.infrared.2007.11.003
24. M. Oane, F. Scarlat, I.N. Mihailescu, The semi-analytical solution of the Fourier heat equation in beam—3D inhomogeneous media interaction. Infrared Phys. Technol. **51**, 344–347 (2008). https://doi.org/10.1016/j.infrared.2007.11.002
25. N.S. Koshlyakov, M.N. Smirnov, E.B. Gliner, *Differential Equation of Mathematical Physics* (North-Holland Publishing, Amsterdam, 1964)
26. T. Visan, D. Sporea, G. Dumitru, Computing method for evaluating the absorption coefficient of infrared optical elements. Infrared Phys. Technol. **39**, 335–346 (1998). https://doi.org/10.1016/S1350-4495(98)00019-X
27. M. Oane, D. Ticoş, C.M. Ticoş, *Charged Particle Beams Processing versus Laser Processing the Mathematics behind Processing Phenomena Particle Beams vs. Laser Processing* (Scholars' Press, Saarbrücken, 2015)
28. M. Oane, A. Peled, R.V. Medianu, Notes on laser processing. Lambert Academic Publishing (2013). ISBN 978-3-659-487-48739-2
29. M. Oane, R.V. Medianu, A. Bucă, A parallel between laser irradiation and relativistic electrons irradiation of solids, in *Radiation Effects in Optical Materials and Photonic Devices* ed. by D. Sporea, A. Sporea (InTech, 2016), pp. 413–430,. ISBN: 978-953-51-2417-7 1
30. Z. Lu, A. Vallabhaneni, B. Cao, X. Ruan, Phonon branch-resolved electron-phonon coupling and the multitemperature model. Phys. Rev. B. **98**, 134309 (2018). https://doi.org/10.1103/PhysRevB.98.134309

31. A.K. Vallabhaneni, D. Singh, H. Bao, J. Murthy, X. Ruan, Reliability of Raman measurements of thermal conductivity of single-layer graphene due to selective electron-phonon coupling: A first-principles study. Phys. Rev. B. **93**, 125432 (2016). https://doi.org/10.1103/PhysRevB.93.125432

Chapter 29
Graphene Oxide—Glass Nanocomposite Obtained by Sintering and Melting

Bogdan Alexandru Sava, Lucica Boroica, Ileana Cristina Vasiliu, Mihail Elisa, Alexandra Trefilov, Ana Violeta Filip, Andreea Matei, Catalin Marius Dinca, Oana Gherasim (Fufa), and Mihai Eftimie

Abstract Composite materials have many obvious advantages, as they can be made to be lightweight, strong, corrosion and heat resistant, flexible, transparent and more according to specific needs. Composites are already used in many industries, like aerospace, automotive components, wind turbine blades, building materials, medical utilities and others. Composite materials' merits and potential assures ample research in the field which is hoped to bring future developments and implementations in additional markets. The presence of graphene can enhance the conductivity and strength of bulk materials and help create composites with superior qualities. Graphene can also be added to metals, polymers and ceramics to create composites that are conductive and resistant to heat and pressure. From GO and a boron-phosphate glass a mix raw material with several % GO was made. The mix was pressed and thermally treated in order to obtain a GO—glass composite. Lithium-zinc-boron-lead-phosphate glass with 3% GO was melted-quenched at 700 °C melting temperature and (Tg + 10 °C) annealing temperature, for 2 h. GO-glass composites were characterized by atomic force microscopy (AFM), Raman and FTIR Spectroscopy.

Keywords Nanocarbon · Graphene oxide · Sol–gel · Glass · Nanocomposite

---

B. A. Sava · L. Boroica · A. Trefilov · A. V. Filip (✉) · A. Matei · C. M. Dinca · O. G. (Fufa)
National Institute for Lasers, Plasma and Radiation Physics, 409 Atomistilor str., 077125 Magurele, Ilfov, Romania
e-mail: ana_filip@ymail.com

B. A. Sava
e-mail: savabogdanalexandru@yahoo.com

I. C. Vasiliu · M. Elisa
National Institute of R & D for Optoelectronics INOE 2000, 409 Atomistilor str., 077125 Magurele, Ilfov, Romania

M. Eftimie
University Politehnica of Bucharest, 313 Splaiul Independenţei, 060042 Bucharest, Romania

I. Visa and A. Duta (eds.), *Solar Energy Conversion in Communities*,
Springer Proceedings in Energy, https://doi.org/10.1007/978-3-030-55757-7_29

29.1 Introduction

Nano-carbon, especially as graphene (G), graphene oxide (GO) or reduced graphene oxide (RGO), single layers of carbon atoms or oxidized carbon atoms in a two-dimensional hexagonal lattice, has been widely studied in the last years due to its amazing properties, including important electrical conductivity, high specific surface area and chemical stability, together with outstanding absorptivity and mechanical properties [1–3]. Many applications of nano-carbon materials included in composites or other complex systems were developed, from bio-applications such as drug delivery or photo-thermal therapy [4, 5] to lasers domain, as passive mode-locking of lasers or fiber lasers [6, 7].

Boron-phosphate glasses, which bring together the high mechanical, thermal and chemical properties of the boron-oxide glasses with the phosphate glass properties from whom the most important are the enhanced luminescence and laser pumping, together with higher level of dopants without quenching [8, 9] have many applications in photonics [10].

Nano-carbon-glass composites are not yet widely investigated, due to several technological problems. Few published works or patents are related to fiber glass-graphene composites [7], graphene nanoplatelets/glass composite coatings [11] or graphene oxide reinforced bioactive glass ceramics composites [12].

The aim of present work is to make a step forward by obtaining nano-carbon NC (as nanographite nG or GO) composites within a rare earth-doped boron-lead-phosphate glass host. Two methods were designed and tested to achieve this, first of them comprise melting the host glass, grinding it at desired granulation, suited by introducing NC, and sintering the obtained composite. The second method is the melting together, at low temperatures, of nano-carbon in doped boron-lead-phosphate glass.

These composites were investigated by using Fourier Transform Infrared – FTIR, Raman spectroscopy and X-ray photoelectron spectroscopy- XPS. Characteristic temperatures and thermal expansion coefficient were obtained by dilatometric measurements.

29.2 Experimental

The samples were obtained in lead-boro-phosphate systems, adding vitreous network modifiers and stabilizers namely Li_2O and ZnO and nano graphite NG. Nominal compositions of the samples, calculated taking into account the raw materials decomposition losses and the estimated volatilities of P_2O_5 and Li_2O of 25 wt% and 15 wt%, respectively, are presented in Table 29.1.

The samples were obtained by melting the raw materials previously processed by an unconventional wet route.

Table 29.1 Samples composition

Sample	Oxide	B_2O_3	P_2O_5	Li_2O	PbO	ZnO	nG	GO
BPP1	Mol%	10	25	10	50	5	0	
	Weight%	4.32	22.02	1.86	69.28	2.53		
BPP1nG	Mol%	10	25	10	50	5	3	
	Weight%	4.3	21.91	1.85	68.91	2.51	0.52	
BPP1GO	Mol%	10	25	9	48	5		3
	Weight%	4.43	22.57	1.71	68.17	2.59		0.53

As raw materials ultra-grade purity reagents: Li_2CO_3, ZnO, H_3PO_4, PbO and B_2O_3 were used. Nano graphite was fabricated at Microtechnology National Institute of Bucharest-Baneasa. Graphene oxide was fabricated at NILPRP Magurele using the Marcano-Tour method.

The wet method of raw materials processing, applied in BPP1 and BPP1GO glass and composite preparation runs according to the following steps:

(i) Gravimetric and/or volumetric measurement of the raw materials;
(ii) Cold homogenization of the raw materials, mechanically, for 15 min;
(iii) Drying the raw materials mixture, with partial elimination of water under constant stirring, up to 120–150 °C;
(iv) Pouring the mixture while it still is fluid, into the melting high alumina crucible.

Melting was done at 800 °C, for 1 h in a Nabertherm furnace, equipped with molybdenum disilicide heating elements, accompanied by mechanical homogenization of the melt by using a thermal shock resistant alumina stirrer at 200–240 rpm.

The melt was cast into high-density graphite molds, previously preheated to reduce the thermal stress due to temperature differences between melt and mold, and annealed at 350 °C, for 3 h. The annealing program was designed using dilatometric data.

The BPP1 glass was grinded, sieved, homogenized without and with 3 mol% nG and pressed at about 2Tf/cm^2 with a hydraulic press. The sintering of BPP1nG samples took place at 400 °C for 4 h.

The thermal expansion coefficient was determined by dilatometry with a horizontally Netzsch DIL 400 PC device. The representative temperatures (strain point, T_S, glass transition temperature, T_g, annealing point, T_A, and dilatometry softening point, T_D) as well as the thermal expansion coefficient ($\alpha_{20}{}^{300}$) of the glass were estimated by processing the results provided by the dilatometry graph. The measurement errors for the thermal expansion coefficient were in the range of $\pm 0.8 \times 10^{-7}$ K^{-1}, supplied by calibration with the SiO_2 standard sample, according to ISO 7991.

For XPS measurements a X-ray photoelectron spectroscope was used, manufactured by Thermo Scientific, 2016, equipped with: Instrument: Escalab Xi+ ; X-ray source: AlKa monochromatic and silver anode upgrade; Analyzer: energy range 0–5000 eV; power steps: for 0–1500 eV step of 6 meV; for 0–3000 eV step of 12 meV; for 0–5000 eV step of 24 meV; System provided with the Ar ion cluster source for contaminants cleaning of the analyzed surfaces and for obtaining the depth profile; System equipped with UV spectroscopy (UPS); The system allows the use of REELS technique (reflection electron energy loss spectroscopy).

FTIR spectra were recorded at RT with a Perkin Elmer Spectrophotometer-Spectrum 100, provided with universal attenuated total reflectance (UATR) accessory, in the range 400–1500 cm^{-1}. The measurement error is of ±0.1%, for 32 scans.

RT Raman spectra were collected with a LABRAM-HR 800 Horiba Jobin Yvon spectrometer, in the 130–1500 cm^{-1} range, using the 514.5 nm line of a cw Ar+ laser as excitation source. The laser power at the sample surface was 2.33 mW, without attenuation filter, objective × 100, exposure time 2 s, accumulation time 5 s, five cycles, resolution 1 cm^{-1}, laser spot 1 μm^2, and grating 1800 gr mm^{-1}.

29.3 Results and Discussion

The sample melted together with graphene oxide in the above presented work conditions crystallized and must be melted in different conditions, which will be subject of a future work. This is the reason why the next investigations were performed only on the base glass and on the nano-graphite-glass sintered composite.

29.3.1 *Thermal Properties*

The thermal expansion coefficient for the base glass is calculated from the dilatometry graph (Fig. 29.1) as $\alpha_{20}{}^{300} = 14.4169 \times 10^{-6}\ K^{-1}$ and the characteristic temperatures are the following: T_S (strain point) = 319.3 °C, T_g (glass transition temperature) = 347.3 °C, T_A (annealing point) = 355.5 °C, and T_D (dilatometry softening point) = 362.5 °C. dL is the elongation of the sample after heating up from 20 to 300 °C and L_0 is the initial length of the sample. The T_A and T_g values were used to design the annealing program, which was identical for the base glass and for the composite obtained using the second method, meaning the melting of the glass together with GO.

29.3.2 *XPS Investigations*

The XPS investigations, shown in Figs. 29.2 and 29.3 and in Tables 29.2 and 29.3, identified the elements in the composite system, including the nanocarbon introduced by the nano-graphite.

The identification of B, P, Li, Pb, Zn, C were qualitatively not quantitatively made using XPS measurements, the difference between the net samples composition and

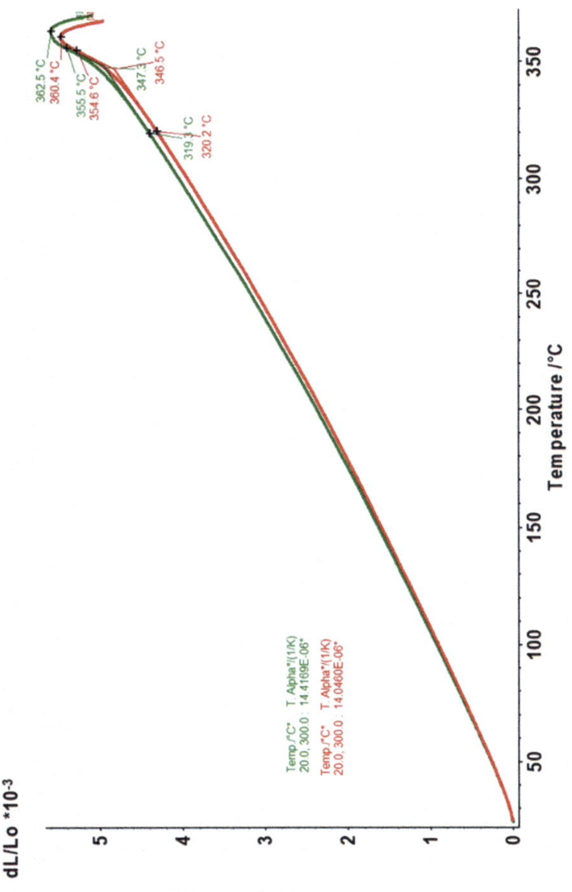

Fig. 29.1 Dilatometry graph of BPP1 (red line) and BPP1GO (green line) glass and composite samples

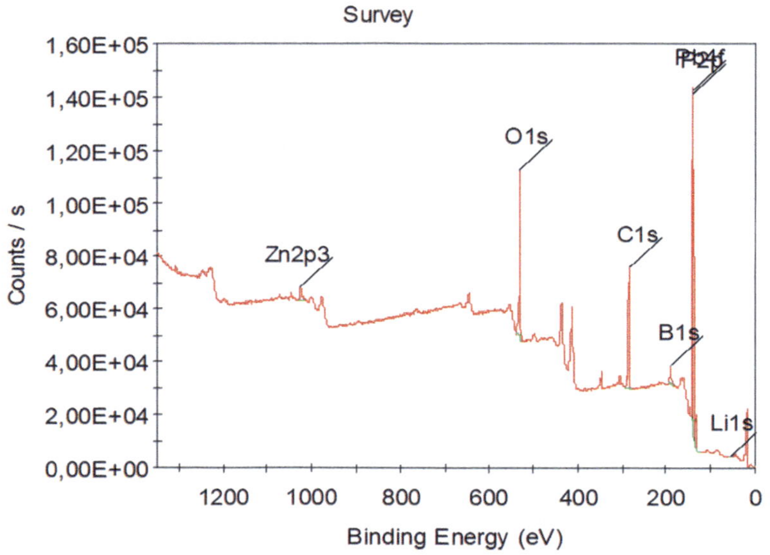

Fig. 29.2 XPS analysis of borophosphate block base glass obtained by melting/casting/sintering technique—sample BPP1

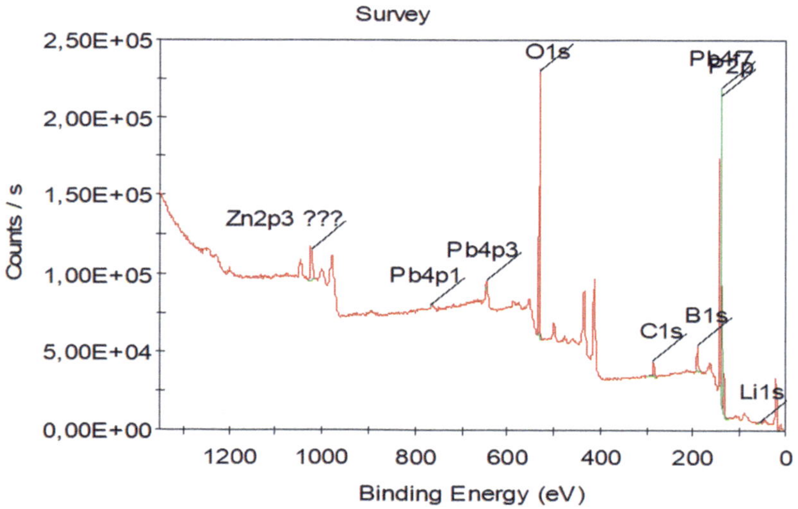

Fig. 29.3 XPS analysis of borophosphate block composites with nanocarbon materials obtained by melting/casting/sintering technique—sample BPP1nG

Table 29.2 Elements in BPP1 sample identified by XPS measurements

Element	Peak BE	FWHM eV	Area (P) CPS eV	Atomic %
Pb4f	138.37	1.41	25,317.56	3.25
O1s	531.16	3.08	16,039.74	27.93
Zn2p	1022.02	3.55	9004.43	2.01
C1s	284.72	2.91	678.38	3.07
B1s	190.73	3.30	1994.08	23.17
P2p1	138.50	1.15	4316.75	40.58

Table 29.3 Elements in BPP1nG sample identified by XPS measurements

Element	Peak BE	FWHM eV	Area (P) CPS.eV	Atomic %
O1s	530.95	3.08	530,160.53	28.04
P2p	137.96	2.18	438,944.82	42.36
Zn2p3	1023.93	6.94	143,403.36	1.45
C1s	284.13	2.75	36,282.59	4.98
B1s	190.12	3.34	55,301.25	19.51
Li1s	54.35	1.33	1624.78	3.39
Pb4p3	644.94	1.96	10,832.00	0.28

the XPS data is attributed to apparatus identification programs and to the possibilities of XPS to identify only surface elements and a few nanometers within the sample.

29.3.3 FTIR Spectroscopy

Figure 29.4 displays the FTIR transmission curve for BPP1 and BPP1nG samples.

Several absorption peaks appear in the FTIR spectra, located at about 581, 650, 725, 954, 1038, 1080, 1263, 1991 and 2104 cm^{-1}. Infrared absorption bands observed can be attributed to the vibration mode P = O stretching vibration of P3 units superposed with $(PO_2)_{as}$ mode in P^2 units (1263 cm^{-1}), to $(P–O^-)_{as}$ vibration in Q^2 and/or (P–O–Zn) linkages (1080 cm^{-1}) [13, 14]. The peak at about 1080 and 1263 cm^{-1} was attributed also to the stretching vibrations of C–O bond [15, 16].

The shift of 1270 cm^{-1} band of phosphate glass to lower wavenumbers, specifically to 1263 cm^{-1}, is attributed to P–O–B connectivity [14]. Bending vibrations of $O_3B–O–BO_3$ links can be identified at 725 cm^{-1} and stretching vibrations of B–O bonds in BO_4 units from tri-borate groups at 1038 cm^{-1} [17]. The peak on 954 cm^{-1} is attributed to symmetric stretching of phosphate groups P-O bond [12]. The bands from 581 and 650 cm^{-1} can be attributed to the vibrations of Zn–O bond and to stretching and torsional modes associated with the vibrational motion of the B-O species [14, 15].

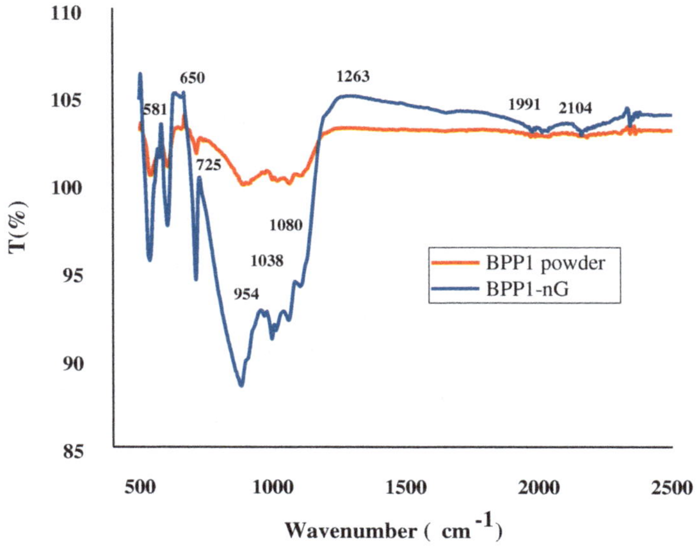

Fig. 29.4 FTIR spectroscopy of BPP1 (red line) and BPP1nG (blue line) samples

29.3.4 Raman Investigations

Raman spectra for BPP1 and BPP1nG together with graphite are presented in Fig. 29.5, for the domain of specific carbon Raman shifts.

The Raman shifts specific for D peak at ~1335 cm^{-1} induced by defects sp^2-carbon bonds, the G peak at ~1582 cm^{-1} that indicated in-plane vibrational mode which involves sp^2 hybridized carbon atoms that comprises graphene sheets [18] are visible in Fig. 29.5 for the investigated samples. The shift of the G peak is visible in the Raman spectrum of BPP1nG sample.

The Raman microscope was also used to take some photos of the samples that showed the NC homogeneous repartition in the glass host. Figure 29.6 shows one of them.

29.4 Conclusions

A new zinc-boron-lead-phosphate glass composite with nano-carbon was obtained and characterized. The glass was melted with or without nano-carbon, grinded, pressed, sintered and characterized. Another sample was melted together with graphene oxide, but this sample crystallized and must be obtained in the future in different conditions and/or with a different low-melting glass composition.

XPS evidenced the nG presence in samples together with all other glass elements.

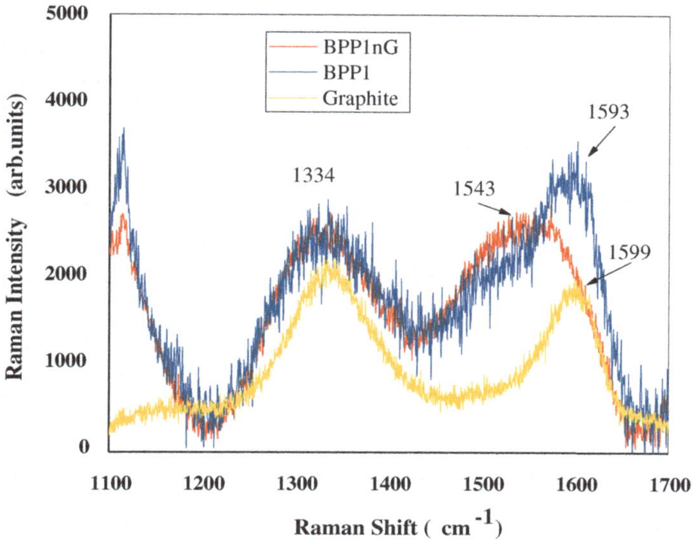

Fig. 29.5 Raman spectra for the samples BPP1 and BPP1nG comparatively to graphite

Fig. 29.6 Raman microscope photo of the sample BPP1nG

The obtained composites were investigated by using Fourier Transform Infrared—FTIR and Raman spectroscopy which demonstrated combined structure including meta-phosphate, boron Q2 units, B–O–P links, together with $[PbO_4]$ groups.

Structural units Q2 and Q3 and also P–O–P, B–O–B and P–O–B links vibrations were evidenced by FTIR spectroscopy revealing a mixed boron-phosphate glass network. The nG is evidenced by the Raman shift at about 1543 cm^{-1}.

Nest research will enlarge the properties measurements and will investigate potential laser applications of these composites.

Acknowledgements UEFISCDI, for the financial support in the frame of the project: PN-III-P1-1.2-PCCDI-2017-0871-contract 47PCCDI/2018, PN-III-P1-1.2-PCCDI-2017-0619-contract 42PCCDI/2018, PN-III-P1-1.2-PCCDI-2017-0387-Contract 80PCCDI/2018, and Core Programs: PN 16N/2019 LAPLAS VI, PN 18N/2019 OPTRONICA VI.

References

1. J.C. Huang, Carbon black filled conducting polymers and polymer blends. Adv Polym Technol **21**(4), 299–313 (2002)
2. M. Ishigami et al., Atomic structure of graphene on SiO_2. Nano Lett **7**(6), 1643–1648 (2007)
3. J. Yan et al., Raman scattering and tunable electron–phonon coupling in single layer graphene. Solid State Commun **143**(1), 39–43 (2007)
4. L. Zhang et al., Functional graphene oxide as a nanocarrier for controlled loading and targeted delivery of mixed anticancer drugs. Small **6**(4), 537–544 (2010)
5. Y. Tao et al., Immunostimulatory oligonucleotides loaded cationic graphene oxide with photothermally enhanced immunogenicity for photothermal/immune cancer therapy. Biomaterials **35**(37), 9963–9971 (2014)
6. M.C. Paul, G. Sobon, J. Sotor, K.M. Abramski, J. Jagiello, R. Kozinski, L. Lipinska, M. Pal, A graphene-based mode-locked nano-engineered zirconia–yttria–aluminosilicate glass-based erbium-doped fiber laser. Laser Phys. **23**, 035110 (2013). https://doi.org/10.1088/1054-660X/23/3/035110
7. US 9.410,246 B2/2016 patent, T.Y. Winarski, Graphene optic fiber laser
8. B.A. Sava, L. Boroica, M. Elisa, O. Shikimaka, D. Grabco, M. Popa, Z. Barbos, R. Iordanescu, A.M. Niculescu, V. Kuncser, A.C. Galca, M. Eftimie, R.C.C. Monteiro, Bismuth and lead oxides codoped boron phosphate glasses for Faraday rotators. Ceram. Int. **44**(6), 6016–6025 (2018)
9. O. Shikimaka, D. Grabco, B.A. Sava, M. Elisa, L. Boroica, E. Harea, C. Pyrtsac, A. Prisacaru, Z. Barbos, Densification contribution as a function of strain rate under indentation of terbium-doped aluminophosphate glass. J. Mater. Sci. **51**(3), 1409–1417 (2015)
10. J.F.C. Carreira, B.A. Sava, L. Boroica, M. Elisa, R. Stefan, R.C.C. Monteiro, T. Monteiro, L. Rino, Structural and luminescence characterization of a Dy/Tb co-doped borophosphate glass. J. Non-Cryst. Solids **526**, 119719 (2019)
11. A. Nistal, E. Garcia, D. Pérez -Coll, C. Prieto, M. Belmonte, M.I. Osendi, P. Miranzo, Low percolation threshold in highly conducting graphene nanoplatelets/glass composite coatings. Carbon **139**, 556–563 (2018). https://doi.org/10.1016/j.carbon.2018.07.030
12. K. Ilyas, S. Zahid, M. Batool, A.A. Chaudhry, A. Jamal, F. Iqbal, M.H. Nawaz, O. Goerke, A. Gurlo, A.T. Shah, I. Rehman, In-vitro investigation of graphene oxide reinforced bioactive glass ceramics composites. J. Non-Cryst. Solids **505**, 122–130 (2019). https://doi.org/10.1016/j.jnoncrysol.2018.10.047
13. S. Ibrahim, M. Abdel-Baki, F. El-Diasty, Zinc borophosphate glasses for infrared-based optical applications. Opt. Eng. **51**, 093401–093411 (2012). https://doi.org/10.1117/1.OE.51.9.093401
14. D. Larink, H. Eckert, M. Reichert, S.W. Martin, Mixed network former effect in ion-conducting alkali borophosphate glasses: structure/property correlations in the system

$[M_2O]_{1/3}[(B_2O_3)_x(P_2O_5)_{1-x}]_{2/3}$ (M = Li, K, Cs). J. Phys. Chem. C **116**, 26162–26176 (2012). https://doi.org/10.1021/jp307085t
15. M. Zamani, M. Rostami, M. Aghajanzadeh, H. Kheiri Manjili, K. Rostamizadeh, H. Danafar, Mesoporous titanium dioxide@ zinc oxide–graphene oxide nano carriers for colon-specific drug delivery. J Mater Sci **53**, 1634–1645 (2018). https://doi.org/10.1007/s10853-017-1673-6
16. D.C. Marcano, D.V. Kosynkin, J.M. Berlin, A. Sinitskii, Z. Sun, A. Slesarev, L.B. Alemany, W. Lu, J.M. Tour, Improved synthesis of graphene oxide. ACS Nano **4**, 4806–4814 (2010). https://doi.org/10.1021/nn1006368
17. F. He, Z. He, J. Xie, Y. Li, IR and Raman spectra properties of Bi_2O_3-ZnO-B_2O_3-BaO quaternary glass system. Am J Anal Chem **5**, 1142–1150 (2014). https://doi.org/10.4236/ajac.2014.516121
18. M. Bayati, H. Peng, H. Deng, J. Lin, M. Fidalgo de Cortalezzi, Laser induced graphene/ceramic membrane composite: preparation and characterization. J. Membr. Sci. **595**, 117537 (2020). https://doi.org/10.1016/j.memsci.2019.117537

Chapter 30
Thin Films Based on ZnO-Graphene Oxide Heterostructures for Self-Cleaning Applications

Dana Perniu, Cristina Bogatu, Silvioara Gheorghita, Maria Covei, and Anca Duta

Abstract Responding to the increasing concern for preserving a clean environment, the development of self-cleaning coatings attracts growing interest, since they already demonstrated efficiency in pollutants removal. Considering the fundamental materials' properties, the target is to develop materials demonstrating enhanced photoactivity under visible light irradiation, corroborated with controlled hydrophilic character. Two major conditions attract the research interest to provide promising, viable solutions: the material should be earth-abundant, non-toxic for humans and environment and the deposition technology should be affordable in terms of costs and energy, and does not lead to environmental risks. This paper presents the development of two types of materials: ZnO, obtained by spray pyrolysis and the ZnO–GO composite, deposited by combined sol–gel and spraying technique. The as-prepared samples proved promising self-cleaning properties. Both hydrophilicity and photodegradation efficiencies were enhanced by a surface pre-activation, consisting of samples' exposure in dark and UV irradiation.

Keywords ZnO · ZnO-graphene oxide · Self-cleaning coating

D. Perniu · C. Bogatu (✉) · S. Gheorghita · M. Covei · A. Duta
Transilvania University of Brasov, Brasov, Romania
e-mail: cristina.bogatu@unitbv.ro

D. Perniu
e-mail: d.perniu@unitbv.ro

S. Gheorghita
e-mail: silviagheorghita@yahoo.com

M. Covei
e-mail: maria_covei@yahoo.com

A. Duta
e-mail: a.duta@unitbv.ro

I. Visa and A. Duta (eds.), *Solar Energy Conversion in Communities*,
Springer Proceedings in Energy, https://doi.org/10.1007/978-3-030-55757-7_30

30.1 Introduction

The continuous need for environmental depollution supports intensive research efforts on finding solutions that can be applied at a large scale, with minimum risks of side-effects related to humans and/or environment and also with minimum costs. Learned from the nature, the self-cleaning concept attracted attention on developing surfaces able to provide protection against chemical/physical or biological pollutants. A self-cleaning surface is meant either to remove the "dirt" by attracting it in the water droplets and rolling these off, or decomposing the "dirt" in less pollutant species in a water film spread over the surface [1]. The two mechanisms are strongly dependent on the surface wettability, assessed by the contact angle developed by the water–vapor interface on the solid surface. A contact angle larger than 90° is specific for the hydrophobic surfaces (and superhydrophobic if the value is higher than 150°), while values less than 90° correspond to a hydrophilic surface. If the contact angle value is less than 10° the surface is superhydrophilic and the water drops spread over the surface forming a water film, able to "host" the contaminant decomposition process, and to "wash" the surface [1]. If the contaminant decomposition process runs under irradiation using a photo-activated catalyst, the surface is characterized as a photocatalytic self-cleaning surface. The most investigated photocatalyst is TiO_2, but others semiconductors (as ZnO) and/or heterostructures involving semiconductors attract attention, especially to surpass the TiO_2 limitations as its wide band gap that allows its photo-activation only under UV radiation (thus limiting its the large-scale applications) or the fast charge carriers recombination, that limits the photocatalytic efficiency. To mitigate these drawbacks, the development of composite structures involving semiconductors and carbonaceous materials, as graphite, graphene and its derivatives—graphene oxide and reduced graphene oxide are extensively studied for photocatalytic applications.

This research focuses on the ZnO semiconductor, known as an efficient photocatalyst, not harmful for humans and/or environment, chemically and photochemically stable that can be prepared using up-scalable methods [2]. However, ZnO has the main drawbacks as TiO_2: the rapid recombination of the photogenerated charge carriers and its wide band gap (E_g = 3.37 eV), limits its large scale applications. Thus, a promising approach is to couple ZnO with graphene derivatives, and graphene oxide (GO) is approached in this paper as a potential filler in the ZnO–GO composite.

The GO selection is based on literature data on its properties and roles in enhancing photocatalytic process efficiency when associated with ZnO [2–4]. It is known that the graphene oxide sheets have their basal planes decorated mostly with epoxy and hydroxyl groups and carboxyl groups located at edges [3, 5]. These polar groups act as anchor sites for ionic species involved in the synthesis and provide GO hydrophilic properties. Thus, GO is a chemically active material, allowing the manipulation in polar solvents to form composite materials. In heterogenous photocatalysis, based on semiconductor-GO composites, the GO (p-type semiconductor) was reported to enhance the VIS absorption if a proper energy bands alignment is formed and, consequently, the process efficiency increases due to a significantly reduced electron–hole pairs' recombination [6–8]. The ZnO–GO conduction and valence bands are well positioned supporting the VIS activation on ZnO.

This paper reports on ZnO and ZnO–GO composite thin films, applied as photocatalytic self-cleaning surfaces. The objective is to develop a (super)hydrophilic surface, with photocatalytic activity, and to investigate the possibility to enhance these properties by photoactivation, as a pre-treatment stage.

30.2 Materials and Equipment

The investigated materials were developed as thin films by spraying the ZnO and ZnO –GO precursors on a glass substrate, in the following structures: Glass/ZnO_SPD and Glass/ZnO_SPD/ZnO–GO. The spray pyrolysis deposition (SPD) was used as technique for the ZnO_SPD thin film development. The zinc acetate dihydrate ($ZnAc_2.2H_2O$, Scharlau) was used as Zn precursor, dissolved in mixed aqueous-alcoholic solvent (1:1, vol), in 0.15 mol/L concentration and the Acethylacetone (Scharlau) was used as additive (1:100, vol). The precursor solution was sprayed (10 sequences, with 20 s break between pulses), on pre-heated (250 °C) commercial glass samples (1.5 cm × 1.5 cm × 0.4 cm), using air as carrier gas (p = 1.2 bar). A CERAN heating plate was used for substrate heating and the ABB IRB Foundy Plus robot for automatized spray deposition. The NORBATHERM oven was further used for thin films annealing at 450 °C, for 5 h.

The structure Glass/ZnO_SPD/ZnO–GO was developed by combining sol–gel process with spray deposition. The sol for the composite (ZnO–GO) layer was prepared using zinc acetate dihydrate ($ZnAc_2.2H_2O$ Scharlau), in c = 0.5 moles/L and GO from stock aqueous dispersion (IMT Bucharest) of 30 mg/mL. The GO ratio was adjusted to obtain a theoretical concentration of 3% (mass) in the ZnO thin film. The precursors were dissolved in isopropylic alcohol (IPA, Scharlau) and monoethanolamine (MEA, Scharlau) as stabilizer. The aged sol (48 h aging duration) was diluted in the ratio sol: IPA = 1:2 (vol) followed by ultrasonication and was sprayed (20 spraying sequences, with 20 s break) on the previously developed layer (glass/ZnO_SPD) pre-heated at 70 °C. The post-deposition treatment, in ambient atmosphere was run at 150 °C for 2 h. The same procedure, without the addition of GO was followed to obtain the structure, glass/ZnO_SPD/ZnO.

The thin films were structurally (by X-Ray diffraction, XRD Brucker, D8 Discover) characterized. The self-cleaning assessment was done by two tests: wettability and photocatalytic activity. The thin film wettability was evaluated by water contact angle (WCA) measurements (OCA 20, Dataphysics). The photocatalytic activity was evaluated as the efficiency of methylene blue (MB, 10 ppm) photodegradation after 9 h (1 h in dark and 8 h under irradiation), following the procedure described in previous work [9, 10]. Different spectral compositions (UV, VIS, UV + VIS) and irradiances (measured with DeltaOHM HD 2102.2 radiometer coupled with two sensors: LP471 A-UV (250–400 nm) and LP471 RAD (400–1050 nm) were used to get insight of the thin films behavior under irradiation. The samples morphology (Scanning Electron Microscopy, SEM S-3400 N-Hitachi and Atomic Force Microscopy, AFM Ntegra Spectra, NT-MDT) and optical properties (Transmittance, T UV–ViS-NIR Perkin Elmer, Lambda 950, Spectrophotometer) were analyzed to discuss the thin films morphology and stability.

30.3 Results and Discussions

30.3.1 Thin Film Synthesis

Two types of samples were prepared, for the examination of the self-cleaning properties: (1) Glass/ZnO_SPD and (2) Glass/ZnO_SPD/ZnO–GO. A supplementary reference sample Glass/ZnO_SPD/ZnO was prepared.

The Glass/ZnO_SPD thin films obtained by the spray pyrolysis deposition, was already proved as photo-active under irradiation [11, 12].

The Glass/ZnO_SPD/ZnO–GO sample development followed a two-steps procedure, aiming at depositing the two layers. Preliminary investigations proved that the the bottom layer, glass/ZnO_SPD, ensures a good adherence of the composite structure to the glass substrate and supports its ordered growth, thus the first step was to deposit this layer. The upper layer, consisting of the composite ZnO–GO, was developed by a method which combines the sol–gel procedure and spray deposition technique. Samples with the upper layer containing pristine ZnO were also developed by using the same procedure as the composite material. The method was designed to simultaneous ensure the film adherence to substrate, the ZnO formation and crystallization, and to prevent the GO decomposition. Five distinct steps were followed for the thin film composite development: (a) sol preparation, (b) sol ageing and gel formation, (c) gel dilution, (d) diluted gel deposition and (e) thin film treatment.

The quality of the final thin film depends on the sol stability. Thus, the role of additives is essential, to keep the relative ratio between hydrolysis and polymerization at a level to ensure the Zn–O chain formation with no precipitation. The literature recommends the mono-ethanolamine (MEA) as a good stabilizer, as it plays complementary roles [13–18]. Zinc acetate has limited solubility in ethanol in the absence of other agents or heating, while in the presence of water it "precipitates" (due to the zinc cations hydratation), giving the appearance of gelation. Since MEA has an alkaline character, it acts as proton acceptor, deprotonating the water molecules, forming hydroxide ions and consequently generating zinc hydroxide involved in the promotion of condensation reactions during the sol–gel process. As a bidentate ligand, MEA supports the chelatization of zinc ions in the formation of oxy-acetate complexes in condensation reactions followed by the formation of Zn–O chains, without precipitation. Thus, MEA plays an essential role both in the complete precursor dissolution and in the sol stabilization.

During the sol formation, the ZnO–GO composite is also expected to be formed, since the zinc (complex) ions, positively charged interact with the negatively charged (or free electron-containing) groups (hydroxyl, epoxy, carboxyl) on graphene oxide [19, 20].

Spraying the aged sol, after dilution followed by thermal treatment at a temperature that is preventing the GO thermal decomposition leads to the final structure that was further investigated.

30.3.2 Thin Film Characterization

Structural characterization. The X-Ray diffraction (XRD), as a powerful investigation, gives a first insight on materials' structure. The XRD spectra for different samples and, as reference, for the pristine GO thin film (deposited by spray technique on glass) are presented in Fig. 30.1. The specific ZnO diffraction peaks in all samples evidence that the crystalline ZnO structure is maintained both in pristine and in composite materials. The GO presence in the composite structure is marked by the peak at 2Θ around 10° but with a slight shift in position suggesting the interaction/intercalation of ZnO particles in GO sheets, also mentioned in the literature [6, 20, 21]. The overall crystallinity degree for the samples varies in the range 30–32%, with no major differences among them.

Thin films self-cleaning properties. The self-cleaning ability of the investigated materials is the result of two simultaneous surface properties: photocatalytic activity and hydrophilicity. On a hydrophilic surface (superhydrophilic), the water droplets spreads over the surface and form a film, creating the condition for pollutant

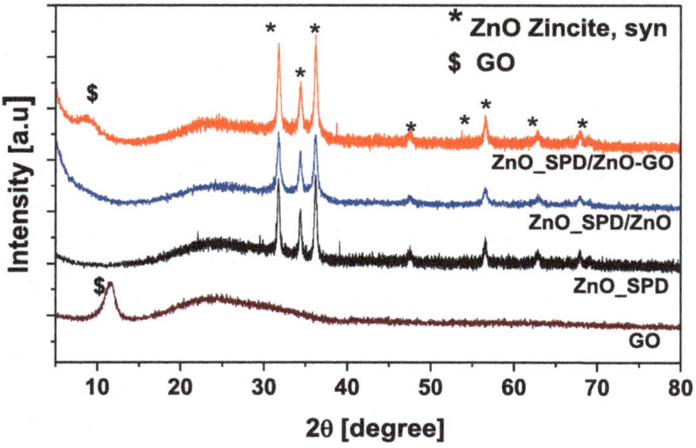

Fig. 30.1 The XRD spectra for the investigated thin films

Table 30.1 Photocatalytic efficiencies of the investigated samples

No	Sample	WCA (°)	η (%) UV + VIS	η (%) UV	η (%) VIS
1	Glass/ZnO_SPD	17.1	40.91	37.47	20.11
2	Glass/ZnO_SPD/ZnO	38.4	14.25	9.87	9.33
3	Glass/ZnO_SPD/ZnO–GO	31.5	36.32	25.05	13.55

photodegradation and hence the removal of the decomposition products from the surface [1, 22].

Thus, a first control parameter is the thin film photocatalytic activity, as its ability to decompose, under irradiation, the targeted pollutant, MB (according to the ISO 10678:2010 standard). In Table 30.1, the photocatalytic efficiencies are presented, together with the water contact angles (WCA) of the thin films. The photodegradation efficiency (η), was assessed under different spectral compositions and irradiance values: UV, ($G = 3$ W/m^2), VIS ($G = 52$ W/m^2) and combined UV + VIS simulating the solar light percentage ($G = 55$ W/m^2).

The data in Table 30.1 reveal the samples hydrophilicity and their photodegradation ability to decompose the MB solution. For all samples the photodegradation efficiency under UV irradiation is rather low, but this can be explained by the very low irradiance value. It is to be noticed the increase in photodegradation efficiency under solar simulated irradiation, and it is quite appreciable, considering the low light irradiance. The ZnO sample obtained by SPD technique is the one for which both hydrophilic character and photdegradation efficiency suggest a promising self-cleaning capacity. For the coatings obtained by combined sol–gel and spray deposition, the pristine ZnO has significant lower photodegradation efficiency and lower hydrophilicity than the composite structure ZnO–GO. It is thus confirmed the role of GO in improving the photocatalytic efficiency of ZnO under VIS-irradiation, most probably due to the efficient photogenerated electron injection from ZnO conduction band to GO, as an excellent electron acceptor, allowing the reduction of electron–hole pairs recombination, a well-known limiting factor of photocatalytic efficiency (as also presented in the literature [1, 6, 20, 23]). Since ZnO is a wide band gap semiconductor, it is activated by the UV radiation. The VIS-activation is explained by the dye sensitization, which can be excited and transfer electrons to the conduction band of ZnO, able to promote the formation of reactive oxygenated species [2]. The VIS-activation is also observed for the ZnO–GO composite material when compared with pristine ZnO, as result of the lowering of the overall bandgap [8, 23].

Further investigations, aiming at understanding the photoinduced surface wettability and variations in its photocatalytic activity will be presented for the more efficient Glass/ZnO_SPD and Glass/ZnO_SPD/ZnO–GO samples.

The surface photo-activation. As known from literature, when the semiconductor oxides (or heterostructures) with photocatalytic activity are irradiated with UV, their wettability is influenced, resulting in superhydrophilic surfaces. In turn, the long dark storage results in increase of the hydrophobicity.[1, 24–26]. Consequently, the Glass/ZnO_SPD and Glass/ZnO–SPD/ZnO_GO samples behavior was investigated after keeping them in dark or exposed to UV irradiation ($G = 8.3\ W/m^2$) for specific periods of 24, 48, 72 h. The samples were also tested after combined activation as a possible solution for surface re-activation. In this respect, the samples were kept in dark for 24 h or 7 days and, after this period were irradiated with UV for 24 h.

The general behavior of the Glass/ZnO_SPD sample, as observed in Fig. 30.2a, is quite regular. The UV irradiation enhances the surface hydrophilicity towards super-hydrophilicity, since the dark exposure promote the decrease of surface hydrophilic character. Generally, this behavior is explained for photoactive semiconductors by the generation of the electron-holes pairs under irradiation, which can react with the surface ions, generating surface defective sites able to host water molecules and/or hydroxyl radicals (formed by water molecule decomposition under UV irradiation) [1, 25, 26]. The result is the increased surface ability for water adsorption, and a lower water contact angle. When kept in dark, due to the lack of induced partially surface charges and also due to oxygen molecule adsorption, the hydrophilic character is significantly reduced. In Fig. 30.2c, the high value of WCA after long dark storage confirms this behavior.

In the case of the sample with composite structure, the behavior is not fully regular. This might be explained by the graphene oxide presence, with the surface polar groups (carboxyl, epoxy), which allows the water molecules surface adsorption (enhanced when the surface is UV irradiated due to ZnO contribution). The graphene oxide also increases the surface roughness, as the results in Fig. 30.3 outline, thus the water adsorption is favored. After a longer exposure under UV radiation, the surface slightly changes towards reduced hydrophilicity, most probably due to the slow replacement of the sorbed water molecules by oxygen molecules from air.

It was thus concluded that the ZnO and ZnO_GO composite surfaces, kept in dark reduce their hydrophilicity, while the UV radiation increases this property. A 24-h irradiation time was considered proper since longer periods do not significantly improve the behavior and induce higher energy consumption. Consequently, it was examined the combined treatment for the samples re-activation, by keeping the samples in dark for specified periods (24 h, and 7 days) followed by UV irradiation for 24 h. The WCA values are presented in Fig. 30.2c. For both samples, the dark exposure brings reduction in hydrophilicity (significant in the case of ZnO_SPD surface), which can be recovered, by exposure to a 24 h of UV irradiation, leading to highly hydrophilic surfaces.

When the surface properties are discussed, the Atomic force microscopy (AFM) analyses give an insight. Figure 30.3 outlines the surface roughness (RMS) values calculated for the samples kept in dark and/or exposed to UV-irradiation. From the RMS data it can be observed that the surface roughness for the sample Glass/ZnO_SPD has not a significant variation as result of dark or UV exposures. From this point of view, the surface seems to be stable, but this conclusion has to be supported by the other investigations (further presented). In the case of the Glass/ZnO_SPD/ZnO–GO sample, the surface roughness decreases in time,

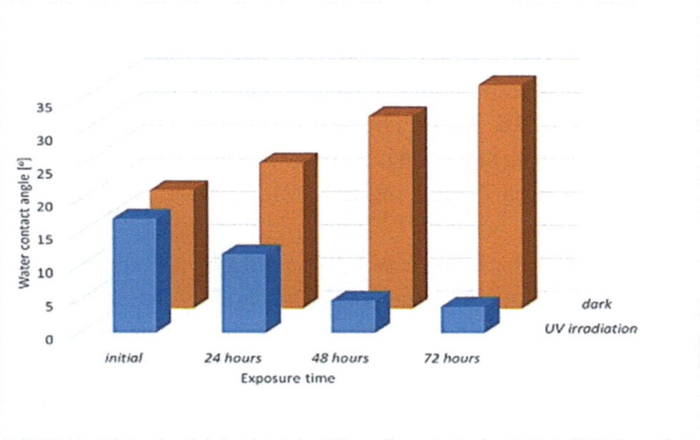

(a)

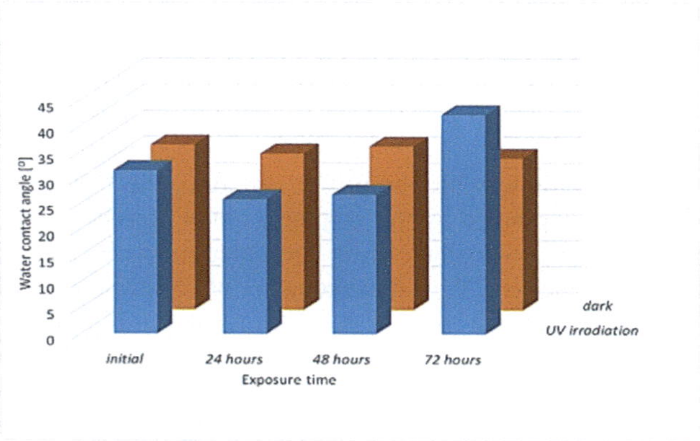

(b)

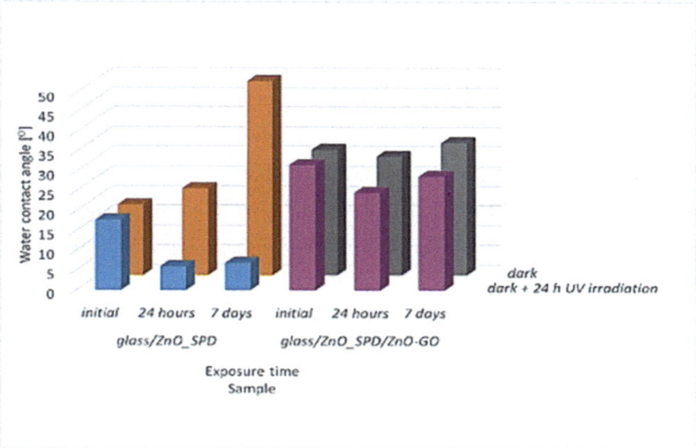

(c)

Fig. 30.2 Water contact angle variation after keeping in dark or under UV radiation for **a** Glass/ZnO_SPD sample, **b** Glass/ZnO_SPD/ZnO–GO sample, **c** combined exposure (dark for 24 h or 7 days + 24 h of UV irradiation)

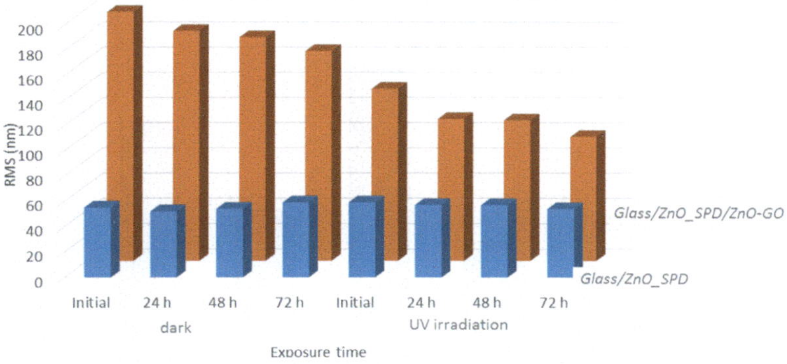

Fig. 30.3 Surface roughness variation

especially after UV irradiation, most probably due to surface chemical bonds re-arrangement. The variations in surface roughness support the changes in surface wettability.

The photocatalytic tests. As fundamental property for self-cleaning behavior of a surface, the photocatalytic activity was investigated.

Learning from the first set of data (Table 30.1) that the as-prepared samples have the ability to decompose the MB solution under solar-simulated radiation, the effect of surface preliminary activation was studied. The photocatalysts pre-treatment consists of dark and UV-irradiation for specific periods: 24 h or 7 days of dark exposure; re-activation 24 h dark + 24 h UV exposure, 7 days dark + 24 h UV exposure. The pre-treated samples were further tested in the photodegradation of MB (10 ppm solution). In Fig. 30.4, photocatalytic efficiencies after 8 h of UV + VIS irradiation are presented. For both samples, the activation for 24 h in dark and 24 h of UV exposure provided higher photodegradation efficiencies. This result, corroborated with the low water contact angle leads to the conclusion that the investigated samples prove the photocatalytic self-cleaning properties, after a preliminary surface conditioning.

Stability tests. In order to get information on the thin film stability, two investigations were run, before and after photocatalysis: surface morphology, by scanning electron microscopy (SEM) and thin film transmittance.

Thin film morphology showed variations in the case of thin films containing the composite material, as presented in Fig. 30.5. The surface maintains its wrinkled morphology, both after UV treatment and after photocatalysis. However, after photocatalysis, additional aggregates, most probably containing the degradation products, are observed. Since the composite surface is highly rough, it can host the photodegradation products, which might be chemically adsorbed by interactions with the polar groups on graphene oxide.

Optical transmittance. Thin films transparency was evaluated from two perspectives: as application property for transparent surfaces and also as control property for thin film stability. The variation in transmittance of the composite sample before

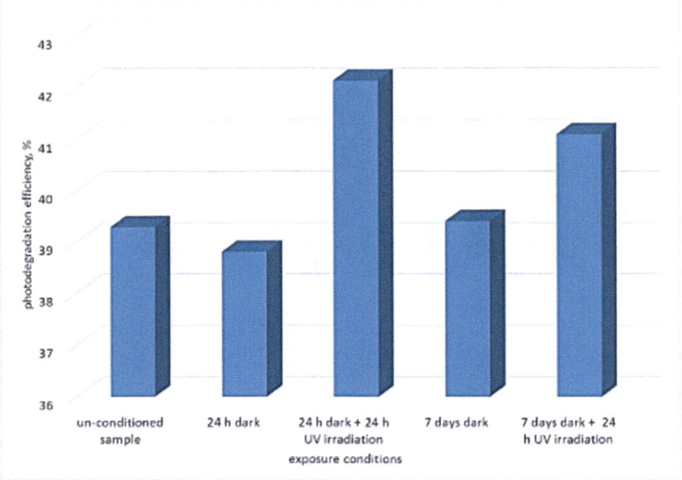

(a)

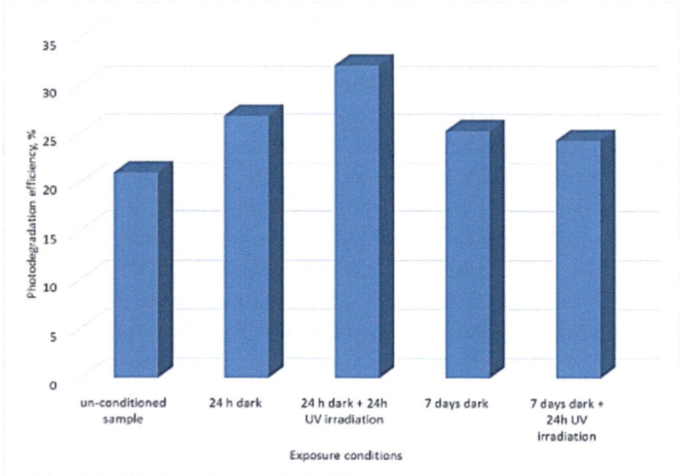

(b)

Fig. 30.4 Photodegradation efficiencies for **a** sample Glass/ZnO_SPD and **b** Glass Glass/ZnO_SPD/ZnO–GO

the photocatalysis, after the UV treatment and after photocatalysis in MB 10 ppm is presented in Fig. 30.6.

The sample transparency is not significantly affected by the UV treatment, as expected, in agreement with the surface morphology results.

The decrease in the sample transmittance is more evident after photocatalysis and may results from adsorption of the colored photodegradation (by)product molecules

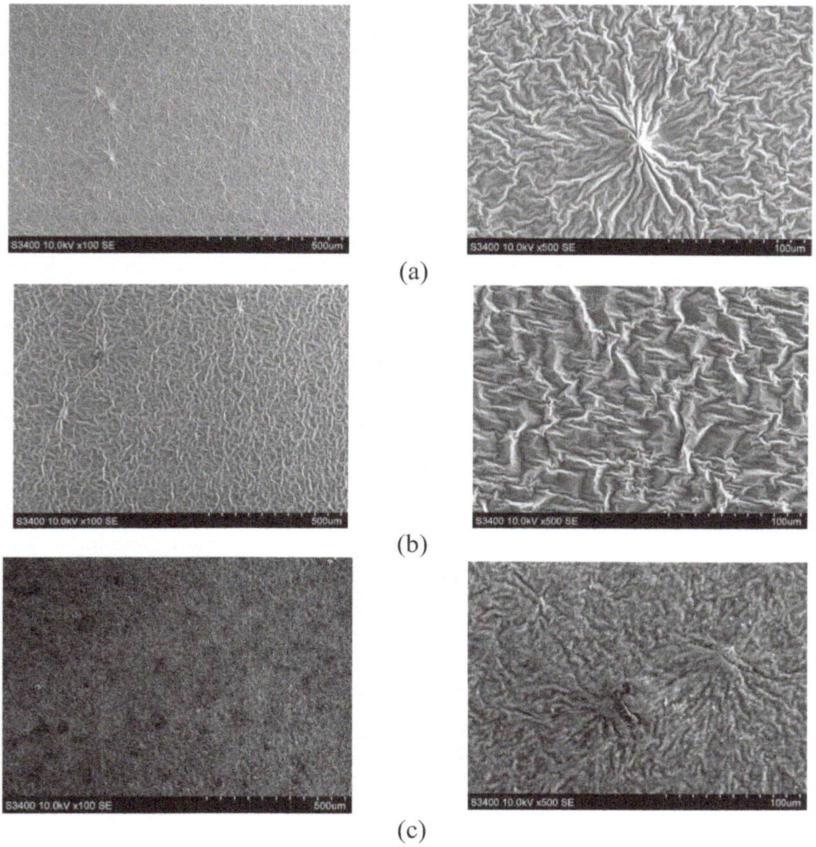

Fig. 30.5 Surface morphology of the Glass/ZnO_SPD/ZnO–GO sample, at different amgnitudes **a** before the photocatalysis and **b** after 24 h UV exposure **c** after photocatalysis

Fig. 30.6 The transparency variation for Glass/ZnO_SPD/ZnO–GO thin film

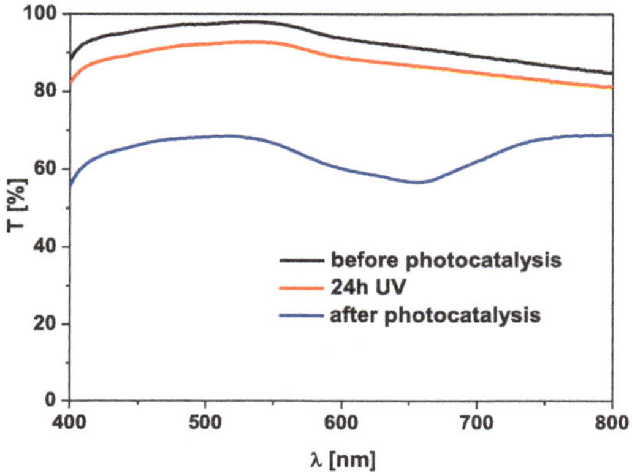

on the film surface. These findings are confirmed by the SEM results. However, the thin films stability can be considered acceptable, but need further optimization for self-cleaning applications, particularly on glazed surfaces.

30.4 Conclusions

This paper presents two types of samples (ZnO and ZnO–GO thin films) developed to provide photocatalytic self-cleaning surfaces, property which corroborates the high hydrophilicty with promising efficiencies of photocatalytic activity. The samples were obtained using techniques that allow the deposition on large surfaces. If the surface allows the high temperature treatment, the spray pyrolysis deposition is a good option, as it leads to stable thin films, with photocatalytic activity under (simulated) solar radiation and hydrophilic character that can be changed to superhydrophilic if UV-irradiation is applied as pre-treatment.

If the substrate does not allow high temperature processing, a combined sol–gel and spray method can be used for developing photocatalytic self-cleaning surfaces. The thin films based on pristine ZnO proved very low photodegradation efficiencies when compared to ZnO–GO composite thin films. The photoinduced pre-activation (after 24 h in dark and 24 under UV radiation exposure) improved the efficiency of the composite thin film and also the surface wettability.

References

1. S. Banerjee, D. Dionysiou, S.C. Pillai, Self-cleaning applications of TiO_2 by photo-induced hydrophilicity and photocatalysis. Appl. Catal. B **176–177**, 396–428 (2015)
2. G. Mamba, G. Gangashe, L. Moss, S. Hariganesh, S. Thakur, S. Vadivel, A.K. Mishra, G.D. Vilakati, V. Muthuraj, T.T.I. Nkambule, State of the arton the photocatalytic applications of graphene based nanostructures: From elimination of hazardous pollutants to disinfection and fuel generation. J. Environ. Chem. Eng. **8**, 103505 (2020)
3. C. Han, N. Zhang, Y.-J. Xu, Structural diversity of graphene materials and their multifaious roles in heterogenous photocatalysis. Nano Today **11**, 351–372 (2016)
4. L. Zhu, Z. Liu, P. Xia, H. Li, Y. Xie, Synthesis of hierarchical ZnO and graphene composites with enhanced photocatalytic activity. Ceram. Int. **44**, 849–856 (2018)
5. P. Kumbhakar, A. Pramanik, S. Biswas, A.K. Kole, R. Sarkar, In-situ synthesis of rGO nanocomposite for demonstration of sunlight driven enhanced photocatalytic and self-cleaning of organic dyes and tea stains of cotton fabrics. J. Hazard. Mater. **360**, 193–203 (2018)
6. P. Raizada, A. Sudhaik, P. Singh, Photocatalytic water decontamination using graphene and ZnO coupled photocatalysis: a review. Mater. Sci. Energy Technol. **2**, 509–525 (2019)
7. K. Qi, B. Cheng, J. Yu, W. Ho, Review on the improvement of the photocatalytic and antibacterial activities of ZnO. J. Allyos Compd. **727**, 792–820 (2017)
8. S. Victor-Roman, E. Garcia-Bordeje, J. Hernandez-Ferrer, J.M. Gonzalez-Dominguez, A. Anson-Casaos, A.M.T. Silva, W.K. Maser, A.M. Benito, Controlling the surface chemistry of graphene oxide: key towards efficient ZnO–GO photcatalysis. Catal. Today, https://doi.org/10.1016/j.cattod.2019.05.049.

9. M. Covei, C. Bogatu, D. Perniu, A. Duta, I. Visa, Self-cleaning thin films with controlled optical properties based on WO3-rGO. Ceram. Int. **45**, 9157–9163 (2019)
10. M. Covei, C. Bogatu, D. Perniu, I. Tismanar, A. Duta, Comparative study on the photodegradation efficiency of organic pollutants using n-p multi-junction thin films. Catal. Today **328**, 57–64 (2019)
11. M. Duta, D. Perniu, A. Duta, Phototcatalytic zinc oxide thin films obtained by surfactant assisted spray pyrolysis deposition. Appl. Surf. Sci. **306**, 80–88 (2014)
12. M. Covei, D. Perniu, C. Bogatu, A. Duta, ZnO thin films with controlled properties deposited by spray pyrolysis, in *ZnO Thin Film: Properties, Performance and Applications, Materials Science and Technologies* ed. by P. Mele. Nova Science Publishers (2019)
13. M. Sharma, R.M. Mehra, Sol-gel derived zinc-oxide films alloyed with cobalt and aluminium. Thin Solid Films **518**, 3725–3730 (2010)
14. P. Hosseini Vajargah, H. Abdizadeh, R. Ebrahimifard, M.R. Golobostanfard, Sol-gel derived ZnO thin films: effect of amono-additives. Appl. Surf. Sci. **285P**, 732–743 (2013)
15. L. Znaidi, G.J.A.A. Soler Illia, S. Benyahia, C. Sanchez, A.V. Kanaev, Oriented ZnO thin films synthesis by sol-gel process for laser applications. Thin Solid Films **428**, 257–262 (2003)
16. D. Nesheva, V. Dzhurkov, I. Stambolova, V. Blaskov, I. Bineva, J.M. Calderon Moreno, S. Preda, M. Gartner, T. Hristova-Vasileva, M. Shipochka, Surface modification and chemical sensitivity of sol gel deposited nano crystalline ZnO films. Mater. Chem. Phys. **209**, 165–171 (2018)
17. T. Demes, C. Ternon, D. Riassetto, H. Roussel, L. Rapenne, I. Gelard, C. Jimenez, V. Stambouli, M. Langlet, New insights in the structural and morphological properties of sol-gel deposited multilayer films. J. Phys. Chem. Solids **95**, 43–55 (2016)
18. A. Gomez-Nunez, S. Alonso-Gil, C. Lopez, P. Roura-Grabulosa, A. Vila, From ethanolamine precursor towards ZnO—how N is released from the experimental and theoretical points of view. Nanomaterials **9**, 1415 (2019)
19. Z. Wu, L. Wang, Graphene oxide (GO) doping hexagonal flower-like ZnO as potential enhancer of photocatalytic ability. Mater. Lett. **234**, 287–290 (2019)
20. P.S. Chauhan, R. Kant, A. Rai, A. Grupa, S. Bhattacharya, Facile synthesis of ZnO/GO nanoflowers over Si susbtarte for improved photocatalysis decolorization of MB dye and industrial wastewater under solar irradiation. Mater. Sci. Semicond. Process. **89**, 6–17 (2019)
21. N. Just, B. Berke, K. Laszlo, L.P. Bakos, A. Szabo, K. Hernadi, I.M. Szilagyi, Preparation of graphene oxide/semiconductor oxide composites by using atomic layer depostion. Appl. Surf. Sci. **453**, 245–251 (2018)
22. G.S. Hikku, K. Jeyasubramanian, J. Jabobjose, P. Thiruramanathan, P. Veluswamy, H. Ikeda, Alkyd resin based hydrophylic self-cleaning surface with self-refreshing behaviour as single step durable coating. J. Colloid Interface Sci. **531**, 628–641 (2018)
23. X. Li, R. Shen, S. Ma, X. Chen, J. Xie, Graphene-based heterojunction photocatalysts. Appl. Surf. Sci. **430**, 53–107 (2017)
24. A. Chakraborty, A. Agresti, R. Pizzoferrato, F. De Matteis, A. Orsini, P.G. Medaglia, Study of structural and optical properties of low temperature photo-activated ZnO-rGO composite thin film. Mater. Res. Bull. **91**, 227–231 (2017)
25. X. Feng, L. Jiang, Design and creation of superwetting/antiwetting surfaces. Adv. Mater. **18**, 3063–3078 (2006)
26. J. Lu, K. Huang, X. Chen, J. Zhu, F. Meng, X. Song, Z. Sun, reversible wettability of nanostructured ZnO thin films by sol-gel method. Appl. Surf. Sci. **256**, 4720–4623 (2010)

Part VI
Sustainable Communities

Chapter 31
Renewable Energy Systems: Current Status and Prospects

Soteris A. Kalogirou

Abstract This keynote presentation examines the current status of renewables in the world. The presentation starts with some facts about the climate change, global warming and the effects of human activities such as the burning of fossil fuels on the climate problem. It then examines shortly the current status of conventional resources of energy, followed by a general outline of the status of renewables in the world, which includes the shares with respect to conventional fuel use for electricity and power and jobs created. Then the basic forms of renewables are examined in some detail, which include solar thermal, both for low and high temperature applications, photovoltaics, hydro power, onshore and offshore wind energy systems and biomass/biofuels. In all these the basic technology is presented followed by the current status as well as the prospects of the technology and new research findings.

Keywords Renewable energy systems · Solar thermal · Photovoltaics · Wind energy systems · Biomass

31.1 Introduction

The purpose of this paper is to show the worldwide status of the various renewable energy systems (RES). This is presented in terms of the total installed capacity of the various types of RES by the end of 2019 and is based on the reports of various international agencies and organizations [1–3]. The paper initially examines the effects of human activities on climate. Today there is a convincing evidence that human activities have a consequence on the climate. The term "global warming" is not now considered as the best term representing what is happening but now the term "climate change" is more representative, as we do not have only warm summers but also severe winters. An analysis is presented in terms of the CO_2 levels in the atmosphere and it is shown that the value of ppm was never before at the current level. It is also shown how major human activities, have affected the amount of the CO_2

S. A. Kalogirou (✉)
Department of Mechanical Engineering and Materials Sciences and Engineering, Cyprus University of Technology, Limassol, Cyprus
e-mail: soteris.kalogirou@cut.ac.cy

I. Visa and A. Duta (eds.), *Solar Energy Conversion in Communities*,
Springer Proceedings in Energy, https://doi.org/10.1007/978-3-030-55757-7_31

in the atmosphere. A quick review of the status of the existing conventional fossil fuels is also presented and their expected depletion, which is based on the existing resources and the current rates of consumption. With current rate of consumption, the oil and natural gas reserves will last for about 56 and 53 years respectively, whereas those for coal for more than 100 years.

31.2 Renewable Energy Systems

The types of RES examined include the main and most important, in terms of capacity, technologies, i.e. solar thermal, photovoltaics, hydro systems, wind energy systems and biomass, biogas and biodiesel. The total amount of renewable power capacity by the end of 2019 is 2588 GW whereas the figure if we exclude hydro is down to 1437 GW [1]. From these, solar PV constitute 627 GW, a 122 GW increase from the 2018 figure, wind power is equal to 651 GW, compared to 591 in 2018, and solar thermal is equal to 479 GWth, compared to 482 GWth for 2018 [1], i.e. there is a slight decrease. The estimated renewable energy share for the global electricity production is 72.7% non-renewable electricity and 27.3% renewable electricity [1]. In each type of RES examined in addition to the total installed capacity the status of the technology is given as well as the outlook in terms of prospects and the current research areas.

A comparison in global renewable energy Levelized Cost of Electricity (LCOE) from 2010 to 2015 shows that almost all renewables have a low and almost equal price in this period whereas there is a substantial decrease in LCOE for solar PV and solar thermal systems; all of them have a lower value compared to conventional fuel electricity production. Another characteristic of renewables is that worldwide more than 10.3 million people are working in jobs related to this industry [1].

In the various technologies, solar thermal power includes systems utilising either the thermal radiation or the light of solar irradiance. The former includes solar thermal systems, which comprise both low temperature systems (mostly for water heating and industrial processes) and high temperature systems (mostly for CSP and high temperature industrial processes), whereas the latter refers to solar photovoltaics. Both have shown a steady increase year-by-year. For solar thermal power the mostly used type of collectors are the glazed water ones. Worldwide the mostly used type is the evacuated tube, whereas in Europe is flat plate collector [2]. Research in this area lead to several new innovations like:

- Use of polymeric materials for the manufacture of solar thermal absorbers to.
  - reduce cost due to lower raw material and manufacturing costs.
  - reduce weight compared to copper or aluminum.
- Improved heat transfer with the use of nanofluids.
- New transparent covers with anti-reflective coatings for high optical transmission, and the use of high vacuum or noble gases.

- Switchable coatings to reduce stagnation temperatures.
- New selective absorber with low emission coatings.
- Temperature-resistant super-insulating materials.

Photovoltaics are breaking every year a new threshold. In 2015 for the first time the installed capacity exceeded the 200 GW limit (actual value: 228 GW), in 2016 the 300 GW limit (actual value: 303 GW), in 2017 the 400 GW limit (actual value: 402 GW), in 2019 the 500 GW (actual value: 505 GW) and on 2020 the 600 GW (actual value: 627 GW). In 2019 an additional of 122 GW were installed, which is equivalent to the installation of more than 50,000 solar panels every hour worldwide. From the 122 GW installed in 2019, 30.1 GW were installed in China (also leading the world in terms of total installed capacity with 204.7 GW); EU, with 16 GW, is second, USA with 13.3 GW is third and India with 9.9 GW is fourth [3].

Hydro energy comes in a wide variety of sizes starting from the very large units that are greater than 10 MW to the pico size of less than 5 kW. The largest application worldwide is the 18 GW scheme at the Three Gorges in China. The four top countries are China with 28%, Brazil 9%, Canada and USA both with 7% [1].

Another area of renewables is the wind power that also shows a year-by-year increase. The current installed capacity is 651 GW with 60 GW installed just in 2019 [1]. Wind has become the least-cost option for new power generating capacity in an increasing number of markets.

The last area of renewables examined is biomass in general, which comprise as main areas in addition to biomass, biogas and biofuels (biodiesel) whereas some people considered waste, like Municipal Solid Waste (MSW), and landfills also as renewables. The big question here is whether to use land for growing energy crops instead of food to satisfy the human needs. The answer to this is to utilize for this purpose land that is not currently used for food production. The biggest challenge here is the transportation sector as more than 90% of transport depends on oil. For example, US consumption with its 200 million cars is 3.5 million barrels of oil per day, which shows the prospects for biofuels. At the end of 2018 the shares of biomass in total final energy consumption and by end-use sector were 88% based on non-biomass and 12% on biomass. The majority of the biofuels produced are ethanol and biodiesel [1].

Other forms of renewables like ocean energy systems, geothermal and hydrogen and fuel cells can also be considered.

Current hot research areas concerning renewables include:

- Increase the efficiency of various renewable energy technologies;
- Design renewable energy components at lower cost;
- Extensive use of RES (many regions, even countries consider transformation into 100% renewables). This will need high shares of renewables, power system transformation and the implementation of storage/integration within a smart energy system; and
- Effective coupling not only for electricity but also for heating, cooling and transportation.

31.3 Conclusions

It can be concluded that although the total installed capacity of RES is small, relative to the total conventional annual fuel consumption, these constitute an important environmental friendly solution to protect the planet with very good prospects of expansion in the coming years. These systems are now becoming cheaper compared to conventional energy sources.

References

1. Renewables 2020—Global status report, REN 21
2. W. Weiss, M. Spörk-Dür, *Solar Heat Worldwide, Global Market Development and Trends in 2019* (Solar Heating and Cooling Program, IEA, 2020)
3. Snapshot of Global PV markets 2020. IEA-PVPS—Annual report 2020, Report IEA-PVPS T1-37: 2020

Chapter 32
How the European Green Deal Promotes Sustainable Energy Research and Innovation

Brigitte Hasewend and Tijana Jokic

The European Union is committed to becoming the first climate neutral continent by 2050. Transitioning to clean energy is key to combating climate change and it plays a central role in the new European Green Deal. Reaching its targets will require action by all sectors of the economy, including investing in environmentally-friendly technologies, supporting industry to innovate, rolling out cleaner, cheaper and healthier forms of transport, decarbonizing the energy sector and ensuring buildings are more energy efficient.

As part of this effort, the European Commission will present new measures aimed at embracing technological advance across all sectors of the energy system. A power sector must be developed that is based largely on renewable resources, complemented by the "rapid phasing out of coal and decarbonizing gas". For this to become a possibility, the deal adds that it is essential to ensure that the European energy market is fully integrated, interconnected and digitalized. The deal adds that increasing offshore wind production will be essential and believes the smart integration of renewables, energy efficiency and other sustainable solutions across sectors will "help to achieve decarbonization at the lowest possible cost".

The formulated objectives in the field of mobility include reducing transport-related emission by 90% by 2050 across all modes of transport, removing subsidies on fossil fuels, effective road pricing and increasing the production of sustainable alternative fuels. Automated and connected multi-mobility intends to be implemented, as well as stricter emission standards for combustion-engine vehicles.

The EU believe about one million public recharging and re-fueling stations will be needed for the 13 million zero and low-emission vehicles expected to be on Europe's roads by 2025.

B. Hasewend (✉)
Director, European Sustainable Energy Innovation Alliance–eseia, Graz, Austria
e-mail: brigitte.hasewend@eseia.eu

T. Jokic
European Sustainable Energy Innovation Alliance–eseia, Graz, Austria

I. Visa and A. Duta (eds.), *Solar Energy Conversion in Communities*,
Springer Proceedings in Energy, https://doi.org/10.1007/978-3-030-55757-7_32

The deal states also changes in the building sector. By 2050, all buildings must be highly energy efficient and decarbonized. This implies that the vast majority of existing buildings must undergo energy renovation providing high-performance building envelops, energy storage facilities and energy management systems. In order to achieve this goal, energy renovation rates have to be accelerated to reach 3% per year. In order to create demand in the housing sector, citizens must become the central actor in any public policy effort to ramp up renovations. Therefore standards must be accompanied by a comprehensive package of financial and technical support and tailored advice to ensure it is a fair deal for tenants, landlords and public investors.

In the current scenario of economic turbulence, with Europe and the whole world suffering from the shock of the COVID-19 pandemic, it is important for the EU to put the green economy and particularly, renewable energy policy at the center of recovery strategies. While governments may be tempted, in the re-opening phase, to take advantage of the historically low oil prices and put their energy transition on hold, we should all be mindful that for the super-wicked challenge of climate change no vaccine is in sight. The European Parliament adopted a resolution urging the Commission to make its upcoming COVID-19 economic recovery proposal fit for the European Green Deal and the Paris Agreement goals. The crisis exit should not return to yesterday's economy, but invest in a clean, modern and healthy economy.

It is clear that comprehensive research is needed in the fight against the climate crisis. The decarbonization of energy flows, the necessary grid expansion and integration of new technologies and renewable energy sources pose major challenges for the system. For this reason research initiatives on sector coupling, energy storage, green gas (hydrogen and methane from renewable energy sources) as well as on the advantages of digitization in decentralized and renewable energy production will be made use of.

As conventional approaches to R&D are not sufficient to manage the transformation, Commission will support the Member States through strategically important investments in R&D, especially where promising industrial policy approaches are available. Industrial partnerships are an important instrument. It is essential for R&D to remain a long-term focus in order to establish Europe as an active and resolute player in the energy transition and to pave the road towards a sustainable economy, consolidating its global position.

From the *ESEIA* perspective, the success of the Green Deal depends to a great extent on the implementation of innovations and as an European alliance, *ESEIA* wishes to support Europe to take the leading role as the most innovative energy region in the world. Supporting the creation of smart energy systems and their market uptake is *ESEIA*'s continuous value creation for a smarter Europe. The *ESEIA* Strategy 2030 supports five European cross-sector Working Groups to get this job done.

Chapter 33
Environmental Impact of a Residential House Using Cardboard Waste as Construction Material

Teofil Câmpean, Nicolae Peiu, Dan-Alexandru Gavrilescu, and Maria Gavrilescu

Abstract Research on waste recycling and developing environmentally friendly materials having good performances and low costs as alternatives to those currently used, in particular in construction and architecture, attempts to minimize the unsustainable use or harmful materials, which also incorporates large amounts of energy. A number of studies demonstrate the use of cardboard waste with sufficient compressive strength offering an alternative for traditional building materials (concrete and steel). In this work the most feasible methods for eco-design and construction of a residential house were analyzed, using corrugated cardboard waste as building material, in the form of old corrugated cardboard panels made without adhesives, by including them between the elements of the supporting structure, made of wood. The analysis shows that the resulted environmental performances recommend the use of recycled corrugated cardboard waste as a construction material with good performance in thermal insulation.

Keywords Eco-designed house · Environmental impact · Insulation · Old cardboard waste

---

T. Câmpean (✉)
Cristofor Simionescu Faculty of Chemical Engineering and Environmental Protection, SC Rondocarton SRL, Cluj-Napoca, Romania and Gheorghe Asachi Technical University of Iasi, Iasi, Romania
e-mail: teofil.campean@student.tuiasi.ro

N. Peiu
Environmental Protection Agency, Iasi, Romania
e-mail: nicolaesw@yahoo.co.uk

D.-A. Gavrilescu · M. Gavrilescu (✉)
Cristofor Simionescu Faculty of Chemical Engineering and Environmental Protection, Gheorghe Asachi Technical University of Iasi, Iasi, Romania
e-mail: mgav@tuiasi.ro

D.-A. Gavrilescu
e-mail: gda@tuiasi.ro

I. Visa and A. Duta (eds.), *Solar Energy Conversion in Communities*,
Springer Proceedings in Energy, https://doi.org/10.1007/978-3-030-55757-7_33

33.1 Introduction

Circular economy is a model for an economy that is designed to function in harmony with the environment, in which biological materials are designed to return safely to ecological cycles, and technical materials are designed to circulate continuously in the economic system. The ultimate goal is to decouple economic growth from resource consumption [1–3]. This presumes that a continuous economic growth will be possible in the context of resource constraints, while avoiding environmental damage by significantly reducing the extraction of virgin materials, eliminating unnecessary and toxic waste, substantial savings in raw material and energy costs [1]. At the same time, the specialized literature has demonstrated the need to integrate eco-design into product development, to guide and encourage product designers (in the sense of any good/product, process or service) to apply the principles of sustainable development and circular economy in design, also considering environmental issues [4–6]. Despite the apparent availability of studies discussing the adoption of practices, methods and tools for the application of eco-design, some researchers have pointed out that implementation is still in its infancy, especially at the company level [7–9].

Recycled or recovered waste can be reintroduced into the economic circuit as a secondary raw material, generating closed loops in accordance with the principles of the circular economy. Even before the circular economy became the subject of an European Commission legislative package, the paper, board and board packaging industry was one of the pioneers in applying the circular economy model at all stages of a product life cycle: design, production, distribution and use, its recovery. According to figures published by the European Union (EU) before World Recycling Day (18 March 2019), the amount of paper and cardboard packaging currently recovered in Europe has touched a record. In the 28 EU countries, the recovery rate for paper and cardboard packaging waste has reached 85.8%, the highest in EU history and the highest of all packaging materials (metal and glass packaging have rates of recycling of 78.3% and 74.1%, respectively) [10]. Globally, Europe continues to be the world champion in the field of paper recycling, followed by North America.

However, recovered paper and cardboard, which are not yet collected, consist mainly of lower grades with a low potential for papermaking. An important finding is related to the continuous decrease in the quality of recovered paper, from the point of view of paper producers. This trend is also manifesting itself on a European scale and is the consequence of the increasing number of recycling of paper and cardboard, which turns them into non-papermaking forms. Solutions for sustainable use can be found for these categories of paper and cardboard, by applying the principles of the circular economy and tools to facilitate the implementation of these principles.

Starting from the realities of the contemporary world and in accordance with European Directives, particularly those concerning the circular economy, this work focuses on the capitalization of recyclable paper and cardboard materials without papermaking potential to make products, economically profitable and with low environmental impact, on the whole life cycle. An analysis of the opportunities and impacts of using recovered cardboard, without papermaking potential, with sufficient compressive strength as an alternative to traditional construction materials has been performed.

33.2 Paper and Cardboard Recycling in the European Context

In the paper industry, the term waste paper is seldom used, the preferred term being *recovered paper*, which better reflects its value and importance as a raw material. Recovered paper (waste paper) is as important as a raw material for making paper as virgin fibers made from wood. The terms recycled paper and secondary fiber are also often used to refer to this waste stream. Recovered paper and cardboard hold second place in the global amount of waste, representing 17%, after the organic fraction [11]. Recycling of paper and board has proven to be a particularly attractive option, with low environmental impact compared to landfilling [12, 13].

At the beginning of the new millennium, recycled fibers are considered an indispensable raw material to meet the need for fibrous material for the paper industry in both developed and developing countries. Thus, by 2020, a waste utilization rate of about 50% is estimated, which will lead to an almost perfect balance between the consumption of virgin fibers and secondary fibers. Under these conditions, the recycling rate of waste will reach approximately 73% in CEPI member countries in 2020 (Table 33.1) [14]. CEPI considers that the maximum waste recycling rate for the paper industry is 78%, as some grades of paper and cardboard cannot be recycled (toilet paper, used paper and cardboard packaging in sanitary facilities etc.) [15]. In Europe, waste consumption is reported for assortment classes and is distributed as follows: waste corrugated packaging (Old Corrugated Containers, OCC); waste paper for deinking (Deinked Paper, DIP), mixed office paper (MOP), other assortments.

Usually, the recycling process involves sorting the fibers, deinking, removal of filling materials, bleaching etc. On the other hand, the repeated use of recovered paper as a raw material can decreases the quality of paper products [16, 17]. According to the current state of technology, there will always be a certain percentage of paper waste that, for economic and technical reasons, cannot be recycled. A favorable option for

Table 33.1 Quantities of recovered paper used for manufacturing paper and cardboard in CEPI member countries in 2017 [17]

Assortment of recovered paper	Quantity used, million tones	Percentage of total, %
Printing papers (including newsprint)	9.125	18.91
Old corrugated containers and other packaging made from paper and board	34.845	72.21
Other paper and board grades	4.278	8.88
Total	48.258	100.00

non-recyclable paper waste is its use in various composites. In this case, no expensive operations such as cleaning and refining the fibers are required. The alternative use of waste paper is interesting for two reasons. Firstly, secondary fibers are a largely unused resource, and secondly, more importantly, the costs of processing this material are significantly lower than those of producing similar wood-based materials [18].

The use of recovered paper as a material resource for composite type products has been investigated by various institutions and companies in the last 20 years. These products were obtained from different types of recovered paper or in combination with particles or wood fibers and were reinforced with various adhesives, organic or inorganic [19, 20].

Cellulose fibers, the main constituent of paper is a highly efficient insulator and could provide the construction industry with an efficient and environmentally friendly alternative to conventional insulation. The advantages of paper and its by-products and, above all, of cardboard is that they are cheap, light and durable, flexible in shape and color, recyclable and "different" [21, 22]. The discovery of opportunities for the efficient use in construction of materials of this nature, renewable and recyclable to a large extent, can be a step forward in the move towards sustainable development.

As a structural construction element, corrugated cardboard has many advantages. In addition to being a relatively low-priced material, it has significant insulating properties (thermal and acoustic), is easily recyclable and can be made from renewable sources. The most important property of corrugated cardboard, as a construction element, is that it has a high degree of structural strength and rigidity [23, 24].

33.3 The Environmental Impact of a Residential House that Uses Cardboard Waste as a Building Material

Proper design of houses and their ancillary systems can bring numerous benefit for the community, contributing to the eco-efficient functions of the building (including surrounding infrastructure, power and heat generation, wastewater treatment, waste management etc.) and a cleaner and resourceful environment. At the same time, a better community environment can benefit buildings by improving living conditions.

In this section it is analyzed the feasible method by which a residential house can be eco-designed and built using corrugated waste as a building material and the environmental impact associated with such a house.

33.3.1 Brief Description of the Residential House

The transition to a circular economy requires changes in value chains, from product design to new business and market models, from new ways of turning waste into a resource to new ways of consumer behavior. This transition implies full systemic change, eco-innovation and eco-design not only in terms of technology, but also in terms of organizations, society, new methods of financing through sustainable policies [25]. The widespread implementation of ecological solutions is intended to help limit many key issues related to environmental protection, such as climate change, declining natural resources, environmental pollution or biodiversity loss [26].

Eco-design plays an important role in reducing the impact on the environment, being essential for the life cycle of processes, products and services, the decisions taken in the design stage being responsible for the sustainable management of a large part of economic resources and those taken from the environment, necessary to support the entire life cycle of a product, process, service. Some specialists estimate that 80% of the environmental impact is predefined in the product design phase [27, 28]. Therefore, designers are interested and motivated to systematically integrate environmental performance into products and processes from the early stages of their design [29].

A house designed in compliance with eco-design requirements should require as few material resources as little energy and help reduce the impact on the environment. McDonough and Braungart [1] stated that if we could imagine buildings as trees, and cities as forests, then they would produce more energy than they consume and purify their own wastewater. Thus, proper design of houses and their ancillary systems can benefit the community, contributing to eco-efficient building functions (including surrounding infrastructure, power and heat generation, wastewater treatment, waste management etc.) and a cleaner environment with resources. On the other hand, a better community environment can benefit buildings by improving living conditions.

The residential house proposed to be eco-designed using OCC is not a temporary house, but one whose life cycle is expected to last as long as that of a traditional house, 50 years, respectively. For this reason, the use of cardboard, in any form as a structural material was excluded from the beginning. Wood, the closest to paper, was naturally chosen as the structural material. In terms of size, a house with a relatively small size was chosen, with a usable area of 85 m^2 and a ground floor height regime.

The outer walls consist mainly of panels, having the core of old corrugated containers (OCC) waste, with thermal insulation role, the exterior face—made of oriented wood chipboard (OSB), and inner face—made of plasterboard. To protect this core of moisture panels, the OCC panels are protected, both inside and outside with two foils with different purposes. Partition walls were not taken into account.

33.3.2 Life Cycle Assessment of a Residential House

In this analysis the use of Life Cycle Assessment (LCA) methodology for assessing the environmental performance of the compared buildings was made based on the provisions of the standards ISO 14,040:2006, ISO 14,044:2006, and ISO EN 15,978:2011 [30–32]. Life Cycle Assessment is an environmental management tool which, according to International Organization for Standardization (ISO) can be considered "a compilation and evaluation of the inputs, outputs and potential environmental impacts of a product throughout its lifecycle" [30].

The life cycle stages of the residential house consist of: **production, construction, usage, and end-of-life.** In this analysis the use of LCA for the calculation of the environmental performance of the construction materials was mainly made based on the calculation rules established in the standard ISO EN 15,978:2011 [32]. **The functional unit refers to the construction and use by a family of four people, and the final disposal**, after a period of 50 years, of a house on the ground floor with an area of about 85 m^2. **The frontiers of the system include construction, usage and end-of-life** (Fig. 33.1). However, production is an important stage because here an analysis can be made on the environmental quality of construction materials new construction materials can be experienced and evaluated.

Therefore, an analysis of production stage can be made from the beginning, addressing the quality of construction materials from the point of view of environmental protection. For example, a comparison of seven thermal insulation materials has been performed (Fig. 33.2). Using a classical environmental impact assessment methodology known as **cumulative energy demand**, a ranking of all cumulative energies required to produce the same amount (1 kg) of different construction materials shows that, if a panel is produced from virgin fibers, the corrugated board is the thermal insulation material whose production needs more cumulative energy consumed for production than the recycled corrugated waste.

The ReCiPe method (final point, total score, hierarchical version (H)) was chosen for impact assessment in the frame of LCA, because it is a combined one, being able to provide results at both midpoint and endpoint levels [33]. Sima Pro developed by the Dutch company PRé, version 8.2.0.0, was used as software for this tool, for which a license was obtained by the Faculty of Chemical Engineering and Environmental Protection within the "Gheorghe Asachi" Technical University of Iași. The endpoint impact categories covered by this tool are: human health, environmental quality and resource availability (with the meaning of damages) (Fig. 33.3). According to the ISO 14,040 (2006) a complete LCA study comprises four major phases (Fig. 33.4):

- *Goal Definition and Scoping*—include the objective and scope of study, system boundaries and the functional unit;
- *Inventory Analysis*—with a detailed compilation of all environmental inputs (resource and energy flows) and outputs (emissions and wastes);
- *Impact Assessment*—involves evaluation of environmental impacts from the inventory and establish environmental performance of the product;
- *Interpretation*—the results of the inventory are interpreted and used in decision and policy making.

LCA (ReCiPe method) was applied during the three stages of life cycle considered in the study: **construction, use, and end-of-life.** All data required for life cycle inventory were collected from standards, handbooks, technologies, visits on building locations, statistics, databases of the SimaPro software and others.

The construction stage is one of the most important stages in the life cycle of the house, although not the most important. This stage is, of course, preceded by

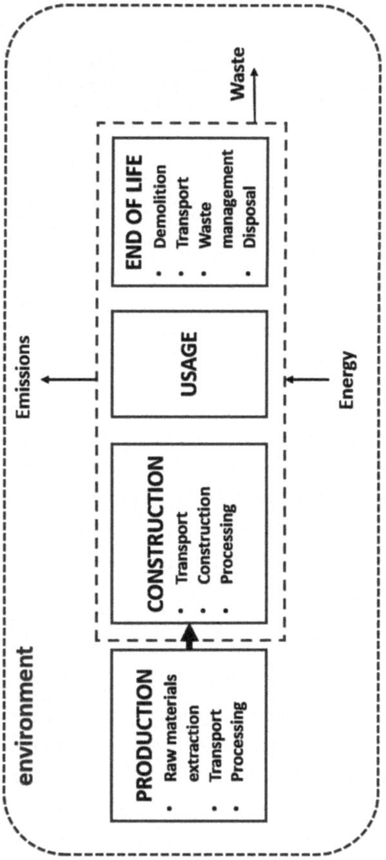

Fig. 33.1 The main stages of the life cycle of a building and the limits of the analyzed system

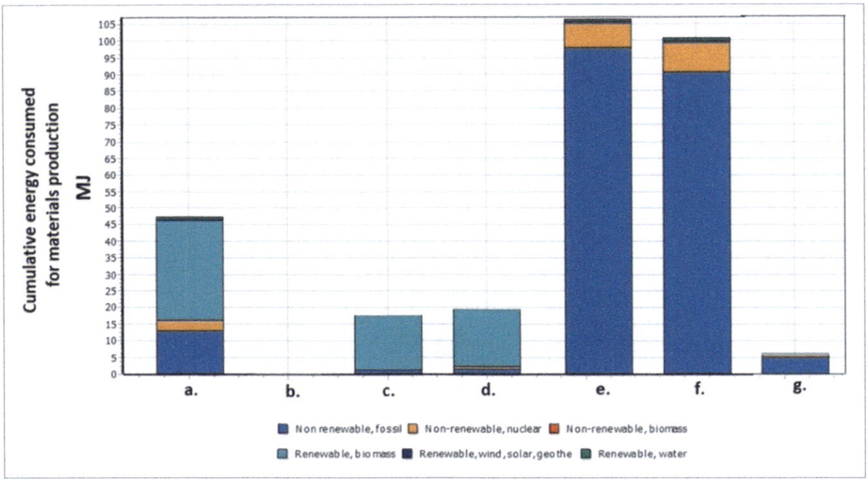

Fig. 33.2 Unit comparison (C.U. = 1 kg) of the main products (construction materials) used for the buildings compared in terms of cumulative energy consumed for their production (MJ); methodology: cumulative (Total) energy demand; total score: **a** virgin fiber corrugated board, **b** recycled corrugated waste, **c** hardwood, **d** softwood, **e** expanded polystyrene PSE, **f** XPS extruded polystyrene, **g** gypsum board

the design, where all the technical details of the future building are developed. According to ISO EN 15,978:2011 [32] this step comprises the following two modules (Fig. 33.1): (i) transporting the raw material from the place of production to the place of construction; (ii) actual construction.

The use/maintenance stage is the longest and most important stage in the entire life cycle of the analyzed house. According to the standard ISO EN 15,978:2011 [32] this step comprises the following seven modules: (i) usage; (ii) maintenance; (iii) repairs; (iv) replacements; (v) renovation; (vi) use of operational energy; (vii) water use. The most important of these modules refers to the operational energy used.

The end-of-life stage is the most difficult to be quantified due to the uncertainty of scenarios evolution and that of waste management technologies resulting from construction and demolition, given the distance within 50 years from the expected period of demolition of the building. According to ISO standard EN 15,978:2011 [32], this step comprises the following four modules: (i) demolition; (ii) transport (to the waste processing site); (iii) waste processing and (vi) final disposal of waste.

Figure 33.5 shows a comparison of the impact on the environment produced in the three stages of the house life cycle, using the ReCiPe impact assessment method. For this analysis, the three impact categories were considered: human health, ecosystem quality and resources availability. In the construction stage, the category of *ecosystem quality* impact has the highest weight, followed by the availability of resources.

In general, the construction activity is considered to be one of the largest contributors to environmental pollution and depletion of resources, producing globally 33%

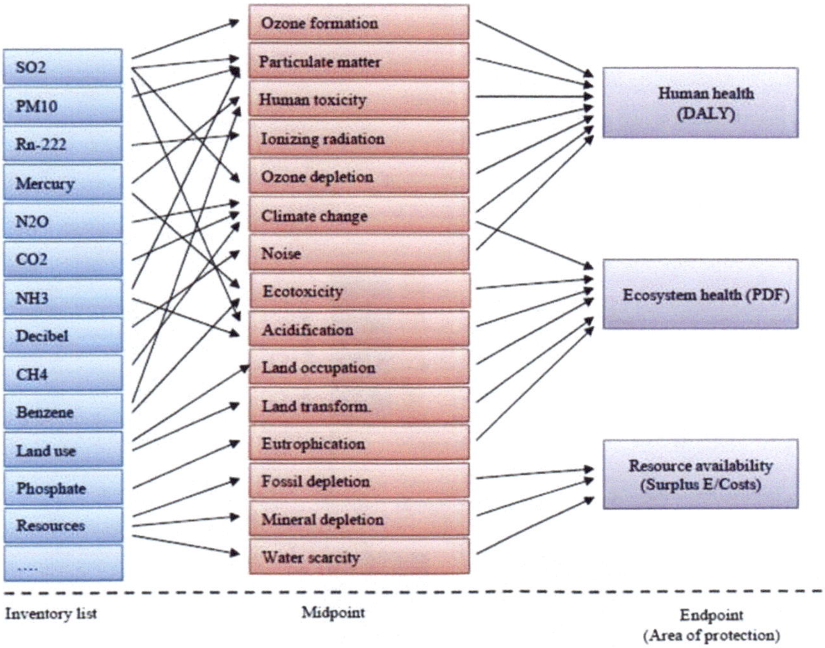

Fig. 33.3 List of impact categories for characterization at midpoint and endpoint level, [34] (under license CC BY-NC-ND, IOP science, https://publishingsupport.iopscience.iop.org/is-permission-required-faqs-using-open-access-content/)

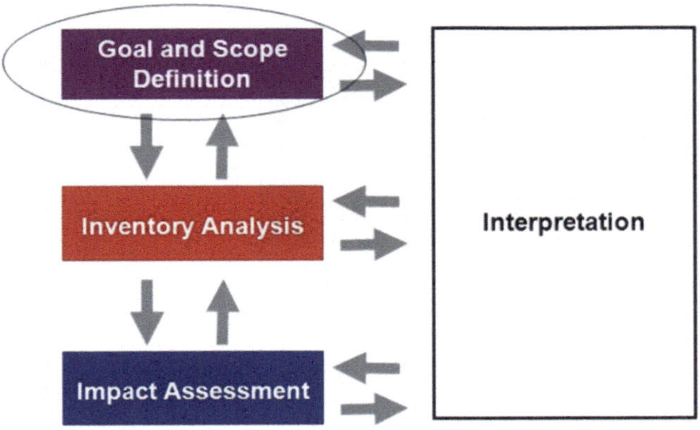

Fig. 33.4 Major phases of life cycle assessment

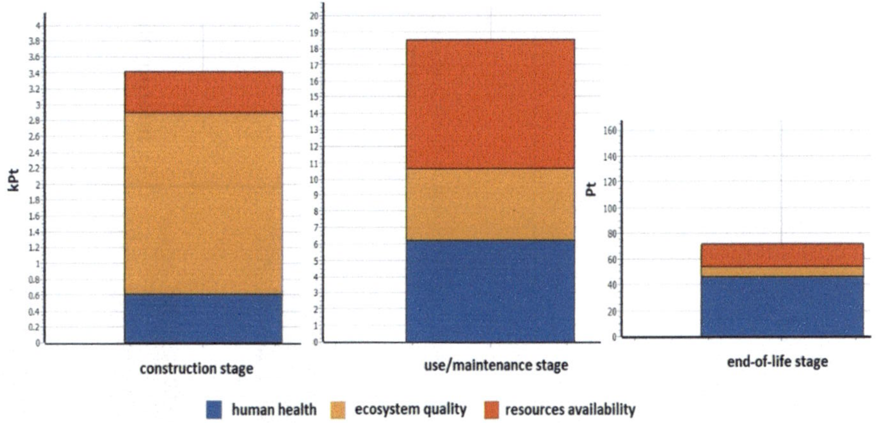

Fig. 33.5 Comparison of the impact produced in the three stages of life cycle of houses (ReCiPe End point methodology hierarchical version, H)

of greenhouse gas emissions and consumes approx. 40% of primary energy [34]. It is expected that this representation will be below the level for the traditional house, because, as can be seen from Fig. 33.2, cumulative energy demand has the lowest value for OCC cardboard waste, used as insulation material in house construction.

33.4 Conclusions

This study aimed at examining the extent to which recyclable corrugated waste without papermaking potential (Old Corrugated Containers, OCC) represents a promising material for the eco-design of a sustainable home (CEP), in terms of environmental performance, analyzed with Life Cycle Assessment (LCA) methodology, a well-structured, standardized and extensive tool used for this purpose especially in Europe, mainly after the launch of the European LCA platform.

The functional unit of the study was the procurement of materials, construction, use and final disposal of a residential house with a period of use of 50 years. ECV focused on all four stages of a building's life cycle: production, construction, usage and final disposal.

Outside the limits of the system analyzed in the production stage, separate investigations were carried out, using Life Cycle Assessment, on the quantities of energy required for the production of corrugated cardboard produced from virgin fiber or cardboard recycled waste. The functional unit chosen for this comparison was 1 kg of the analyzed construction material. Research has shown that the material represented by virgin fiber corrugated cardboard requires one of the highest energy consumption along with extruded or expanded polystyrene. Instead, old corrugated containers (OCC) has the lowest energy consumption, which implicitly means the lowest impact on the environment.

Comparisons made separately for the construction, usage and end-of-life stages show that the human health, environmental quality and resource availability impact categories have different weight in the three stages. In the construction stage, environmental quality has the highest weight. In the usage stage, resource availability is has the highest weight. Human health is the impact category with the highest share in the end-of-life stage.

In the future works, the analysis will be extended, to allow comparisons of environmental impacts and costs generated by a traditional house and a house made of virgin fibers cardboard, with the eco-innovated residential house, which use waste corrugated cardboard without papermaking potential as an insulation material.

References

1. W. McDonough, M. Braungart, A world of abundance. Interfaces **30**, 55–65 (2000)
2. Ellen MacArthur Foundation: Towards the Circular Economy: *Economic and Business Rationale for an Accelerated Transition*, Report vol. 1, (2012). Online at: https://www.ellenmacarthurfoundation.org/business/reports/ce2012
3. F. Preston, A global redesign? *Shaping the Circular Economy, Energy, Environment and Resource Governance, Chatham House, London* (2012). Online at: https://www.biblioteca.fundacionicbc.edu.ar/images/d/d7/Bp0312_preston.pdf
4. F. Brones, M.M. Carvalho, E.S. Zancul, Ecodesign in project management: a missing link for the integration of sustainability in product development? J. Cleaner Prod. **80**, 106–118 (2014)
5. D.C. Pigosso, E.T. Zanette, F.A. Guelere, A.R. Ometto, H. Rozenfeld, Ecodesign methods focused on remanufacturing. J. Cleaner Prod. **18**, 21–31 (2010)
6. S. Sihvonen, J. Partanen, Implementing environmental considerations within product development practices: a survey on employees' perspectives. J. Cleaner Prod. **125**, 189–203 (2016)
7. E.A. Dekoninck, L. Domingo, J.A. O'Hare, D.C.A. Pigosso, T. Reyes, N. Troussier, Defining the challenges for ecodesign implementation in companies: Development and consolidation of a framework, J. Cleaner Prod. **135**, 410–425 (2016). Online at: https://doi.org/10.1016/j.jclepro.2016.06.045
8. M. Rossi, M. Germani, A. Zamagni, Review of ecodesign methods and tools. Barriers and strategies for an effective implementation in industrial companies. J Cleaner Prod. **129**, 361–373 (2016)
9. A.G. Silvius, M. Kampinga, S. Paniagua, H. Mooi, Considering sustainability in project management decision making; an investigation using Q-methodology. Int. J. Proj. Manage. **35**, 1133e1150 (2017). Online at: https://doi.org/10.1016/j.ijproman.2017.01.011
10. Packaging Europe, Paper and cardboard recycling reach record high across Europe, (2019). Online at: https://packagingeurope.com/paper-and-cardboard-recycling-have-reached-record-high-acros/
11. D. Hoornweg, P. Bhada-Tata, *What a Waste—A Global Review of Solid Waste Management, Urban Development and Local Government Unit, World Bank*. (Washington DC, 2012). Online at: https://openknowledge.worldbank.org/handle/10986/17388
12. A. Björklund, G. Finnveden, Recycling revisited—life cycle comparisons of global warming impact and total energy use of waste management strategies. Resour. Conserv. Recycl. **44**, 309–317 (2005)
13. A. Villanueva, H. Wenzel, Paper waste—recycling, incineration or landfilling? Rev. Existing Life Cycle Assessments Waste Manage. **27**, 29–46 (2007)
14. CEPI, CEPI key statistics 2017, (2018). Online at: https://www.cepi.org/system/files/public/documents/publications/statistics/2018/210X140_CEPI_Brochure_KeyStatistics2017
15. EPRC, *European Declaration on Paper Recycling*, 2016–2020, European Paper Recycling Council—EPRC, (2018). Online at: https://www.cepi.org/publication/european-declaration-paper-recycling-2016-2020
16. L. Groom, S. Shaler, L. Mott, Physical and mechanical properties of virgin and recycled wood fibers, In: *Processing of Environmental Quality in Wood Processing*. (Atlanta, 1994), pp. 20–35
17. O. Kordsachia, Remarks on the waste paper problem. Holz-Zentralblatt **81**(82), 1327–1330 (1992)
18. A.H. Grigoriou, Waste paper–wood composites bonded with isocyanate. Wood Sci. Technol. **37**, 79–90 (2003)
19. G. Gerischer, Structural particleboard from waste material. Holzforschung **31**, 129–133 (1977)
20. A. Krzysik, J.A. Youngquist, R.M. Rowell, J.H. Muehl, P. Chow, S.R. Shook, Feasibility of using recycled newspapers as a fiber source for dry-process hardboards. For. Prod. J. **43**, 53–58 (1993)
21. J.F. Latka, Paper in architecture. Research by design, engineering and prototyping, PhD Thesis, Delft University of Technology. (Delft, The Netherlands, 2017)

22. J. Schonwalder, Cardboard as building material, Tentech BV—Engineering consultancy of lightweight structures utrecht, faculty of architecture T.U. (Delft, 2016). Online at: https://www.kcpk.nl/algemeen/bijeenkomsten/presentaties/20160202-9-julia-schonwalder
23. A. Russ, J. Schwartz, S. Bohacek, H. Lubke, V. Ihnat, A. Pazitny, Reuse of old corrugated cardboard in constructional and thermal insulating boards. Wood Sci. **58**, 505–510 (2013)
24. S. Secchi, F. Asdrubali, G. Cellai, E. Nannipieri, A. Rotili, I. Vannucchi, Experimental and environmental analysis of new sound-absorbing and insulating elements in recycled cardboard. J. Build. Eng. **5**, 1–12 (2016)
25. COM 614, *Closing the Loop—an EU Action Plan for the Circular Economy, Communication from the Commission to the European Parliament, the Council, the European Economic and Social Committee and the Committee of the Regions.* (European Commission, Brussels, 2015)
26. A. Ociepa-Kubicka, P. Pachura, Eco-innovations in the functioning of companies. Environ. Res. **156**, 284–290 (2017)
27. T.G. Navarro, S.C. Rizo, M.J.B. Ceca, D.C. Ruiz, Designing for the future—application and management of knowledge and information in engineering design, in *DS 35: Proceedings ICED 05, the 15th International Conference on Engineering Design*, 15–18 August. (Melbourne, Australia, 2005)
28. M. Dufrene, A methodological framework to support integrated ecodesign for companies: requirements and conceptualization towards a software platform, PhD Thesis, Université Grenoble Alpes. (France, 2015)
29. M.D. Bovea, V. Pérez-Belis, A taxonomy of ecodesign tools for integrating environmental requirements into the product design process. J. Cleaner Prod. **20**, 61–71 (2012)
30. ISO 14040, *International Organization for Standardization, Environmental Management—Life Cycle Assessment—Principles and Framework.* (2006). Online at: https://www.iso.org/iso/catalogue_detail.htm?csnumber=23151
31. ISO 14044, *Environmental Management—International Organization for Standardization, Life cycle Assessment. Requirements and Guidelines.* (2006). Online at: https://www.iso.org/iso/catalogue_detail?csnumber=38498
32. ISO EN 15978, *Sustainability of Construction Works. Assessment of Environmental Performance of Buildings.* Calculation method. (2011). Online at: https://shop.bsigroup.com/ProductDetail/?pid=000000000030256638
33. M. Goedkoop, R. Heijungs, M. Huijbregts, A. De Schryver, J. Struijs, R. van Zelm, *ReCiPe 2008 A life cycle impact assessment method which comprises harmonized category indicators at the midpoint and the endpoint level.* Ministerie van VROM, Den Haag, 1st edn. (2013). Online at: https://www.leidenuniv.nl/cml/ssp/publications/recipe_characterisation.pdf
34. M. Wolfova, A. Estokova, M. Ondova, A. Monokova, Comparing of the external bearing wall using three cultural perspectives in the life cycle impact assessment. IOP Conf. Ser. Mater. Sci. Eng. **385**, 012064 (2018). https://doi.org/10.1088/1757-899X/385/1/012064

Chapter 34
National Mechatronic Platform—The Foundation of a Smart and Sustainable Society Development

Vistrian Mătieş, Călin Rusu, and Vlad Tămaş

Abstract The paper aims to explain details related on the innovative potential of mechatronics as environment for smart education, organizational learning and sustainable development, to enhance the pillars of a Smart and Sustainable Society. At the EU level these pillars are defined as smart governance, smart economy, smart mobility, smart environment and sustainable development, smart living and smart people. Mechatronic platforms as complex technical systems integrating elements of mechanical engineering, electrical-electronic engineering and computer science engineering are the basic infrastructure of these environments. The National Mechatronic Platform (NMP) is conceived as a national mechanism of a network structure to activate the human and material resources at all the levels and to ensure the systemic approach of the complex problems regarding smart education, organizational learning and sustainable development in Romania. Based on this technical and scientific support, the Platform for Smart Education, Organizational Learning and Sustainable Development will be developed. Specific details are also presented in the paper. The proposed solutions are useful to get the main goals of the UN 2030 Project too. As it is known, Mechatronics (the backbone), Cyber Physical Systems (CPS) and Internet of Things (IoT), are the triangle of the twenty-first century technologies, the foundation of the 4.0 industry

Keywords Smart society · Sustainability · Mechatronics · Smart education · Organizational learning

34.1 Introduction

The concept of sustainable development emerged as a response to a growing concern about human society's impact on the natural environment. The concept was defined in 1987 by the Brundtland Commission, as "development that meets the needs of

V. Mătieş · C. Rusu (✉) · V. Tămaş
Technical University of Cluj-Napoca, Cluj-Napoca, Romania
e-mail: calin.rusu@mdm.utcluj.ro

V. Mătieş
e-mail: vistrian.maties@yahoo.com

I. Visa and A. Duta (eds.), *Solar Energy Conversion in Communities*,
Springer Proceedings in Energy, https://doi.org/10.1007/978-3-030-55757-7_34

the present without compromising the ability of future generations to meet their own needs" [1].This definition acknowledges that while development may be necessary to meet human needs and improve the quality of life, it must happen without depleting the capacity of the natural environment to meet present and future needs. The sustainable development movement has grown and campaigned on the basis that sustainability protects both the interests of future generations and the earth's capacity to regenerate.

The underlying concepts of sustainable development are defined by Agenda twenty-first, which is the Action Program for the twenty-first century adopted by governments at the United Nations Conference on Environment in 1992 [2]. The Summit confirmed its resolve to promote the three pillars of sustainable development: economy, society and environment, as interdependent and mutually reinforcing concepts. The UN Agenda twenty-first includes 17 goals and 169 targets. The concept of smart and sustainable society is highlighted in the Fig. 34.1

Education is an essential tool for achieving a more sustainable world. Education for Sustainable Development (ESD) promotes the development of the knowledge, skills, understanding, values and actions required to create a sustainable world, which ensures environmental protection and conservation, promotes social equity and encourages economic sustainability. The aim of ESD is to enable people to make

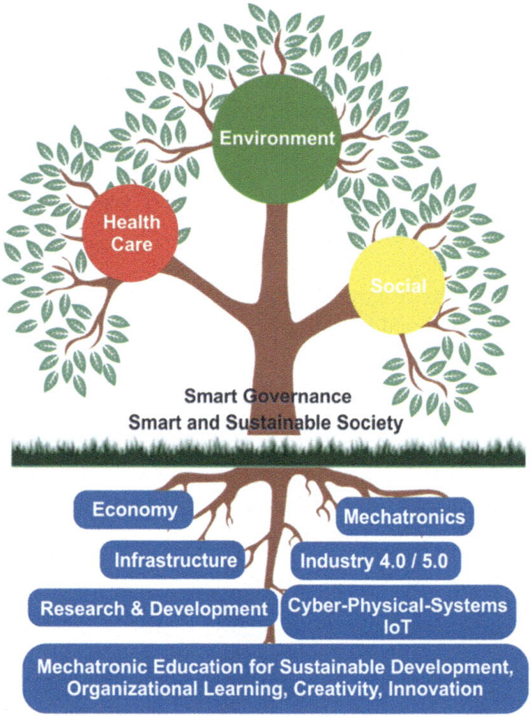

Fig. 34.1 The smart and sustainable society concept

decisions and carry out actions to improve our quality of life without compromising the planet. It also aims to integrate the values inherent in sustainable development into all aspects and levels of learning [3].

There are a number of key themes in ESD and while the dominant focus is on environmental concerns, it also addresses themes such as poverty alleviation, citizenship, peace, ethics, responsibility in local and global contexts, democracy and governance, justice, human rights, gender equality, corporate responsibility, natural resource management and biological diversity.

To promote ESD, the United Nations Decade of Education for Sustainable Development, 2005–2014, (DESD) was adopted by the UN General Assembly with the United Nations Educational, Scientific and Cultural Organization (UNESCO) designated as the lead agency for promotion throughout the decade [4].

The goal of the decade, as outlined by UNESCO, was to integrate the principles, values and practices of sustainable development into all aspects of education and learning. This aims to encourage changes in behavior that will create a more sustainable future. One of the most important aspects of the DESD is the recognition that ESD must engage a wide range of stakeholders from government, private sector, civil society, non-governmental organizations and the general public [4].

The complex problems related on ESD are to be solved in the context of the world crisis in education. The crisis is caused by the delay of the educational technologies by comparing with technological development. The crisis is recognized at the UNESCO level too. The establishing the International Commission for Education in the twenty-first century and Delors's report to UNESCO confirms that. The initiatives for ESD promoting at UNESCO level are fine, but the suggested solutions are not connected to the twenty-first century technologies.

As it is known, the triangle of the 21st century technologies includes: Mechatronics (the backbone), Cyber-Physical Systems (CPS) and Internet of Things (IoT). These technologies are the foundation of the 4.0 industry [5–9].

Mechatronics as the "Science of intelligent machines" and also "Environment for smart education, organizational learning and sustainable development" ensures the scientific foundation and efficient tools for promotion the smart education for sustainable development in order to get the goals included in the UN Agenda twenty-first. The NMP is the result of the research activities along of a more than a quarter of century (1990-present), of the specialists of the Mechatronic Departments of the Technological Universities from Romania that established the branch of mechatronics in engineering. The NMP is conceived as a national mechanism of a network structure to activate the human and material resources at all levels and to ensure the systemic approach of the complex problems regarding smart education, organizational learning and sustainable development in Romania. Based on this technical and scientific support, the Platform for Smart Education, Organizational Learning and Sustainable Development will be established. The specific details are also outlined in the papers: [8, 10–12].

34.2 The Innovative Resources of Mechatronics for Smart Education and Sustainable Development

34.2.1 Why Was Mechatronics Born in Japan?

The word mechatronics was coined by Japanese in the beginning of the eighth decade of the last century to describe a new technology fusion." It is an advanced technology to create energy-saving, resource-saving and high intelligent systems by integrating mechanics, electronics and software" [13]. In the paper [13] are analyzed the relationships between mechatronics and Japanese lifestyle from an industrial, technological, economical, historical and cultural point of view. The link is described by the following keywords: average, blend, clean, delicacy, education and flexibility [13–15].

Mechatronics was born in Japan as a result of the industrial policy for post-war rehabilitation. The frame was established by four laws: In 1956—was established the law for mechanical industry development. In 1957—was established the law for electrical industry development. In 1964, the production of the integrated circuits began on industrial basis. In 1971—was set-up the law for electromechanical industry development. At that stage Japanese realized their own weakness. Japan was obliged to organize a drastic change of the industrial structure. Engineers were strictly requested from the administration to save energy, material and cost [13–15]. Under such a serious condition, energy and resource saving products with high intelligence have been developed. Japanese engineers could turn a misfortune into a blessing. Japan has succeeded to strengthen its industrial technology owing to a shortage of natural resources. In 1978—the law for information handling machine development was established instead of the previous laws. In 1978 the Mechatronic Education was initiated at the newly founded University of Toyohashi. The spirit and layout of the program has been described by the major professor at 1984 World Conference on Engineering Education [13, 16, 17]. The legal frame above mentioned was the starting point of Mechatronic Revolution in Japan [16].

It is important to note that Prof. Eduard Deming, the pioneer and promoter of the Total Quality Management concept was credited as one of the inspirations for what has become known as Japanese post-war economic miracle of 1950–1960. Then Japan rise from the ashes of war on the road to becoming the second largest economy in the world through processes partially influenced by the Deming's ideas. As a recognition of his contribution to Japan development he was honored in 1951 with the establishment of the Deming Prize. In 1960, at the end of his stage of cooperation with Japan's Government Deming awarded on behalf of Emperor Hirohito Japan's Order of Sacred Treasure, Second Class. The citation on the medal recognizes Deming's contributions to Japan's industrial rebirth and its worldwide success [18].

Mechatronic philosophy has spread around the world very quickly. The decisions adopted at the national and international level to promote mechatronics in education, research and technological development stimulated the initiatives to develop projects and programs in the field. Two very representative initiatives are outlined below:

In 1985, The USA Department of Commerce has elaborated a comparative report on mechatronics in the USA and Japan [19]. The report was the support for the decision of the National Science Foundation to finance the National Mechatronic Educational Program. The project was managed by Stanford University.

In 1986 the word mechatronics acquired its citizenship in the vocabulary of the European Community. The Industrial Research and Development Advisory Committee (IRDAC) recognized mechatronics as one of the major needs to be met by the European Research and Educational Programs [14–16]. It specifies the definition as follows:" Mechatronics is a synergistic combination of precision mechanical engineering, electronic control and system thinking in design of product and processes. It is an interdisciplinary technology that both draws on the constituent's disciplines and includes subjects not normally associated with any of those above". In 1988 a European Integrated Course in Mechatronics started as a result of a high-level transnational cooperation of academics and industrialists [16]. It integrated three modules respectively offered at T.H. Aachen (with emphasis on CAD-CAM), K.U. Leuven (System Approach, Robotization) and Institute of Technology Cranfield (precision mechanics and measurement). It was specially designed for Engineers intending to supplement their traditional education with recently technologies integrated in a system's vision. In 1989 it was enlarged including two additional modules.

These decisions stimulated initiatives at national and international level to develop projects and programs to promote mechatronics in education, research and technological development too [8, 10].

34.2.2 *Mechatronic Technology.*

Mechatronics was born as a result of technology development. Technological flow to mechatronics integration is shown in Fig. 34.2 [13–15]. The backbone of mechatronics is mechanical technology that developed towards mechanization. The progress in the field of electronic technology, the integrated circuits development,

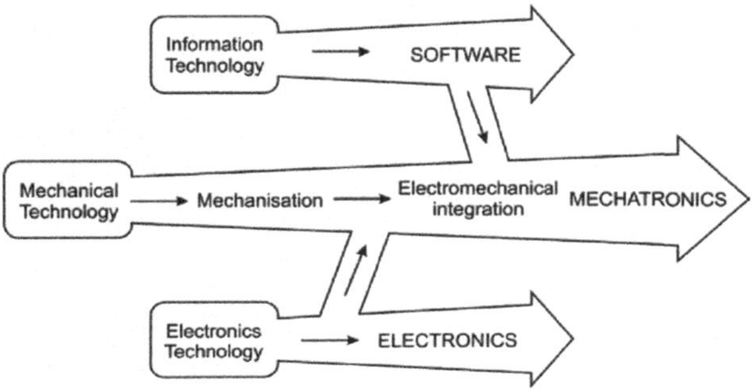

Fig. 34.2 The flow to mechatronic integration

allowed the integration of electronics into mechanical structures. This is the first step towards integration: electromechanical integration.

In the next step, by integrating microprocessors into previously built electromechanical structures they can retrieve information about their internal status and about the state of the environment. By processing the information, they can make decisions about how the system behaves.

The concept of mechatronics is highlighted in Fig. 34.3. The integration of the three main fields of engineering is based on Integronics principles and methods [8, 20, 21]. Integronics is defined as "a science of the integration processes and of a hyper integrated system" [21]. In the structure of a hyper integrated system everything is linked to everything and everything depends on everything (eg. human body). Details related on the integration process in the nature and technology are outlined in the papers [8, 21, 22].

Mechatronics makes possible the integrated approach of material, energy and information. Information is the basic component of the mechatronic technology.

This position of information in relation with material and energy is motivated by the following arguments: satisfaction of the mind of human beings is caused by information; only information can increase added value of all things; information means culture [8, 13, 16, 22, 23].

Based on Fig. 34.4, a comparative analysis of the three components of mechatronic technology can be made [8, 13, 16, 22, 23]. The analysis considers the origin of the resources, the supply, the demand and the meaning of life from the three elements' point of view.

Certainly, by integrating a large amount of information in the structure of intelligent products their functional performance will increase.

On the other hand, material and energy resources are preserved in this way. But, by consuming less material and less energy, the pollution decreases resulting

Fig. 34.3 Mechatronics concept

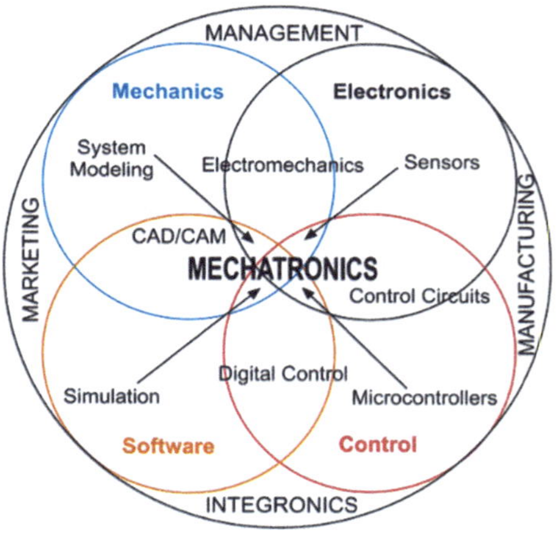

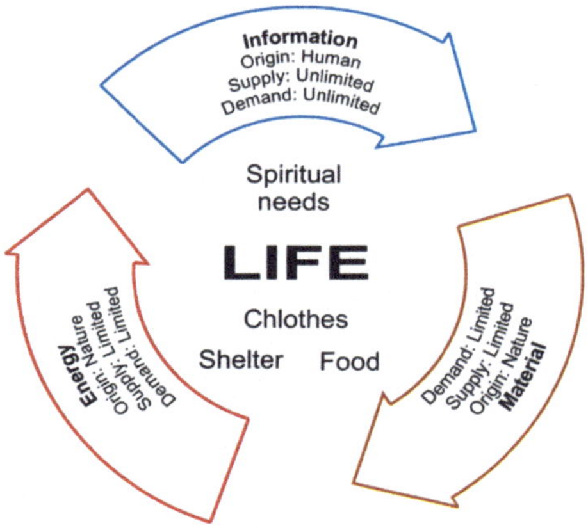

Fig. 34.4 The relationship material—energy—information

new valences of mechatronic technology: it is a non-dissipative and less polluting technology.

The information links integrated in the structure of technical systems requires small quantities of material and energy increasing flexibility, efficiency and reconfigurability [8, 13, 16, 22, 23].

More than, the value of information does not depend so much on quantity as it does on freshness, since human mind always demands new stimuli. Also, the information can be multiplied.

The value of material and energy depends on integration, and the information value depends on differentiation. Mechatronic technology launched the challenge of sensitive information. The style, color, design in general of every machine. transmit information that stimulates the senses of the human being.

The signal is the means of physical presence of information. The signals are generated by sensors (artificial sensing organs) integrated into the structure of smart machines and systems. The sensors materialize the perception function in the structure of an intelligent system. Microcontrollers materialize the brain functions and the actuators (the execution elements) are the artificial muscles.

34.2.3 Mechatronic Education

The mechatronic principles in education focus on the systems thinking and skills for teamwork/team learning developing.

Systems thinking is holistic, it attempts to derive understanding of parts from the behavior and properties of wholes, rather than derive the behavior and properties of wholes from those of their parts. "Disciplines are taken by science to represent different parts of the reality we experience. But disciplines do not constitute different parts of reality; they are different aspects of reality, different points of view. Any part of reality can be viewed from any of these aspects. The whole can be understood only by viewing it from all the perspectives simultaneously" [24].

In the context of globalization and internationalization promoting the smart educational technologies for sustainable development is a major need [5, 6, 8, 12, 22, 23]. The basic problems for promoting these educational technologies are: the system thinking development; team work/learning; organizational learning; integral education concept promotion; the understanding of integration—complexification process in the nature, technology and society; the role of information in the integration complexification process: information carriers, information link, information kinematic chains, information field, sensitive information etc.

In the last two decade were launched challenges for smart cities development. Based on EU standards the mains pillars of the smart city's platforms are smart governance, smart economy, smart mobility, smart environment and sustainable development, smart living and smart people [26]. The smart city is a city of learning, of creativity and innovation. The smart city learns through their citizens that are integrated into organizations (family, schools, universities, companies etc.) [12, 26–28]. In the last decade, were launched challenges for evolution to Smart Society development too.

Mechatronic education ensures flexibility in thinking and action. In this context, mechatronics education meets the requirements of a smart education, providing the necessary skills for pupils, students or adults for intelligent integration, smart organizations, smart communities etc. Organizations and communities become intelligent and therefore competitive, by learning. Competence is the bridge between man and organizations (institution) and beyond, between organizations and the community [22, 29].

Mechatronic platforms are the basic infrastructure of the environments for smart education and sustainable development As complex technical systems, these platforms integrate into their structure elements of mechanical engineering (mechanisms, mechanical transmissions etc.), electrical engineering elements (actuators, sensors, microcontrollers, filters, amplifiers etc.) and control engineering elements (control algorithms, dedicated software, human–machine interfaces etc.). The specific structure of the mechatronic platforms facilitates the understanding of the integration-complexification process in the nature and technology, as well as the role of information and information links in this process [8, 10].

34.2.4 Mechatronics in Practice and Engineering Education

For the engineering practice, mechatronics marked the shift from traditional sequential engineering to simultaneous, concurrent engineering. Therefore, the concepts of integrated design and design for control were developed. The details on the integrated design methodology are presented in the works [8, 23, 25, 30, 35]. Thus, it is necessary from the conceptual design phase to consider the problems regarding the integration—interfacing processes, the information links as well as the integration of the control functions into the product structure. In this way, the conventional functions made by the mechanical components are transferred to the electronic control and software components. This increases the constructive and functional performance of products and systems [8, 14, 23].

The approaches in the field of mechatronics require advanced knowledge from multiple engineering fields (see Fig. 34.5). Being the result of an integrated design process, mechatronic systems are superior to any product achieved through a sequential (classic) design methodology [8, 14, 16, 20, 23, 25].

The integrated approach promoted by mechatronics is essential for developing and manufacturing of Cyber-Physical Systems (CPS) [7] as well as for harnessing the Internet communication potential through Internet of Things (IoT) approaches [9].

34.3 Organizational Learning

As it was underlined by UNESCO, one of the most important aspects of the DESD is the recognition that ESD must engage a wide range of stakeholders from government, private sector, civil society, non-governmental organizations and the general public [4]. Schools and universities are the main actors to launch initiatives on this direction. For that, they must get the status of learning organizations.

The concept was launched by Prof. P. Senge from MIT, in 1990. In 1997 at MIT was established the Society of Organizational Learning [27, 28].

Senge defines the Learning Organization as "an organization where people continually expand their capacity to create results they truly desire, where new and expansive patterns of thinking are nurtured, where collective aspiration is set free, and where people are continually learning how to learn together".

The organizations that will truly excel in the future will be the organizations that discover how to tap people's commitment and capacity to learn at all levels in an organization [27, 28]. Traditionally, productive organizations have been viewed as center of work. However, the Learning Organization sees the productive organization as not only a center of work, but also a center of learning. Schools and universities are also learning organizations.

P. Senge defines the disciplines of the learning organization as: Personal Mastery, Mental Model, Shared Vision, Team Learning and System Thinking. The disciplines are very similar with the four pillars of a profound knowledge as they are defined by Prof. Eduards Deming in the book [8]. These pillars are: System thinking, Changes management, Psychology and Theory of knowledge.

The discipline consists on a body of theory and technique that must be studied and mastered to be put into practice. A discipline is a developmental path for acquiring certain skills or competencies.

Senge sees systems thinking at the heart of his "learning organization" models, where all members of organization develop an understanding of the whole rather

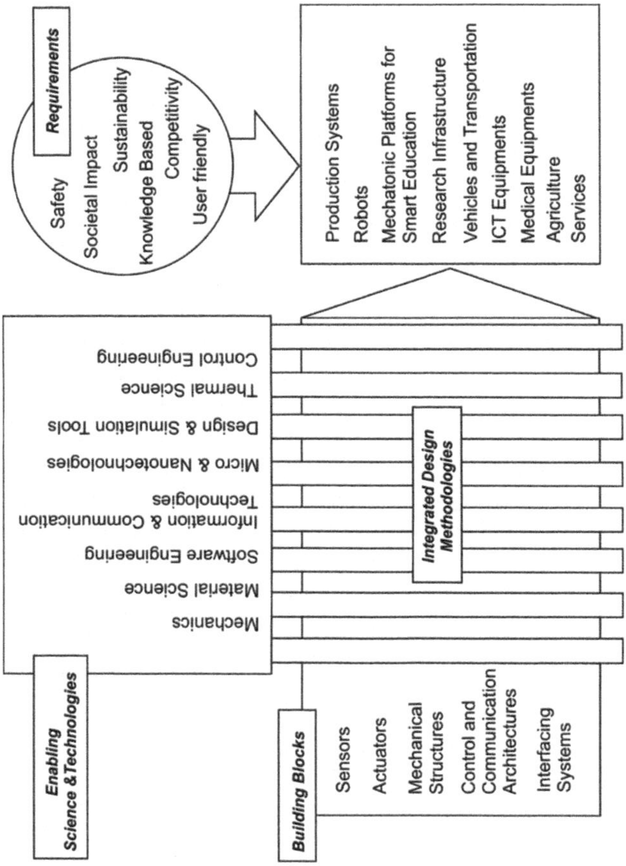

Fig. 34.5 The integrated design concepts

than just fractional parts of organization in terms of structures, processes, thinking and behavior.

Analyzing the scientific content of the five disciplines of the learning organization as it is outlined in the books [27, 28] we can understand that mechatronics is true environment for smart education, organizational learning and sustainable development. The mechatronic platforms are the basic infrastructure for such environments. The specific approaches are appropriate, since kindergarten to adult education [8, 12, 13].

34.4 Romanian National Mechatronic Platform

Mechatronic philosophy developed in Romania since 1991 by establishing the branch of mechatronics in engineering in the main technological universities from the country [8, 10]. In 1999 was established the National Council for Technological Education and Innovation. The Council elaborated the National Program for Mechatronic Education. The goals of the program were oriented to promote the mechatronic philosophy at all the levels in education, research activities, industry and services too.

As a result of cooperation at the academic level in education and research activities, along of a more than a quarter of century (1991–present) the National Mechatronic Platform was developed [8, 10–12].

It is conceived as a "national mechanism of a network structure which aims to activate material and human resources on a local, regional and national level, and also to ensure the systemic approach, in a holistic way of dealing with complex problems regarding smart education, organizational learning and sustainable development" [8, 10, 12].

At that stage (the pilot stage) the platform integrates seven Regional Centers of Mechatronics established on the structure of the Mechatronic Departments of the technological university's partners in the Flexform project [10, 12]. The regional Center established at the Technical University of Cluj-Napoca is the coordinating one. For practice and experiments at all the levels in education and training activities the mobile lab of mechatronics and portable mechatronic platform were developed. The portable platform for mechatronic education makes possible experiments everywhere and every time, being very efficient to stimulate initiative and creativity. For mechatronic portable platforms are very efficient solutions based on Lego modules including sensors, actuators, microcontrollers and elements for mechanical structures.

The mobile lab of mechatronics was established since 2010. It is used for research activities out of university (technical state of equipment's evaluation in factories, pollution parameters registration etc.) and for demonstrations in schools. The specific equipment's are mounted on a dedicated platform based on Mercedes Vito (see Fig. 34.6).

The automobile platform is itself a very representative mechatronic system. The applications developed are: study of the integration—complexification process and

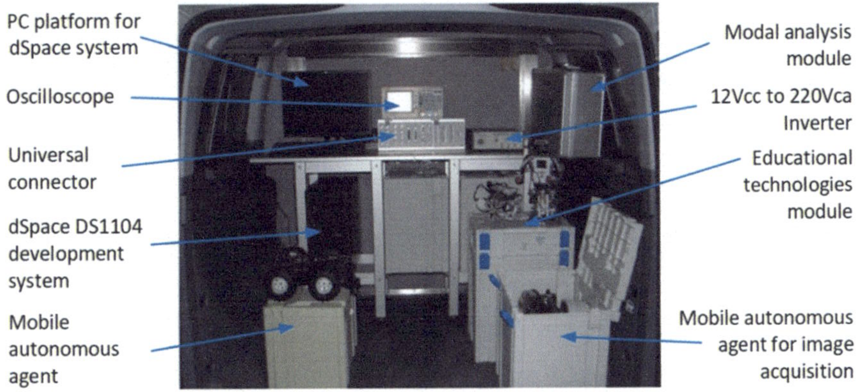

Fig. 34.6 The mobile laboratory of mechatronics

the role of information links; study of the main mechatronic modules integrated in the structure of a mechatronic automobile; study of the interfacing problems; understanding the importance of a system thinking developing to design and manufacture the complex systems. The anatomy of a mechatronic car is very similar with human body anatomy. Details related on the structure and applications based on mobile lab of mechatronics are integrated in the paper [8].

In the last years, 11 universities in the country established mechatronic departments. Also, the National Institute for Research and Development on Mechatronics and Measurement Technique is integrated in the structure of the National Mechatronic Platform [8, 12]. It is important to note that since 2010, every year, at the academic level is organized the scientific meeting "Mechatronic Education Days" [8, 12]. The program of the event includes the National Mechatronic Olympiad too. The competition is organized based on the rules of the International Olympiad of Mechatronics. The International Olympiad of Mechatronics is organized every two years under the aegis of the International Vocational and Training Organization (IVTO).

The partnership of our university with CIRET (International Center for Transdisciplinary Research and Studies) since 2007 and with ATLAS (Academy for Transdisciplinarity Learning and Advanced Studies) since 2010, were very fruitful and helped us to discover the innovative potential of mechatronics for education and research activities too. CIRET (founded in Paris in 1987) and ATLAS (founded in Texas, USA in 2000) are the main vectors of promoting transdisciplinarity at the international level [31, 32]. In the context of globalization and internationalization, being transdisciplinary is a major need [32]. But in order to be transdisciplinary we must learn transdisciplinarity. Transdisciplinarity and integral education are the basic component of smart education for sustainable development too. The basic problems for transdisciplinarity learning are outlined in the paper [36].

The trans—thematic identity of mechatronics, based on complexity concept is demonstrated in the works [8, 33–35]. On the other hand Russel Ackoff, the pioneer and promotor of the systems thinking concept says that the effective research is not disciplinary, interdisciplinary or multidisciplinary it is transdisciplinary [24].

Now, The National Mechatronic Platform is the scientific foundation of The National Platform for Smart Education, Organizational Learning and Sustainable Development. Based on this scientific support was launched the Project: Smart Romania: The Country of The Learning Communities [12]. The project aims to stimulate the evolution of schools and universities to the status of Learning Organizations. On this way, a state of spirit will be developed at the level of communities to support the effort for sustainable development.

34.5 Conclusions

Mechatronics, the twenty-first century technology, the integration philosophy, science of intelligent machine is also the scientific foundation of the smart and sustainable educational technologies development. Transdisciplinarity and organizational learning are the basic components of these educational technologies too. The evolution of the schools and universities to the status of learning organizations is of great importance for getting the status of smart schools and smart universities too. Mechatronic platforms are the basic infrastructures of these educational environments. Mechatronics makes possible the integrated approach of material, energy and information, opening new horizons in all the fields of activities. Also, mechatronics launched new openings for innovation, through technologies integration. The flexibility in thinking and action, the result of mechatronic education, are essential to stimulate initiative and creativity. Flexibility and reconfigurability of the mechatronic technical systems are the result of the integration in their structure of information links. The triangle of the twenty-first century technologies (Mechatronics, CPS and IoT), fundamentals of industry 4.0, defines the scientific and technical platform for developing the main pillars: economy, society and environment of the smart and sustainable society, according with UN Agenda 21.

References

1. G. Brundtland, (edit.) *Our Common Future: The World Commission on Environment and Development*. (Oxford University Press, Oxford 1987).
2. United Nations Agenda 21. Available at: https://www.un.org/esa/sustdev/documents/agenda21/index.htm. Last Accessed 21 Feb 2020
3. United Nations *Report of the World Summit on Sustainable Development*, Johannesburg. Available at: https://daccessdds.un.org/doc
4. UNESCO *International Implementation Scheme for the Decade in Brief*. Available at https://portal.unesco.org/education/en/ev.php. Last Accessed 21 Feb 2020
5. M.E. Auer, R. Langmann, (eds.), Smart Industry and Smart Education. in *Proceedings of the 15th International Conference on Remote Engineering and Virtual Instrumentation*. (Springer Publishing, 2019)
6. M.E. Auer, K. Sun Kim, (eds.), *Engineering Education for a Smart Society. World Engineering Education Forum and Global Engineering Council 2016*. (Springer Publishing, 2018)
7. V. Gunes, S. Peter et al., A survey on concepts, applications, and challenges in cyber-physical systems. KSH Trans. Internet Inform. Sys. **8**(12) (2014)
8. V. Mătieș (ed.), *National Machatronic Platform* (Fundamental of educational and training programmes in the knowledge society. TUPress, Cluj-Napoca, Romania, 2016)
9. O. Vermeșan, P. Friess (eds.), *Internet of Things from Research and Innovation to Market Deployment* (River Publishers, Aalborg, Denmark, 2014)
10. V. Mătieș (ed.), *Mechatronic Platforms for Education and Research* (Todesco Publishing, Cluj-Napoca, Romania, 2009)
11. FlexFORM–Program for flexible training on mechatronic platforms (POSDRU-ID-64069) Project Homepage, www.flexform.ro. Last Accessed 21 Feb 2020
12. Smart Romania: The country of the learning communities, Project Homepage, www.smarteducation-clujnapoca.ro. Last Accessed 21 Feb 2020
13. M. Kajitani, What has brought mechatronics into existence in Japan? In: *Proceedings of the 1st France-Japan Congress of Mechatronics* (Besancon, France, 1992)
14. J. Buur, *Mechatronics Design in Japan* (Technical University of Denmark, Institute for Engineering Design, 1989)
15. J. Buur, *A Theoretical Approach to Mechatronics Design* (Institute for Engineering Design, Technical University of Denmark, Ph.D.Thesis, 1990)

16. J. Peters, H. Van Brussel, Mechatronic revolution and engineering education. Eur. J. Mech Eng **34**(1), 5–8 (1989)
17. K. Yamazaky, T. Suzuki Hoshi, Methodology of education and R&D in mechatronics. Int. Appl. Eng. Ed. **1**(1), 35–41 (1985)
18. E. Deming, *The New Economics for Industry, Government and Education*. (MIT Press, 1993)
19. V.D. Hunt, *Mechatronics: Japan's Newest Threat*. (Chapman and Hall, New York)
20. F. Jacob, *The Logic of Living* (Hermes, Paris, 1970)
21. A. Restian, *The World Unity and Sciences Integration* (Integronics. Scientific and Cultural Publishing House, Bucharest, Romania, 1989)
22. I. Vlașin, *Research Regarding Valorization the Innovative Potential of Mechatronics for Smart, Competence-based Education Development*. Ph.D. Thesis. (Technical University of Cluj-Napoca, Romania, 2018)
23. C. Rusu, *Contributions Regarding the Study of the Mechanisms for Mechatronics*. Ph.D. Thesis. (Technical University of Cluj-Napoca, Romania, 2005)
24. https://thesystemsthinker.com/a-lifetime-of-systems-thinking/
25. I. Vișa, A. Duța (eds.), *Nearly Zero Energy Communities. Proceedings of the Conference for Sustainable Energy (CSE)*. (Springer Publishing, 2017)
26. R. Giffinger et al., *Smart Cities: Ranking of European Medium-Sized Cities* (Vienna University of Technology, Centre of Regional Science (SRF), 2010)
27. P. Senge, *The Fifth Discipline* (Doubleday Publishing Group, Ney York, The Art and Practice of the Learning Organization, 2006)
28. P. Senge, *School That Learn. A Fifth Discipline Field Book for Educators, Parents and Everyone Who Cares about Education* (Three Publishing House, Bucharest, 2012)
29. I. Vlașin, *From Mechatronics to Smart Education* (TU Publishing, Online Library, Cluj Napoca, Romania, 2018)
30. C. Lăpușan, *Contributions Regarding Integrated Design and Interfacing of Mechatronic Systems*. Ph.D. Thesis. (Technical University of Cluj-Napoca, Romania, 2010)
31. B. Nicolescu, The need for transdisciplinarity in higher education in a globalized world. Transdisciplinary J. Eng. Sci **3**, 11–18 (2012)
32. B. Nicolescu, R. Yeh, A. Ertaș (eds.), *Being Transdisciplinary* (ATLAS Publishing, Texas, USA, 2019)
33. S. Berian, *Research Regarding Transdisciplinary Potential of Mechatronics*, Ph.D. Thesis. (Technical University of Cluj-Napoca, Romania, 2010)
34. S. Berian, V. Matieş, *Transdisciplinarity and Mechatronics* (Curtea Veche Publishing, Bucharest (in Romanian), 2011)
35. I. Pop, V. Mătieș, Transdisciplinary approach of the mechatronics in the knowledge based society. in *Advances in Mechatronics*, (Intech Open Access Publisher, Croatia, 2011)
36. V. Mătieș, I. Vlașin, V. Tămaș, Transdisciplinarity, mechatronics and organizational learning. in *Proceedings of the Atlas Conference 2018, volume-being Transdisciplinary*, (ATLAS Publishing, USA, 2019)

Chapter 35
Rethinking Architectural Spaces for Solar Energy Better Use

Anda-Ioana Sfintes and Radu Sfintes

Abstract Using renewable resources nowadays is not only a reality but also a necessity. However, when designing architectural spaces we are still rather considering a sole main function or related functions, thus making the building easier to manage. In this paper we wish to present a vision for adapting representative buildings in a community in order to respond to sustainability issues (from social sustainability to economic and environmental sustainability), highlighting ways of producing and better using solar energy. In this context the building itself can also become a promoter by informing about (while also making use of) renewable energy systems and principles. From sports grounds to museums, we shall see how different functions can be accommodated and integrated in a single area, leading to using resources locally produced in real time and in that space instead of producing, storing, sharing energy on longer distances. In this scenario the consumer is drawn towards the place of production and consumption, having his needs met while also having something to gain at various levels—from chances for personal development to chances of social interaction.

Keywords Architecture for the community · Sustainable architecture · Multifunctionality

35.1 Introduction

Architecture nowadays has to respond to a vast array of challenges. Beyond being a spatial response given to the needs and realities of a specific time and place, it becomes iconic in promoting and fostering key concepts of contemporaneity; "globalization, speed, technology, sustainability, and more are transforming how design is practiced" [1, p. 1] but also how architecture is being used and experienced. More than that, architects have to acknowledge the impact their architecture has upon society and

A.-I. Sfintes (✉) · R. Sfintes
Center for Architectural and Urban Studies, "Ion Mincu" University of Architecture and Urbanism, 18-20 Academiei Street, 010014 Bucharest, Romania
e-mail: anda.sfintes@uauim.ro

I. Visa and A. Duta (eds.), *Solar Energy Conversion in Communities*,
Springer Proceedings in Energy, https://doi.org/10.1007/978-3-030-55757-7_35

environment whilst responsibly considering the complexity of present and future reality [2], possibilities of growth, development, adaptability.

The issue of sustainability, with all its well-known aspects—social sustainability, economic and environmental sustainability—but also with its secondary aspects like cultural, political, aesthetic, functional sustainability etc. is central (if not even mandatory) in any high-quality contemporary approach.

In this paper we wish to present an architectural vision for adapting representative buildings in a community, considering the potential, difficulties and implications of using solar energy as a keystone in reframing an architectural program. While the potential is obvious—that of producing and using renewable energy sources, the difficulties taken into consideration refer to the possibility of placing the necessary systems—area, orientation, but also site ownership and interest. The implications are being addressed through issues of consumption optimisation, scale of the impact and related effects.

Evolving around the Romanian overdue/weak implementation of sustainable thinking and the potential presented by existing facilities in Bucharest like *Complexul Cultural Sportiv Studentesc Tei*, we explore a scenario of sustainable transformation linking sports venues as green infrastructure and multifunctionality.

Romania is still struggling with responding to basic sustainability challenges like waste management, water quality, nature conservation and air quality [3], various infringements being active. These facts not only highlight the need for overstepping these challenges, but also the effort that should be put in meeting other sustainable aspects through any means.

In this regard, the proposal also assumes urban regeneration through qualities which follow centrality issues, hybridization, participative and inclusive activities, enhancement of economic and social aspects, ecological responsibility [4]. Thus through interventions in existing spaces (in our case a facility which tackles both issues of public and semi-public spaces anyway) notions of urban space as object of consumption (needed to be exploited in order to further activate the area) are being put together with "forms of ethical consumption" like eco-friendly architecture [5].

35.2 Benefits of Sports Facilities

The health and healthy living aspects assumed by sports are far from being the only type of benefits they bring. Doing sports leads to learning or enhancing life skills like problem-solving, decision making, dependability, teamwork, self-management, to develop abilities like work under pressure, set goals, handle success and failure etc. Watching sports (by attending sporting events) has its own positive outcomes by being anything between a leisure activity and a way of life (for die-hard fans). It is an occasion for meeting up with friends, family or the community of fans, for fostering these relationships, for experiencing joy or sorrow etc.

At a larger scale, they also drive social change through the impact they have upon education, international relations, and the natural environment [6]. Taking as example the issue of sustainability, we can name the role of sports organizations and facilities in acknowledging, informing about, promoting a sustainable way of life. For example, "[e]ducational interventions such as *green games* can inform fans on recycling, green tailgating, carpooling, and environmental issues specific to that region" [6, p. 69].

Getting closer to our case study, further we shall discuss some aspects which will help us grasp the potential presented by the specificity of the site—*Complexul Cultural Sportiv Studentesc Tei*. We will consider the benefits of a sports facility being also a public space with green infrastructure elements and the various sustainable criteria which can be met in regard to this context.

35.2.1 Sports in the Context of Public Spaces and Green Infrastructure

Urban green areas play different roles in the life of a city, no matter their scale. When it comes to sports, people can easily appropriate almost any kind of setting, from paved alleys along green areas (which can be used for jogging, riding bicycles, skateboarding, roller-skating), to rough terrain (for off-road walks and cycling) and lawns (for playing badminton, volley, frisbee) etc. Green spaces become casual occasions for relaxation and exercise, having a positive impact upon the quality of living and becoming a buffer for the more negative effects of urban life [10].

Even more positive outcomes arise from considering a public space part of Green Infrastructure (GI) which is "a planned or managed spatial structure and network of interconnected environmental features, natural areas, open spaces, and landscapes" [7, p. 3917/1]. This assumes the implementation of various systems capable of sustaining Green Infrastructure roles like protecting ecosystem state, functioning and services, protecting biodiversity, protecting societal health and wellbeing as well as supporting the development of a green economy and sustainable land and water management [9]. For an underdeveloped community, GI can play different roles beyond its ecological aspects: it can raise attractiveness, productivity and employment opportunities, it can foster economic growth while also improving the physical, social and environmental conditions [7]. From a social and cultural perspective, it emphasises the educational role played by nature in the more casual context of leisure activities while also having an impact upon physical as well as mental health; it helps acknowledge, manage, adapt to and participate in environmental and ecological issues [7].

35.2.2 *Sports Facilities in Service of Sustainability*

Because of their great public span, sports facilities, organizations and services became involved in sustainable initiatives, feeling the (real or virtual) pressure of society or just anticipating their role in social change. Thus they concentrated their efforts in two directions: "to reduce the ecological footprint of sports, and to use sports as a means to raise environmental awareness" [12, p. 10]. Most of the entities referred only to basic and easier to apply measures like saving electricity and waste management, but more important in this context are the use of both formal and informal environmental systems, the latter speaking more of commitment, engagement and responsibility beyond the main scope of the entities. This approach has the benefit of also triggering public commitment to environmental protection [12].

To a larger and more programmatic scale, sports facilities can be included in strategies of Sustainable Urban Planning and Development (SUPD), addressing issues of economic importance (creating new jobs, promoting a circular economy and regional/local interdependencies as well as informal economies), the environment and available resources (promoting and even adopting clean energy sources, addressing the ecology and biodiversity of the site, waste management), the society (by promoting wellbeing, involving and empowering communities, addressing issues of right to the city, social innovation and justice) and so on [2]. The built environment is the one that often makes visible and/or possible the aspects mentioned above by accommodating various functions and making use of various systems, but also through its image (use of materials, display of the systems use) and aesthetics (buildings which facilitate appropriation).

Further we shall briefly discuss the three aspects of sustainability—economic, environmental and sociocultural sustainability—in regard to sports facilities.

Environmental Sustainability. Environmental Sustainability refers to "the efficiency of land use through the conservation of energy and of natural terrain, the minimization of pollution, creation of parks and green areas, securing of open space, and the maximization of land use efficiency through multi-dimensional and complex development" [8, p. 1249/5]. Thus it assumes use of natural resources including local plants, but also making use of natural conditions and manipulating the built space in order to respond to climatic needs—proper natural ventilation and lighting, heating and cooling through shading systems and natural air flow, rainwater management through permeable paving, collecting systems etc.

Many of the sports venues are accompanied by large open spaces in order to accommodate large numbers of people attending events and/or large green spaces through the number of green play fields or as a way of reducing the carbon footprint of the facility. These open and green spaces can sometimes be appropriated by nearby communities for outdoor activities, thus becoming public places.

Economic Sustainability. "The characteristics of economic sustainability can be summarized as recycling, low cost, local resources, and flexible space utilization" [8, p. 1249/19] which can easily be fit into sports facilities. Recycling is not only an environmental friendly initiative but it also leads to low costs in the context of circular economies. As facilities which bring together large numbers of various users, waste management becomes mandatory. Considering also the scale of sports facilities in general, recycling (including use of recycled materials), accompanying the use of local resources, can be an essential cost-saving method. The use of local resources can be part of urban planning and landscaping, architectural design, but can also refer to consumption of local products, thus spanning to the whole life of the facility. Reusing water as well as making use of environmental factors (climate, orientation, air flows) to reduce energy needs are other important challenges. All of the above should be implemented considering also reduced maintenance and maintenance costs. Flexibility of space is one of the methods that can help in this regard, assuring continuous use as well as easy adaptation to changing needs. When it comes to sports, we can easily think of multifunctional play fields besides the multifunctionality of architectural spaces. This issue will be further addressed in the next section.

Socio-cultural Sustainability. The idea of socio-cultural sustainability evolves around issues of rehabilitation, identity and social participation, all related to the social and cultural realities of the local context. While preserving local values, customs and social norms, socio-cultural sustainable developments are expected to facilitate access, integration (of different ages, people with different social statuses, with different ethnic origins etc.), to assure gathering places and encourage active and passive participation, involvement, interaction between users of the facility and local community. Thus, space should also be easily appropriated by the community. These facts make mobility and amenity key issues in conceiving sustainable interventions [8].

In the case of sporting events, the phenomena mentioned above easily occur but special attention should be paid to framing or managing such encounters (to avoid, for example, hooliganism and negative impacts of stranger flows into the community).

Given the ever-changing nature of communities, this type of sustainability also assumes adaptability and flexibility in order for the space to keep responding to contemporary needs and developments.

Many sports organizations have taken seriously their social role amplified by the great number of (professional and amateur) trainees and players, fans and viewers. Their sustainable initiatives easily inspire positive social change among a vast public while also educating about how sustainable measures can be implemented or adopted in everyday life by every individual [6].

35.3 Benefits of Multifunctionality/Multifunctional Benefits

Multifunctionality nowadays is often incorporated in architectural and landscape proposals, thus assuring the good use of spaces as well as a continuous and flexible adaptation that can keep up with changing needs and trends. As we have seen in the previous section, multifunctionality is also a tool of sustainable design from an environmental and social perspective. It signifies overlapping functions and it can refer to both "multifunctional use of space as the interrelation between functional, social, and morphological possibilities, in which activities can be integrated and coexist simultaneously; multifunctionality over time of the object as its potential and useful transformation, guaranteeing new lives to it" [4, p. 139].

The latter aspect attracted much attention around sports in the last years as the development of Olympic facilities demanded integrating multifunctional views in order to avoid desertion after games end or as stadiums started to stand out as big scale resilient facilities. The former aspect is closely linked to key aspects of contemporaneity like connection and networking, cooperation, inclusion, equality, social justice, empowerment. We can speak of a twofold adjustment—of the space to the demands of the society and vice versa, as the space, through its attributes and affordances, can promote social change.

Multifunctionality unfolded especially along public space proposals because of its social, political and economic position. As it gathers large and various numbers of people, pertaining to various communities, public space was and is the perfect scene for fostering dialogue and participation, for contesting values, views, citizenship etc. through triggering mechanisms which assume "interaction between the cultural, social, landscape, environmental, economic, and institutional dimensions of a context" [4, p. 138].

Multifunctional buildings can link ecological, economic, socio-cultural, historical and aesthetic functionalities [10], the benefits of multifunctionality residing in spatial outcomes like openness, flexibility, adaptability, hybridity, as well as in environmental and economic sustainability concerning especially management of resources. All these benefits of multifunctionality also lead to multifunctional benefits as they give back to society by creating appropriate living conditions, by fostering local production and responsible consumption, by leading to appropriation and meaningful experiences, by accommodating identity development and expression etc.

Circling back to grasping aspects of our case study context, we can add up to these benefits those emphasized by the development of Green Infrastructures: improvement of the built environment, ecosystem protection, development of economic capacities, emergence of new (including informal) educational opportunities [7].

It should be noted though that, by operating at the intersection of so many factors and less controllable variables, the benefits and outcomes cannot all be anticipated, fact which also applies to possible negative impacts. In order for multifunctionality to lead in a certain extent to positive impact it should be based on in-depth interdisciplinary researches that take into consideration at least regional as well as local social, cultural, political, economic realities. On the other hand, by operating with multifunctionality unplanned outcomes can also arise, based on people's actual response and organic unfolding and development of uses.

35.4 Adapting Existing Facilities—A Proposal

Romania lacks sustainable high-standard buildings, even more so sustainable high-standard rehabilitations that would consider the complexity of meeting all aspects of sustainability. To some extent this is caused by the fact that most of the major projects are funded by private stakeholders having no interest for public good and without being at least stimulated by the state to include in their developments aspects of public good. Issues of administrative dialogue, coordination and capacity also block possible initiatives in this regard coming from private stakeholders. When it comes to public funds, the priorities/national demands are set on building infrastructure and hospitals. However, Romania could engage in solving secondary issues by strategically attracting and directing European funds.

The case study we propose—regeneration and enhancement of *Complexul Cultural Sportiv Studentesc Tei* (Tei Student Cultural Sports Complex)—is based upon its characteristics, its position and discrete openness towards the community despite passing, along the years, through limited and unessential modernisation. As it stands on the shore of one of Bucharest's lakes forming the city's series of lakes and being part of the green belt accompanying it, the Complex has the potential of being integrated in the development of a Green Infrastructure. Its public ownership, as well as its main functions which cover both education and sports make it a valuable candidate as a future pilot project of sustainable and multifunctional transformation.

35.4.1 Context

Complexul Cultural Sportiv Studentesc Tei (Fig. 35.1) is a public institution under the patronage of the Ministry of Youth and Sports. Hence, public ownership already highlights the engagement in extending public good and in fostering the public interest.

The complex gains its budget from its own revenues and subventions from state budget. While there is no entrance fee, some of the (basically modernized) play fields can be rented for professional and amateur matches and others can be used freely. Besides the fields (configured for rugby and football, volleyball, basketball, tennis, pinball but also appropriate for various smaller sports), the complex attracts, through its large alleys and natural context, runners, cyclists, roller-skaters, skateboarders etc.

Despite the poor conditions of some of the fields, by being one of the very few such free entrance sports facilities, they are used by a large community inhabiting the residential areas nearby. Some of the free features can even be considered important triggers in testing various sports. For example, the presence and free use of rugby training equipment led some into embracing the sport at least as amateurs.

Another aspect worth mentioning is the informal use of the complex, not only as a sports facility but as a local park and leisure site. Activities from parents walking

Fig. 35.1 Aerial view of the Complex. *Image Source* Google Maps

newborns and toddlers to people sitting in the grass, sunbathing in the grass or on the uncovered stands around the rugby field are quite common.

To the aspects mentioned above we can add, as justification for the appropriateness of the site for being transformed into a sustainable facility, its formal and informal educational function. The complex is used by students for training and competitions, but the organization also arranges various activities and sessions dedicated to youths on subjects like health, culture, art, entrepreneurship, volunteering.

A last important feature to be considered is the location. The complex is part of a series of open and green spaces along the Bucharest lakes which still lack a coherent strategy for embracing their potential as Green Infrastructure elements. The complex has a long, accessible but unexploited shore to Tei Lake in the east. Furthermore, the existence of and relationship between open, free and built spaces create numerous opportunities of transformation, adaptation and implementation of sustainable systems. In this article we shall highlight the possibilities of better using solar energy in such a context after evaluating the multifunctional potential of the site.

35.4.2 Services at the Community Scale—Overview of Possibilities

Complexul Cultural Sportiv Studentesc Tei is at a walking distance from the habitation area of Tei neighbourhood (in the south–east). The site is also easily accessible by public or sustainable transportation from more upper-class habitation areas like Floreasca (in the south–west) and Aviatorilor (in the north–west). An element of contrast, the office area in the north highly developed in the last 10 years, assures the possibility of further variating the services envisioned for the complex (Fig. 35.2).

Given this still basic identification of functions around the site and the communities they accommodate, we can come up with a few proposals of urban regeneration and sustainable transformation. It is obvious that a real proposal should be based upon a thorough radiography of the area and a deep understanding of the needs and values of potential users. Anyway, here our scope is that of overviewing a larger palette of viable options, based on the reading and interpretation of the site and its evolution, as well as on issues which should be addressed in a sustainable perspective, in order to highlight the potential of the site.

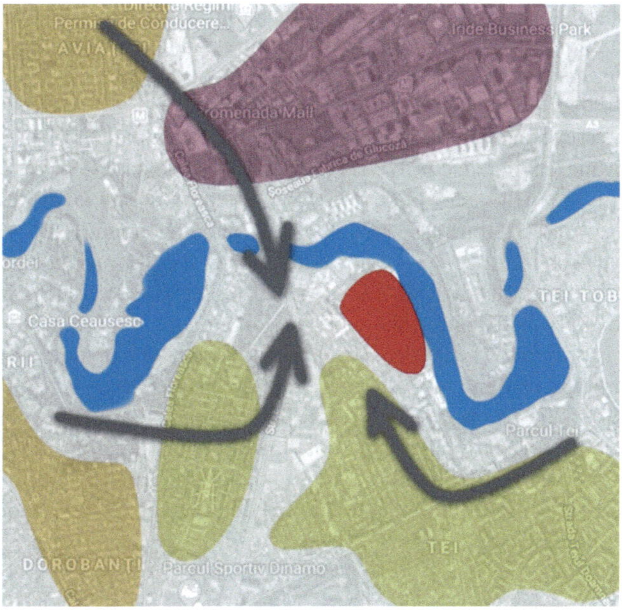

Fig. 35.2 Functions around the complex and main access routes. In red—the site, blue—series of lakes, yellow—habitation areas, orange—upper class habitation areas, purple—offices and corporation headquarters

First, the Complex can be exploited as part of a new quartier center, serving both habitation and work areas. Derived from the main functions of the complex—education and sports—and its current forms of appropriation, issues of health, healthy living, sustainable living, and responsible consumption can be addressed. Such services could foster interactions between the various inhabitants and enhance economic and social aspects. Participative and inclusive activities could underline informal forms of education and foster creativity. Activities of production could also be proposed at a certain scale, keeping in balance the proportion of main and secondary activities.

As the site is placed at the intersection of old and new areas, with different functions and with communities more or less constituted, fostering the discovery and development of collective identities would play an important role in use and appropriation. In this regard, including a community museum that would grow through the active involvement of the public could be beneficial.

The multifunctionality envisioned for such a place would assume drawing users towards using the services and facilities on long periods of time. The inhabitants of the nearby dwelling areas could spend here an entire day, caught in a variety of activities (from leisure and production, to education and work). Transforming the site into a landmark—place to be consumed—can have a greater call upon those working in the Pipera office district who could use the site rather after work or in weekends. However, the biggest challenge would be finding and tuning the motivations that would lead to a continuous use of the Complex and that can be made by assuring access to useful services, facilities, and systems that otherwise wouldn't be available as well as "through the creation of a relationship system able to expand and to «contaminate» the urban environment that needs to be revitalized" [4, p. 145]. Through inclusiveness the issue of bringing in the same place people of different ages, with different interests etc. would be addressed, thus assuring an equilibrated occupancy.

The main idea is that of using renewable energy locally produced in real time, avoiding storing and sharing a big part of it on longer distances, fact which would need thorough analysis for optimizing production and consumption. In this scenario the consumer is drawn towards the place of production and consumption, having his needs met while also having something to gain at various levels—from chances for personal development to chances of social interaction. A related effect of such an approach would be popularizing the idea of ethical and sustainable consumption and energy saving. Eco-friendly architecture and landscape planning are in this context the first exemplifying features that can lead to social change and further engage the public in sustainable living.

35.4.3 Integrating Solar Energy Use

Further, we shall highlight ways of integrating and better using solar energy in the context of extending the current function and usability of *Complexul Cultural Sportiv Studentesc Tei* by transforming it into a multifunctional and sustainable public space while still unfolding it around the ideas of sports and education, but in an even more opened and accessible way.

Many of the systems proposed would be visible or at least people would be informed about the production-consumption cycle and the sustainable outcomes in order to acknowledge the importance of such systems and value them.

Various systems can be either integrated into the new and old buildings or proposed along the open spaces. In the former case, solar and photovoltaic panels are the already classical solutions that come to mind. However, their development in the last years makes them easier to integrate and use as architectural elements, thus conferring aesthetic identity to the buildings they are part of (research is done in playing with color and shapes). They can be positioned on terraces as well as on facades but special attention has to be paid to orientation for maximizing efficiency. Other Building-integrated photovoltaics systems (BIPV) which can be used refer to adding semitransparent PV modules to windows (with the negative effect of diminishing to some extent visibility) or using solar glass bricks which produce and also store energy, acting as radiant walls.

Complexul Cultural Sportiv Studentesc Tei has the advantage of a small built/unbuilt ratio as well as of being surrounded by low rise buildings, facts which give plenty of liberty in finding the best orientation for various systems. Panels for example can also be used for covering less common structures like the stands around the main field or covering structures along alleys or above benches as sun-blinds and protections from rain.

Urban furniture can produce and use solar energy in various ways and for various reasons—for example for charging devices. Other systems which can be implemented in the open space include fixtures like tall lighting sources or even solar brick pavers.

Last but not least, everything should be sustained by using all sorts of low-energy equipment and, as we said before, by informing the public, raising awareness and fostering sustainable thinking.

35.5 Conclusions

The success of a sustainable project relies not only on the quality and complexity of the systems applied, but also on the public values that it references which make the project appropriable. Thus, the implementation of a sustainable approach in a public space, even more so existent, taking into consideration all the aspects of sustainability—environmental, economic and social—has the potential of influencing the sustainability of the whole city [8]. In our case study, *Complexul Cultural Sportiv Studentesc Tei* could also have been included in the development of the city's Green Infrastructure comprised of two of its most important natural features: the series of lakes and the green belt accompanying it.

Transforming such a place and making it part of an urban regeneration scheme should exploit to the maximum its possible outcomes as "[t]he contemporary city can become an experimental center of progress and creativity, also thanks to the adoption of new drivers for sustainable growth, innovation, creation and dissemination of knowledge and information" [4, p. 139].

The proposal discreetly framed in this paper spoke of multifunctionality revolving around sports and education, of sustainable systems (detailing ways of integrating and better using solar energy), of impacts upon communities and benefits. All these elements also advocate for environmental diversity sustained by architectural and landscape interventions. A high-quality design is based upon a thorough research, reckoning and understanding of possible stakeholders which will further inform ways of designing dynamic, adaptable, flexible buildings. It will take into consideration possibilities of engagement with the built/unbuilt space, affordances. When it comes to the environment, it will emphasize and relate to the natural conditions, making people aware of the ecological specificities [11].

Here we only envisioned possibilities and potentialities based on specific elements and a specific context. The approach can only be considered a triggering basic research but it can be taken further in various directions, underlined throughout the entire article: possibilities of urban regeneration of the area considering *Complexul Cultural Sportiv Studentesc Tei* a central point, possibilities of developing the city's Green Infrastructure along which the Complex can become a landmark, possibilities of sustainable development of the existing area (considering all sustainable aspect or less), possibilities of adapting the context and making it multifunctional etc.

References

1. S. Augustin, C. Coleman, *The designer's guide to doing research: applying knowledge to inform design* (Wiley Inc, 2012)
2. L. Balula, J. Seixas, Contemporary city and plural knowledge: reframing urban planning, in *Architecture and the Social Sciences—Inter—and Multidisciplinary Approaches between Society and Space* ed. by M.M. Mendes, T. Sá, J. Cabral (Springer, 2017), pp. 69–84
3. European Commission, in *The Environmental Implementation Review 2019. Country Report—ROMANIA* (SWD(2019) 130 final) (2019)
4. E. Ginelli, G. Pozzi, G. Lazzati, D. Pirillo, G. Vignati, Regenerative urban space: a box for public space use. in *Regeneration of the Built Environment from a Circular Economy Perspective* ed. by S. Della Torre, S. Cattaneo, C. Lenzi, A. Zanelli (Springer, 2020) pp. 137–147
5. R. Jaffe, A. de Koning, Social life in public space, in *Introducing Urban Anthropology* (Routledge, 2016), pp. 55–68
6. T.B. Kellison, S. Trendafilova, B.P. McCullough, Considering the social impact of sustainable stadium design. Int. J. Event. Manage. Res. **10**(1), 63–83 (2015)
7. D. Kim, S.-K. Song, The multifunctional benefits of green infrastructure in community development: an analytical review based on 447 cases. Sustain. **11**(14), 3917 (2019). https://doi.org/10.3390/su11143917
8. S. Kim, H. Kwon, Urban sustainability through public architecture, in *Sustainability*, **10**(4), 1249/1–21 (2018)
9. Science for Environment Policy. *The Multifunctionality of Green Infrastructure* [In-depth report] (2012)
10. W. Shi, H. Woolley, managing for multifunctionality in urban open spaces: approaches for sustainable development. J. Urban. Manage. **3**(1–2), 3–21 (2014). https://doi.org/10.1016/S2226-5856(18)30081-5

11. M.A. Steane, K. Steemers *Environmental Diversity in Architecture* (Spon Press, 2004)
12. S. Trendafilova, B. McCullough, M. Pfahl, S.N. Nguyen, J. Casper, M. Picariello, Environmental sustainability in sport: current state and future trends. Glob. J. Adv. Pure. Appl. Sci. **3**, 9–14 (2014)

Chapter 36
From Theory to Practice and All the Way Back. Designing with The Sun

Ana-Maria Dabija

Abstract "The Sun does not realize how wonderful it is until after a room is made."—Louis Kahn. The sun is also a building material; not by itself but in the architectural concept, the sun—light gives value to the space. Encapsulated in some building products, the light and energy of the sun are used to provide active or/and passive gain; sun control is a field of architectural approaches and there are products that provide protection against sun glare and heat. Professional education develops the knowledge to use the sun or protect against it, from the concept to the detail. At least in theory. In practice, a poor or incomplete design generates a chain-reaction: problems that are difficult to repair and that generate other problems that can be sometimes solved only with huge costs. As in the case of all building products, if the specifications are not respected, the product will fail in practice. In any activity in life, the better the "story"/the "script" is analyzed, planned and carried out considering as many angles as possible and integrating specialists from other connected fields, the less bad surprises may occur. A building will never be perfect. It is an art to identify the smallest compromise (however, the compromise is always there). Teaching on how to do things right is fine but the dimension of what is the meaning of "to do right", can also be understood while tackling failed systems. A failure, analyzed and understood, brings light not only into the specific system but also in the broader field of knowledge and emphasizes and strengthens the theory. The paper presents several examples, including cases of failures, in relation with the sun and light and draws conclusions about the importance of tackling the design of the building considering the biunivocal relation between the architectural object and the natural and anthropic environment.

Keywords Architecture · Design · Education · Solar energy

A.-M. Dabija (✉)
Center for Architectural and Urban Studies CSAU, Ion Mincu, University of Architecture and Urbanism, Bucharest, Romania
e-mail: am.dabija@uauim.ro

Renewable Energies Commission, Romanian Academy, București, Romania

I. Visa and A. Duta (eds.), *Solar Energy Conversion in Communities*,
Springer Proceedings in Energy, https://doi.org/10.1007/978-3-030-55757-7_36

36.1 Introduction

The legend of the "burning mirrors" that Archimedes used to repel the Roman attack in 213 B.C. is well known. Anthemius of Tralles was probably the first to launch the story, about three centuries after the event. According to Kryza [1], Anthemius declares that Archimedes used burning mirrors: *"Archimedes cannot be deprived of the credit of the unanimous tradition which says he burnt the enemy fleet with the rays of the sun*"; he "*gave a signal and a few hundred soldiers standing along the waterfront raised their highly polished bronze shields to reflect the sun against a target on the sails of the nearest ship. The concentrated heat from these shields, acting as mirrors, was so intense that the dark sailcloth smoldered and exploded into flames within seconds. Panicked, the Roman fleet retreated, giving the Greeks time to strengthen ranks and foil the invasion*". In the same book [1] Kryza also describes an experiment regarding Archimedes' approach, carried out in 1992 by a group of scientists of the Leicester University. The "mirrors" were plywood boards covered with sheets of polished metal that transmitted their weight to the ground through a pivot (that was also granting the possibility of orienting the panels) and taking into consideration the physical capacity of the soldiers to handle the reflecting panels as well as the climate and the solar radiation of Sicily, they determined that the temperature of 500 °C could have easily been produced on a target situated at around 50 m from the shore, if some hundreds of such mirrors focused on it. The results of the experiment were that the mirror could not have burned the ship but could have injured Marcellus, as his uniform would have made him easy recognizable.

While the story is beautiful (but probably untrue), there are other true stories, in the history of architecture of the past century that are worth mentioning, as examples of good practices or failures that teach us a lesson.

36.2 Glass and Sun

The story of glass is probably one of the most fascinating in the building materials' history. From its multi-millennial life span, the invention of float glass has implications on the solar control glass, as different coatings can be applied evenly on perfectly flat surfaces.

Solar control glazing was developed to overcome indoor overheating due to the lack of solar devices, on the south and west facades of the first generation(s) of curtain walls. According to Button et al. [2], the new performances that the building skins ought to comply with in the 1920s were introduced by Le Corbusier in the curtain walls of the La Cite de Refuge, in Paris. A part of a larger assembly that was commissioned by the Salvation Army, the building was erected between 1929 and 1931 and the sealed glass facade embodied all the means that represented the environmental control of the period [3]: "*The original glass sheating […] was the physical manifestation of a totalizing system of environmental control. The intended parts were an internal system of forced air, both temperature—controlled and humidified, called'exact respiration' (la respiration exacte) and a hermetically sealed, double glazed curtain wall (mur neutralisant) with a continuous cavity for the circulation of temperature-controlled air.[...] the mechanical and material system would maintain a constant temperature while providing maximum light and purified air. Like a bio-mechanical skin, the wall would mediate between the internal organs and external conditions*".

From design to execution changes may occur, mainly aiming to decrease the costs of the buildings. It was also the case of the La Cité de Refuge: the double glazing as well as the central cooling plant were eliminated. Consequently, during the summer the interior space was overheated by the exterior fresh air pumped inside. The problem increased as there was no return air system. To diminish the heating effect of the interior space, operable windows were installed (much to the protest of the architects) in 1935 and, after the second World Was, Le Corbusier designed a composite façade, preserved to this day (Fig. 36.1) that included fix and operable panels and a shallow brises-soleil system.

Having this building as an example, the glass industry—encouraged by the designers but also by visionary writers [4], aimed at developing a glazing that included solar control characteristics as a result of body-tinting glass. Light transmission was not the sole characteristic for performance; the solar factor became just as important.

Reflecting glass panels were highly used in the 1970s. Coated with metallic alloys, the reflexive glass panels proved to be a double-edged sword: the light was directed in the environment from where it emerged. While during the day the reflection of the sunrays avoided extra-heating of the interior spaces, in the evening, when the artificial light became necessary, the glass reflected the interior like a mirror. Opaque

Fig. 36.1 Le Corbusier & Pierre Jeanneret—Cité de Refuge, as seen in 2019 photo Iantomferry (CC BY-SA 4.0)

to the exterior and at the same time transparent to outside viewers, the inner spaces were transformed into genuine showcases.

Due to the glare, car accidents were reported, as the disturbing reflections blinded the by-passers and the drivers and led, about 25–30 years ago, to special provisions that forbade the use of reflective glass at the street level. Today the glass coatings are less aggressive and a balance of the light transmittance and the solar factor of the glass provides a fair solution for the users of the buildings as well as for the by-passers.

However, glass is not the only reflective building material used for façades cladding.

36.2.1 Light Reflection as Architectural Design Tool

Reflective materials can be considered architectural means of passive design, to maximize the level of light in a dark space. Carefully oriented, the reflecting surfaces guide the light into the dark spaces. The technique is not new as it has been used since the Antiquity; it was rediscovered after the energy crisis of the 1970s and reiterated again today, in the context of the efforts to diminish the use of fossil fuels as energy resources.

In this context, the ground floor of Torre Glòries (former Torre Agbar), the office building in Barcelona designed by arch. Jean Nouvel represents an interesting example, as the window jamb extensions are oriented to lead the sunrays inside the building, with the aid of reflective metal sheets (see Fig. 36.2).

Fig. 36.2 Jean Nouvel & Fermín Vázquez, Torre Glòries, groundfloor

36.2.2 From Energy to Architecture and Back

36.2.2.1 From Energy to Architecture

Archimede's experiment—reality or legend—found a peaceful field of application in research, sometime in 1949, when Felix Trombe[1] built in Mont Louis, France, the first solar furnace in the world [5], with the purpose of providing high temperatures for producing alloys with a non-pollutant energy source: the sun. The result was an assembly that used the double reflection provided by 1420 mirrors. Called "Four Solaire"—Solar Furnace—it is the translation into French of the Roman Heliocaminus. In principle the system uses a heliostat (a flat mirror that rotates according to the movement of the sun) that reflects the sun beams to a fixed parabolic concentrator mirror [6].

After the success of the experiment in Mont Louis, Trombe designed between 1962–1968, a larger scale solar furnace in Font-Romeu-Odeillo-Via, in fact the largest assembly of this type in the world [7]. The location, in the Pyrénées-Orientales, was selected due to the air quality and considering that the number of sunny days/year is close to 300. The solar furnace of Odeillo is 54 m high and 48 m wide and consists of a field of sixty-three heliostats, located on eight terraces, that direct the sun-rays onto a large concave mirror of about 2000 m^2, that is part of the building façade of the research facility.

The mirror focuses the enormous amount of sunlight onto an area "roughly the size of a cooking pot" [8], in fact an area of 40 cm diameter [9] which reaches temperatures above 3,000 °C (Fig. 36.3). Altogether the 9600 mirrors concentrate the energy of "10,000" suns.

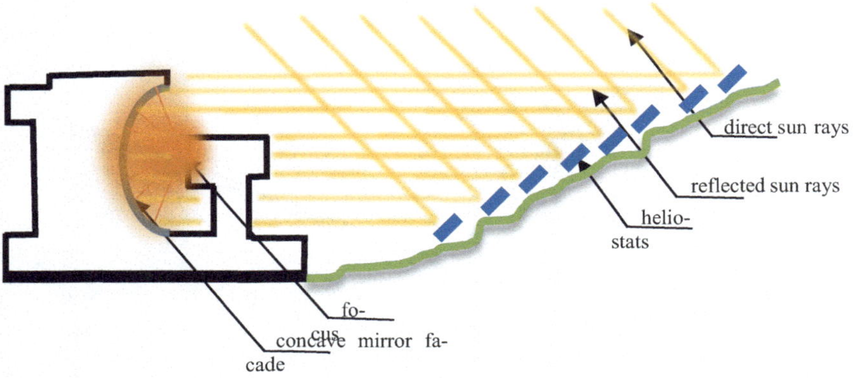

Fig. 36.3 Scheme of the Font-Romeu-Odeillo-Via solar furnace

[1] famous for the "Trombe walls", a passive solar design strategy, developed in the late 1950, in fact, a previous invention of Edward Morse, almost a century before.

The aim of building such a complex was to produce high temperature for the metallurgic industry (bronze and aluminum products) or for firing ceramic products, with a smaller impact on the environment (Fig. 36.4).

During the time, the research field initially referring to the study of materials at very high temperatures, extended to applications in the aviation and aerospace industries.

While in the case of the Solar Furnace of Odeillo the heliostats were directing the solar rays to the concave façade, in the One Central Park Towers in Sydney, Australia, heliostats direct the sunrays to provide daylight in darker areas of the assembly (Fig. 36.5), shifting the field of application from the scientific laboratory

Fig. 36.4 Four Solaire Font-Romeu-Odeillo-Via in 2010. Photo H.Zell (CC BY-SA 3.0)

Fig. 36.5 Arch. Jean Nouvel (Patric Blanc—green facades, Kennovation—heliostats) One Central Park, Sydney. Photo Daniel Neubauer (CC BY-ND 2.0)

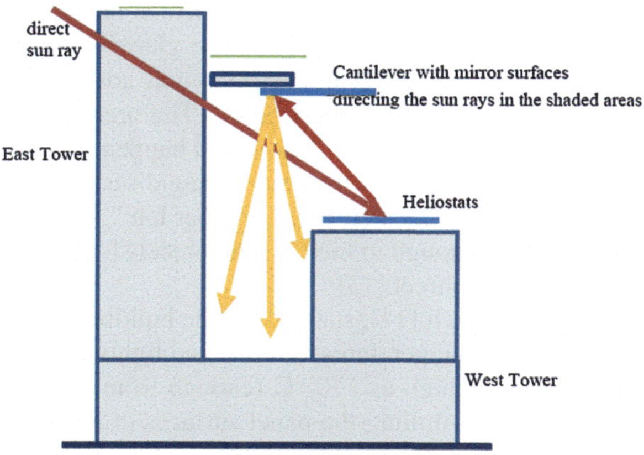

Fig. 36.6 Scheme of the One Central Park heliostats and mirror assembly

to the civil architecture and urban landscape.

The One Central Park complex is composed of two residential towers, of 116 and 64.5 m, located on a compact multileveled retail basis that sums 16,000 square meters [10]. The West tower roof is provided with 42 motorised heliostat mirrors that direct the sunrays to the cantilever soffit covered with 320 reflective mirror panels [11], thus re-directing the natural light in the areas shaded by the East, higher tower (Fig. 36.6).

Designed by Jean Nouvel (the architect of the Torre Agbar), the green walls are the creation of botanist Patrick Blanc,[2] as well as with Kennovations, the designers of the heliostats-mirror system.

Such spectacular projects must consider all the aspects of the natural and built environment, from the climatic data of the site to the influences of the already existing constructions. Omissions may lead to failures that can be addressed with difficulty and high expenses.

36.2.2.2 Back to the Design

While in the Torre Agbar the light is directed into the interior space, otherwise rather dark as the openings are of small dimension, reflections of the metallic building products lead to glare and overheating of the neighborhood of an iconic building in Los Angeles: the Walt Disney Concert Hall.

Built between 1999 and 2003, the remarkable city-scale sculpture owes its design to architect Frank Gehry. The organic, fluid shapes of the envelope—where roofs slide into façades and curves and angles alternate in a masterly proportion—were also the source of a huge scandal, as the shining stainless steel shingles that covered the convex-concave surfaces concentrated heat spots where, allegedly "reports of

[2]Jean Nouvel and Patrick Blanc collaborated for the design of the living façade of Musée du Quai Branly—Paris, in 2005.

bystanders roasting hot dogs in the reflected heat quickly became legendary" [12]. Drivers were affected by the blinding glare; tenants in the residential buildings complained by the growing demands and costs for air conditioning, due to the increased environmental temperature. The architect's answer was "The reflection wasn't my fault. I told them that would happen. I was taking the heat for all that. It made the list of the ten worst engineering disasters in the decade. I saw it on television, the History Channel. I was number ten." [12]. The temperature at the ground level was high enough to melt plastic objects [12] or to drive the pedestrians on the other side of the street [13].

Further research [14], states that "the building apparently increased surrounding ground surface temperatures to 60 °C and lightweight freestanding surfaces reached temperatures as high as 170 °C (enough to melt some plastics.)" The solution to the problem was dulling the panel surfaces in a two phases industrial sandblasting process that eliminated the metal gloss while providing an aesthetic appearance at the finalized process (Fig. 36.7). The cost of the intervention that re-established the thermal and visual comfort balance of the neighborhood was of $90,000 in 2005, that increased with one third the original costs of the building. The lesson that must be learned is that "specular surfaces must be used with great care" [15].

It is also the case of a famous London building, the 'Walkie Talkie' (voted in 2015 "the ugliest and most hated building in Britain" by the readers of the Mail Online [16]). It is not the shape, however, that produced a scandal, nor the wind tunnel created by the emplacement of the building on the site, nor the roof garden that allegedly did not meet the specifications but the concave glass façade that, like a lens, focalizes the sun-rays and projects them on a spot on the street, increasing substantially the environmental temperature. The building, designed by architect Rafael Viñoly, is

Fig. 36.7 Arch. Frank Gehry, Walt Disney Concert Hall in 2009 (after sanding the building envelope) Photo: Andrew M Butler (CC BY-NC 2.0)

Fig. 36.8 Arch. Rafael Viñoly: 20 Fenchurch Street, London Photo Glynne Hather (CC BY-SA 2.0)

located in the business district of London. Designed in 2006, the construction began in 2009 and was finalized in 2014 (the roof top garden was opened in 2015). The initial design was changed, by down-scaling (from 200 to 160 m), as concerns about its visual impact on the St. Paul's Cathedral and the Tower of London were expressed. The original design seems to have had a brise-soleil system incorporated but, as it was built during the economic crisis of 2008–2010, the shading devices were cut off (Fig. 36.8). Its shape, that resembles the communication device, is where the name originates but the building earned a more ironic nickname, because of the effects that the Southern façade had on the environment: the Fryscaper. Even from the construction phase it was observed that the glass reflections of the (curved) Southern façade heated spots of the sidewalk and street up to 91 °C. According to the press other 'casualties' included smoldering, singeing or blistering different types of materials [17], the top of the list being melting of parts of a convertible car that parked on the specific spot, by a "beam of light so hot it melted parts of a Jaguar sports car and sparked a fire at a local barber shop" [18].

To remediate the damages, the façade was provided with a permanent sun protection, in fact the world's largest sunshade. In an interview to Architectural Record in 2014 [19], the architect stated that "*It wasn't even within my range of decision-making. Our client commissioned a building that got approval, and then was sold to somebody else. And the new owner redesigned the building. So my name is on it, but it's not my building.*"

However, similar causes (concave facades that affect the environment) produced similar focalization of the sun rays in other Viñoly building, previously designed to the 20 Fenchurch Street Walkie-Talkie, in Las Vegas, the Vdara hotel (Fig. 36.9) that sends the "death ray" into the swimming pool. Complaints have been made but the problem was solved by the hotel management: they provided umbrellas [20].

Fig. 36.9 Arch. Rafael Viñoly, the Vdara Hotel, Las Vegas. Photo: Sven Brendel (CC BY 2.0)

36.3 Discussions

Architecture is an integrating and multidisciplinary profession. Today, even more than in the past, the implications of human actions—and a building is the result of human action—can be verified by mock-ups and models, by specific software or by other laboratory devices and tools. Less than in the history, phrases like "However, the "script"—the functional requirements—are formulated by the specialists and the changes must be carefully considered by the entire team: owners, specialists, users." are valid. Contemporary performant building materials have specific characteristics and a "user manual" that the designers must be aware of. Although coming from another field—Medicine—the Hippocratic Oath is just as applicable: *primum non nocere* (first do no harm).

References

1. K.T. Frank, *The Power of Light: the Epic Story of Man's Quest to Harness the Sun* (McGraw-Hill, New York, London, 2003), pp. 43–46
2. D. Button, J. Colvin, J. Cunliffe, C. Inman, K. Jackson, G. Lightfoot, P. Owens, B. Pye, B. Waldron, *Glass in Building a Guide to Modern Architectural Glass Performance* ed. by D. Button, Brian Pye (Butterworth Architecture, 1993), pp. 10
3. D. Gans, *The Le Corbusier Guide*, 3rd edn (Princeton Architectural Press, 2006), pp. 41–45
4. https://hts3.files.wordpress.com/2010/12/scheerbart-glass-architecture.pdf. Accessed Apr 2020
5. https://web.archive.org/web/20090523033436/four-solaire.fr/historique.htm. Accessed Apr 2020

6. https://www.promes.cnrs.fr/TOUT-PUBLIC/Les-fours/eng-lesfours1.htm Dec 2006. Accessed Apr 2020
7. https://anglophone-direct.com/mont-louisfont-romeu-odeillo-via/. Accessed Apr 2020
8. https://www.atlasobscura.com/places/worlds-largest-solar-furnace. Accessed Apr 2020
9. https://www.gsenergy.eu/odeillo-solar-furnace/. Accessed Apr 2020
10. https://www.jeannouvel.com/projets/one-central-park/. Accessed Apr 2020
11. https://good-design.org/projects/one-central-park-heliostat-and-reflector-system/. Accessed Apr 2020
12. https://www.thoughtco.com/gehry-responds-to-concert-hall-heat-178089. Accessed Apr 2020
13. M. Schiler, E. Valmont, Microclimatic impact: glare around the walt disney concert hall, in *Solar World Congress, Joint American Solar Energy Society/International Solar Energy Society Conference* (Orlando, 2005)
14. J.Y. Suk, Post-treatment analysis of the glare remediation of the Walt Disney Concert Hall (ProQuest Dissertations Publishing, 1443880, 2007) p. 6, in https://search.proquest.com/openview/cf88c408ab9ba9f0f43fcdcb22e3490c/1?pq-origsite=gscholar&cbl=18750&diss=y. Accessed Apr 2020
15. M. Schiler, E. Valmont, Urban environmental glare: the secondary consequence of highly reflective materials, in PLEA—The 23rd Conference on Passive and Low Energy Architecture (Geneva, Switzerland, 2006)
16. https://www.dailymail.co.uk/news/article-3219883/London-s-Walkie-Talkie-skyscraper-voted-ugliest-hated-building-Britain-branded-gratuitous-glass-gargoyle-graffitied-skyline.html. Accessed Apr 2020
17. https://www.bbc.com/news/magazine-239446793 Sep 2013. Accessed Apr 2020
18. https://www.mississauga.com/news-story/4067822-london-s-fryscaper-draws-crowd-on-hot test-day/Sept. 6, 2013. Accessed Apr 2020
19. F.A. Bernstein, Newsmaker: rafael viñoly, Architectural Record, in https://www.architecturalrecord.com/articles/3077-newsmaker-rafael-vi-241-oly. Accessed Apr 2020
20. https://amp.businessinsider.com/the-vdara-death-ray-hotel-is-still-burning-people-in-las-vegas-2016-6. Accessed Apr 2020

Chapter 37
A Framework for Implementing Sustainable Design Strategies in Romanian School Buildings Projects

Oana Mihăescu

Abstract All debates on the Romanian educational system should consider the need to create a sustainable built environment. However, as long as educating our younger generations is one of society's core purposes, building new schools remains one of our essential challenges. Over 80% of the existing Romanian school buildings were built before 1970. Box type buildings, designed to restrain learning experiences to indoor environments hardly meet the needs of today's generations, already compromising the needs of future ones. Almost every rehabilitation or extension project of a school building aims minimal safety or comfort, while restricting to a minimal budget. Schools' users are completely absent from any dialogue with the design teams and educational architecture practice in Romania is not grounded on research. Research is a dynamic phenomenon, providing complex and sometimes contradictory data, which encourages the continuous testing of new ideas. Beside healthcare, education is the most important architectural program that should be approached through Evidence-based Design (EBD). This paper emphasizes the necessity of adapting our building legislation based on systematic Post-Occupancy Evaluations (POE). The primary objective is creating evidence: what works, what should be avoided, what and how could be improved in planning and designing sustainable educational facilities. One of the main results consists in proposing a framework for implementing the POE method in Romanian schools. School should not be a capsule where education takes place. Our public school buildings should transform into community landmarks, which can provide significant learning experiences and promote sustainable ways of understanding education.

Keywords Sustainable school buildings · Post-Occupancy evaluation · Evidence-based design

---

O. Mihăescu (✉)
Ion Mincu University of Architecture and Urbanism, 010014 Bucharest, Romania
e-mail: mihaescu.t.oana@gmail.com
URL: http://orcid.org/0040744633343

I. Visa and A. Duta (eds.), *Solar Energy Conversion in Communities*,
Springer Proceedings in Energy, https://doi.org/10.1007/978-3-030-55757-7_37

37.1 Our National Challenge: Motivation for Education

37.1.1 School Spaces Remained Far Behind the Educational Process

The gap between the anchorage into the past of the Romanian school architecture and the current context of complex school functioning scenarios is rather significant.

In the last 23 years since the publication of currently in effect **NP 010–97** *Normative regarding design, construction and utilization of school buildings and high school buildings* major changes took place throughout the legislative, pedagogical approach, building solutions and users' needs.

A school project is however approached as a collection of inherited templates, which require no additional research. The Ministry of National Education had been working on standard designs for school projects and the tenders are still about commissioning works that simply adapt these standardized functional designs.

At most 2% of the school buildings in Romania were built after 1989, while over 80% of them have been finalized before 45 years ago. School architecture should therefore outgrow the stagnant level it currently finds itself in: functional standard shaped design sets amount to school building type projects, remaining indifferent to the specificity of an educational climate, to dialogue with its users, to elements of sustainable construction and to principles of Education for Sustainable Development.

The current research inquire system is grounded on the following question: *What will happen when Romanian schools, conceived as for the 1950′will no longer be rehabilitated to comply with the renewable standards of 2050′?*

Unlike the evolution of kindergartens or universities, the institution of school in Romania continues to be considered an environment of theoretical lectures, with frontal teaching, with an identical content for all students, having 50-min classes, during which time all students must keep quiet and still. The register, the blackboard and the chalk, the bell, the break, the homework, the test, the grade, the fence, the bars on the ground floor windows, the door keeper's booth or the asphalt schoolyard stay emblematic labels, which have deeply imprinted everyone's perception.

The well-known and debated PISA evaluation results (Programme for International Student Assessment) published in the last years depict a worrying situation for Romania. Those results have demonstrated a precarious level of knowledge and understanding of environmental issues for students between 15 and 16 years old. The percentage of students in the studied age group is a significant indicator of a country's future adult population, a population who would be able to develop enough knowledge and understanding in order to be up for the future environmental challenges.

The point is not to accuse the Romanian educational system of staying behind, stuck in principles the social reality has outgrown, it is about pinpointing certain typologies and dominant practices and placing them under question marks. This process grounds itself on the idea that the fluctuation in evolution stems, for the best part of it, from a lack of research on the relation between the dynamics of pedagogy and architectural responses. For this reason, we will continue to delve in the ways through which the needs of the learners can be better catered for by rethinking the spatial configurations. A rethinking based on a deep understanding of the role school plays in the development of the human being.

The chart in Fig. 37.1 outlines a timeline of the most significant legislative landmarks regarding school planning and design at national level. It can be noticed that the national strategic objectives for the following years are currently in disharmony with the approach to spatial configurations for school buildings.

Given the two decades disparity, any new norm concerning school design in Romania ought to be well grounded on a research program based on evidences collected from relevant case studies from existing schools.

37.1.2 Building Sustainable New Schools: Learning from Others Experiences

The traditionalist education model is based on underappreciating the child's capabilities, based on subordination, ready-made items, experience and knowledge division, memorizing, dependence and procrastination. On the opposite end, the scope of education may be defined as a continuous remodelling of the human being through intrinsic motivation, into physical, intellectual, affective and ethical integrity.

In the chapter *How to Prevent Obsolete Schools* Raymond Bordwell [2] discusses the negative aspects of schools that discourage learning: stagnation, stillness, boredom, inactivity and outdated conditions in which the educational process takes place in many schools (classrooms with a single outlet, with poor mechanical ventilation or with direct sunlight from only one direction).

The interpretation key could be the following: to understand and own your past in order to transcend it. A school project involves a conscious assumption of the historical heritage of the experiments of time and space use in the educational process, in order to create new learning environments. A radical transformation of schools is a complex, long-term process, which must be based on significant analysis: both on the operating conditions of the existing schools and on the possible needs of the future schools' users.

According to statistical studies, between 1950 and 1970, a new school was opened in the United Kingdom every day [3]. In a study conducted in 1960 on the existing German school buildings, it was concluded that any inclination of the balance towards an educational reform was limited by the existence of a good building stock of old schools [4]. This observation can be applied today to the Romanian existing school building stock. Our national reality is quite different: that of *an old building stock of outdated schools*. If we follow the warnings in the specialized literature analysing school evolution, the danger today would be to design "new old schools" [5].

School should not be a neutral container of spaces but an active agent that helps modelling students' experiences and promotes new ways of understanding education.

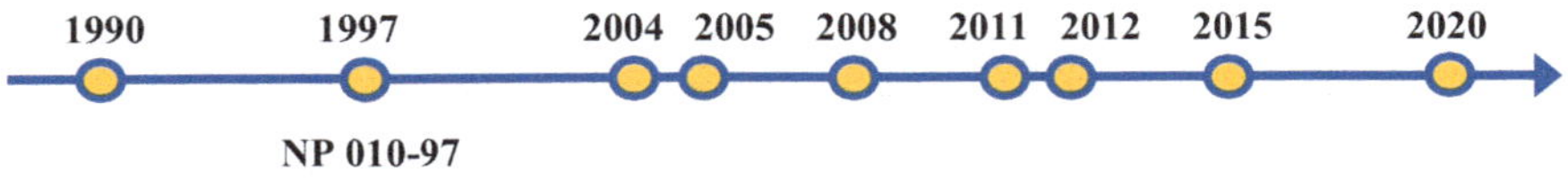

- **Before 1990:**
  - **1959 – 1965**
    - about 1000 new schools were built, with 26000 classrooms;
    - a school type project was developed within the Bucharest Design Institute (IPB): "Project type school with 24 classes", "U"-shaped plan, Arch. G. Cristea.
  - **1965 – 1970**
    - new standard design projects for secondary schools with 8, 16 or 24 classroom and for high schools with 8, 12, 16 or 20 classrooms;
    - programs for schools with special profiles: professional, sports, music, choreography schools; student houses, canteens and clubs have been built;
    - standard projects for education, easily adaptable to various locations, are considered to be "the only rational solution, with high technical and economic efficiency" [1];
    - 82% of the existing school buildings in Romania have been built before 1970;
    - in 1970 there were registered 15595 schools in Romania: 14 127 secondary schools, 577 theoretical high schools, 226 vocational high schools, 463 professional schools, 131 technical schools and 71 special schools;
    - 63% of the Romanian children left school after the eighth grade.
- **1990 – 1997**
  - The National Institute of Statistics registered 15726 units of primary and secondary education in 1990 and 16610 on 1997;
  - alternative pedagogies to the traditional education syste appear and the first schools are built: in 1990 *Waldorf* educational alternative and *Curative Pedagogy* were recognized; in 1993, *Montessori* pedagogy; in 1994 the alternative education program *Step by Step*; in 1995 *Freinet* pedagogy; in 1996 the alternative program *Jena Plan*;
  - 884 new educational units in the first 7 years after the fall of the communist regime, but this is mainly due to constructions started before 1989
  - **NP 010-97** *Normative regarding design, construction and utilization of school buildings and high school buildings*
- **2004** *The Law no. 272 from 21/06/2004 regarding protection and promoting of the rights of children' rights*
- **2005** UNECE Strategy on Education for Sustainable Development
- **2008** National Sustainable Development Strategy *Romania 2013-2020-2030*
- **2011** The Law of National Education
- **2012** NP 051-2012 *Normative regarding accommodation of civil buildings and ur ban space to the individual needs of people with disabilities* = redefines the concept of accessibility – a first step towards *universal design*
- **2015** The Ministry of National Education registers 18799 schools in Romania
- **2020** 18597 schools in Romania: 12062 rural schools and 6535 urban schools

Fig. 37.1 Legislative and design landmarks with significant impact on national school planning

In our opinion, sustainable design for school buildings should be a natural design approach for any architect [6].

Following our previous research [7] we published a series of recommended design measures for architects [8]. We defined and analysed 25 major interconnected elements of a sustainable school building that encourage a high quality of the educational process [9]. These are: *Site, Transportation/Mobility, Safety & Security, Energy, Lighting, Indoor Air Quality, Acoustics, Water Management, Waste Management, Materials & Resources, Structural Systems, Modularity & Prefabrication, Accessibility/Universal Design, Image/Awareness/Local Footprint, Interior Design & Furniture, Orientation & Signage, Health, Nutrition & Physical Activity, Operations & Maintenance, Building services, BMS, Smart/Intelligent Buildings, Information and Communications Technology, Costs & Financing, New or Rehabilitated, Innovation.*

Existing schools amount to a nucleus of resources that must be borne in mind whenever planning new sustainable school building programs [10]. We support the systematic conducting of evaluation and research programs for contemporary educational facilities, with the intent of creating evidence on *what works, what should be avoided in planning and designing schools, what and how could be improved.*

This research aims for a design approach revival of Romanian school architecture. Following this study, architects may benefit from new open paths, landmarks, and a research toolkit for generating new results.

37.2 Post-Occupancy Evaluation and Evidence-Based Design of School Buildings

37.2.1 Post-Occupancy Evaluation (POE) of Educational Facilities

Introduced in 1988 by architect Wolfgang F. E. Preiser, Post-Occupancy Evaluation (POE) is a research process of a building's functioning after a certain time from its occupancy. POE integrates quantitative analysis, through objectively monitoring the building's performance, with qualitative research, of users' values, needs, complaints and expectations.

This cross-disciplinary research combines technical design answers with the approach of the social dynamics context within the respective built environment. The impact of POE is reflected on health, productivity and users' performances as well as on the operations and maintenance of systems, equipment and building services.

Bearing in mind the barriers identified throughout the analyzed case studies (at international level), we ought to draw the attention upon a risk factor at a national level. This mainly consists in understanding POE as a process of external evaluation, as an audit or as a certification mean for a new building with the intent of simply demonstrating that it functions according to pre-determined performance criteria.

A POE is not limited to a data gathering exercise. It entails more than simply examining the degree of conformation with fundamental criteria through objective means in an occupied building. Instead, it demands a lot of efforts and time, but the achieved benefits can also be plenty.

On the short term, leading a POE has the advantage it actively involves the administrators and users of a building, who become more responsible where operations or maintenance is concerned. This will thus lead to a more efficient distribution and use of resource and costs allowed. On the long term, it generates data, new solutions and criteria which can be implemented in future projects.

The building occupant's ability to operate changes in their environment as well as changes concerning their personal condition with the purpose of improving comfort is called *adaptive opportunity*. This key-factor entails both positive and negative attributes, materialized in users' complaints. A *complaint* (written, over-the-phone, e-mailed or in conversation demand) is a perception of an underperforming building's element or piece of equipment. A systematic analysis of such complaints may provide new significant perspectives over building related problems.

37.2.2 Evidence-Based Design (EBD) and School Architecture

Research is a dynamic phenomenon, which provides complex data, sometimes contradictory, which encourages the continuous testing of new ideas. And as more research on the environment gets published, certain design decisions may be questioned. Design therefore, always gets its information from research.

Evidence-based Design (EBD) is a design method based on self-evident data. The goal of EBD is to promote learning from past projects' experiences and to apply all this new knowledge in new projects, in the context of continuous evolution. The EBD approach is a step further from the Performance-Based Building Design or Building Performance Evaluation, as all decisions made concerning the built environment are made following research after the building has already been occupied.

After healthcare buildings, schools are part of the most important architectural program that should be approached through EBD. Numerous studies performed over the past few decades demonstrate correlative and causal relations between the schools' built environment and a wide range of variables connected to its users: health, performance, absenteeism degree, dropout degree, timing for competences acquisition, attention span etc. Moreover, operations and maintenance of a school building strongly influences its functioning in time, and the contained educational process.

In a report published in 2013, ARACIP (the Romanian Agency of Quality Insurance in the Primary and Secondary Educational System) explains the cause of European teaching policies' inefficiency. In the list they put forward, they highlighted the "Lack of evidence based policies, the absence of detailed information concerning the foundation and incidence of the problems, as well as the absence of a systemic approach of data collection, monitoring and analyzing. Only where complex systems of data collection, scanning and correlation are in place, the real causes of problems became apparent and the risk of generating inefficient policies significantly decreased" [11]. ARACIP recommends a common goal for all decision factors in the national education system: elaboration, putting into practice and supporting coherent, comprehensive and evidence-based strategies" [12].

Therefore, "evidence" means more than "information", "data" or "knowledge". Studying several examples published in specialized literature, visiting a recently finalized project or leading a single POE is not synonymous to generating evidence. And a randomly obtained feedback from building' users is not a substitute for the validation of relationship between users and the built environment. In order to state the design is evidence-based, all identified hypotheses must be tested and verified.

Today's educational architecture practice in our country is not grounded on research. Taking into account the level of pioneering of sustainable buildings altogether, as well as the disparity between the legal frameworks highlighted in the beginning, we will consider the formal replication of architectural solutions without seeing the big picture, as the highest risk factor.

37.3 Methodological Framework for Implementing POE in Romanian Schools

The built environment of an educational facility bears a great responsibility towards those who learn within. An integrated approach through cross-disciplinary teams and dialogue is necessary. This research puts forward a tool kit for conducting a Diagnostic POE in Romanian school buildings and describes in stages the techniques, instruments and activities detailed in view of future implementation.

Firstly, we will resort to a direct incursion into the thorny side of the problem: the costs. 15 years after the launch of the term "post-occupational evaluation", the author of this concept, architect and university professor Wolfgang Preiser published in 2002 the study *Continuous quality improvement through post-occupancy evaluation feedback* [13]. Preiser specifies in this article examples of costs allocated to US schools in recent years: Indicative POE—US $ 8,000; Investigative POE—US $ 25,000; Diagnostic POE—$ 90,000.

- **Indicative POE**: entails the analysis of execution drawings and indexing the degree of conformation with the fundamental requirements; a tour of the building, accompanied by marking all observations (both positive and negative aspects) and photographs; interview with the building administrator.
- **Investigative POE**: a step further from Indicative POE, it compares the existing situation with the results obtained from POEs of similar facilities and with results covered by specialized literature; additionally, Investigative POE also employs questionnaires; the evaluation may last for several weeks or months, as it is necessary to cover the variations between seasons, between moments during occupation hours and days of the week (because of the fluctuating use of spaces).
- **Diagnostic POE**: involves multiple tactics and instruments (inquiries through questionnaires, in-depth interviews, observations, physical measurements,

discussing the results with buildings' users within focus groups and workshops); the collected data is comparatively analyzed with that coming from similar facilities. The data derived from this study features a high degree of validity and generalization, worthy of use in forecasts. The results of a Diagnostic POE have the potential of being turned into design directions for certain architectural programs.

According to the UK's Commission for Architecture and Built Environment (CABE), rigorous, in-depth post-occupational assessments (Diagnostic POE) that lead to relevant results require significant financial and time resources, approximated by CABE at £ 10,000 for a school [14].

A model worthy to be followed is provided by the public authorities of Scotland. They require that 1% of the value of the investment capital in the construction works for new school buildings be used to carry out POEs and to communicate the results obtained and of the lessons learned [15]. At the same time, in Scotland it is compulsory to carry out POEs for new school buildings since 2006. In addition, in the last years POEs are also conducted before the projects of extension or rehabilitation of existing school buildings.

The only known benchmark at the national level is the research project *EDU-Quality* (*Optimization of Indoor Environmental Quality & Energy Efficiency in educational facilities using a multi-criteria decision approach*), realized within the Faculty of Building Services Engineering from Technical University of Civil Engineering Bucharest [16] and financed by UEFISCDI. This project had a budget for 2013–2014 of about 98,000 EUR. *EDU*-Quality aimed at the experimental evaluation through measurements and questionnaires of the comfort level of the indoor environment in 20 classrooms belonging to several types of schools in the country.

Such high costs can discourage an organization from investing in conducting a POE. However, it is important to mention that the savings after implementing the results obtained after conducting a POE can reach a multiple value of 3–10 times the value of the investment.

Certainly, a research with a low budget can be conducted through the POE method on relevant case studies of school buildings in our country. However, in order to obtain relevant data under pioneering conditions, the active involvement of a motivated transdisciplinary team is required. If the research will be an integral part of a national or European program supported by the authorities, the educational institutions can see real opportunities for evolution. A directly interested school management, together with actively engaged school users can compensate for financial barriers.

Participatory design experiences [17] have shown that the main source of user satisfaction with the built environment is not the degree to which a person's needs have been met, but the individual's sense of being able to influence design decisions [18]. The phenomenon of "identification with a building" contributes to a sense of well-being which, in the case of schools, adds value to the pedagogical process.

Regarding possible workshop directions, our proposed methodological guide starts from the ludic activities experimented by the German research group Die Baupiloten. Led by Susanne Hofmann, this architectural group from Berlin has developed a series of methods and tools for communication between architects and users, summarized in the work *Architecture is participation*, published in 2014. Moreover, our toolkit proposes other creative exercises, useful for a research team of architects conducting a POE in a school, involving students, teachers, parents and administrative staff.

The specificity of our proposed POE model consists of applying a double analysis filter: *What sustainability components can be implemented within the school?* and *How can we find the best solutions to turn the school into an interactive 3D textbook that encourages Education for Sustainable Development?*

A Diagnostic POE model for the Romanian school buildings developed by this research uses the following seven data collection techniques: Documents Analysis, Observation, Questionnaire, Focus group, Interview, Workshop, Monitoring. The orientation route of our Methodological Framework aims at creating the work plan specific to a certain case study and proposes the following main stages:

- Putting together the research team and the general work plan;
- Collecting preliminary data;
- Initial contact;
- Informal visits on school site;
- Documents analysis;
- Observation: walking tours; photography; filming; traces analysis;
- Questionnaires—adapted to each target group of users: students, teachers, administrative staff and community members.
- Focus groups
- Interviews
- Workshops: creative exercises and playful activities, based on five main work tactics adequate to certain age groups of students or other target public categories (students, teachers, school administrative staff, parents, local decision factors): mapping; sorting; building; exploring; negotiation.
- Monitoring entails the use of certain technologies and incorporating data sensors which allow for tracking the equipment, building services and systems performances as well as the quality of the microclimate in various spaces, at a certain point in time or over a certain time frame:
  - incorporating fixed data sensors, connected to displays and manual measurements with mobile devices; will be monitored: air quality—the level of air pollutants (from both inside and outside the building): the level of gas particles, concentration and distribution of particles; volatile organic compounds, water quality; comfort (temperature, relative humidity, the movement of air currents); electrical energy balance; levels of artificial and natural lighting; level of noise and acoustic performance; vibrations; openings air-tightness performance; functioning of IT equipment, electronic and household appliances in classrooms, offices;
- Data interpretation—reading, annotating, classifying, connecting, corroborating, comparing, elaborating the research report;
- Results communication—feedback towards the school community;

- Results dissemination—towards other interested institutions and the wide public.

During POE, the research team needs to develop a series of instruments destined for working with school users as well as internal instruments, necessary to collect data and elaborate reports: instruments for working with school users—*Questionnaire; School building evaluation scale; School building evaluation form; Journal for using the school environment; "The wishing box"*; instruments for internal using (tools destined for filling-in/management within the research team)—*Approval map; Invitation; Presentation summaries; List of participants—focus group; Individual interview guide; Interview guide—focus group; The top of "to keep" elements and The top of, "to change" elements; Activity evaluation form (focus group, workshop); Focus group Report; Room evaluation file; Monitoring Plan.*

This "toolkit" needs permanent adaptation: the developed framework is therefore, simply a first stepping stone, which will lead to something perhaps differently built altogether, according to the organizing plans of each study case.

37.4 Recommendations for Further Research Implementation

A sustainable school building transformed into an interactive 3D textbook of Education for Sustainable Development has the power to strengthen motivation for learning experiences. The initial stages of planning for the architecture typology of school buildings therefore get improved once the design act has been embedded in the Post-Occupancy Evaluation, valid link between users and architects.

In order to put the methodology into practice and benefit from the proposed "toolkit", we offer a series of featured recommendations:

- Implementing pilot research projects, supported by national or European financing schemes. Subsequent to the dissemination of the results, benefitting from higher trust, a competition could be organized between schools, the prize of which would be conducting an in-depth Post-Occupancy Evaluation for the winning facility. Therefore, POE would be regarded by the institution as a well-deserved and assumed investment rather than a burden or a risk factor.
- Introducing POE as mandatory criteria for the actual design team who is going to take on a project of school rehabilitation or on a project for building a new school for an existing scholar community. Therefore, the decision factors can be assured from the design phases that the real users' needs will be accurately met.
- Organizing architectural competitions for the schools which have already been evaluated. This way, POEs may be conducted by one research team, while the architectural competition may stay open for other teams to register too. The theme of a project as such ought to be extremely well structured, its intent being to familiarize the competing teams with the results of POE and with the vision of the respective school community. For example, a series of visits on the schools' sites,

accompanied by observation tours and presentations may constitute a mandatory requirement during the registration phase of the competition.
- Focusing on the integrated design with the aid of instruments which help the connection between specialties: BIM—Building Information Modelling).
- Contracting local or regional companies, crafters, organizations or consultants over the course of the design and construction phases and actual involvement of current or future school users in the process of design and construction.
- Among the requirements, building site operations will be taken into account, conceived and detailed from the onset of the project, bearing in mind the school's program as well as the fact that the site itself may become an object of study.
- Once the construction of the sustainable school building is finalized and it is handed over, the design/construction team should continue to honor its commitment towards its users, taking it upon itself to ensure consultancy at least during the first year after occupancy. Although the international examples have demonstrated shorter consultancy programs, 3–6 months, we believe that in Romania it would be extremely necessary to track the users' getting familiarized to the school building over the entire year and during all four seasons.
- The architects' role is thus extended, towards a stake connected to communication: the design team involved in the construction or rehabilitation of a school building should coordinate the creation of various visual materials, guides, textbooks, manuals of operations or maintenance, practical training classes, samples accompanied by information about the producers of every material or equipment.
- Launching new school construction pilot-projects in rural Romania: maintaining a tight collaboration with the local communities and keeping the banner high as opposed to attaining a minimal threshold. Accompanied by adequate services for staff, these new schools would attract well trained teachers and would become magnetic poles of the region.

37.5 Conclusions

Within the choreography of designing sustainable school buildings, the architects' roles are multiplied. Additionally to managing the already complex system of parameters and constraints which are inherent to any project, a responsive architect will understand that a school building project ought to be approached as a team sport, to begin with. The responsibility load on the architect's shoulders is increasing: a responsive architect can no longer look the other way and can no longer go with the gut feeling in the design process when there is relevant and accessible evidence at hand.

In the same way that specific legislative requirements for school buildings is essential for a team of architects in charge of designing a project for a new school, so is the acquisition of knowledge about the educational process. Creating a receptive design culture means not only being up to date with the latest research in education, but understanding how this information can be applied to a specific project and how it can be integrated into a particular educational facility.

Romania has been a part of UNECE Strategy on Education for Sustainable Development since 2005, taking upon itself to substantially change the educational curricula in the direction of sustainability. However, most of Romanian teachers presently lack the methods, techniques and work instruments and what is more, they are not trained for this direction at all. To add to the distress, the physical built environment of schools where they teach hardly ever encourages their initiatives.

Thus, our proposed Methodological Framework may well be regarded as a starting point for architects—the leaders of a POE—to become teachers of sustainability for the school teachers.

Grounding the design process on evidence from research creates a positive effect. Not only in the clients' perception of architects, but also in the clients' motivation to invest in quality design services. In other words, the EBD approach could bring back the guild of architects to the high level of trust, status and respect it held in Romania in the interwar period.

An increase of architects' worthiness goes hand in hand with an increase in their responsibility. Just as healthcare architecture is a form of treatment, just as the indoor office environment became a tool for good performance in the workplace, so could sustainable school buildings become a three-dimensional representation of the educational curricula, embedded in an architecture that encourages learning experiences.

References

1. C. Lăzărescu et al., Arhitectura Românească Contemporană (Bucureşti, Editura Meridiane, 1972), p. 117
2. W.C. Brubaker, R. Bordwell, G. Christopher, *How to Prevent Obsolete Schools* (In Planning and Designing Schools. McGraw-Hill, New York, 1998)
3. Department for Education and Skills, Building Bulletin 95, Schools for the Future. Designs for Learning Communities. (The Stationery Office, London, 2002)
4. CLEAPSS Home page, https://www.cleapss.org.uk/. Accessed 15 Apr 2020
5. C. Burke, I. Grosevenor, School, London: Reaktion Books Ltd (2008), p. 86
6. O. Mihaescu, Spaţiul şcolii— motivaţie pentru educaţie: către o regândire a arhitecturii şcolii din România in "Tendințe contemporane în arhitectura pentru educația preuniversitară", volum (coord.: Augustin Ioan), Editura Paideia, (2016), pp. 63–88
7. *Id.*: Elemente de sustenabilitate în arhitectura şcolilor in, Peisaj—Arhitectură—Tehnologie—Ambient", Bucharest: Editura universitară , Ion Mincu" (2011), pp. 275–288
8. *Id.*: Concluzii pentru elaborarea unui ghid de bune practici în arhitectura pentru învăţământul preuniversitar in "Tendințe contemporane în arhitectura pentru educația preuniversitară", volum (coord.: Augustin Ioan), Editura Paideia, (2016), pp. 89–164
9. *Id.*: Education through Architecture. Sustainable Schools in Romania. An Evidence-based Design Approach through Post-Occupancy Evaluation Method (unpublished PhD Thesis), Bucharest: CSUD CSD, Ion Mincu" University of Architecture and Urbanism (2016)
10. *Id.:* How School Buildings Can Teach Sustainability in ICAR 2012. (RE)writing history", Bucharest: Editura universitară, Ion Mincu" (2012), pp. 116–117

11. Agenţia Română de Asigurare a Calităţii în Învăţământul Preuniversitar: Propuneri privind strategia națională de îmbunătățire a calității în sistemul de învățământ preuniversitar (2013), p. 3. https://www.edu.ro/index.php/articles/21132. Accessed 29 Jan 2016
12. Agenţia Română de Asigurare a Calităţii în Învăţământul Preuniversitar: Propuneri privind strategia națională de îmbunătățire a calității în sistemul de învățământ preuniversitar, (2013), p. 4. https://www.edu.ro/index.php/articles/21132. Accessed Jan 29 2016
13. W.F.E. Preiser, Continuous quality improvement through post-occupancy evaluation feedback. J. Corp. Real Estate. **5**(1), 42–56 (2002)
14. Commission for Architecture and The Built Environment: Creating excellent primary schools. A guide for clients. CABE, London (2010), p. 119. https://webarchive.nationalarchives.gov.uk/20110118095356, https://www.cabe.org.uk/files/creating-excellent-primary-schools.pdf. Accessed Jan 30 2016
15. Sci-Network: Post Occupancy Evaluation. An insight into current POE-related policies and practices by European public authorities to drive sustainable construction and innovation, (2012), p. 6. https://www.sci-network.eu/fileadmin/templates/sci-network/files/Resource_Centre/Reports/SCIN_POE_final_report_-_June_2012.pdf. Accessed Jan 19 2016
16. T. Catalina, Optimization of Indoor Environmental Quality & Energy Efficiency in educational facilities using a multi-criteria decision approach (EDU-Quality) (2014). https://www.ieq-energy.com/pdf/Raport_stiintific_2014.pdf. Accessed Jan 30 2016
17. H. Sanoff, Multiple views of participatory design in, METU J. Fac Archit, Middle East Technical University Faculty of Architecture **23**(2), 131–143 (2006). https://jfa.arch.metu.edu.tr/archive/0258-5316/2006/cilt23/sayi_2/131_143.pdf. Accessed Jan 30 2016
18. Hofmann, S.: Architecture is Participation, Berlin: Jovis, (2014) p. 9

Correction to: Emerging Issues and Challenges with the Integration of Solar Power Plants into Power Systems

Fazel Mohammadi and Mircea Neagoe

Correction to:
Chapter 11 in: I. Visa and A. Duta (eds.), *Solar Energy Conversion in Communities*, Springer Proceedings in Energy, https://doi.org/10.1007/978-3-030-55757-7_11

The original version of this chapter's first reference had wrong URLs as CrossRef and Google Scholar & reference 108 had incorrect title of the paper and wrong Google Scholar, CrossRef. These have been corrected.

The correction chapter has been updated with the changes.

The updated version of this chapter can be found at
https://doi.org/10.1007/978-3-030-55757-7_11

I. Visa and A. Duta (eds.), *Solar Energy Conversion in Communities*,
Springer Proceedings in Energy, https://doi.org/10.1007/978-3-030-55757-7_38

The manufacturer's authorised representative in the EU is Springer Nature Customer Service Centre GmbH, Europaplatz 3, 69115 Heidelberg, Germany. If you have any concerns regarding our products, please contact ProductSafety@springernature.com

Printed and bound by CPI Group (UK) Ltd, Croydon, CR0 4YY
30/09/2026
02225697-0016